MARCEL HUET au

formerly
Director of the Research Station
of Waters and Forests of Belgium
Professor at Louvain University

TEXTBOOK
OF FISH CULTURE
Breeding and Cultivation of Fish

Second Edition

In collaboration with J A TIMMERMANS
Hydrobiological Section, Research Station
of Waters and Forests of Belgium

Translated by HENRY KAHN
from the fourth French edition of 'Traité de Pisciculture'
published by *Editions Ch. De Wyngaert, Brussels*, 1970
This book was awarded the Adolphe Wetrems Prize (1971)
by the Royal Belgian Acadamy of Sciences

Fishing News Books Ltd
Farnham, Surrey, England

First printing 1972
Second printing 1975
Third printing 1979
Second edition 1986

British Library CIP Data

Huet, Marcel
  Textbook of fish culture: breeding and
  cultivation of fish.—2nd ed.
  1. Fish culture 2. Fish ponds
  I. Title II. Timmermans, J. A. III. Traité
  de pisciculture. *English*
  639.3'11   SH159

  ISBN 0 85238 140 9

Published by
Fishing News Books Ltd
1 Long Garden Walk
Farnham, Surrey, England

Printed in Great Britain by
Adlard & Son Ltd, Dorking, Surrey

FOREWORD

"Traité de Pisciculture" was first published in 1952. It was concerned principally with the breeding and rearing of salmonids, cyprinids, a few special forms of fish breeding in the temperate regions of Europe and the breeding and culture of tilapias then developing in Central Africa. It sold out rapidly and was republished in 1953. The third edition published in 1960 was brought up to date particularly with regard to fish breeding in Africa.

The present fourth edition, now translated into English and published under the title of "Textbook of Fish Culture—Breeding and Cultivation of Fish", has been entirely and completely revised and considerably enlarged. Instead of being devoted mainly to fish breeding in temperate climates, it also discusses the principal types of fish culture practised in both fresh and brackish water all over the world. It includes the rearing of more than 100 different species of fish. The author and his collaborator took advantage and made use of knowledge acquired over the past 20 years on study missions and journeys to many European countries, Africa, North America, the Far East and the Middle East. They also took an active part in the "First World Symposium on Warm Water Pond Fish Culture" organized by the Food and Agriculture Organization (United Nations) in Rome in 1966, which confirmed at world level and over a wide field all that is known about fish culture in warm water ponds.

Since the end of the second world war great progress has been made in different techniques – all very important to the development of fish culture in fresh water.

As an example, note should be taken of the progress made in the breeding of species such as carp and Asiatic cyprinids, which was formerly more or less achieved only by natural reproduction. The success attained by introducing artificial spawning after hypophysation and the artificial incubation of eggs permitted the introduction of these species into countries far from their natural growth areas. This dispersion has also been helped by the greater transport facilities now available over long distances, not only of eggs and alevins or fry, but also of fish of all sizes. This development of air transport has been greatly aided by the use of polythene bags under pressurized oxygen instead of the more normal receptacles.

Over the past few years the increasing use of dry concentrated food has spread and is progressively replacing the use of fresh food. There is also a tendency now for pelleted food to be used generally – not only in salmonid culture. It is also being introduced and developed by farms rearing other fish such as eels, cyprinids and silurids. The intensification and regularity of feeding are also being helped considerably by the employment of controllable automatic distributors.

For these reasons the present edition differs considerably from previous editions. The author and his collaborator have, however, retained the essentially practical character which makes it accessible not only to specialists but also to amateur culturists. They hope it will be welcome to readers.

The author must thank, in particular, his collaborator, Dr. J. A. Timmermans of the Research Station of Waters and Forests (Belgium), who drafted the passages devoted to fish diseases and who revised, very thoroughly, the chapters devoted to the control of invasive aquatic vegetation, the manuring of ponds and the feeding of fish. He also corrected, with the author, the entire manuscript and the printed proofs.

Thanks must also be given to others who were kind enough to provide original material and photographic documents and who helped with certain passages. In particular Dr. Cl. F. Clark, Columbus, Ohio; Dansk Orredfoder firm, Brande; Dr. J. Dobie, Saint-Paul, Minnesota; M. D. E. van Drimmelen, Utrecht; Prof. A. R. Fuji, Tokyo; Prof. P. Ghittino, Torino; Dr. J. P. Gosse, Brussels; Gov. C. Halain, Brussels; Dr. D. Hey, Cape Town; Prof. M. Mojdehi, Teheran; Dr. C. J. Spillmann, Paris; and Prof. E. Woynarovich, Budapest.

The translation into English has been done with care and sympathy by Mr. Henry Kahn, to whom the author wishes to express his sincere thanks.

The author also warmly thanks Dr. D. A. Conroy, Fish Pathologist at the Zoological Society of London, who checked the English translation, especially the accuracy of the scientific and technical terms.

M. HUET

PUBLISHER'S NOTE

This second edition of the English version of "Traité de Pisciculture" by M. Huet is being published after the author's death in 1976.

Although the text of the earlier edition is still valid, it has been revised and completed, mainly concerning the cultivation of fish in floating cages and in heated water by the author's collaborator J. A. Timmermans.

It is hoped that this edition will play the same role as the previous edition and prove to be a useful and complete instrument for all those who plan to use fish culture for the exploitation of water resources.

CONTENTS

Chapter V

SPECIAL TYPES OF FISH CULTIVATION FOR RESTOCKING

Section I – Cultivation of pike

Section II – Cultivation of coregonids

Chapter VI

BREEDING AND CULTIVATION OF PERCIFORMES

Section I – Cultivation of perch *174*

Section II – Cultivation of pike-perch

Section III – Cultivation of walleye

Section IV – Cultivation of black bass and other centrarchids

Section V – Cultivation of tilapias and other African cichlids

Chapter XI

BIOLOGICAL MEANS OF INCREASING PRODUCTION

Chapter XII

MAINTENANCE AND IMPROVEMENT OF PONDS

Section I – Control of excessive aquatic vegetation

Chapter XIII

ARTIFICIAL FEEDING OF FISH

Section I – Artificial feeding in general

Section V – Conservation of fish

General appendix
LISTS OF PRINCIPAL CULTIVATED FISH

BIBLIOGRAPHY

"Il n'est pas de petite surface d'eau convenable qui ne puisse être utilisée pour la production du Poisson".

L. Léger

PRELIMINARY NOTES

I Aim of Fish Culture

THE aim of fish culture is the rational rearing of fish including, notably, the control of growth and breeding.

Rearing of fish is not only concerned with quantitative growth, however, but also with improving the quality of the product. Reared fish are intended either for food or for restocking open waters such as running water, stagnant water, natural and artificial lakes and ponds.

Fish culture is principally practised in ponds. It permits the supervision and regulation of reproduction, feeding, quantitative growth and control of the size of the fish as well as the stocking and maintenance of the ponds instead of leaving this to nature.

Among other advantages is the exploitation of ponds resulting in the development of land which otherwise would have remained unproductive, either because of too much water or because it was just marsh land. It also contributes to the production of proteins which can provide a serious addition to food particularly in inter-tropical regions as well as in many temperate regions. For the latter, the fish are needed for artificial restocking because of over fishing, pollution and navigational and other works carried out on and around the water.

Generally ponds are only a small proportion of the total fresh water surface of any country. But this is compensated for by their high productivity per acre. In effect natural production can be greatly increased by the possibilities which intensive exploitation presents.

There is a difference between "fish culture for restocking" which has as its aim the production of alevins and young fish, and "fish culture for food" which extends fish farming to the production of fish for consumption. The latter concerns, for example, the rearing of trout or carp intended for food. Fish culture for restocking, on the other hand, can have economic or recreational aims. It is economic when it is practised for the purpose of improving the return on commercial fishing; such as, for example,

repopulation of young salmon or whitefish fry (coregonids). It is recreational when its aim is, above all, to increase fish population for angling. This is the reason for the restocking of brown trout rivers carried out in the form of eyed eggs, alevins or of young fish. However, the differences between these two kinds of restocking fish culture are not always clear.

There is also a distinction between complete and restricted fish farms according to whether the farming starts with the production of the eggs to full-size fish for food or to be used as breeding stock, or whether only one stage of farming is followed.

As for the types of farms, these may be classified as extensive, semi-intensive and intensive fish culture according to whether the farming is based on natural feed only, or whether it includes, additionally, more-or-less complete artificial feeding. Extensive farming means obtaining a quantity of fish corresponding to natural productivity; intensive farming seeks to produce a maximum quantity of fish in a minimum of water.

The particularities of the cultivation of ornamental fish reproduced by reason of their behaviour, of their forms or vivid colouring is not developed in this work.

II Evolution of Fish Culture

The practice of fish culture is very old. Egyptian bas-reliefs show fishing scenes and the conservation of fish raised in ponds. The Romans reared fish in fish ponds (Fig. 1). For centuries the people of the Indo-Pacific regions and above all the Chinese have reared fish.

Throughout China, Indonesia, Vietnam, and Cambodia, fish culture is expanding under the pressure of growing population; traditional methods are being modernized and used for an increasing number of species among which the most important is the "Chinese carp" group.

In India and Pakistan fish culture is ancient in certain provinces and is developing. "Indian carp"

1

Fig. 1   Old circular shaped and divided fish breeding ponds dating from Roman times. Lago di Paola, Sabaudia, Lazio, Italy.

are the principal fish used. In the Philippines fish culture is carried out in brackish water and is in full development. The same is the case with intensive culture of eels in Japan.

In the Middle Ages, in Central and Occidental Europe, fish culture developed along with the monasteries. At first it was limited to the production of fish for food, carp in particular, but it has been profoundly modified since the nineteenth century. Better food conditions in these regions led, in some at least, to a reduction in the consumption of pond grown cyprinids while in other regions, especially in Central and Oriental Europe, the modernization of rearing methods, the introduction and general use of fertilizers as well as the growing use of artificial feeding gave cyprinids culture a new lease of life. Real farming on an industrial scale began. Simultaneously, the discovery of artificial salmonid reproduction and the development of rearing techniques for these fish as well as improvement in the means of communication gave a great lift to salmonid culture. On the other hand the depopulation of open waters due to the increase in the number of anglers able to move about easily and using perfected tackle, as well as the harm done by pollution and the industrialization of water

courses for navigation or by the hydro-electric industry, led to the growth of fish culture for the restocking of open waters. Thus fish culture modified its image in Europe and is now orientated towards the production of fish for consumption, or towards fish for restocking.

In North America fish culture has developed considerably since the start of this century. The aim is to produce fish for food, principally trout or fish for restocking which, according to the desired purpose can be cold water fish – different salmonids – or warm water fish, principally black bass.

The origins of fish cultivation in Central Africa are far more recent. Only at the end and, above all, after the second world war, were efforts made to introduce and develop fish cultivation there. The debut was almost spectacular after all the possibilities of rearing tilapias were discovered but was slowed down, for the most part because of political upheavals in certain areas. It is certain that the development of fish cultivation in these parts is really needed as a contribution to the food supply which is chronically lacking in animal proteins. This same problem, equally important, exists in the inter-tropical regions of other continents.

Fish culture is only beginning in Latin America

Fig. 2 A two-centuries-old pond with a dated monk. Overijse, Belgium.

and most of the Middle East, with the exception of Israel where it is well established and is developing rapidly.

Since the end of the second world war three factors have developed simultaneously and have modified considerably the image of fish cultivation while at the same time having an influence on its development all over the world. These factors are: facilities offered by modern forms of transport for fish, the constant extension of artificial reproduction of farmed fish, and the development of the use of concentrates for feeding.

The transport of fish, above all over long distances, has spread and been simplified at the same time thanks to the increasing use of aircraft as a means of transport as well as the advantages offered by the use of oxygen filled, hermetically closed polythene bags, with an addition, when necessary, of a tranquillizer in the water. For these reasons the transport of eggs, fry, young fish and even full grown fish offers no unsurmountable difficulties and the dispersion of different species is no longer limited. This explains why the most important species used in fish culture are more and more dispersed over the entire world and are found wherever they are capable of achieving satisfactory productivity. Examples of this dispersion of fish suitable for rearing in a number of regions all over the world include rainbow trout, carp, certain tilapias and grass carp.

The cultivation of many economically important species has been helped greatly by the growing use of artificial fertilization and incubation. The artificial spawning and fertilization of many species of fish to which this technique could not be applied formerly is now possible thanks to hypophysation. This is particularly the case with common carp and Asiatic carps. The eggs of the fish obtained in this way are generally incubated artificially which is a guarantee of success. For a long time this was used only for salmonids. Since the discovery and generalization of these methods the species of fish mentioned have now been introduced in many countries all over the world.

Another development is the intensification in fish cultivation and the use of artificial food based on concentrates distributed in pelleted form. This food is easier to prepare, to conserve and distribute than traditional food. In a number of cases it can be prepared under very economical conditions. First used for the cultivation of salmonids, pelleted food is now on the way to being used for other fish cultivation such as for cyprinids, eels and catfish.

There is therefore, taking place at the same time, an extension of fish culture and the standardization of the methods used due to the fact that the same species are cultivated throughout the world more and more in suitable waters and by the same techniques. Artificial fertilization and feeding with pelleted food are now used for both salmonids and cyprinids which formerly were reared by different methods.

III   Requisite Conditions for Fish suitable for Fish Culture

Fish cultivation is suitable for only a restricted number of species. In order that a fish be useful for pond cultivation it must: (1) withstand the climate of the region in which it will be raised; (2) its rate of growth must be sufficiently high; (3) successfully reproduce under conditions of rearing; (4) if possible accept and thrive on abundant and cheap artificial food; (5) prove satisfactory to the consumer; (6) support high population density in the ponds; (7) be resistant to disease. Very few fish meet all or even most of these requirements.

1   Adaptation to climate

This is an essential condition which limits the use of both cold and warm water species. Salmonids and especially trout which are cold water fish, cannot tolerate warm water because of its low oxygen content. Rearing is limited, therefore, to temperate regions or to mountain water which is sufficiently cold in tropical climates. Conversely, certain fish from tropical climates, African tilapias, mud carp (*Cirrhina molitorella*) of Southern China which one might be tempted to introduce into temperate regions with cold winters, should be eliminated because they cannot tolerate the climate unless it is decided to winter them in restricted surroundings which are artificially heated, such as is the case with tilapias in some places in the southern part of the United States.

2 Rate of growth; feeding

Fish intended for food must reach full size rapidly. This means that small species are not suitable for rearing, even if they reproduce well in ponds and even if they accept the diet. For this reason fish such as the three-spined stickleback (*Gasterosteus aculeatus*) in Europe and many species of *Haplochromis* in Africa, which do not even grow to 10 cm (approximately 4 in) in length, are not suitable for cultivation despite the fact that they reproduce well in ponds. The same applies to small species of tilapias, *T. sparmanni* for example.

It is best for a fish being raised to have a short feeding chain, in order to reduce as far as possible the loss of energy resulting from the passage of one link of production to the next.

In this regard the best fish are herbivorous, plankton-eaters, microphagous and fish which feed off detritus. These fish also present another advantage; they tolerate other species and thereby permit the mixing of different species.

3 Reproduction under conditions of cultivation

In order to assure an easy and constant supply of fish for rearing ponds, it is best if the fish reproduce in captivity and that they can do this without needing special conditions which have to be fulfilled; and, finally, that they give a high return in eggs and fry. A fish which does not fulfil these conditions cannot really be domesticated. For example, certain African *Barbus* and *Citharinus* grow easily in ponds but they do not reproduce and for that reason are only reared occasionally.

But there are important exceptions to this rule and the most evident is without doubt Chinese carp (*Ctenopharyngodon* and others) the rearing of which has been largely carried out in ponds despite the fact that they do not reproduce. For centuries the fry were caught in open water and transported over long distances. New methods now being perfected give every reason to believe that it will be possible to obtain regular reproduction in ponds.

While it is desirable that reared fish should reproduce in ponds, it is also true that if their reproduction is too prolific it will create other irritating problems; this is the case with tilapias.

Relatively late reproduction helps to avoid over population and dwarfing which is the result; it also permits maximum benefits to be obtained in the growth rate of young fish, which is higher than that of adult fish.

4 Aptitude to artificial food

To obtain a high production rate it is necessary that reared fish accept an abundance of cheap, artificial food. The use of species fulfilling this condition is of particular interest since it permits very high production rates reaching and passing even 10 tons per hectare (about $2\frac{1}{2}$ acres) per year.

5 The taste value of fish for eating

It is absolutely essential that reared fish should suit the taste of the consumer. For example, certain catfish introduced into Europe (e.g. *Ictalurus melas*) which breed and grow easily in ponds are considered most unsuitable because consumers do not like them. On the other hand certain species of relatively slow growth and feeble reproduction are highly appreciated. A typical case is that of the gourami (*Osphronemus goramy*) which is reared in Indonesia and other Far Eastern countries.

6 Population density in fish farming

One requisite is the possibility of assuring a sufficiently dense population. The best species are those which are social and gregarious. Certain fish, such as pike, will only accept a dense population to a certain age, 6 or 8 weeks. Over that age they devour each other with the result that, at a later stage, the economic return is not so good.

7 Resistance

Reared fish must be resistant to disease and accept handling and transport without difficulty.

IV Plan of the Book

The present edition of *Textbook of Fish culture* has been divided into four parts of unequal importance. It includes 16 chapters.

The first part deals with *ponds and fish suitable for cultivation*. As fish culture is practised for the most part in ponds, the first chapter is devoted to their construction and layout. The second chapter deals with natural food for cultivated fish.

The seven chapters forming the second part of the book describe the principal *techniques and methods of breeding and raising fish*.

The aim is to produce either fish for eating or fish for restocking. The first two chapters, which may be considered among the most important, are devoted to the cultivation of salmonids and cyprinids which are typical examples of fish breeding and fish cultivation in cold water and in warm water respectively. The summer temperature is the measure of the difference between the two types of fish cultures – not exceeding 20°C for salmonids and exceeding 20°C for cyprinids. The purpose of salmonid cultivation is the rearing of several species of trout, salmon, grayling, and char with a view to producing young fish for restocking or fish for eating (rainbow trout principally). The cultivation of cyprinids concerns the breeding and rearing of carp and other cyprinids with a view to producing fish for eating and, from time to time, for restocking. The next chapter covers fish cultivation for restocking based on artificial fertilization and incubation. It deals with the breeding and cultivation of pike which is particularly important in Europe; of coregonids present in Europe and North America; the cultivation of sturgeons, particularly important in the Caspian area; and the culture of silversides in Argentina. The book then discusses the breeding and cultivation of perciforms; the perch is of only limited interest but the raising of pike-perch is particularly important for restocking in Europe; the rearing of black bass is well developed in certain relatively warm waters – above all in the United States. As for the cultivation of African cichlids, this has assumed considerable dimensions not only in Africa but also in the Far East and in other parts of the world with warm waters. The breeding and cultivation of silurids, of less importance, is examined in the following chapter; certain representatives of this group are raised in North America,

in Europe and in the Far East. One chapter is devoted to certain special forms of fish breeding and cultivation: eels, intensively raised in Japan; rearing in rice fields, particularly developed in the Far East, as well as fish culture in brackish water important above all in the Philippines and Indonesia. The last chapter deals with breeding and cultivation peculiar to Europe, to North America, the Indo-Pacific region, to Africa, the Near and Middle East and to Latin America. In the appendices to the chapters devoted principally to the methods of fish breeding and cultivation, are given certain particulars concerning restricted forms of cultivation.

Six chapters make up the third part of the work devoted to the *control of and increase in production*. The latter depends first of all on the use of natural food which develops in ponds; the first of these chapters deals with the evaluation of natural productivity in ponds. Natural productivity can be sensibly increased by the use of different biological methods such as: the choice of species, the mixing of species and age groups, successive or simultaneous production, etc.; they are studied in the next chapter. Non-biological methods used for increasing production are examined in the two following chapters. The first deals with the maintenance and improvement of ponds. This considers successively: the control of invading aquatic vegetation, the upkeep and restoration of pond bottoms, liming, and fertilizers. The second covers feeding and artificial food for cultivated fish. The methods differ according to whether it is a question of salmonids or cyprinids or herbivorous fish. The following chapter deals with the total productivity of ponds and lays down rules determining the rational stocking rate for rearing ponds. A last chapter completes this third part; it describes briefly the principal enemies and diseases of fish.

The fourth part of the work deals with operations when the rearing is completed: *cropping or harvesting* principally by means of draining the ponds, sorting out and storage of fish and finally transport to their final destinations, where they will be eaten or used for the restocking of running or stagnant water.

Part One

PONDS AND FISH SUITABLE FOR CULTIVATION

Chapter I

CONSTRUCTION AND LAYOUT OF PONDS

I Generalities

THE farming of fish is generally practised in ponds. In current language the word pond has several meanings. In fish cultivation a pond is a section of fairly shallow water used for the controlled farming of fish and laid out in such a way that it may be easily and completely drained.

This definition excludes waters which cannot be drained (pools, natural ponds and lakes) in which fish are caught with lines, traps or nets. This book does not include these.

It is not possible to establish artificial ponds anywhere. To construct a pond under good conditions it is necessary to have enough water of good quality and it is also necessary that the topographical characteristics should be satisfactory. To ensure this the following demands are indispensable: *topographical characteristics, the amount of water available* and *the quality of the water.*

The size and the surface of the ponds can vary considerably. The smallest cover only a fraction of an acre; the largest can cover several acres and, exceptionally, dozens and even hundreds of acres. The small and medium size ponds are more easily handled and, proportionally, the most productive.

The principal points which must be considered when constructing a pond are the choice of ground and the preparation of the layout; the construction of the dike, the drainage system, the water inlet, the by-pass and the overflow (Fig. 3).

Depth will be chosen to avoid the invasion of emergent vegetation but it must not be excessive so that submerged plants can grow and develop. The ideal depth varies between 0·75 and 2 m (about 2½ and 7 ft).

It should be possible to drain the pond dry easily, rapidly and completely which is achieved by means of a series of drains or ditches comprising one principal ditch and several secondary ditches (Fig. 33). This network of ditches should terminate at a monk (Fig. 24), the construction of which is where the pond is deepest. It should include a screen in order to avoid the escape of the fish, and also a series of small wooden boards which allow the level of the water to be regulated at will.

If possible each pond should be provided with a water inlet, and an outlet independent from those of other ponds.

It is necessary to assure that legal regulations covering this point are not broken.

The present chapter is concerned with normally shaped ponds. The particulars of spawning and wintering ponds used in the cultivation of cyprinids as well as raising ponds for salmonids are given in chapters relating to these types of farming.

II Classification of Ponds

According to their water supply, ponds are divided into:

1. *Spring water ponds* (Fig. 8), with water supply from a spring in the bottom of the pond, or in close proximity to it or supplied by ground water.

2. *Ponds supplied with rain water or by runoff water* (Fig. 7). After being drained dry this pond fills up more or less quickly according to its size and to the amount of rainfall.

3. *Ponds supplied by a water course* (Fig. 4). These are classified as: (*a*) barrage ponds and (*b*) diversion ponds, according to whether all the water course or only a part crosses the ponds.

Ponds supplied by springs and by rain water are known as dam or barrage ponds. They are sometimes installed with a by-pass for evacuation of excess water (Fig. 20).

Diversion ponds are subdivided into "linked ponds" and "parallel ponds". In both cases the by-pass channel evacuates unnecessary and excess water, but in the case of linked ponds (Fig. 9) the allowed water flow crosses all the ponds, while with parallel ponds (Fig. 10) each one has its own individual water supply and its own individual outlet.

Ponds may also be classified by the species of fish raised in them (cyprinid ponds, salmonid ponds) or according to their usage (brood ponds, spawning ponds, nursery ponds, fattening ponds, storage ponds, etc.).

III Choice of Ground – Topographical Conditions

The laying down of ponds is often best on sandy and marshy ground not economically suitable for other forms of exploitation.

To be of economic value the pond must give a return above that which the land itself would give. Further, the economic advantages are doubled by the very aesthetic and hygienic improvement a pond brings in place of an unhealthy swamp.

The land should not be too porous for it is only in small ponds for trout and other intensive raising that one can expect complete impermeability. Elsewhere it could be too burdensome. One cannot build ponds on moving soil.

The dike must always rest on solid and watertight ground. This must be selected in such a way that the cost of installing the dike or dikes should be as low as possible and amortization from the earnings of the pond feasible.

Regarding the topographical characteristics, the ground should be neither too hilly nor too flat. The ideal is a slight depression which can be easily put under water by constructing a single transverse dike (Ex. valley of the Ry des Gattes at Philippeville, Belgium) (Fig. 7).

If the ground is too broken it would be necessary to build a very high and strong dike across the valley in order to place only a limited surface under water. Further, the draining of trickling water (which is always dangerous for a dike) would be difficult if not impossible (Ex. valley of the Kiliba, Zaire) (Fig. 5). Under these circumstances the construction of ponds is not recommended and in any case is practically impossible.

If the ground is too flat and it is necessary to dig the pond completely or surround it entirely with dikes it would be better to abandon the idea. The evacuation of the water and the draining dry of the ponds would be very difficult and could only be carried out with the use of pumps which produces difficulties (Ex. "watering" between Abcoude and Amsterdam, Holland) (Fig. 6).

The examination of the topographical characteristics must, then, begin with the cross section of the valley to be laid out. One can reduce cross sections of valleys to four principal kinds, the first of these being subdivided into three sub-types (Fig. 11). The interest each type presents for the construction of ponds is set out (Fig. 12).

Many of them are satisfactory excepting those with extreme profiles such as untruncated V-shaped profiles on the one hand and totally truncated V-shaped profiles on the other. In between these it is possible to construct ponds the surface of which increases as the valley widens. But at the same time as the surface of these ponds grows so the temperature of the water increases and in temperate Western Europe the fish raised will be of less economic value, so really what is gained on the swings is lost on the roundabouts.

1 Valleys not truncated horizontally

Untruncated V-shaped valleys; V-shaped valleys slightly and obliquely truncated; or rounded off V-shaped valleys. (Valleys of types 1a, 1b and 1c.)

(a) The untruncated V-shaped valleys (type 1a) at an acute, right or slightly obtuse angle are of no practical interest for the construction of ponds because they are too closed in. The construction of ponds here would call for high dikes in order to place limited surfaces under water (Ex. valley of the Kiliba, Zaire) (Fig. 5).

The advantages offered by untruncated V-shaped valleys at an obtuse angle increase at the same time as the size of the angle increases (Ex. valley of the Ruisseau de Mont, at Achouffe, Luxembourg prov., Belgium) (Fig. 13). If the angle is strongly obtuse it is often possible to construct barrage ponds and sometimes even with a by-pass channel for the evacuation of flood water (Fig. 4, arrangement 2). Such ponds are costly to install and should be reserved, if possible, for the intensive production of fish of high economic value.

(b) V-shaped valleys slightly and obliquely truncated (type 1b) permit the construction of diversion ponds. As these valleys are often rather narrow the ponds laid down generally have a limited surface (Ex. valley of the Ruisseau de Valire Chevral at Achouffe, Luxembourg prov., Belgium) (Fig. 14). In order that these ponds should be economically viable it is necessary to raise quality fish (salmonids), and if it can be done it should be an intensive form of farming.

(c) Valleys with a rounded off V-shape (type 1c) are generally found in less hilly regions than those mentioned above. If the water courses are not very important these valleys are generally suitable for barrage ponds or diversion ponds, the diversion being built at the foot of a valley slope (Ex. valley of the Ruisseau de Gémioncourt, at Sart-Dames-Avelines, Brabant, Belgium) (Fig. 15).

Note – In valleys belonging to this group (type 1), it is possible to build, without too much difficulty, linear contour ponds along the sides of the valley. Frequently these ponds exist already as mill or saw-mill races, or as irrigation channels. These linear ponds can be exploited as raising channels for trout fingerlings intended for restocking and they also play an important part in the development of salmonid running water culture (Ex. valley of the Ruisseau de Valire Chevral, at Achouffe, Luxembourg prov., Belgium.) (Fig. 172).

2 V-shaped valleys slightly and horizontally truncated
Water course flowing at the foot of a valley slope (type 2 valleys).

It is relatively easy to build a series of diversion ponds here. The water course flowing at the foot of one of the valley slopes serves to evacuate excess water and drained off water. At the foot of the other slope, a by-pass is built which serves as supply channel; the ponds are situated between the two in the alluvial plain (Ex. valley of the Bocq, below Mohiville, Condroz, Belgium) (Fig 16).

3 Strongly truncated V-shaped valley
The water course winding in the alluvial plain (type 3 valleys).

If the water course is not very important this is an excellent location for building diversion ponds. It is indeed possible to build two series of ponds – one on one side, and the other on the other side of the evacuation channel. It is then necessary to install two supply channels at the foot of the valley slopes.

The ponds can also be fed by springs or lateral supplies such as brooklets or brooks which flow down from secondary valleys (Ex. valley of the Dyle at Neerijse, Brabant, Belgium) (Fig 17).

4 V-shaped valleys very strongly or totally truncated (type 4 valleys)
If the alluvial plain is slightly inclined the building of diversion ponds offers no difficulties (Ex. Fish farm of Vaassen, Apeldoorn, the Netherlands) (Fig. 18).

On the other hand if the plain is not inclined the location has little or no interest for fish cultivation.

Absence of sloping ground would make supply and evacuation of water very difficult or even impossible (Ex. "watering" between Abcoude and Amsterdam, Holland) (Fig. 6).

IV Quantity of Water Necessary

Besides favourable topographical conditions, it is equally necessary to have a sufficient supply of good quality water for the construction of ponds. Feebly flowing brooks cannot supply important fish farms. But it is difficult, on the other hand, to build ponds, even diversion ponds, in valleys through which strong water courses flow, such as streams or large rivers. These are liable to flood and this would be a grave and permanent danger to the security of the dikes and the ponds.

This means that normally fish farms should be installed in valleys with water courses of small or average importance, such as brooks, streamlets or small streams. Only exceptionally are ponds built in valleys served by important water courses.

What is important is a sufficient but not excessive quantity of water. First of all it is necessary to assure enough water to compensate for losses through seepage, infiltration and evaporation. This same quantity of water must be replaced by an adequate flow to maintain the level, in addition to enough water to meet the breathing requirements of the fish. This is far higher in cold water fish than in warm water fish. But once the amount of water to meet this is achieved it must not be increased and all surplus water must be removed.

The quantity of water needed for fish cultivation depends on the species of fish being raised and the quantity of fish held in a determined volume of water. Certain species need a lot of fresh and well aerated water such as for example, salmonids and all so-called cold water fish. Other species require less oxygen and are happy with small quantities of water. In this category are, above all, carp and tench. The same goes for eels and in general for all fish raised in high temperature water.

Calculating the necessary quantity of water for a fish farm, if intensive raising is the aim (a large number of fish in a restricted space and depending on artificial food), it is necessary above all to base calculations on the breathing requirements of the fish. If the aim is extensive farming (a relatively small number of fish raised on natural food in a fairly large pond), then it is first of all necessary to compensate for loss caused by evaporation and infiltration. The water level must be kept constant and the temperature within the limits imposed by the requirements of the species.

Loss from infiltrations will depend on the care taken when constructing the dikes. Loss through evaporation differs throughout the year and also according to climatic and local conditions. In Europe, carp ponds are estimated to lose an average of 1 litre (2 pt) per second per hectare (2½ acres). But at times this quantity has been reduced to 0·5 litre per second and even less. In tropical climates, on the other hand, the amount varies according to regions. Evaporation can reach 2·5 cm (about 1 in) per day and if so requires an inflow of 3 litres (6¼ pt) per second per hectare (2½ acres) to compensate for the loss. In Indonesia, Buschkiel used from 6 to 12 litres (approximately 1½ to 3 gal) per second per hectare to keep his cyprinids ponds in adequate supply.

Water needs are high for trout if intensive cultivation is practised but the size of the ponds is small. In Europe, it is estimated that 5 litres (1¼ gal) per second per hectare (2½ acres) is the minimum for extensive salmonid cultivation, a flow of 10 litres per second (just over 2½ gal) for semi-intensive, and a flow of 100 litres per second (approximately 26 gal) for intensive cultivation. Ponds used for intensive cultivation rarely exceed 5 ares (1 are is one-fortieth of an acre) while for semi-intensive farming ponds do not exceed 15 ares. Precise details on this is given in Chapter III, section I.

V   Quality of Water Supply

The question of the quality of the water supply for ponds is dealt with in Chapter XI: Natural productivity in ponds.

VI   Installation Planning

There is no general rule for planning ponds. When the intention is extensive fish farming, it is not always possible to determine in advance the shape the ponds will take. The dike is adapted to the layout of the ground and the shape of the pond is really of no importance.

As far as production is concerned the principle is to build a dike which, calling for a minimum of transport of soil, gives the greatest water surface possible. The size of the pond depends therefore on the ground and the slope.

The slighter the longitudinal slope, the greater is the surface secured underwater by the building of a single dike of a predetermined height. When the ground has a steep longitudinal slope it is only possible to build small surface ponds.

The depth of the water will always be somewhat shallow, for if it is exaggerated without a proper reason because of the nature of the farming, the result will be a decline in productivity. The temperature of ponds which are too deep rises only with difficulty. Drainage will also be complicated because of the enormous mass of water. But if the ponds are too flat they will be invaded by emergent vegetation and will become unproductive.

Normal depths vary according to the species being farmed. However, as the fish grow, the larger fish will need greater depths. An average depth from 0·75 to 1·25 or 1·50 m (approximately 2½ to 4 or 5 ft) is generally recommended.

In Europe the average depths for carp and trout are as follows:

1. Raising carp: (a) nursery ponds: 0·50 to 0·75 m (approximately 2 to 2½ ft); (b) carp of 1 year and over: 1 m (about 3½ ft).

2. Raising trout: (a) nursery ponds: 0·75 m (2½ ft); (b) ponds for trout of 1 year and over: 1·50 m (5 ft).

This applies to the depth of the water but not to the total depth. To obtain this it is necessary to add the difference between the level of the water and the top of the dike. The depths given for the pond are average. At the monk – that is the deepest spot – about 40 cm (approximately 15 in) must be added.

The depth of the water at the banks must never be less than 0·50 m (19 in) in order to avoid invading emergent vegetation.

In type 1 valleys and particularly type 1c (V-shaped rounded off valleys), the valleys often narrow and this itself indicates the ideal spot for building the dike.

In order to find out the approximate surface of the pond, rudimentary levelling is carried out. A peg is driven into the ground, the top reaching the level that will be considered the surface of the water. A lath is placed on the peg horizontally. If it is spun on the peg to point up-stream in all directions then where the lines of sight meet the ground line will fix the limits of the pond.

If a more precise levelling is required the profiles, lengthwise and crosswise, should be traced. One establishes a base following the bottom of the valley and after that the axis pegs are placed more or less close together depending on the contour of the ground. At these points the cross sections profiles are traced perpendicularly (Fig. 19).

Once this is done examination will show whether it is possible to construct one or several ponds. But this will depend on the type of farming and the nature of the ground. If the pond is to be built on deep clay ground the dike can be several metres high, but if the ground is porous the height will be limited.

The nature of the ground is tested by digging testing holes. On calcareous soil, moorland or cracked shale only small ponds can be constructed.

If an important spring is going to be used for supply, then tracing the water level above that of the spring can be dangerous. By way of precaution it is a good idea to build a barrage at the spring itself to see if water is not lost when the level is raised.

When several ponds are built it is useful to see that each one has its own supply and drainage systems, for in this way the ponds are independent and can be emptied at will. Also the danger of spreading epizootics is limited.

It is almost always possible to lay out diversion ponds by evacuating flood water from the side, and also water not needed for normal supply. Often enough this is done by correcting the water course which acts as an evacuation channel (Fig. 9).

When ponds are not diversionary they are called barrage ponds (Fig. 8). It is not only necessary to provide a monk to regulate the level of the water and to permit them to be drained, but also an overflow for evacuating flood water which cannot be drained off by the monk. When this system is followed, however, there is considerable risk of the dike breaking and it is necessary therefore to build a very solid one. It is not possible to drain a pond completely or at least to leave it dry for long periods; also it is difficult to avoid the presence of wild fish in the pond, except when it is a spring-fed pond.

The diversion pond system is by far preferable to the barrage system.

The topographical characteristics will decide whether the ponds should be linked like a "rosary" or in "parallel".

If the valley is somewhat narrow and has an accentuated longitudinal slope then the "rosary" system is best (Figs. 9 and 20), as the water of one pond will then spill over into the next. However, each pond can also be supplied with water separately.

If the flow of water through the valley is light it may be necessary at times to adapt the "rosary" disposition, even if the topographical conditions otherwise permit the construction of parallel ponds.

If the valley is wide, the longitudinal slope slight, and the flow of water sufficient, then parallel ponds should be built (Fig. 10) each with its own supply and evacuation system. A supply channel is built and also a collecting channel by which the water is drained off and by which water not needed for supply is drawn off. Sometimes it is preferable to build two series of ponds on either side of the rectified water course which serves as a collector (Fig. 4).

VII   Laying out the Bottom of a Pond

Grass is only removed if it is superabundant.

The bottom of the pond must be covered with a network of drains or evacuation ditches of "fish-bone" design which will permit complete and easy dry draining (Figs. 21 and 33).

A principal longitudinal ditch is necessary as well as lateral secondary ditches. The bottom of these ditches must be at least 0·50 m (19 in) wide and the sides sloping 1 in 1½.

The minimum slope of the principal ditch must be 1 in per thousand but it is necessary to aim for from 2 to 3 in per thousand or more. The slope of the secondary ditches is 5 in per thousand.

In certain cases when it is a matter of ponds of relatively small size, only the principal ditch is dug (dish-shaped pond) or replaced by a peripheral ditch (crested pond) (Fig. 378).

When a pond is emptied it should be completely drained and all the fish removed. During construction it is necessary to eliminate all the holes and marshy spots. Peat pockets should be removed and replaced by gravel, sand or clay.

When the ground is compact the secondary ditches are only 10 m apart (33 ft), while the separating space can be as much as 50 m (55 yds) when the soil is light. Their bearing depends on local conditions but in any case all the depressions must be dried out at the time of emptying.

If there are springs outside the drain network, the water is drained off by secondary ditches but never by means of drainpipes in which the fish might hide.

If the principal ditch is dug in resistant soil it should remain firm, but if it is dug in sand or soft clay, the water might become troubled at the time of emptying and the fish could then be asphyxiated. In this case it is wise to lay down a lining of stone or bricks.

After emptying, the ditches are always cleaned out.

In front of the monk the principal ditch must be very carefully built. It is necessary to widen and deepen it over a length of from 2 to 3 m (about 7 to 10 ft) while the area immediately next to the monk must be protected by faced earthwork. This section will be at least 10 to 20 cm (4 to 8 in) below the principal ditch. Nevertheless it must also be above the draining pipe in order to permit the pond to be drained dry (Fig. 24).

Care must be taken to ascertain that the monk is placed at the deepest part of both the principal ditch and the pond. If a depression is made in front of the monk, supposedly for collecting the fish when emptying, then all the water therein must be removed and this creates difficulties.

VIII   The Dike

The dike is a very important part of a pond – in fact a badly made dike is almost irreparable. However, the use of aluminium complex hydrated silicates which swell strongly in water have given good results for filling and consolidating cracks in the dike (Arrignon, 1964).

Dikes must be built with the greatest care. The two principal qualities are solidity and water tightness.

If the ground on which the dike is to be built is sandy, gravelly or marshy then it will be necessary to dig down to solid watertight foundations. Start by removing the grass and top soil, peaty lumps and organic matter from the site. If the soil dug up does not appear to be watertight then a trench must be dug until watertight foundations are found.

If the earth removed is satisfactory for constructing the dike so much the better – this is sometimes possible with the earth which has been removed to deepen the pond.

If sand is used then the width of the dike must be doubled or a clay core from 40 to 50 cm thick (15 to 20 in) should tie the dike into the watertight ground. But otherwise such a core would be superfluous. (Fig. 22 shows transversal sections of dikes with clay cores).

Sandy clay is the best material to use. Do not use turf, humus or peaty earth. Eliminate stones, wood or any other materials which might soften

or rot. If the soil is too compact then it can be made usable by mixing it with humus. If pure clay is used it must be covered with some other material in order to avoid cracking when it dries.

The dike should be built in layers of 20 cm (8 in) and each must be rammed down. Dry earth must be moistened. Never build on frozen ground.

For the earthworks and transport use bulldozers, mechanical excavators, narrow gauge rail trucks or wheelbarrows according to conditions and the importance of the work (Fig. 23).

In principle the width of the dike at the top should equal its height but should never be less than 1 m (1 yd) wide. A wider top is quite feasible and will permit the passage of vehicles. The dike must be thick and solid enough to support the transport of material, fish and equipment necessary for the operation of the fish farm.

The dike must be some 30 cm (12 in) above surface of water for small ponds and 50 cm (20 in) for large ponds.

Sinking must be taken into account. This can reach as much as one-tenth of the height so for this reason the final height at the construction is generally greater than the original specification.

The slope for the outside angle of the dike should be 1:1 to 1:1½ and on the inside about 1:2. The inside slope can be reduced to 1:1 for small ponds or if heavy earth is used. But it can also be increased to 1:4 for large ponds especially if the waves are strong or very light earth is used.

When the dike is finished it can be covered with turf using that which was dug out and put aside previously. If no turf is available then the dike should be covered with a little earth and grass seed sown.

If the dike is constructed carefully then the slight oozing of water which might be noticed at first, disappears little by little once the pond is filled with water and the dike becomes watertight.

Down-stream from the dike and all along its side a drainage ditch is dug to draw water off from the dike. It should be about 30 to 40 cm (12 to 15 in) deep.

Trees should not be planted on the dike because the roots can cause water to infiltrate.

During construction the emplacement where the monk will be erected should be left free (Fig. 29).

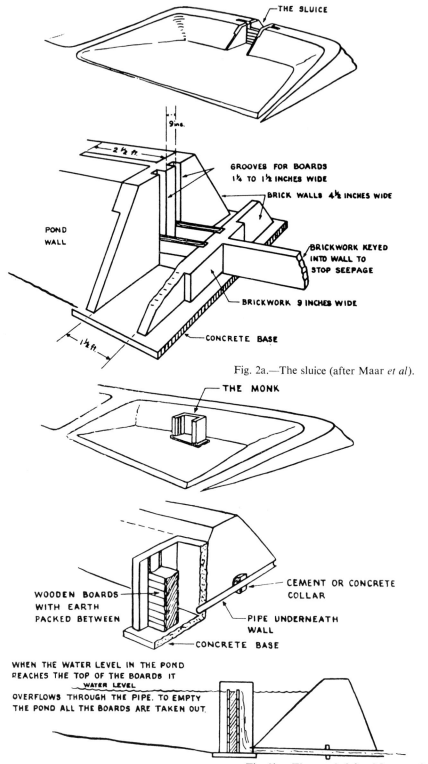

THE SLUICE

9 ins.

2½ ft.

GROOVES FOR BOARDS
1¼ TO 1½ INCHES WIDE

BRICK WALLS 4½ INCHES WIDE

POND
WALL

BRICKWORK KEYED
INTO WALL TO
STOP SEEPAGE

BRICKWORK 9 INCHES WIDE

1½ ft.

CONCRETE BASE

Fig. 2a.—The sluice (after Maar *et al*).

THE MONK

WOODEN BOARDS
WITH EARTH
PACKED BETWEEN

CEMENT OR CONCRETE
COLLAR

PIPE UNDERNEATH
WALL

CONCRETE BASE

WHEN THE WATER LEVEL IN THE POND
REACHES THE TOP OF THE BOARDS IT
WATER LEVEL
OVERFLOWS THROUGH THE PIPE. TO EMPTY
THE POND ALL THE BOARDS ARE TAKEN OUT.

Fig. 2b.—The monk (after Maar *et al*).

IX Draining Installation – The Monk

Many different draining systems exist. Formerly what were known as draining plugs were used. These allow no margin at the time of emptying and are only to be recommended for small ponds.

Among the modern outlet systems, special mention must be made of the sluice and the monk (Maar *et al*, 1966). The *sluice* (Fig. 2a) is a construction which is an integral part of the dike and is mainly formed by two lateral and parallel walls. The *monk* (Fig. 2b) is a construction placed in front of the dike and formed by two lateral and parallel walls, plus a back wall. In fact these two systems are very similar. Only the monk will be described in this book.

The monk has two important functions. When the pond is being filled, it controls the level of the water and prevents escape of the fish. When it is being emptied it permits progressive draining of the ponds.

The emptying device comprises a horizontal channel or drainage pipe running the full length of the foot of the dike, and also a vertical branch or so-called monk, quadrangular in shape and at least 30 cm wide (1 ft) open on one side in front (Figs. 27 to 30). The monk should reach at least 40 cm (16 in) above the level of the water.

On the lateral sides (sometimes slightly inclined backwards, Fig. 26) grooves are cut to take the screen and the boards and it is thanks to these the level of the water can be controlled. There are two or three grooves, the first either entirely or partially used for the screen.

One or two series of boards are put in place but if there is plenty of water then one only is sufficient. If the pond is fed with rainwater and it is necessary to retain all the water – in order to avoid a leak – two series of boards are used between which clay is packed to assure perfect watertightness (Fig. 24).

If complete proofing against leaks with a single series of boards is required, the following method permits the filling in of small chinks between the boards or between them and the grooves. When the pond is filled it is a good idea to throw some fine cinders into the water just in front of the boards where there might be leakage. The current will carry these cinders and within seconds they will have closed the chinks themselves.

The soil must be well rammed down at the spot where the monk is placed. Eventually the monk will be given a concrete foundation resting on piles.

The top of the monk must be just in front of the top part of the dike so that its lower half or somewhat higher part of the back side of the monk is embedded in the dike. In order to avoid the drainage section of the monk being obstructed in the event of the dike crumbling near the monk, it can be given two wing walls built obliquely as a protection (Fig. 31).

The entire monk can also be built totally out from the dike. This makes it less accessible and prevents thieving by draining the pond. But in this case a movable footbridge must link the dike and the monk.

The monk can be built of strong wood, provided it has no knots. It can be built of planks from 4 to 5 cm (1½ to 2 in) thick. In Europe such a monk would last about 25 years (Figs. 29 and 30).

Often the monk is built in concrete (Figs. 27 and 28) and sometimes in brick. If it is in concrete it should be reinforced and "U"-shaped iron rods used as grooves should be included as they are smoother than if cut directly into the concrete or brick, thus making the placing of the screen and the boards far easier. The iron grooves must not only be sunk into the sides of the monk but also into the bottom.

Monks either in concrete or wood can be prefabricated (Fig. 26).

In order to obtain the exact fitting of the boards which are between 20 and 30 cm (8 to 12 in) high they are sometimes bevel edged. Each board is provided with a hook for lifting (Fig. 24).

A screen stops the fish escaping while permitting the water to flow. Normally the screen is installed in the whole of the first groove. But it is also possible to use a smaller screen placed above the boards. In this case one groove fewer is necessary, but it should be remembered that the screen can be stopped up quickly with plants carried by the water. This is also possible even when the screen is slipped into the first groove entirely.

To overcome this inconvenience the screen can be replaced by a screened case which will increase the surface screened. It is also possible to install the screen in the lower part of the first groove with the boards above it. When this system is used the cold water at the bottom of the pond can be evacuated.

Another method also permits the removal of the bottom cold water as well as of the toxic gases which may accumulate there while leaving the first groove entirely occupied by the screen. This is very useful, for a large screen is preferable to a small one which can be choked very quickly and in this way slow down the flow of the water.

This method (Herrguth type monk) (Fig. 25) calls for a monk with three grooves, the first holding the screen and the second a series of boards, the lowest of which is provided with a large notch. The top board is above the water level which is determined by the top board of the series placed in the third groove. In this groove the boards go right down to the bottom. There is a free space between the two series through which an ascending current of water flows.

The screen must be made to suit the sizes of the fish. For large fish only bars are necessary. For small fish the screens are made in perforated sheet metal with holes according to the sizes of the fish (for fry about 2 mm in diameter). As the fish grow so the screens have to be changed and screens with bars have to be installed. These facilitate the passage of floating vegetation, dead leaves, twigs, etc. which the water carries. Exchange screens must always be ready. Screens must have similar measurements especially in breadth so that interchange is easy.

The screen bars of the monk, the water inlets, the overflow and the fishing out device should be vertical rather than horizontal as this makes cleaning easier.

The monk can also be provided with a safety padlocked device (a cover or flattened-out bar) as a protection against thieves.

The horizontal draining pipe can be made of stoneware or concrete (beware of acid water) at least 20 cm in diameter (8 in) but never more than 40 cm (15 in). If necessary two pipes can be used.

In order to prevent theft the best size is one which takes at least one night to drain the pond.

In large ponds a valve is sometimes installed on the drainage pipe immediately behind the vertical branch of the monk. When the pond is being emptied this valve can be used to regulate the speed of flow and reduce it so as to allow the fish to adapt themselves to a supportable pressure. This is more manageable than removing or replacing the boards in the grooves of the monk.

In order that the pond might be dried completely, the average level of the bottom of the pond should be around 30 to 40 cm (12 to 15 in) higher than the drainage pipe. Further, never drive the latter deeper than is necessary in order to avoid water flow-back from the outlet ditch.

Take care that the drainage pipe is on solid ground. The pipes must never be bent. If they are, then they are very difficult to clean out when clogged.

Normally the drainage pipe gives onto the fishing-out device behind the dike. This is either in stone, in bricks or in concrete, and it is here that the fish are harvested. The size of this fishing-out device should be in proportion to the size of the pond. At its down-stream end grooves are cut in the walls in which to slip screens. These permit the evacuation of the water and the retention of the fish. How these fishing-out devices are laid out is described in Chapter XVI devoted to harvesting fish.

When there is no fishing-out device the fish are harvested in an enlarged catching ditch in front of the monk. Stones should be placed under the outlet of the drainage pipe to avoid erosion.

X The By-pass Channel

In every case where diversion ponds can be built, it should be done in order to avoid the high cost of installing a weir and having to cope with its numerous inconveniences.

More water in a pond than is necessary must be avoided and, above all, flood water and an excess of water brought about through rainstorms must be prevented. Even if there is only danger from an excess of water at certain times of the year, diversion ponds are still customarily built.

When the ponds are laid out either side of the water course, or on one side only, the bed of the course, more or less rectified, serves for the evacuation of excess water and of water which has to be drained out of the ponds. It is necessary to build one or two supply channels (according to whether there are one or two series of ponds) at the foot of the valley slopes (Fig. 41).

The individual water inlets of the ponds are tied into the supply channel. If there is a row of linked ponds the supply channel can be suppressed and only a general inlet need be kept.

If the ponds are situated in the bottom of the valley and cover the original bed of the water course, then excess water can be evacuated by the by-pass channel which should be sufficiently wide. However, such a solution can only be considered if the water course supplying the ponds is small.

According to each case it can be seen that the by-pass channel can serve either as a supply channel or for the removal of excess water. In any case the arrangement is such that it serves the holding tanks in which the fish are kept at the time of emptying (Fig. 3).

The by-pass channel should be dug out of the ground and not constructed from soil. The bed of the channel must be built higher than the water level of the pond.

XI   The Water Inlet

All ponds must be installed with a water inlet with the exception of ponds fed by springs which have a regular though in no way excessive supply. In the case of ponds fed by rainwater they should only be provided with a water inlet if there is an excess of run-off water which will have to be removed by a by-pass. This is essential to avoid sudden influxes of water.

A well-constructed water inlet must fulfil the following conditions: (1) it must assure a regular and regulatable supply of water for the pond; (2) it must prevent the escape of fish, especially the possibility of their escaping into the feeding water course; (3) it must keep out undesirable fish which might come in through the water fed into the pond. Above all, care must be taken to stop the entry of wild fish into nursery ponds for they could include voracious species. To protect the fry, a single or double screened case can be installed at the mouth of the water inflow of each pond.

Water inlets can be constructed in different ways according to whether a general inflow for a group of ponds or an individual inflow for one pond is wanted.

1   General water inlet

This is installed at the head of a group of ponds and over the supply channel in the case of diversion ponds in parallel series; or at the head of the up-stream pond if the ponds are linked in a row.

The best general water inlet is the one known as an inlet with sunken horizontal screen and with constant flow. This type meets the three conditions outlined above.

It is very important to establish a constant and regular flow. This is relatively easy with diversion ponds, for then the excess water is evacuated by the by-pass. Constant flow is not possible in barrage ponds.

To ensure a regular and controllable flow into the ponds a vertical sluice is installed. This has a metal handle the height of which can be regulated by means of a series of punched holes (Fig. 42). A

series of superimposed boards which slip in and out of two vertical grooves can also be used. An opening is left either beneath the vertical sluice or beneath the bottom board or between two of the boards (Fig. 36), through which the water can flow at the desired speed and quantity. Excess water is diverted to the by-pass which evacuates it.

The horizontal sunken screen meets the two last of the three requirements mentioned above. This is the best way of stopping the intrusion of undesirable fish brought by the current. The holes in the screen, which is generally of perforated sheet metal, are adapted to the sizes of the fish to be kept out. In order to stop these fish getting in round the sides of the screen, the latter rests on suitable supports (Fig. 36).

To reduce the chances of the screen becoming choked by filamentous algae, vegetation, debris, etc. brought by the current, the inlet is sometimes built against the current (Fig. 43).

The advantage of a sunken horizontal screen is that it stays much cleaner than a vertical screen or an oblique screen. The latter gets choked far too quickly and has to be cleaned frequently, which is not always possible. A choked vertical or oblique screen over which the water flows is just a sunken obstacle really and is insufficient to stop the escape of the fish from the pond or keep out intruders.

For the horizontal screen to function normally it must be sunk and covered with 10 cm (4 in) or more of water. The fact that the screen is sunk, some of the debris, leaves, twigs, etc. brought by the water remain on the surface and do not stick to the screen. This means there is always a section which is free. According to Diessner-Arens, a horizontal screen of about 1 m² (just over 1 sq. yd.) will permit 1 m³ (1⅓ cu. yd.) of water to pass per minute.

The horizontal screen is completed by two small vertical walls (Fig. 36). The up-stream wall descends to the bottom of the water inlet and stops the passage of fish likely to pass beneath the screen; the top of the down-stream wall is well above the highest water level reached in the supply channel and so stops the passage of the fish over it. If the screen and the two vertical walls are assembled they can be pulled out of the water together for cleaning (Fig. 38).

In order to sink the horizontal screen a dam must be built in the water supply channel more or less abreast of the water inlet (Fig. 34). The top of the dam should be at a slightly higher level than that of the horizontal screen so that the latter is sunk. As waste accumulates behind the dam, it must be cleaned away from time to time. If the dam is made in hard materials it should have a plug at the bottom to permit periodic cleaning out. It is easier to build the sill of the dam at the level of the bottom of the supply channel and to lift it with movable beams which slip in and out of two grooves cut in the pillars of the dam (Fig. 35).

There are other devices for controlling the water supply which meet the demands given above. These generally consist of cylindrical rotating screens with an automatic cleaning device (Vibert, 1956). They are fairly common in California (Fig. 40) and are very efficient, but they are expensive. The most recent models are electrically controlled (Richardson, 1965).

In certain cases a horizontal screen placed on the bottom of the supply channel can be used (Fig. 39). Beneath the screen is a chamber from which the inlet pipe of the pond starts. Such a device is only possible, however, if the difference in the level between the supply channel and the pond or ponds to be supplied is sufficiently high and if the supply current is sufficiently strong to remove, regularly, the leaves and twigs which may choke the bottom screen.

2   Individual water inlets
These can be built in many different ways and are much simpler when compared with general water inlets. Individual inlets are used for diversion ponds in parallel series (Fig. 41).

If the ponds are branched on to the supply canal in which there are fish which must be kept out of the ponds, then each pond must have its own screen. A horizontal sunken screen of reduced size compared with the general inlet can be used, but it is also possible to use an oblique screen or a vertical screen because the danger of obstruction is much less than at the head of the installation.

If there are no undesirable fish in the water supply the inlet need only meet two conditions. They are to ensure a regular flow of water and to prevent the escape of fish from the pond. Several devices are able to do this. The constant flow can be ensured by a sluice or by a series of boards of the same type as those described above (Fig. 42). The escape of fish from the pond can be prevented by a horizontal screen (Fig. 45) or by a fall from a sufficient height (Fig. 44).

When the difference in level is sufficient a good supply of water into the pond is obtained by a fountain jet (Fig. 46). The water arrives by a pipe curved towards the top or cut open at its top side; the water spurts up and then falls in a thin spray. This system has the double advantage of aerating the water well and of hindering the escape of the fish. The same result is obtained with a screen or a horizontal perforated plaque fixed to the end of the supply pipe. There is a large variety of these different devices and all help the aeration of the water supply (Haskell *et al*, 1960).

XII   The Weir

In order to evacuate flood water, or excess water through storms, an overflow is indispensable for barrage ponds, which do not have a by-pass channel (Fig. 3).

The overflow should not be in the middle of the dike but at the side, at a shallow point, in order to avoid a rupture in the dike. If necessary, there can be two overflows, one at each end of the dike.

The ditch for evacuating the water from the overflow is tied into the ditch which drains the pond.

The overflow can be in stoneware, concrete or cast iron, or it can be an aqueduct or an open channel.

Concrete pipes risk attack by acid water. Stoneware pipes do not but they are not very resistant. It is necessary to protect them with a masonry collar about the thickness of half a brick.

The overflow (Fig. 47) should have a minimum fall of 20 cm (8 in). The width of the overflow should be proportional to the size of the pond and the quantity of water it evacuates. Its width and level should be so calculated that the level of the water, when in flood, is some 40 cm (15 in) below the top of the dike.

It is a good idea to build a device into the overflow which will allow a beam or plank to be thrust into it if necessary and constitute a reserve of water, or to halt the flow of water momentarily. On the other hand one or two sluices installed in the top of the overflow will accelerate the flow. The sluices and the beam device can be seen in Figure 48.

A screen is placed up-stream from this arrangement. It can be semi-circular or wedge-shaped. It is placed in such a way that foreign matter floats on the surface, so leaving the water a free passage below. It must be perfectly stable. Its bars should be spaced from 10 to 15 mm apart and they should be 15 mm in diameter. Each time the pond is drained the bars must be given a protecting coat. If the fish are small a trellis should be placed in the groove *a* (Fig. 47).

Appendix: Family Ponds

In the tropical regions of the Far East the breeding of fish in family owned ponds has been widespread for a long time. It is a matter of breeding for food carried out in the same way as raising poultry and goats. These varied farms are intended to transform domestic refuse into animal proteins.

In Europe, where family fish breeding does not occur, this transformation was and still is carried out in certain areas by keeping pigs. In Central Africa, shortly after fish breeding started, first as a communal operation, efforts were made to promote fish breeding by families. However, while in the Far East the family owned ponds are generally individual and separated small stretches of water, in Africa they are frequently found in compound groups of some relative importance.

An area of around 1 to 5 ares (1 are equals one-fortieth of an acre) seems a worthwhile recommendation for family ponds. If the size is less than a fortieth of an acre then they often produce little, even if intensive breeding is practised. If the size is over one-eighth of an acre they are too large for the owner to give thought to intensive breeding.

One capital consideration to be taken into account when setting up family ponds is the location of the pond and the distance between it and the owner's home. Small individual ponds have no real value unless intensive breeding is practised. This kind of breeding means frequent feeding – at least daily feeding of the fish. The method is only practical or feasible then if the owner does not have to take a long daily walk with the food. In the Far East (Fig. 49), the owner's home is often on the very edge of the water. This is rarely so in Africa where villages are situated in the hills, while the fields, the water sources and the ponds are in the valleys. In any case it is best that the ponds should be situated near the paths leading to water sources or brooks, or to the fields. Often these family ponds form groups (Fig. 50), the importance and working of which depend on the number of owners, the topographical conditions and the water supply.

If the ponds are grouped together in a complex of some importance, it is essential that the water supply channel should be suitably installed and above all that its slope is neither too steep nor too gradual – just enough for the water to flow. It is necessary that the individual water inlets be well placed.

Pond dikes (Fig. 51) must be sufficiently wide and solid, but not excessively so. Too much earth should not be used. Also the dike must not be too narrow for then it would be easily eroded or undermined by the water. The lighter the earth the wider the dike. After construction both the top and the slopes of the dike should be covered with grass in order to avoid cracks. The grass must be cut regularly and can be thrown into the pond to feed the fish. The best cover to use is paspalum grass.

The depth of the ponds should be sufficient but not excessive. In general, recommended depths for the individual ponds is between 0·50 and 1 m (18 in and 3 ft) with an optimum of about 0·75 m (2½ ft). However, in regions where temperatures vary considerably or where there is a risk of a steep drop in temperature, depths should be 3 ft and even more in order to avoid low temperatures which might harm the fish.

The bottoms of family ponds do not call for any special arrangement. It is only necessary to provide a gentle and continuous slope to the lowest point of the pond where it will be emptied.

Devices for supplying and evacuating family ponds should be as simple as possible. Whilst for large ponds it is indispensable to provide more or less complicated water inlets, overflows and emptying devices built with strong materials, family ponds should rule out complicated constructions and hard materials. Simple and inexpensive systems should be used. One of the best is to use bamboo pipes both for the supply and evacuation of water (Figs. 52 and 53). Where it is not possible to find bamboo, a pipe made in sheet metal or even a hollow tree trunk will do. A ditch cut across the dike is not good enough, however, for this could lead to the dike crumbling or breaking.

Generally, there is no emptying device built in hard material such as a concrete or brick monk prolonged by a drainage pipe. But a wooden monk can be used. Often a simple emptying device is sufficient, consisting of a bamboo pipe closed with a plug inside the pond. In the simplest cases, there is no emptying device at all, for this is done by making a hole in the dike. After emptying, the opening is plugged up before the pond is again put under water.

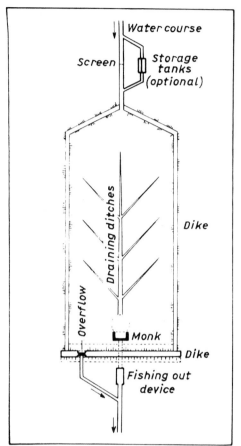 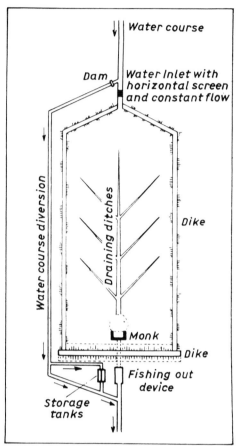

Fig. 3   Schematic drawing of ponds. Left: Barrage pond. Right: Diversion pond.

I.   Barrage Ponds (diagram 1: barrage ponds) – No by-pass; ponds overflow successively one into the other. This is not generally advised. It might be used in places where there is no danger of a rush of water such as ponds fed by springs. Example: Fig. 8.

II.   Diversion Ponds – By-pass channel for the evacuation of excess water.

A.   Linked Ponds in a Row – A by-pass channel evacuates excess water. Ponds overflow successively one into the other. This must be adopted for valleys with rather steep longitudinal slopes or when the water flow is insufficient.

1.   No supply channel (diagram 2: linked ponds without autonomous and individual water supply). A single by-pass for evacuating excess water. Disposition not permitting individual and autonomous supply nor the isolation of ponds. Example: Fig. 9.

2.   A supply channel (diagram 3: linked ponds with autonomous water supply). Two by-passes:

one for supply, the other for the evacuation of excess water and drainage water. This is preferable to the preceding method and permits the individual supply of each of the ponds. In addition the overflow of water from one pond to another is possible.

B.   Parallel Ponds – Two by-passes: one for supply, the other for evacuation. Supply channel for feeding the pond with water; by-pass channel for the evacuation of excess water and for emptying. Each pond has its own independent supply and evacuation. Ideal disposition, realizable in valleys with a gentle longitudinal slope so long as the flow is sufficient.

1.   One series of ponds (diagram 4: a series of parallel ponds) can be built in relatively narrow valleys.   Example: Fig. 41.

2.   Two series of ponds (diagram 5: two series of parallel ponds) can be built in wide valleys. Example: Fig. 10.

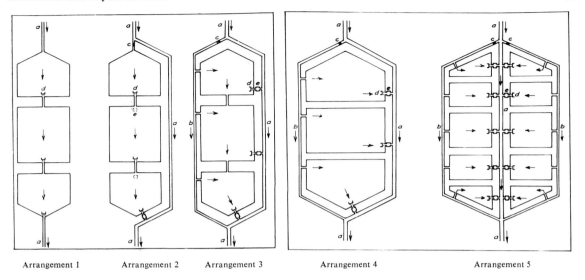

Arrangement 1   Arrangement 2   Arrangement 3   Arrangement 4   Arrangement 5

Fig. 4 Schematic drawings of principal arrangements of ponds. From left to right: Arrangement 1, Barrage ponds; Arrangement 2, Linked ponds without autonomous and individual water supply; Arrangement 3, Linked ponds with autonomous water supply; Arrangement 4, Parallel ponds. One series of ponds; Arrangement 5, Parallel ponds. Two series of ponds.

Fig. 5 Broken ground unsuitable for the installation of ponds. Deeply embanked valley, V shaped and untruncated (type 1a), with an obtuse angle. Kiliba valley, a tributary of Lake Tanganyika, Zaire.

Fig. 6   Ground too flat and unsuitable for the construction of ponds. Polders and drainage systems between Abcoude and Amsterdam, Holland. V shaped valley completely truncated (type 4); an unbroken and absolutely flat alluvial plain. Unwise to build ponds here; pumps will be necessary for the introduction and evacuation of water.

Fig. 7   Site well suited for the installation of ponds. Valley dips slightly (type 1c). Formation of a pool due to trickling water on impermeable ground. Valley of the Ry des Gattes, Philippeville, Belgium.

Fig. 8   A series of linked barrage ponds, fed by springs in a dale. No diversion channel (Arrangement 1). Degreyse ponds, Kisangani, Zaire.

Fig. 9   A series of linked diversion ponds (Arrangement 2). There is no supply channel. The water supply crosses all the ponds successively. On the right the water course acts as an evacuation channel. Fish farm at Achouffe, Belgium.

Fig. 10 (above)   Parallel diversion ponds (Arrangement 5). A water supply channel and an evacuation channel. Sewage purification ponds of the city of Munich, Germany.

Fig. 11 (right)   Different types of cross sections of valleys. From top to bottom: (left) Type 1a, Untruncated V shaped valley; (right) Type 1b, V shaped valley slightly obliquely truncated; Type 1c, V shaped valley rounded off; Type 2, V shaped valley slightly and horizontally truncated, the water course skirts the foot of one of the flanks of the valley; Type 3, V shaped valley strongly truncated, the water course winds in the alluvial plain; Type 4, V shaped valley very strongly or totally truncated.

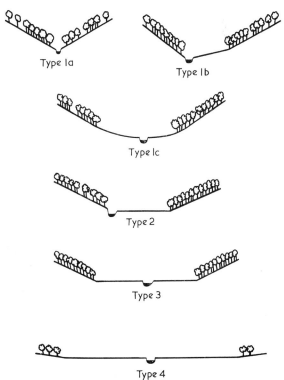

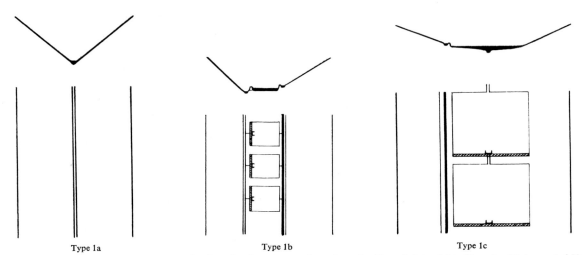

Fig. 12   Utilisation of different types of valleys for the construction of ponds. From left to right: Type 1a, Untruncated V shaped valley. Not recommended often for the building of ponds except if the angle is strongly obtuse (Fig. 13); Type 1b, V shaped valley slightly and obliquely truncated. Building of small diversion ponds (Fig. 14); Type 1c, Rounded off V shaped valley. Building barrage ponds(without displacing the water course) or a series of linked diversion ponds (with displacement of the water course) (Fig. 15).

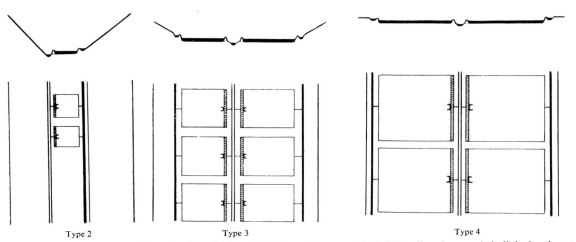

From left to right: Type 2, V shaped valley slightly and horizontally truncated. Building diversion ponds in linked series or in parallel. One series of ponds (Fig. 16); Type 3, V shaped valley strongly truncated. Building diversion ponds. One or two series of ponds in parallel (or linked together) (Fig. 17); Type 4, V shaped valley very strongly or totally truncated. Building of diversion ponds often possible. One or two parallel or linked series of ponds. (Fig. 18).

Fig. 13   An untruncated but obtuse angled V shaped valley (type 1a). Valley of Ruisseau de Mont at Achouffe, Wibrin, Belgian Luxembourg. Site only suitable for small surface ponds in which fish of high commercial value such as trout fingerlings for restocking are reared.

Fig. 14   V shaped valley slightly and obliquely truncated (type 1b). Valley of Ruisseau de Valire Chevral at Achouffe, Wibrin, Belgian Luxembourg. Small surface diversion ponds in linked series in which fish of high value, such as trout, are reared.

Fig. 15   Well rounded off V shaped valley (type 1c). Valley of Ruisseau de Gemioncourt at Sart-Dames-Avelines, Brabant, Belgium. Undulating relief, suitable for the construction of large barrage ponds, for the rearing of cyprinids.

Fig. 17   V shaped valley strongly truncated, the water course winds in the alluvial plain (type 3). Valley of the Dyle, a small stream between Wavre and Louvain, Brabant, Belgium. Very well situated for the construction of large diversion ponds for cyprinids. These ponds are fed either by the principal water course, or by a spring or by a small lateral tributary stream. The photo shows the large pond of Neerijse, fed by a rheocrene spring at the foot of the left flank of the valley; the spring first feeds a watercress bed which could be replaced by a salmonid farm.

Fig. 16   V shaped valley slightly and horizontally truncated, the watercourse skirts the foot of one of the flanks of the valley (type 2), in this case the right flank. Valley of the Bocq brook, down-stream from Mohiville, Condroz, Belgium. Well situated for building diversion ponds; the water supply channel can be installed at the foot of the left flank of the valley.

Fig. 18   V shaped valley, totally truncated (type 4). Cyprinid ponds of the Vassen fish culture station at Apeldoorn, the Netherlands. The alluvial plain is slightly broken. Easy construction of large and shallow diversion ponds for the rearing of cyprinids.

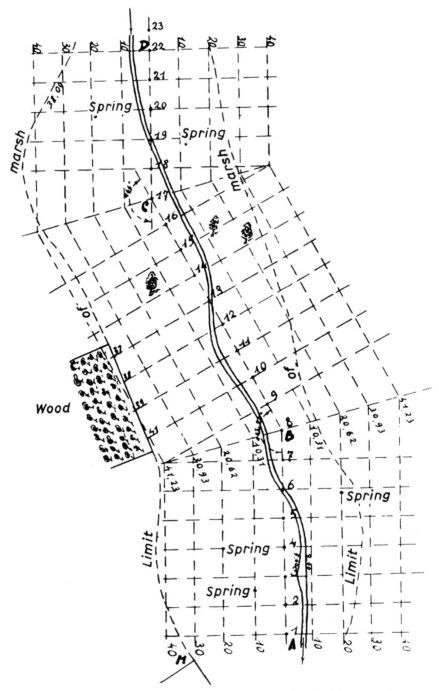

Fig. 19   Pegging of a valley before the construction of ponds (after Evrard).

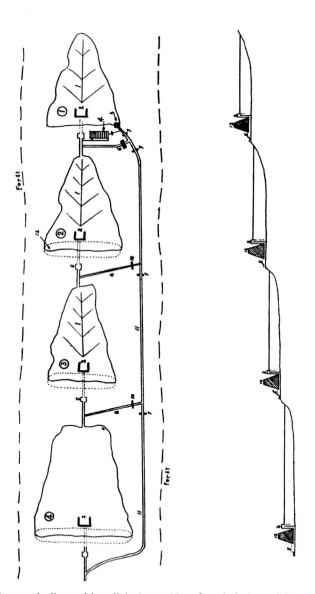

Fig. 20 Diversion ponds disposed in a linked row. Use of a relatively straight valley with a rather steep longitudinal slope.

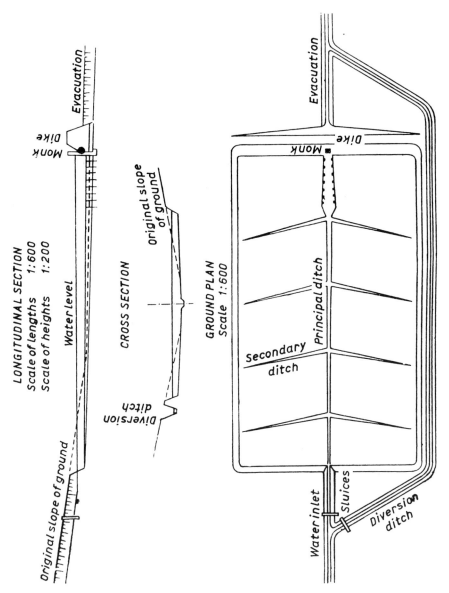

Fig. 21   Groundplan, longitudinal and cross sections of a pond (after Maier-Hofmann).

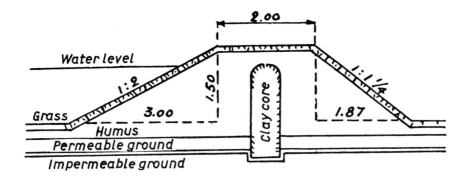

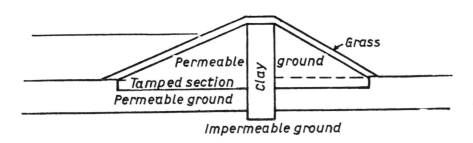

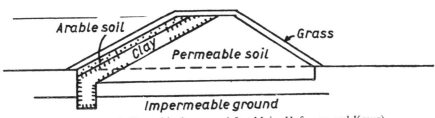

Fig. 22   Cross-section of pond dikes with clay cores (after Maier-Hofmann and Kreuz).

Fig. 23 Preparation for new ponds. Construction of the dike using an excavator.

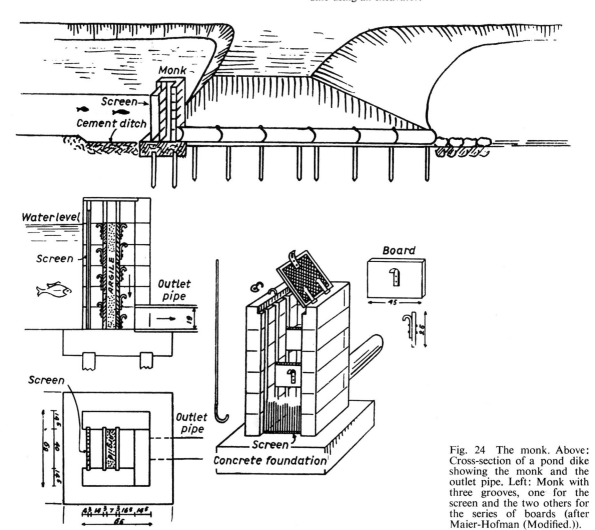

Fig. 24 The monk. Above: Cross-section of a pond dike showing the monk and the outlet pipe. Left: Monk with three grooves, one for the screen and the two others for the series of boards (after Maier-Hofman (Modified.)).

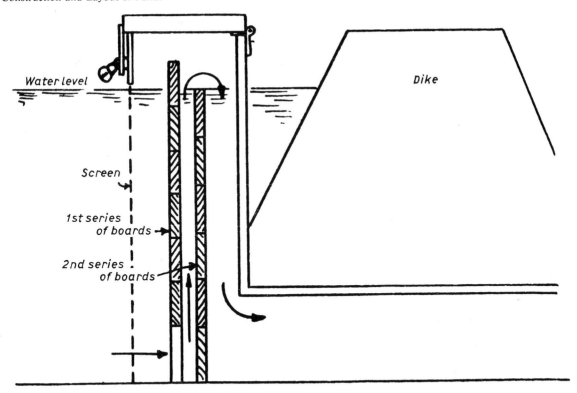

Fig. 25 Herrguth type monk permitting the evacuation of water near the bottom (after Diessner-Arens).

Fig. 26 Prefabricated drainage system: wooden monk slightly inclined and outlet pipe. Fish farm at Ligneuville, Belgium.

Fig. 27

Fig. 28

Figs. 27 and 28   Concrete monk. Three grooves: one for the screen, two for the boards. Fish culture station at Linkebeek, Belgium.

Figs. 29 and 30   Wooden monks. The outlet pipe is seen behind. Fish culture station at Linkebeek, Belgium.

Fig. 29   Fig. 30

Fig. 31   Concrete monk with wing-walls to prevent collapse of the dike. Fish farm at Mirwart, Belgium.

Fig. 32   Double monk with three grooves. Fishing-out device in concrete is seen in front. Fish farm at Bokrijk, Belgium.

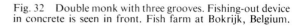

Fig. 33   Principal and secondary draining ditches of a pond. The monk is seen on the left. Bussche ponds, Lubumbashi, Zaire.

Fig. 34   General view of a water inlet with a small overflow on the right, and a sunken horizontal screen on the left. Ponds at Flamizoulle, Belgium.

Fig. 35   Collective water inlet for a group of diversion ponds. On the right a dam divided in three parts crosses the water course. The beams are removable. On the left, a horizontal sunken screen at the head of the water supply channel for the ponds. Fish farm at Mirwart, Belgium.

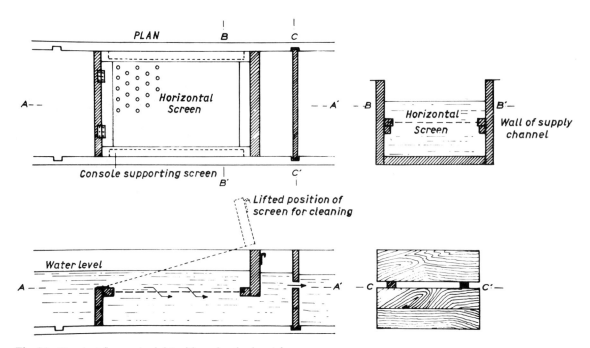

Fig. 36   Constant flow water inlet with sunken horizontal screen.

Fig. 37 Water inlet with horizontal sunken screen raised for cleaning. Diversion channel at top right; below left, inflow of water towards the pond. Keyberg ponds at Lubumbashi, Zaire.

Fig. 38 Horizontal screen removed from water for cleaning. The up-stream side of the screen is on the left. Fish farm at Mirwart, Belgium.

Fig. 39 Sunken horizontal screen. Water inlet for a linear pond for rearing trout fingerlings. Vevey, Switzerland.

Fig. 40 Rotating cylindrical screens equipped for automatic cleaning, photographed looking down-stream Munsel Creek, Florence, Oregon, U.S.A.

Fig. 41 Parallel diversion ponds. On the right the supply channel to which water inlets for the ponds are branched. Keyberg ponds, Lubumbashi, Zaire.

Fig. 42 Water distribution sluices on the general water supply channel for the ponds. Fish culture station of Jonkershoek, South Africa.

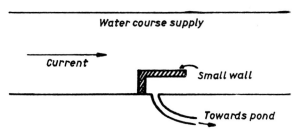

Fig. 43 Schematic drawing showing water inlet against the current.

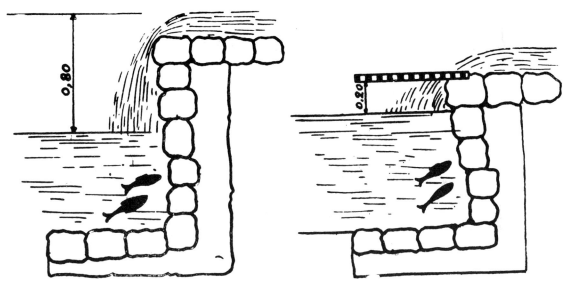

Fig. 44 Up-stream waterfall of a pond (after Evrard).

Fig. 45 Horizontal screen up-stream of a pond (after Evrard).

Fig. 46 Spouting fountain water supply system in a pond. Fish farm at St. Georges, Pas-de-Calais, France.

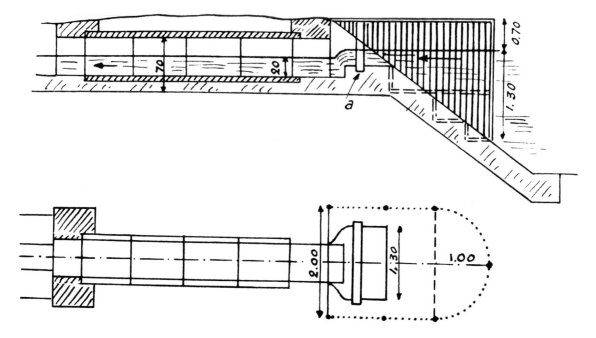

Fig. 47   Plan of barrage pond weir (after Evrard).

Fig. 48   General view of the weir of a barrage pond. Pond of Oise, Seloignes, Belgium.

Fig. 49 Small family pond at Tjisäät, Soekabumi, Indonesia. The pond has many uses: water reserve for certain domestic uses and fish culture. The latrines overhang the pond.

Fig. 50 Individual ponds grouped together in the heart of a valley. Nyakadaka, Kivu, Zaire.

Fig. 51 A group of individual ponds each one covering about 1 are (40th of an acre) at Katuishi, Luluabourg, Zaire. The fish can be fattened with draff from a neighbouring brewery.

Fig. 52 Fountain water supply in a pond achieved by splitting the top of the bamboo pipe. Soerabaya, Indonesia.

Fig. 53 Water control by bamboo screen. Soekabumi, Indonesia.

Chapter II

NATURAL FOOD AND GROWTH OF CULTIVATED FISH
SECTION I
FEEDING AND GROWTH OF FISH IN GENERAL

I Digestive System of Fish

THE digestive system of fish includes the same essential organs as in the higher vertebrates.

The mouth

Situated at the anterior end of the head, it is normally a horizontal opening which can close. It is placed either above, at the tip, or below. In lampreys, the mouth is a sucker and cannot close.

The oesophagus or gullet is situated at the back of the mouth. Both on the right and left sides there are a series of apertures known as branchial apertures, separated one from the other by branchial arches of which there are four on either side.

These branchial arches have a multitude of branches, called gill lamellae, which help the fish to breathe. They are also provided with small protuberances on either side. These are called branchiospines, the number, relative density and length of which vary according to the species. The branchial arches and branchiospines together on the one hand, and the branchial apertures on the other, form a kind of filter through which the water passes but which retains the food to be carried to the oesophagus. From the degree of fineness of the filter it is possible partly to deduce the kind of food a fish eats: for voracious fish the filter is large; for fish which eat small animals it is narrow, and is very fine for plankton-eating fish (Fig. 57).

The mouth generally includes a certain number of teeth. A basic distinction must be made however between voracious and non-voracious fish.

The first (Fig. 54) have numerous teeth often turned inwards. Normally they are set out in

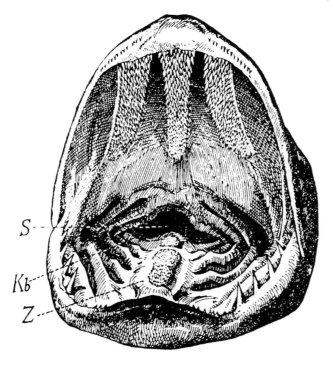

Fig. 54 Strongly dentated mouth of a voracious fish (pike) (after Walter, 1913).

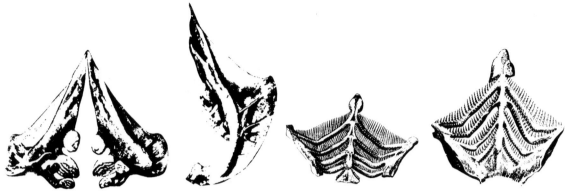

Fig. 55 Pharyngeal bone of carp showing pharyngeal teeth (after Rousseau, 1915).

Fig. 57 Branchial system of two species of whitefish (coregonids) with different feeding habits. The plankton feeding whitefish, on left, has a plexus of long, thin and tight branchiospines; the whitefish which feeds off larger nutritive fauna, on right, has shorter, thicker and fewer branchiospines (after Grote-Vogt-Hofer in Wunder).

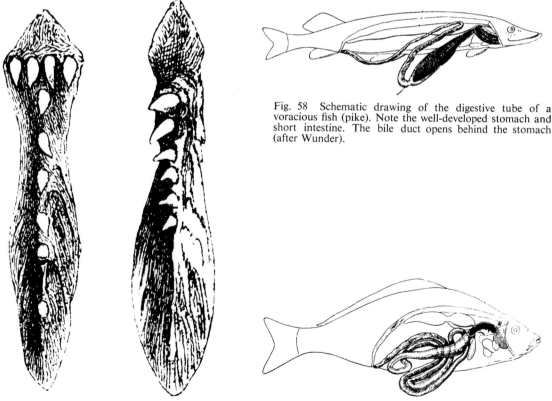

Fig. 58 Schematic drawing of the digestive tube of a voracious fish (pike). Note the well-developed stomach and short intestine. The bile duct opens behind the stomach (after Wunder).

Fig. 56 Vomer of salmon (right) and sea trout (left) showing vomerine teeth (after Rousseau, 1915).

Fig. 59 Schematic drawing of the digestive tube of a non-voracious fish (carp). Note the absence of a stomach and the relatively long intestine. The gall bladder opens immediately behind the gullet (after Wunder).

several rows on the jaws and also on the tongue which is rudimentary and only moves slightly. They are also found on the vomer (vomerine teeth, Fig. 56). These teeth have no roots. They fall out easily but grow again. Used teeth are replaced by other teeth. The teeth of voracious fish are not used for mastication but simply for the retention of prey.

Cyprinids, which are not voracious, have no teeth in their mouths proper but only behind. These are called pharyngeal teeth (Fig. 55). They are strong and large and disposed in one or several rows. They are used for mastication and cutting the food up before it descends into the stomach.

On the outside, the lips of many fish carry barbels.

The oesophagus is very short and dilatable.

The stomach of fish is easily extensible and this permits them to gorge themselves with food. This characteristic is marked in carnivorous fish (salmonids, perches, pike) but it is not very distinct in others (cyprinids, etc.) Voracious fish, trout in particular, have a distinctly defined stomach with a short intestine. In non-voracious fish, such as carp, there is no stomach, only a relatively long intestine.

In voracious fish (Fig. 58) the caeca or pyloric caeca at the limit of the stomach and intestine secrete powerful digestive enzymes. The number of pyloric caeca varies: perch have three, salmonids have up to 150. The stomach of voracious fish has an acid reaction (pH below 5·0); it produces pepsin.

In other fish (Fig. 59) the intestine starts almost immediately behind the oesophagus. The absence of pyloric caeca, however, sharply reduces the digestive action. There is no pepsin secretion and the reaction is rather alkaline (pH = 6·7 to 7·7). Digestion therefore is not so easy excepting when the necessary pepsins are brought to the fish by the natural food they eat: that is pepsins already contained in the stomachs and intestines of the animals eaten. It is for this reason that carp do not profit by traditional artificial food which they are given unless the natural food providing indispensable enzymes is an important part (about 50 per cent) of the total amount of food consumed. A different situation exists, however, if carp are given a complete diet including all the necessary elements, but at the moment this is still rare.

The intestine of fish is of variable length but relatively short when compared with that of other vertebrates.

In voracious fish digestion starts in the stomach and terminates in the intestine. Only when it is well advanced does the food pass from the stomach into the intestine. This means the food will be well digested and it is possible, in consequence, to give these fish just about as much as they can swallow.

It is not so with other fish which, because of the absence of the pylorus, can swallow such quantities of food that the intestine becomes completely filled. The actual digestion of the food starts at a determined point of the digestive tube and provokes the excretion and elimination of all food, digested or not, situated at the back. This can mean a loss of food.

In practice, for purposes of fattening, voracious fish (trout) can be given as much food as they can swallow, while for other fish, e.g. cyprinids, it is recommended that only small quantities be given at a time, in order to fill the intestine only partially: so it is necessary to feed them often.

Several other organs complete the digestive apparatus. The principal of these are the liver, the spleen and the pancreas.

II Growth of Fish in Ponds

Schäperclaus classes as follows the nutritive needs of fish considered from the points of view of rearing and physiology.

	Physiological angle		Fish cultivation angle
Total nutritive requirement	Maintenance of the body		Conservation of the body weight
	Construction	Replacement of worn parts of the body	
		Growth of the body; Constitution of reserves	Growth

The fact of satisfying the food requirements necessary for maintaining the body and replacing worn parts, results finally in conserving the body at a predetermined degree of development. The first of these needs is the pure and simple maintenance of the body, the second the replacement of worn or eliminated parts by cutaneous and digestive secretions, etc.

Fish farmers include under the general term "growth", the production of a new and larger mass including on the one hand increase in size or growth and, on the other, the formation of well localized reserves.

The growth of fish depends on a great many factors, some of internal origin, hereditary and relative to the speed of growth, the ability to benefit from food, and resistance to disease. The others are of external origin, designated on the whole as vital conditions, notably temperature, the quantity and quality of food, the composition and purity of the chemical medium (oxygen content and absence of poisons), necessary amount of life space (according to whether it is extensive or restricted so that growth is rapid or slow), etc.

The principal rules which determine the growth of fish are:

1. The maintenance need depends on the activity of the fish and its relative size. In the first case, all stress or troubles which interfere with the normal life of a fish (transport, capture, etc.), result in an increased activity which can cause the using up of three times the normal energy necessary for breathing. As for the second point, the smaller the fish the greater the relative surface of its body and the higher its vital demands, notably for oxygen. Consequently, from a practical point of view, the total weight of small fish of one species carried in a similar receptacle should be smaller than that of large fish of the same species.

2. For any one single species receiving sufficient food, the speed of growth is relatively that much greater as the size of the fish is smaller. Age is of secondary importance. That is to say a fish which is small because it does not receive sufficient food can grow rapidly and suddenly in the same way as a younger fish if it is finally given as much food as it needs.

The relative precocity of sexual maturity grows in proportion to the rapid development of the fish, but manifests itself at almost normal age even when the fish is dwarfed through lack of food and space.

3. Under normal conditions the relation of growth requirements to those of maintenance lies between 1/1·5 and 1/3·2.

4. Total energy demand and consequently the necessary quantity of dissolved oxygen required depends on the species. Under given conditions, the needs of carp are 2·5 times higher and those of rainbow trout are six times higher than the requirements of tench.

5. The "normal" growth of fish varies considerably from one species to another. Further, great differences exist for the same species. This is due to the strain and to individual particularities springing notably from the hereditary ability, first of seeking and finding food, and next of assimilating it. On any one fish farm it is not rare to find from the first stage of cultivation (say 6 months for example) fish whose increased weights vary once, twice or even more.

The "normal" size of a fish is difficult to establish for it depends essentially on the individual fish and the amount of food at its disposal. It is not possible to determine the age of a fish by its size. Thus, young trout of one summer (6 months) can measure 6 cm ($2\frac{1}{2}$ in) ($\pm$ 3 grm) while others under particularly favourable growth conditions can reach 20 cm (8 in) (75 grm), or a weight proportion of 1:25.

The assimilation of food by fish can occur in bursts rather than regularly. Very resistant to hunger, living on their interior reserves, they can withstand complete deprivation of all food without any appreciable trouble. If food is reduced more and more, growth will of course slow down or stop, yet without being harmful to the health of the fish or without affecting its subsequent growth. It is even possible to reduce the quantity of food in such a way that it no longer covers the maintenance needs. The fish will then lose weight. But it will start to develop when it is fed again. Resistance to hunger springs partly from the ability to assimilate directly the organic substances dissolved in the water. These are absorbed by the gills in very small quantities but continuously. Constantly entering the blood stream, they are an appreciable element, insufficient to feed the fish on its own but sufficient to give the body the ability of maintenance and resistance, sometimes for quite a long time, despite the absence of food.

6. The vital activity of fish and their growth in particular, depends on temperature as opposed to vertebrates living at a more or less constant temperature. Fish have a variable body temperature which depends on the external medium. Their bodies are capable of functioning at different temperatures and each species is adapted to a thermal zone, possessing an optimum within more or less narrow limits.

NATURAL FOOD OF FISH IN PONDS

I Biological Cycle in Ponds

Knowledge of the biological cycle which takes place in ponds is the basis of fish farming when natural foods are used exclusively or only partially. This is generally the case excepting when artificial food is used principally or exclusively.

1 The general biological cycle in fresh water

A fish is the final result of a complex biological cycle.

The origin of this cycle (Fig. 60) is in the mineral nutrients in solution in the water. They come from soluble substances in the ground, with which running or stagnant water is in contact, or from substances carried to the water by exogenous

detritus and also by rainfall.

By means of the sun's heat and light, the green vegetation is able to transform this inorganic matter and carbonic acid in solution in the water into organic matter which forms vegetable tissue (higher plants and lower plants: planktonic algae and biological cover).

"Biological cover or periphyton" is composed of vegetable and animal microscopic life living on the most diverse aquatic supports: stones and higher vegetation as well as mud. "Plankton" is formed by all the microscopic organisms: plants (phytoplankton) and animals (zooplankton), which swim or live suspended in the aquatic mass without being able to fight against the currents.

Living or dead (in the form of more or less fine

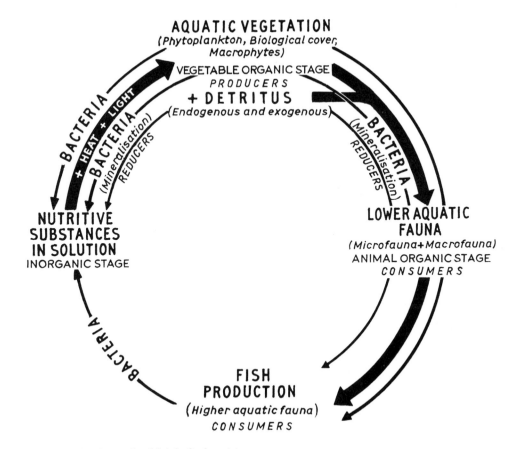

Fig. 60   Production cycle of fish in fresh water.

detritus) the plants are consumed by numerous small animal organisms. These then serve as food for larger water animals, which in turn (as well as the smallest of them) and also certain types of vegetation, both living or dead, are eaten by fish.

The chain or cycle of natural fish production includes the following links: mineral nutrients, plant production, intermediate animal consumption and production, leading to the final production which is the fish. The last stage is the reduction which takes place thanks to bacteria which by a mineralization mechanism permits the return in solution of all the dead components of organic matter – vegetable and animal – and their re-integration into the biological cycle.

Fish production in relation to the quantity of food present, depends in the final analysis on the production of vegetation. The latter depends on physical factors: light and heat. It is also subject to the law of minimum, according to which the quantity of vegetable produced is dependent on the nutrients found proportionally in the smallest quantity. Calcium is one of the mineral nutrients most subject to variations. Phosphorus, nitrogen and without doubt other elements also exist in insufficient quantities. These deficiencies justify the liming and fertilization of ponds (Ch. XII).

2 The biological cycle proper to ponds

The biological cycle proper to ponds is nothing more than the general biological cycle of fresh water but with certain particularities. It is, in sum, the biological cycle of a lake restricted to the single biocenosis of the littoral zone which is an independent biocenosis. It includes producers: vegetation (lower vegetables, planktonic algae and biological cover; higher vegetables); consumers: nutritive aquatic fauna at first, feeding off living vegetables and detritus, then fish; then reducers which are the bacteria.

In a pond the nutritive fauna is composed of: (1) zooplankton, small animal organisms living and swimming in water without direct contact with the soil or fixed aquatic plants; (2) animal organisms living among the biological cover on the bottom, plants and other submerged substrate; (3) living organisms in the bottom itself (mud, slime or sand) which may or may not be covered with plants (Ex.: larvae of *Chironomus plumosus*, *Tubifex*, nymphs of *Ephemera* sp.).

As far as the feeding of fish in ponds is concerned the normal and most important organisms are those which live on and in the bottom of the pond or among the aquatic plants. Planktonic organisms are important, above all, for young fish and plankton-eating fish.

If it is desirable to increase the production of cultivated fish by giving them greater quantities of natural food, it is not possible to do so directly. It is first necessary to increase the vegetable growth on which the animals feed. If it is desired to improve the biological cycle of the ponds then, evidently, the weakest link in the chain must be reinforced. To this end fertilizers, such as phosphorus compounds, have shown themselves to be effective. It is also necessary that the fish should live under the best possible conditions: sufficient oxygen content, alkaline reaction, and the absence of toxic substances. This is reached by rational stocking and adequate upkeep of the ponds.

The general classification of ponds is as follows: oligotrophic, eutrophic, and dystrophic.

An oligotrophic lake or pond is an aquatic medium or habitat poor in essential mineral nutrients such as calcium, phosphorus, nitrogen. The production of organic matter is not abundant. The water is generally clear and blue if it is deep. In its depths the dissolved oxygen content is high. A eutrophic water is rich in those nutritive elements which determine an abundant production of organic matter. The water is generally alkaline and favours the development of plankton, which gives a green or brown-green colour. In its depths the dissolved oxygen content is reduced, and at some periods (summer and winter) may be non-existent. Dystrophic water is rich in humic matter present above all in a colloidal form. The water is yellow to brown in colour and the pH is acid. The medium is not very productive; the aquatic vegetation is only slightly developed. Water derived from peaty marshes presents these characteristics. There are also intermediary stages between the three principal types of aquatic medium.

As for the origin of the nutritive substances, ponds are autotrophic or heterotrophic according to the origin of the substances present in the pond itself or coming in from outside (Ex.: ponds fed with sewage).

II Nutritive Aquatic Fauna in Ponds

A Characteristics of nutritive fauna in ponds

Following the point of view taken certain characteristics can be accepted, according to Schäperclaus:

1. From a general point of view the organisms which constitute the nutritive fauna develop in regularly exploited ponds, that is to say those which are periodically and alternatively dried and then put under water. They react to that alteration. According to the period and the length of time the pond is dry, which will differ according to the categories of the ponds (spawning ponds, nursery ponds, wintering ponds), the kind of fauna varies considerably. It varies also according to the average temperature and the flow of the water, lending itself to the cultivation of cyprinids or to salmonids. In any case the fauna is rather uniform in character, given that the medium is less varied than in a lake or in running water.

2. According to the type of food eaten the following fauna are singled out:

(*a*) consumers of higher plants (gammarids, many Trichoptera larvae with movable cases;

(*b*) consumers of microscopic vegetation; consumers of planktonic algae (zooplankton); consumers of biological covers (certain worms, crustaceans, chironomids larvae, Ephemeroptera nymphs);

(*c*) consumers of detritus (zooplankton, *Gammarus*, *Asellus*, worms and chironomids larvae living in mud with the exception of Tanypodinae);

(*d*) voracious fauna: eaters of plankton (*Bythotrephes*); consumers of organisms present among the vegetation and on the bottom (free-living larvae of Trichoptera, nymphs of Ephemeroptera and Odonata, larvae of *Sialis*, Hemiptera, Coleoptera and Tanypodinae; water-mites).

Among this last category of organisms certain serve as food for fish (Ephemeroptera nymphs, *Sialis*, Trichoptera and Tanypodinae larvae); others compete with the fish for food (water-mites, small Hemiptera); others are the enemies of fry (large nymphs of Odonata, larvae of Coleoptera, Hemiptera).

3. Considering their demands for dissolved oxygen and their resistance to organic pollution the following are singled out according to classification of the biological analysis in polluted water:

(*a*) sensitive organisms (nymphs of Ephemeroptera, Plecoptera and Odonata, larvae of Trichoptera);

(*b*) organisms rather resistant (different molluscs, *Asellus*, microcrustaceans, larvae of *Sialis*, Hemiptera, Coleoptera and certain chironomids larvae);

(*c*) very resistant organisms (*Tubifex*, certain chironomids larvae).

4. According to demands regarding water alkalinity the following are singled out:

(*a*) forms resistant to strong variations of pH and able to support low alkalinity: certain chironomids larvae, Ephemeroptera nymphs, certain crustaceans (*Cyclops*);

(*b*) forms more or less tied to a single and predetermined pH or better still to a minimum of alkalinity (gammarids and certain molluscs).

B Development of nutritive pond fauna

1 Origin of nutritive fauna

Organisms found in ponds arise from various sources:

(*a*) from supply water or from pools of water which have not dried out (in general all aquatic organisms);

(*b*) soil in which organisms have hibernated (molluscs, certain larvae of chironomids);

(*c*) those encysted organisms which resist dryness and cold and which hibernate at the bottom or can be carried by the wind (Copepods, Cladocera, Rotifera, leeches);

(*d*) eggs laid by adult insects in ponds recently put under water (Ephemeroptera, Odonata, Diptera, etc.);

(*e*) adults passing from an aquatic medium to another (Hemiptera, Coleoptera).

Populating a pond is carried out quickly and the work terminated only a short time after the pond is put under water. One must not, therefore, be afraid of drying out a pond in the fear that this will kill off the fauna. What would really be lost would certainly not exceed the lack of control of the cultivated population and the bad decomposition of the mud which would result.

2 Multiplication of the nutritive fauna

The speed at which aquatic organisms multiply depends largely on the annual generations. Wundsch singles out the following:

(*a*) species with pluri-annual generations (large larvae of Coleoptera, Odonata, Ephemeroptera, Trichoptera);

(*b*) species with one generation per year (the most part of Ephemeroptera, Trichoptera, Odonata, Hemiptera, molluscs, large larvae of chironomids);

(*c*) species producing several annual generations (small larvae of chironomids and culicids, gammarids);

(*d*) species which produce multiple generations annually (Cladocera, rotifers, etc.).

The number of generations is influenced strongly by both atmospheric and food conditions.

In the same order it is possible to single out, according to the period of maximum development of a species:

(*a*) predominantly spring species: Cladocera;

(*b*) predominantly pre-summer species: Diptera (chironomids in particular), molluscs;

(*c*) predominantly summer species: Ephemeroptera, Trichoptera;

(*d*) predominantly post-summer species: Odonata.

3 Variations in development

The development of aquatic fauna in ponds and its variations depend on the following rules:

(*a*) The evaluation of both the quality and quantity of aquatic nutritive fauna in ponds is calculated by the methods employed for determining the biogenic capacity (Huet, 1949).

(*b*) The more a pond is stocked the greater is the amount of natural food consumed. In other words the greater the density of the population so the greater the chances of their finding and consuming the natural organisms.

(*c*) Nevertheless, even when the fish stock is much greater than what is generally considered normal all the natural food will be far from eaten. But such stocking will lead to the stopping of growth by limiting the quantity consumed by each fish (Fig. 61).

When the number of fish is too high there is still no need to fear that there will be a fall in the number of nutritive organisms, for the natural ability of most species to multiply is so high that this need not be feared. However, the greater the fish population the more the fauna will be consumed before they reach average size let alone full size. Under these circumstances the nutritive population will be diminished in gross volume. Finding the fauna will be more difficult and the calories they offer will be insufficient, or only just what is required to compensate for the search for such food. The result will lead to stagnation of growth.

It has been found that more food is produced in a well stocked pond than in a similarly productive pond with fewer fish. This phenomenon is similar to that which determines an increase in the production of cultivated fish as a consequence of intermediary fishing.

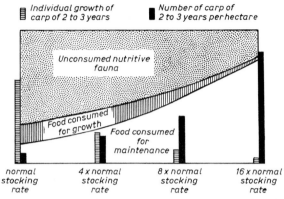

Fig. 61 Individual growth in ponds and density of stock. Schematic illustration showing relationship of growth to density of stock in fish culture, the individual growth of carp, the consumed part and the unconsumed part of nutritive fauna, and the utilization of the food consumed (after Schäperclaus, 1933).

C Principal constituents of nutritive fauna in ponds

The principal constituents of aquatic nutritive fauna found in ponds belong to the following zoological groups: rotifers, worms, crustaceans, molluscs, insects (Figs. 62 to 112).

Rotifers. Rotifers (Figs. 62 to 64) are very small measuring from 1/50 to 2 mm; they live no longer than from 2 to 3 weeks but can exist in prodigious quantities. They are found in great numbers in the planktonic fauna as well as on the bottom, particularly where the vegetation is rich. Rotifers are eaten by fry and small fish but above all by the aquatic animals on which the fish feed. According to Schäperclaus, Pauly singles out the following forms in eutrophic ponds: (1) forms periodically dominant (*Conochilus*, *Brachionus*, *Anuraea*), (2) permanent but not abundant forms (*Asplanchna*, *Synchaeta*), (3) abundant but irregular forms (*Triarthra*, *Polyarthra*, *Rattulus*, etc.).

Worms. Only a few worms (Figs. 65 to 70) play an important part as food for fish. Among the principal are to be found the oligochetes. These vary markedly in size; while some are almost microscopic, most of them measure from 8 to 15 mm while others reach 30 cm (12 in). In the littoral zone, particularly among vegetation, Naididae are found: *Stylaria* and *Nais*. These are small animals, both transparent and colourless. In the mud, particularly when it is rich in organic matter, important colonies of *Tubifex tubifex* Müller (= *T. rivulorum* Lamark) are found. It is a reddish worm from 25 to 85 mm long (between 1 and 3½ in) and 1 mm in diameter. It makes a kind of tube for itself out of mucus and mud particles which it sinks into the ground. The anterior part of the body settles in the tube while the rest of its body waves above the ground in a breathing movement.

The leeches (*Herpobdella*, *Glossiphonia*) are not generally eaten by fish.

Molluscs. Lots of molluscs live in ponds (Figs. 71 to 79) feeding off detritus and vegetation.

Gasteropods are the most important molluscs for fish. They live, above all, among submerged plants. The shell is twisted in a kind of spiral and is sometimes closed by an operculum. Among those which must be mentioned are *Limnaea* (*L. stagnalis*, length 35 to 60, diameter 16 to 27 mm; *L. ovata*, l. 15 to 33, d. 10 to 22 mm; *L. auricularia*, l. 15 to 35, d. 14 to 28 mm); several *Planorbis*, *Bythinia tentaculata*, l. 8 to 12, d. 4·5 to 7 mm; *Vivipara vivipara*, l. 18 to 30, d. 14 to 25 mm; *Valvata piscinalis*, l. 5 to 6, d. 5·5 mm; *Physa fontinalis*, l. 8 to 12, d. 5 to 9 mm.

Lamellibranchs have shells formed by two symmetrical valves and are not often eaten by fish, excepting the very small species, *Pisidium*, l. 2·5 to 10 mm and *Sphaerium*, l. 7 to 25 mm.

Crustaceans. Crustaceans are one of the most important of the zoological groups which are food for fish. The lower crustaceans are the principal plankton food eaten by fry and young fish. They also form the principal food for plankton-eating fish. The higher crustaceans are consumed by adult fish.

Many entomostracans or lower crustaceans (Figs. 82 to 89) are found among the fauna on the bottom and among the vegetation. The most common and important of these are the Cladocera, small crustaceans from $\frac{1}{4}$ to 3 mm in length. *Eurycercus lamellatus*, *Sida crystallina*, are among the most abundant; while *Alona*, *Simocephalus*, *Chydorus* are less abundant. *Daphnia pulex* and *D. magna* are particularly abundant in recently filled small ponds or those manured by organic matter. There are also many entomostracans in planktonic forms: *Bosmina longirostris*, different *Diaptomus*, *Daphnia longispina*, *Diaphanosoma brachyura*, *Polyphemus pediculus*, *Ceriodaphnia*. The latter is a typical representative example of pond plankton while *Bythotrephes* are never found and belong to the lacustrine plankton.

Ostracods about 0·5 to 2·5 mm in length belong to fauna inhabiting the bottoms. They are rarely to be found in large number. *Cypris* spp.

Copepods, rarely longer than 5 mm, play an important part among the littoral and pelagic fauna of ponds. Along with the Cladocera they are the principal food of a whole series of fresh water fish. *Cyclops*, *Diaptomus* spp.

Among the malacostracans or higher crustaceans the most abundant is the fresh water shrimp (*Gammarus pulex*, Fig. 81), found in trout ponds and in running water while the water louse (*Asellus aquaticus*, Fig. 80), is found in calm muddy water or in water which has plenty of plant life and is even slightly polluted by organic matter.

Insects. Insects form one of the most important food organisms for cultivated fish in ponds. They are generally eaten as larvae or nymphs: Ephemeroptera, Plecoptera, Odonata, Megaloptera, Trichoptera, Diptera, Hemiptera and Coleoptera are found as both larvae and adults. The most important nymphs and larvae for fish are the Ephemeroptera and the chironomids.

Ephemeroptera (mayflies) (Figs. 90 to 92). This is one of the most important groups on which cultivated fish feed. The nymphs, either burrowed or swimming, live among the submerged plants and on the bottom. The principal genera of mayflies in the ponds are *Cloëon*, *Caenis* and *Baetis*. The adults are of great importance as exogenous food.

Plecoptera (stoneflies) (Fig. 93). The same role is played but it is of less importance. *Nemura* is the principal genus found in ponds (nymphs from 6 to 9 mm).

Odonata (dragonflies and damselflies) (Figs. 94 and 97). Dragonfly larvae (Libellulidae, Aeschnidae, Agrionidae are found (living among aquatic plants and on the bottom and in the mud of all ponds. The fish do not care for them much and the smallest of them compete for food. The large ones are enemies of the fry.

Megaloptera (alderflies) (Fig. 98). The larvae of *Sialis* measure from 20 to 25 mm and are found in ponds with rich mud bottom and organic matter.

Trichoptera (caddisflies) (Figs. 95, 96, 99 to 103). Caddisfly larvae, many of which are protected by a plant or mineral case into which they can withdraw, are rather common to all ponds, especially salmonid ponds. In cyprinid ponds, Leptoceridae and Phryganeidae are very common but not so numerous if the ponds are drained dry annually and for prolonged periods.

Hemiptera (water-bugs) (Fig. 104). These are numerous in many ponds, but those appreciated by the fish are rare with the exception of *Corixa*

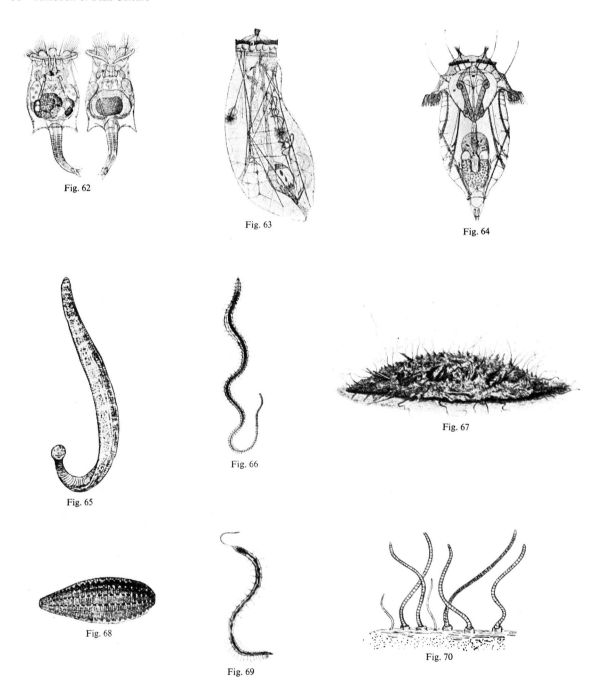

Fig. 62

Fig. 63

Fig. 64

Fig. 65

Fig. 66

Fig. 67

Fig. 68

Fig. 69

Fig. 70

ROTIFERS AND WORMS
Fig. 62 *Brachionus entzii* Fr.   Fig. 63 *Asplanchna priodonta* Gosse.   Fig. 64 *Synchaeta pectinata* Ehrbg.   Fig. 65 *Herpobdella octoculata* L.   Fig. 66 *Tubifex tubifex* Müll.   Fig. 67 Colony of *Tubifex* with *Lumbriculus.*   Fig. 68 *Glossiphonia complanata* L.   Fig. 69 *Stylaria lacustris* L.   Fig. 70 Detail of a colony of *Tubifex tubifex* Müll.   Fig. 62 according to Francé, 1910.   Figs. 63, 64, 66, 67, 68, 69 according to Wesenberg-Lund, 1939.   Figs. 65, 70, according to Wundsch, 1926.

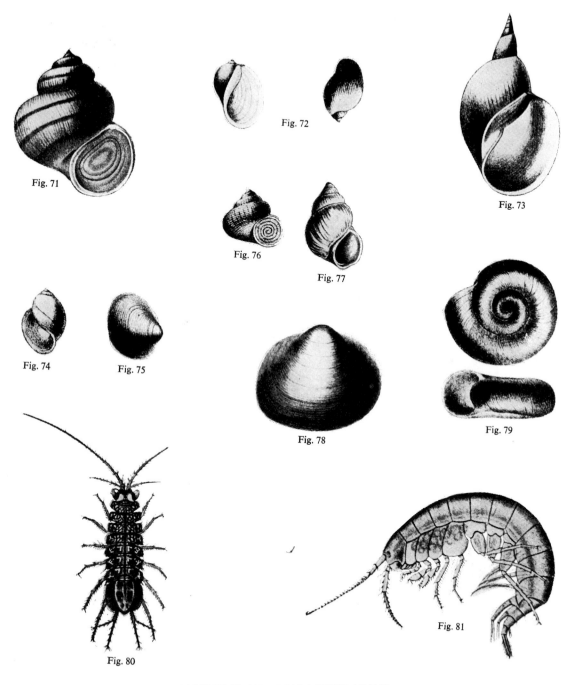

Fig. 71

Fig. 72

Fig. 73

Fig. 76

Fig. 77

Fig. 74

Fig. 75

Fig. 78

Fig. 79

Fig. 80

Fig. 81

MOLLUSCS AND MALACOSTRACANS

Fig. 71 *Vivipara vivipara* L. Fig. 72 *Radix (Limnaea) ovata* Drap. Fig. 73 *Limnaea stagnalis* L. Fig. 74 *Physa fontinalis* L. Fig. 75 *Pisidium casertanum* Poli. Fig. 76 *Valvata piscinalis* Müll. Fig. 77 *Bithynia tentaculata* L. Fig. 78 *Sphaerium corneum* L. Fig. 79 *Planorbis (Coretus) corneus* L. Fig. 80 *Asellus aquaticus* L. Fig. 81 *Gammarus pulex* De Geer. Figs. 71, 72, 73, 74, 75, 76, 77, 78, 79 according to Geyer, 1927. Figs. 80, 81 according to Wesenberg-Lund, 1939.

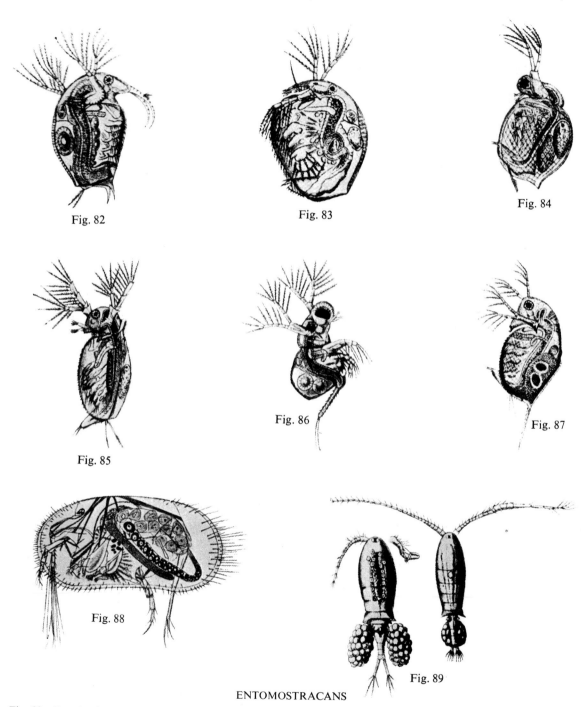

ENTOMOSTRACANS

Fig. 82   *Bosmina longispina* Leyd.   Fig. 83   *Chydorus sphaericus* Müll.   Fig. 84   *Ceriodaphnia reticulata* Jur.   Fig. 85 *Sida crystallina* Müll.   Fig. 86   *Polyphemus pediculus* L.   Fig. 87   *Daphnia longispina* Sars.   Fig. 88   *Cypris reptans* Baird.   Fig. 89   *Cyclops strenuus* Fisch. and *Diaptomus coeruleus* Fisch. Figs. 82, 83, 84, 85, 86, 87, 89 according to Lampert, 1925. Fig. 88 according to Wesenberg-Lund, 1939.

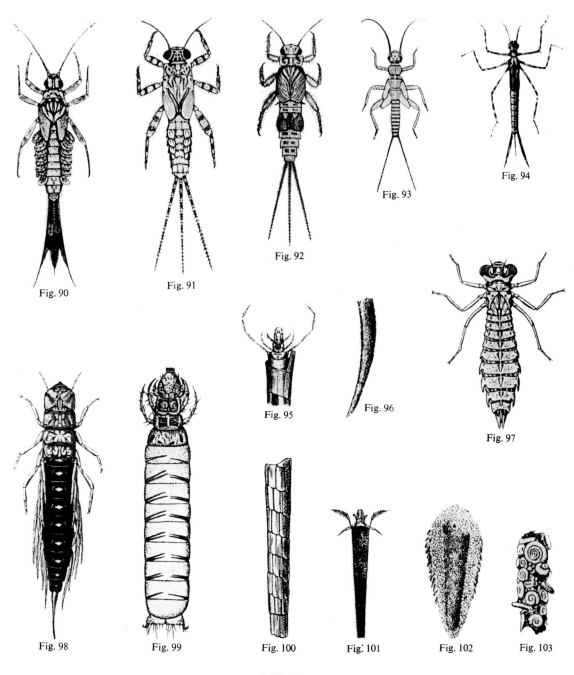

Fig. 90

Fig. 91

Fig. 92

Fig. 93

Fig. 94

Fig. 95

Fig. 96

Fig. 97

Fig. 98

Fig. 99

Fig. 100

Fig. 101

Fig. 102

Fig. 103

INSECTS

Fig. 90 Nymph of *Cloëon dipterum* L. Fig. 91 Nymph of *Ephemerella ignita* Poda. Fig. 92 Nymph of *Caenis macrura* Steph. Fig. 93 Nymph of *Nemura marginata* Pict. Fig. 94 Nymph of *Calopteryx splendens* Harr. Fig. 95 Larva of *Triaenodes bicolor* Curt. emerging from case. Fig. 96 Larva case of *Leptocerus aterrimus* Steph. Fig. 97 Nymph of *Aeschnidae.* Fig. 98 Larva of *Sialis lutaria* L. Fig. 99 Larva of *Stenophylax* sp. Fig. 100 Larva case of *Phryganea grandis* L. Fig. 101 Larva of *Setodes interrupta* Fab. Fig. 102 Larva case of *Molanna angustata* Curt. Fig. 103 Larva case of *Limnophilus flavicornis* L. Figs. 90, 91, 92, according to Schoenemund, 1930. Figs. 93, 94, 95, 99, 101 according to Wesenberg-Lund, 1943. Figs. 96, 100, 102, 103 according to Rousseau, 1921. Figs. 97, 98 according to Miall, 1934.

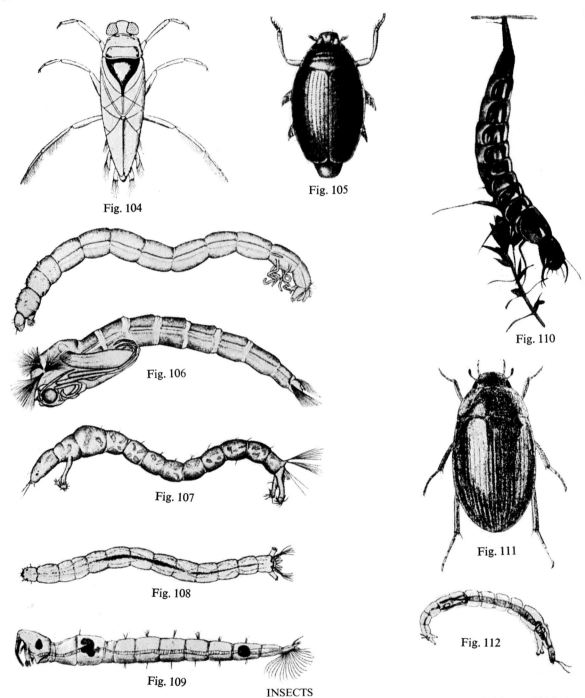

Fig. 104

Fig. 105

Fig. 106

Fig. 107

Fig. 108

Fig. 109

Fig. 110

Fig. 111

Fig. 112

INSECTS

Fig. 104 *Notonecta glauca* L. Fig. 105 *Gyrinus natator*. L. adult. Fig. 106 Larva and nymph of chironomid of the *plumosus* group: *Chironomus tentans* Fabr. Fig. 107 Larva of *Tanypodinae: Pentaneura monilis* L. Fig. 108 Larva of *Orthocladiinae: Metriocnemus knabi* Coq. Fig. 109 Larva of *Corethra plumicornis* F. Fig. 110 Larva of *Dytiscus marginalis* L. Fig. 111. *Hydrobius fuscipes* L. adult. Fig. 112 Larva of *Corynoneura celeripes* Winnertz. Fig. 104 according to Miall, 1934. Figs. 105, 110, 111 according to Wesenberg-Lund, 1943. Figs. 106, 107, 108 according to Johannsen, 1937. Fig. 109 according to Lampert, 1925. Fig. 112 according to Karny, 1934.

of which there are several species. Other genera are either competitors for food or a danger to the fry (*Notonecta, Nepa*).

Coleoptera (water-beetles) (Figs. 105, 110, 111). In the form of larvae and adults, beetles are fairly abundant in ponds (Dytiscidze, Gyrinidae, Hydrophilidae). They are not appreciated as food by the fish but they compete for food. Larvae of many species, and certain adults also, are enemies of both fry and small fish. The adults fly easily and pass from one pond to another rapidly, populating those ponds recently put under water. They vary between 2 and 45 mm in length.

Diptera (twin-winged flies) (Figs. 106 to 109, 112). The most important two-winged flies are the chironomids (midges). Their larvae alone form one of the principal sources of food for fish. Apart from the plankton-eating species, all fish eat chironomids and many make it their principal form of nourishment. Certain larvae, red in colour, are also well known as "blood worms" to anglers. Chironomids larvae are oblong, cylindrical, and more or less wormlike in shape. They are from 2 to 20 mm long according to the species, but most are from 6 to 12 mm. In colour they are red, rose, green, yellow and white. They live free or build cases for themselves. They are found in abundance everywhere, in trout ponds as well as in those of cyprinids. They live among the vegetation and in the mud. The larvae of Ceratopogonidae and multitudes of Chironomidae (Tanypodinae, Diamesinae, Orthocladiinae, Tanytarsariae, Chironomariae) are very difficult to identify.

Beside chironomids, other two-winged fly larvae are found, particularly *Corethra plumicornis*. This is a transparent planktonic larva. There is also *Simulium* where there is an obvious flow of water.

III Natural Food of Principal Cultivated Fish

Food eaten by fish differs considerably according to the species and their ages, bearing in mind that differences between the species develop with age. There are three feeding phases for fish according to age.

The first phase: resorption of the yolk sac

Fish absorb no food. The larger the yolk sac the more slowly it is absorbed. The higher the temperature the faster it is absorbed. Fry hatched at the end of the winter (salmonids) have larger sacs than those at the end of the spring (cyprinids).

Second phase: fry stage

This phase lasts about 6 months. A short time before the yolk sac is absorbed, fry start eating. All species eat more or less the same food: Cladocera and algae of the biological cover. Then follow other crustaceans and chironomids larvae. Little by little, as the fish grow older, the food requirements change and resemble more and more these of adult fish.

Third phase: adult stage

It is possible to distinguish those fish which eat plants, those which are consumers of small aquatic fauna, and those that are voracious types. In general, however, these distinctions are not absolute, for most fish eat different kinds of food though preferring one food in particular.

1 The herbivores

This type is rare and fish which live almost exclusively on plants are not very numerous. Among cultivated fish of European origin they are practically non-existent. Typical herbivores include *Tilapia rendalli* of African origin and *Ctenopharyngodon idella*, of East Asiatic origin. When farming is intensive, for non-voracious fish such as carp for example, vegetable food can be used as the basis of artificial food.

2 Consumers of small aquatic fauna

These are the principal sources of natural food for most cultivated fish. Phytophilous, benthic, planktonic and exogenous foods can be singled out.

Phytophilous food is that which develops either on or among submerged aquatic vegetation. Benthic food, or food at the bottom, develops on the bottom but also burrows in the bottom as well as just above the bottom or thereabouts.

Apart from the herbivorous and the voracious species, most cultivated fish such as carp, tench and trout also belong to this category. Generally fish falling into this group eat all nutritive organisms of the phytophilous and benthic fauna. However, well defined preferences for different nutritive fauna exist. The Mormyridae and Siluridae are typical benthic food eaters. Some, which find food on the bottom, only eat detritus. They are the pelophagous species of which *Citharinus gibbosus* is typical. There are also malacophagous species such as *Haplochromis mellandi* which eat molluscs principally. There is however general repulsion for certain animals with tastes of their own such as Hydrachnids (or water-mites) or those possessing a strong chitinous carapace (Coleoptera).

If most of these eaters of small aquatic fauna live partially on plankton and even make it their principal food when young, very few species eat it exclusively when fully grown. Coregonids and certain tilapias belong to this category.

No cultivated fish live exclusively off exogenous food. However, among those species which have a marked preference for this kind of food, trout must be mentioned.

3 Voracious fish

These are fish which live principally off other fish. When still young they eat, principally, small aquatic fauna. Later on their voracious characters become more or less marked. After a few weeks certain fish, such as pike, become exclusively voracious. Others, trout and perch among them, are happy enough with small fauna but will eat other fish given the chance. Under these circumstances their growth is greatly accelerated.

The eating habits of a species depend largely on the mechanical ability to catch and consume prey, and also upon the physiological ability to seek their food out. These faculties differ considerably from one species to another and are notably distinct between the voracious and the non-voracious types.

1. Voracious species (trout, pike and perch) have a terminal mouth with teeth. They swallow their prey at once without cutting it up or masticating it. From this fact, and because of the shape of their mouths, these fish leave all that exists on the bottom and in the mud of the bottom. They are only interested in that which swims above the bottom or the vegetation and that which moves in the water.

2. The non-voracious species (cyprinids) have more or less wide mouths which allow them to dig over the muddy bottom. Their barbels are very helpful for seeking out prey. Having filled the mouth with the nutritive organisms contained in the mud and some of the mud itself, the latter is ejected in part. Thanks to the presence of pharyngeal teeth the food is partly cut up before being swallowed.

As an example, here are some precise facts on the eating habits of three important fish that are cultivated: carp, trout, and pike. They are very different one from the other.

Carp

Carp fry live off plankton and small animals found on the vegetation or near it and as well as on the bottom. The principal constituents are insect larvae and littoral Cladocera. Older carp also eat insect larvae principally (especially chironomids), and small crustaceans (Cladocera, Copepoda), found among the vegetation, on and in the bottom. Occasionally they also eat plants. In the same way as the fry, big carp also eat plankton.

Trout

When young, brown trout feed exclusively off small prey. Only if they are tightly cultivated (hatching troughs, intensive nursery ponds) does cannibalism develop at an early age among trout, particularly brown trout.

As they grow older trout eat larger prey, notably large caddisfly larvae which are swallowed with their cases. If it is possible to catch small fish (sculpin, minnow) then they benefit from them. But the natural food for large trout differs considerably when they are in running water or when they are in ponds.

Pike

This fish is the most voracious to be found in temperate fresh water. Except when it is very young and lives off small aquatic prey (crustaceans, insect larvae), it eats fish almost exclusively. It can swallow prey 20 or 30 per cent its own weight and even more than that. When pike are mixed with cyprinids, the latter must always be much larger than the pike.

Part Two

TECHNIQUES AND METHODS OF FISH CULTIVATION

Introduction

FISH cultivation is spread over the entire world. Methods are diversified and systems differ according to the species cultivated, the aims (fish for eating or repopulation) and the techniques and methods employed.

These last terms are used in a definite sense by Lemasson and Bard (1966). A "technique for the cultivation of fish" is a procedure applicable to a specified operation employed for the cultivation of fish; for example the artificial fertilization of fish or their artificial feeding. A "fish cultivation method" is a combination of an ensemble of fish cultivation techniques: the raising of common carp in different age categories for example. A "system of fish cultivation" is the employment of a method of fish cultivation under specified ecological and socio-economic conditions such as the farming of carp in Central Europe or the raising of milkfish in tambaks in Indonesia.

The cultivation of fish is of interest for only a relatively small number of species. The reason for this is that only a few fish meet the required conditions for cultivation. These conditions were given in the "preliminary notes".

The first and most important distinction to be made between the different types of fish cultivation is between cold water and warm water cultivation. In practice the temperature separating the two types of cultivation is 20°C. In reality it would be better to speak of cultivation in water with rather low temperature and cultivation in water with rather high temperature. However, for the sake of simplification and for conventional reasons the expression "cold water" and "warm water" will be used. The cultivation of salmonids is adapted to cold water while the cultivation of cyprinids and others is adapted to warm water. Apart from this fundamental difference of distinct temperatures, the two groups of cultivation also use different reproduction and artificial feeding techniques but

these differences, which have been very marked up till now, are tending to lessen.

The cultivation of cyprinids is the cultivation of carp and other cyprinids in water around and over 20°C, the optimum being between 20 and 30°C. A relatively high temperature is necessary not only for growth but also for reproduction. Natural but controlled reproduction is practised; control is notably used for choosing parent fish and the arrangement of the spawning ponds. However, there is now some development towards artificial fecundation techniques for cyprinids. In general, raising cyprinids rests first of all on the use of natural food found in ponds. This means the population must not be very dense and the cultivation of cyprinids should extend over rather large areas. The ponds are shallow and water supply is not important or, maybe, no water is supplied at all. The general trend is to intensify the cultivation of cyprinids by practising fertilization and artificial feeding. In this way intensive cultivation of cyprinids in small ponds with constant renewal of water can be carried out. In most regions cyprinids are generally cultivated to be sold and eaten.

The cultivation of salmonids includes, in varying degrees, several important species of the salmonid family. The aim is either the production of young fish for restocking free waters or the production of marketable fish for the table. In the last case rainbow trout are the principal fish. The cultivation of salmonids is carried out in cold water, that is water with a temperature not above 20°C. Such waters are found either in the mountains or near abundant springs at low altitudes in temperate regions. The optimum for the growth of salmonids is between 20 and 10°C and temperatures equal to the last or lower are indispensable for reproducing the species. In fish cultivation, the reproduction of salmonids is always artificial. Production of salmonids rests, sometimes, on natural food but most often on artificial feeding in its intensive and even very intensive forms. The ponds used are small but the renewal of the water is always very important.

The cultivation of cyprinids and salmonids constitute the two most important types of fish farming. There are others which differ somewhat but which, nevertheless, approach either according to the techniques and methods used. According to each case, it can be cultivation in cold water, though more likely in warm water; it can also be cultivation for restocking or for eating; it can be based on natural reproduction or on artificial fertilization; it can be extensive or intensive cultivation.

In this way the cultivation of pike, the cultivation of coregonids and sturgeons are principally for restocking and artificial fertilization is used. This is developed in temperate climate regions. The cultivation of perciformes is more common and more varied. In Europe, pike-perch are cultivated principally for restocking. In the United States black bass are cultivated principally for angling. In the tropical regions of Africa, principally, but also in the Far East and on other continents the cultivation of cichlids (tilapias and others) has developed regularly over the past 20 years. Other forms of cultivation such as eels and silurids have developed more slowly.

The first two chapters of the second part of this work are devoted to salmonid cultivation and cyprinid cultivation. Cultivation for restocking using artificial fertilization and incubation follows. After that comes the cultivation of perciformes, followed by silurids and then such special forms of cultivation as eels, cultivation in rice paddies and in brackish water. The last chapter is devoted to the specifics of regional fish cultivation.

Chapter III

BREEDING AND CULTIVATION OF SALMONIDS

or

FISH CULTURE IN COLD WATER

THE purpose of salmonid cultivation is the production of fertilized eggs, of fry, of fish of one and two summers, most of the latter being intended for eating. The principal operations group successively: artificial fertilization and incubation, rearing fry and fattening trout intended for eating. Fish of one summer reach marketable size in the autumn of their first year. They are known as summerlings or fingerlings. Trout for eating are generally sold during their second year.

Salmonid cultivation intended for the table is principally practised with rainbow trout and occasionally with other species such as brown trout and brook trout.

Salmonid cultivation for restocking uses the species given above as well as other salmonids such as Atlantic salmon. Pacific salmon, Danube salmon and Arctic char. This cultivation also takes in coregonids and grayling but only according to special methods which will be examined separately (Chapter V, sections II and IV).

In a word, the cultivation of salmonids is practised in cold water throughout the world. It spread with the introduction of such principal species as brown trout and rainbow trout in salmonid waters of all continents. These fish are found in Europe, North and South America, Asia, Africa and in Oceania.

The reasons for this wide dispersion are the facility and advantages of artificial reproduction of the fish and also the possibility of fattening most of them by feeding.

The principal steps towards artificial reproduction (fertilization and incubation) such as is practised now, were as follows: (1) the discovery or rather the rediscovery of wet, artificial fertilization around 1842; (2) the discovery of dry fertilization around 1857; (3) the use of incubators with water circulation from bottom to top.

The principal advantages of artificial reproduction and incubation are as follows: (1) a great number of fertilized eggs are obtained; (2) the protection of eggs and fry against many natural enemies; (3) the ease of repopulating open water; (4) making possible the repopulating of certain water courses of plains which do not have natural spawning grounds; (5) the ease with which large numbers of fingerling trout can be produced.

The principal stages of artificial salmonid cultivation are as follows: (1) the harvesting of sexual products (eggs and milt) and artificial fertilization; (2) artificial incubation and hatching until the yolk sac is absorbed; (3) the production of 1 year fish; (4) the fattening of 2 year fish.

SECTION I

REQUISITE CONDITIONS FOR SETTING UP A SALMONID CULTIVATION FARM

I   Sites for Salmonid Fish Farms

In salmonid cultivation the first and principal question is water. It is essential that there should be abundant, clear, pure, fresh water, always renewable and protected against heating in the summer season in order to assure a sufficient dissolved oxygen content. These hydrographical conditions limit the number of sites where such a fish farm might be possible.

The most important biological characteristic of salmonids and one on which the whole arrangement and exploitation of their cultivation depends, is their requirement for breathing. Salmonids must have water which is rich in dissolved oxygen and always, with certain exceptions, 9 mg per litre of dissolved oxygen or almost double that required by cyprinids.

The choice of a satisfactory site for salmonid cultivation rests on technical considerations (topographical conditions, quantity and quality of water) and commercial conditions (availability of food for fattening and marketing possibilities). It is essential the latter be good for those farms producing trout for the table. They are less important but by no means negligible for the cultivation of fry and fingerlings for restocking. As for technical needs a sufficient quantity of good quality water is necessary and topographical conditions must be favourable. These last conditions were considered in a preceding chapter (Chapter I).

1 Quality of the water

The water must be pure and fresh – a physico-chemical analysis is indispensable.

As regards chemical composition, calcareous water is best because it is richer in natural food and is the most productive. This is an important point for those farms which produce fingerlings or second year trout without artificial food, and also where brood fish are kept. Water from granite sources is pure but not so rich in natural food. Marsh and peaty water is acid, poor and should, in principle, be rejected because it is only slightly productive. Also beware of magnesium, ferruginous and selenetic water – one containing calcium sulphate.

The chemical analysis of the water can be completed, with advantage, with a biological study (see Chapter IX). If carried out with proper care then the latter should be sufficient. If the biogenic capacity is average (IV to VI) or high (VII to X), then, *a priori* the water should be sufficiently productive.

Chemically, it is essential that the water should have a sufficient oxygen content at all times and in all seasons. The dissolved oxygen content needed by most salmonids for breathing is 9 mg per litre (pt). But this minimum level is generally only possible in renewed water with a temperature below 20°C.

The oxygenization of water depends on different factors but is always related to the temperature of the water. The higher the temperature the less the dissolved oxygen. The relation of temperature to the oxygen content in pure water is given in Chapter XII (Fig. 382).

Only exceptionally should water exceed 20°C in salmonids ponds. It is also indispensable to take note of the temperature of the water during the hottest season of the year – and it is the temperature of the pond water that must be tested and not the water supply. The fact that there are trout in the water course is not a sufficient guarantee that the ponds are right for the production of salmonids, and this would most certainly not be the case when rather large ponds are established at low altitudes and supplied with water coming from a small brook.

Under certain conditions, which occur in Spain (Calderon, 1965), it is possible to cultivate rainbow trout or even brown trout in water with a temperature higher than 25°C. Rainbow trout have been raised in water where the maximum temperature at 6 o'clock in the afternoon in the months of July and August have reached between 26·5 and 27°C. But to be successful under such conditions there must be: (1) an abundance of submerged plants (*Potamogeton*, *Ceratophyllum*, *Elodea*) covering about 40 per cent of the bottom; (2) good sunshine on the ponds, such as is found in Mediterranean countries; (3) a relatively small flow of water during the day permitting over-saturation in dissolved oxygen.

Generally it is best to rule out water which is too warm, but water which is too cold should also be avoided. The water should not have a temperature below 10°C in summer. Trout in it will remain quite healthy but their growth will be poor.

Provided the water is renewed sufficiently, temperatures between 15 and 17°C are best for the growth of most salmonids, excepting the very young, for which the water should be colder: 10°C and less.

2 Quantity of water

The quantity of water, along with its quality, is of prime importance in relation to the type of fish farm. It is absolutely indispensable that a sufficient quantity of water is assured particularly during the heat of the summer.

The quantities given below can serve as a basis for calculation when temperatures are below 10°C during incubation, and not above 15°C during the hatching and rearing.

For incubation: $\frac{1}{2}$ litre per minute for 1,000 eggs;

For rearing fry from 0 to 3 months: 1 to 3 litres per minute for 1,000 fry;

For rearing fingerlings from 4 to 8 months: 4 to 8 litres per minute for 1,000 trout fingerlings;

For fingerlings of 6 to 12 months: 6 to 12 litres per minute for 1,000 fingerlings.

It is necessary, on the average, for 1,000 fry and fingerling trouts to have 1 litre (pt) of water per minute per month of the age of the trout.

For trout intended for consumption, Léger has proposed a litre-kilo formula: 1 litre of water per minute will permit the rearing of 1 to $1\frac{1}{2}$ kg (3 lb) of trout if the temperature is fresh (about 15°C).

Little has been published on the subject of the quantity of water necessary for the different types of salmonid cultivation and writers do not always agree. Flow given in seconds per hectare ($2\frac{1}{2}$ acres) go from 20 to 1,400 litres (pt). In the first place the wide differences of data published are due to the variable intensity practised by different farms. Another reason is the conditions which are also variable between farms. Temperatures, for example, can differ from average to the extreme.

It is agreed that for small ponds with an average depth of 1 m (3 ft) supplied with abundant fresh water it is possible to raise 10 to 15 kilos of trout (20 to 30 lb) per square metre (square yard) or per cubic metre of water. It is even possible to force the number to 30 kilos per square metre and even more than that, but it is not at all certain that establishments which reach that kind of density can give a sufficient guarantee for hygienic conditions.

Referring again to the Léger formula (1 litre of water per minute permits the production of 1 to 1·5 kilos of trout), it can be seen that for a density reaching 15 kilos per square metre a flow of 10 litres per minute per square metre would be necessary. This means 1,000 litres per minute for a pond covering 1 are (one-fortieth of an acre) in which 1,500 kilos or one and a half tons of trout were being raised and 100,000 litres per minute or 1,666 litres per second for an ensemble of ponds covering 1 hectare in which 150 tons of trout were being raised. These figures look pretty high.

Some writers in German (Kostomarov, 1961; Schäperclaus, 1961) give the following quantities as being necessary: 100 litres per second per hectare (1 litre per second per are) for extensive raising, 200 to 300 litres per second per hectare for semi-intensive raising and 300 to 500 litres per second per hectare for intensive raising. The figures permit, according to the individual case, the

renewal of the water in the pond from once to five times per day. They do not give precise findings as to the relative stocking corresponding to each type of farming. These evaluations or estimates seem good for intensive culture but too high for extensive rearing.

Concerning intensive raising, the evaluations given by these writers are: 300 to 500 litres per second per hectare. This approaches the water quantity effectively used on large Italian fish farms where there is an abundant supply (one to several cubic metres per second) of fresh water of excellent quality with temperatures from 9 to 16°C. For example, 400 litres per second per hectare are used for farms stocking 100 to 150 tons of fish per hectare. These last evaluations may be considered good for reference. They permit, with 500 litres per second per hectare (5 litres per second per are: 0·05 litre per second per square metre = 3 litres per minute per square metre) the raising of 150 tons of fish (1·5 tons per are or 15 kilos per square metre). Young fish would require greater quantities of water and evaluations given above can be taken as a guide.

The amount of water available will decide, above all, whether the farm can undertake extensive, semi-intensive, intensive or condensed intensive production.

From this angle, care should be taken not to conclude that a salmonid farm of any type can be set up because the trout exist naturally in the water course of the supply. At a low altitude, running water with only a feeble flow can still be suitable for trout but stagnant water overheats in the valleys during the hot season, and at that time of the year is only good for cyprinids and their accompanying voracious fish. Apart from the hot season, and if at the same time the ponds are fished by anglers, it is possible to add or even replace them with rainbow trout. Such ponds are to be found in low and middle Belgium and in Holland. It is also possible that the flow might be sufficient for extensive cultivation but not for intensive cultivation.

3 Origin of the water

Water from springs and running water can be used.

The three principal types of springs are: rheocrene, limnocrene and helocrene springs (Fig. 113).

1. Rheocrene springs: These are running springs. On emerging the water runs down towards

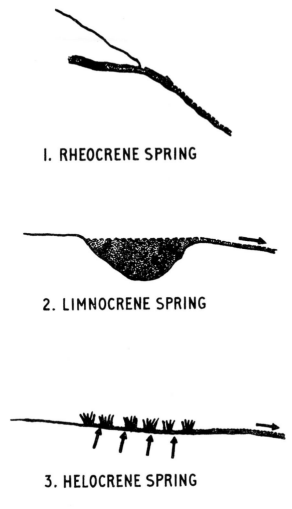

1. RHEOCRENE SPRING

2. LIMNOCRENE SPRING

3. HELOCRENE SPRING

Fig. 113 Schematic drawings of principal types of spring.

the valley immediately and the water course takes on its sloping profile. Fine waste is carried by the water and the soil is stony and gravelly.

2. Limnocrene springs: These springs are in a depression. The water first fills the natural depression before taking on its sloping profile. As it spills over, a water course is formed. The beds of these springs are often muddy or sandy and rich in water plants.

3. Helocrene springs: Marsh springs. The water seeps through a more or less thick layer of earth and transforms the ground into a marsh soaked like a sponge. The water is subject to acute changes of temperature. It is often rich in humic acids and

organic matter. Its pH is low and its oxygen content feeble. Limonite often forms. This water is not to be recommended for fish cultivation.

Water from rheocrene and limnocrene springs is clear. The temperature is not very variable but there is a lack of oxygen at the point of emergence; less than 50 per cent saturation for limnocrene and 70 to 80 per cent for rheocrene springs. The latter are the best but still not oxygenized enough. In any case it is not advisable to tap the water at its immediate source but to let it flow and make contact with the air for at least several yards. If the slope is insufficient the water is spread out over a shallow and wide space.

If the water of rheocrene springs is tapped at a sufficient distance from the point of emergence, in order that it can become oxygenized, it is perfectly suitable to use for incubation and hatching. Beware of vauclusien or resurgent springs which have variable temperatures and disturbed water particularly after heavy rainfalls.

The temperature of rheocrene springs is practically constant and equal to the annual average temperature of the region. In temperate Western Europe rheocrene springs have an average temperature between 7 and 10°C according to the altitude.

Running water (brooks and streams) runs the risk of becoming disturbed or affected by rain and storm often considerably. Temperature is variable but normally it is saturated with oxygen.

As far as it is possible the ideal is to supply incubation halls with spring water, and the ponds for fingerling trout, trout for eating and brood fish with water from streams.

In this way, the construction of expensive filters and their operation hazards are avoided and the water available is not too cold for the ponds.

On the other hand, by mixing spring and stream water it is possible to modify the temperature of the water in order to slow down or accelerate the development of the eggs and the fry. For the incubation of fry intended for repopulating open water it is necessary to cool spring water, the temperature of which is too high. Too warm water would accelerate incubation and the fry would be liberated in water which is too cold and swollen by spring flood. Spring water is cooled by allowing it to run free in the open air. This also favours oxygenation. Or it can be held in a freshening pond located between the source of the spring and the incubation hall. The aim is to reduce the temperature to around

4 to 5°C, at least if the intention is to produce fry for restocking open water or stocking nursery ponds where the young fish, unfed, will become fingerling trout of one summer for restocking.

It is indispensable to have complete control of the water to be used. In consequence, for a spring it is most important to watch out that it is not reduced in a way likely to influence its flow. For a water course it is necessary to own both banks and it is essential to see it is protected from all risks of pollution. Further, the farm must be protected from flood water from the water course. These two precautions relative to the dangers of pollution and floods are very important and must be given careful consideration.

4 Commercial considerations
For fish farms planned for the production of fry and trout fingerlings the technical aspects come before the commercial considerations. It is the reverse for farms producing fish for eating.

For the latter, supplying food and selling the marketable product is of primordial importance. An industrial salmonid farm must be located so that it can easily receive all the necessary food for fattening the trout and, on the other hand, be near to important commercial markets able to offer a remunerative price for the fish.

At one time it was indispensable to be near or to have easy access to a fishing port or an important town where enough food could be bought at a reasonable price (salt water fish, spleen and abattoir waste) required to feed the trout. This is much less important now that it is possible to replace fresh food by dry feeding compounds which, apart from the cost of transport which is not in any way high, can be delivered anywhere at a uniform price.

However, easy communications and proximity to centres of restocking for trout fingerlings or for disposing of trout for eating (densely populated areas and tourist centres) are indispensable for disposing of the products of fish cultivation.

5 Conclusion
If full consideration is given to what is written above it should be evident that, contrary to previous opinion, salmonid cultivation is no longer tied to the mountains. Little by little this conception has been modified. Provided the conditions here stated are possible, notably those concerning the quality and quantity of the water, then a farm for the cultivation of salmonids can be installed in a plain close to the sea. For there it is quite possible to find fresh and abundant water and, more easily than in the mountains, also find important supply sources of cheap food for fattening. This is the case, notably of Danish salmonid farms which enjoy a high reputation, and also French salmonid farms in the Pas-de-Calais.

II General Arrangement of a Salmonid Farm

Industrial salmonid cultivation set up to produce fish for eating is one of those agricultural industries which gives the highest production and the best return per acre. The size of most farms varies between 0·5 and 2 hectares of pond surface. Industrial salmonid cultivation covering 5 hectares and beyond are rare. If there is sufficient water it is possible to start a fish farm for intensive production by using artificial food.

The layout and installing of a farm for industrial salmonid cultivation varies according to local topography and the aims of the exploitation. That is to say whether the objective is the production of fish for restocking open water, the production of trout for eating or whether it will include both. A farm can include everything or only part. In the first case it will have ponds for parent fish, an incubation and handling building, nursery and fattening ponds and finally storage ponds. In the second case it will include only those ponds necessary for one or two of the rearing stages. It can also be arranged in such a way that the production of eggs and trout fingerlings cover the needs of the farm only, or more than is required in which case this surplus can be sold.

The nursery ponds and the incubation shed must be so located to permit the supply of absolutely fresh and pure water. As far as possible the works and warehouse should not be far from the fattening ponds in order to reduce as much as possible the transport of food. The holding ponds should be nearby as well, so as to enable them to be watched easily. They will hold trout to be delivered for the table, and also the brood fish ready for reproduction when the time comes. These ponds need only be small but there should be an abundant flow of water across them.

During the different stages of rearing, trout are the kind of fish which like to face the current;

installations should be long and narrow: nursery tanks, ponds for rearing trout fingerlings and the fattening ponds.

In order to maintain a sufficient dissolved oxygen content in trout ponds, which should remain at a fairly low temperature and where abundant water is renewed, ponds should be narrow and long so that the current can flow across

the full length and its movement be felt. Wherever possible these long ponds should be built from north to south. In the summer the water will not be heated so quickly by the sun and in the spring it will be less exposed to the east winds.

Round ponds with a circular current have assumed more and more a greater importance for they meet the same current requirements.

SECTION II

PRINCIPAL CULTIVATED SALMONIDS

I European Salmonids

1 Brown Trout (*Salmo trutta fario* L.) (Fig. 117)
Brown trout are indigenous to mountain waters of Central and Western Europe. It was the first fish to be artificially reproduced and reared. It initiated this method of fish cultivation.

It is a fish very varied in colour and form. Sea trout (*Salmo trutta* L.) and lake trout (*Salmo trutta lacustris* L.) are closely related to brown or river trout and they are all considered as constituting a single species though differentiated in various forms. The first have silver flanks but no red-orange spots which brown trout always have.

The size of brown trout depends on the medium and the richness of its food. Sea trout and lake trout can reach 3 ft and longer and weigh 10 kilos (20 lb) but brown trout rarely pass 18 in.

Apart from the three forms given above, other forms, more or less widely dispersed geographically, can be singled out. Among the principal of these are (Ladiges and Vogt, 1965) "*Salmo trutta macrostigma*" A. Dum. which exists in salmonid water courses tributaries of the Mediterranean; different varieties have been established and "*Salmo trutta marmoratus*" should be mentioned among others. It is also necessary to mention "*Salmo trutta labrax*" Pallas for the basins of the Black Sea and the Sea of Azov and "*Salmo trutta caspius*" Kessler found in the basins of the Caspian Sea.

Brown trout is a stenothermal cold water fish which needs fresh water and does not withstand high temperature variations. They like pure, running water rich in oxygen. They spawn in the autumn or at the beginning of winter, that is to say from November to January in the northern hemisphere.

At that time they leave their normal habitat: the sea, lakes, streams or brooks to swim upstream and spawn on gravel-bedded shallows of fast current waters of the trout zone. Brown trout swim up their own streams to the neighbourhood of the sources.

They are carnivorous and their favourite food is relatively large living prey. In open waters they live principally off animals such as crustaceans, insect larvae and small fish which are swallowed in one gulp. They are fish living near the bottom, which differentiates them from rainbow trout which live more on the surface.

Brown trout lend themselves to rearing for restocking. This gives good results if they are fed natural food and reared in water with summer temperatures not exceeding 20°C. But their raising can also be carried out with artificial food.

Brown trout do not fatten so well as rainbow trout. They are more difficult to feed and benefit less from the artificial food they receive. Their growth therefore is slower. Brown trout are also more timid and more demanding when it comes to both the quantity and quality of water. Nevertheless their adaptability should not be underestimated on condition that acclimatization starts when they are young. Brown trout are less sensitive to whirling disease but are more subject to furunculosis than are rainbow trout.

If they are adequately fed the flesh of brown trout is of excellent quality, and if their food includes fresh water shrimps then their flesh becomes the colour of salmon flesh.

Brown trout from Western Europe were introduced to other continents and released in suitable salmonid waters at the end of the nineteenth century.

2 Atlantic Salmon (*Salmo salar* L.)

Salmon differ from trout not only because of their vomerine teeth but also because of the caudal fin which is more or less distinctly scalloped on the outside, especially in young fish: parr and smolts. In colour Atlantic salmon resemble sea trout. They can grow more than 1 m (3 ft) in length.

The Atlantic salmon is a migrant. It grows at sea and this stage can last from one to three years. Then it swims upstream, often over considerable distances inland to spawn in the waters in which it originated on the gravel shallow spawning grounds of the grayling and trout zones. Young salmon remain in fresh water for one or two years and then migrate to the sea.

In Europe the Atlantic salmon swim up the salmonid rivers and tributaries of the Atlantic, from the north of Portugal to the Arctic ocean including the North Sea and the Baltic. Formerly very abundant in many waters, it is now in reduced numbers because of the construction of dams and also because of pollution. The Atlantic salmon is also found in tributaries of the North American Atlantic.

Atlantic salmon are exclusively cultivated for repopulating areas by means of eggs, fry and second year fish. This is particularly practised in Sweden.

3 Danube Salmon (*Hucho hucho* L.)

This is a continental salmon indigenous to the Danube basin but not found in its lower reaches. It lives in running water of the upper barbel zone and migrates over short distances upstream at the spawning season. Just like the Atlantic salmon, it can grow to over 3 ft in length.

The introduction of Danube salmon has been tried in several countries in Western Europe, notably in France and Sweden. It is raised for restocking.

4 Arctic Char (*Salvelinus alpinus* L.)

This name groups an important number of fish living in the cold water lakes of the northern slopes of Europe, Asia and North America. In Europe it is essentially a lake fish populating the sub-Alpine and northern lakes. It lives in the deep waters.

Arctic char is, at the same time, both a beautiful and good fish. It can weigh over 4 kilos (8 lb). It spawns in the autumn in the same way as other European salmonids. Its cultivation is confined to repopulation.

II American Salmonids

5 Rainbow Trout (*Salmo gairdneri* Richardson) (Fig. 118)

Rainbow trout originated from the North American Pacific area and were imported into Europe in 1880.

Salmo gairdneri lives in water courses of the coastal range from the south of Alaska to the south of Oregon and California. It is possible to single out the fish in different forms: the continental which does not migrate to the sea, and those which do. Generally the continental type which do not migrate are known as "Rainbow Trout" and those which do are known as "Steelhead Trout".

There is an astonishing variety of types of *Salmo gairdneri*, even more than those mentioned for European brown trout. Originally the principal types could be more or less distinguished, for example by the number of scales on the lateral line and by their colouring, but they have been subject to so much cross breeding during the course of artificial cultivation that it has become practically impossible to identify them and to obtain eggs which correspond to primitive strains.

In Europe, at the beginning, several strains were introduced. Among them, *Salmo shasta* Jordan was probably the first to be introduced. This originated from the Sierra Nevada where it populated the waters of Mount Shasta at the limits between Oregon and California. Its form is continental. *Salmo irideus* Gibbons inhabits the running waters of coastal ranges of the same region and is a migrant.

Rainbow trout have no red spots. The skin is covered with numbers of small black star-shaped spots. The adults have an iridescent reflecting rose-coloured band on their flanks, which is particularly apparent in the male at the time of reproduction. It spawns later than brown trout, from January to May. However, varieties have been obtained which start spawning in December and even before.

Rainbow trout are salmonids best suited for raising industrially and for the production of trout for the table. The reasons why they are preferred are as follows:

1. They adapt themselves easier than brown trout. They lend themselves more easily to domestication in general and to taking artificial food in

particular. They withstand higher temperatures better and also reduced amounts of oxygen. In sufficiently renewed and deep water they can withstand temperatures between 20 and 22°C and momentarily as high as 24°C and over.

2. They are more resistant to certain diseases, notably furunculosis.

3. They develop rapidly: incubation is shorter and their growth faster. They are less carnivorous and destroy fewer young fish. They also make the most of the food on the bottom, seeking out worms and molluscs living there. Their growth is rapid so that if they are well fed they can reach 100 grm (almost 4 oz) and more in one year, to 250 to 300 grm (almost 9 to 11 oz) in two years (30 to 35 cm). They can measure from 40 to 45 cm (15 to 17 in) in three years.

On the other hand they are more liable to suffer from whirling disease and, according to their origin, they show more or less pronounced tendencies to migrate.

It is said that their flesh is of inferior quality compared with that of brown trout. It should not be forgotten that the flesh of trout, no matter what the species, depends a great deal on the quality of the water in which they live and, above all, on the food found there or that which is fed to them. The season in which they are eaten must also be taken into account. Too often brown trout taken from running water which has natural food of excellent quality are compared with rainbow trout fattened on a fish farm with food of only mediocre quality. To get an exact idea it would be necessary to taste fish taken from the same water, at the same time and fed and cooked in the same way. Taken from fresh and clean water and having been fed with choice food, rainbow trout, eaten out of spawning season, has a very delicate flesh.

To conclude, rainbow trout are the best, easiest and most attractive fish to raise for the table. They are less difficult to feed, less demanding with regards to temperature and the quality of the water. Consequently they are to be found in cold water fish farms throughout the world where they have become the principal salmonids for eating. For repopulating open water in Europe, the brown trout is preferred, except for water which is intensively fished by anglers. These waters can very well be restocked with rainbow trout.

6 Cutthroat Trout (*Salmo clarkii* Richardson)

These are indigenous to the Rocky Mountains where they live more inside the inland waters than the rainbow trout, at least in the United States. They are found from the north of California to British Columbia and Alaska. They carry one oblong red spot either side of the throat. Their skin is marked with white circled black spots, large and more accentuated than for rainbow trout. It is possible to find local types. They are raised on fish farms in the same way as rainbow trout with which they cross.

7 Pacific Salmon

There are five species of salmon which grow out at sea, swimming up the waters of North America rivers entering the Pacific Ocean along the United States coast to Alaska. They are *Oncorhynchus gorbuscha* (Walbaum) (pink salmon), *O. keta* (Walbaum) (chum salmon), *O. kisutch* (Walbaum) (coho salmon), *O. nerka* (Walbaum) (sockeye salmon) and *O. tshawytscha* (Walbaum) (chinook salmon). These salmon can also be found in Siberia as well as a sixth species called *O. masou* (Brevoort) (masu salmon). Biologically, they are on the whole the same as Atlantic salmon. They are cultivated for restocking.

8 Brook Trout (*Salvelinus fontinalis* Mitchill) (Fig. 119)

Brook trout originated from the coastal slopes of the North American Atlantic. They are to be found in cold and pure water from Labrador in the north to the Alleghany Mountains in the south. They were introduced into Europe in 1889. Very popular at the beginning, their cultivation is now diminishing.

These salmonids are easily recognizable by the green stripes on a yellow ground which covers the top and the sides, the back, the dorsal and caudal fins. The belly is vermilion red. The pectoral, ventral and anal fins also have black, white and red edges. These fish spawn early in the autumn starting at the beginning of November.

Brook trout are very difficult with regard to the temperature of water which must not be too high (no more than 18°C and preferably 12 to 14°C). It must also be perfectly pure. They are able to live in ponds fed by springs with water constantly fresh but less rich in oxygen than running water.

They benefit more from eating different prey within their reach than brown trout. Growth is rapid and at 2 years they can reach 250 to 300 grm (9 to 11 oz) under satisfactory raising conditions.

The flesh is firm and delicate and in water rich in gammarids or fresh water shrimps, it has an attractive salmon colour. Later it becomes stringy. These qualities, including the limited demand for dissolved oxygen and the fact that they reproduce naturally in certain streams and ponds, explain why they enjoyed popularity for a certain time.

Unfortunately they appeared to be very sensitive to pollution and infectious diseases, especially furunculosis and fin rot. Cultivation is still justified in small ponds fed naturally by spring water which cannot be contaminated and whose temperature remains low in summer. In such water rich in vegetation and food, fresh water shrimps in particular, brook trout give good quality specimens

of rapid growth. They can measure 40 cm (15 in).

9 Lake Trout (American) (*Salvelinus namaycush* (Walbaum))

The American lake trout is a fish which lives in the deep and cold lakes of North America from the State of New York to British Columbia, reaching Labrador and Alaska near the north. This voracious fish can grow very large and can pass 50 lb. Its colours are not so bright as those of other salmonids. It was introduced into Europe in certain cold water lakes, the high Swiss mountains for example. Its cultivation is for repopulation.

SECTION III

ARTIFICIAL REPRODUCTION OF SALMONIDS

The discovery of artificial reproduction without doubt permitted the great strides taken by salmonid cultivation. Thanks to this it is possible to obtain great quantities of fry indispensable for raising in ponds in which trout farming for the table and for repopulating many salmonid rivers and streams is practised.

The artificial reproduction of salmonids includes different successive stages. These are artificial fertilization, incubation and rearing. In order to do all this it is necessary to have a sufficient quantity of parent or brood fish of good quality.

I Choice and Rearing of Brood Fish

A Number of brood fish

1 Absolute quantity

The quantity of brood fish necessary is predetermined by the number of fry that are wanted, which in turn decides the number of eggs taking into account a normal percentage of loss. To establish the number of brood fish for a given quantity of eggs it is necessary to base the calculations on the fact that a female weighing about 1 kilo (2 lb) gives an average of from 1,500 to 2,000 eggs. For females which weigh less the actual number of eggs will be smaller while being proportionately higher (up to 2,700 for a kilo). For

very small females, weighing 150 grm (5½ oz) and less, the number of eggs, both actual and relative diminish, falling even lower than 1,500 eggs per kilo.

Salmonid eggs have a normal diameter of between 3·5 and 5 mm. The diameter depends on the size of the female and not on its age. It grows according to the size of the mother.

2 Relative proportion of the sexes

Fewer males than females are necessary: half or even one-third is sufficient. In effect, in the course of the reproduction period, the male can give milt several times; between three and eight times provided that the successive spawnings are separated by two weeks or better still 70 degree days. Often the milt of two males is used to fertilize the eggs of four females.

In order to excite the sperm production, the males are placed, at this time of reproduction, in tanks or ponds situated downstream from similar tanks or ponds holding females. If they are left with the latter then the males will fight violent and mortal battles between themselves.

In order to obtain 1,000,000 eggs, 750 kilos (over 14 cwt) of parent fish will be needed.

3 Differentiation of the sexes

At the time of spawning the differentiation of the sexes is easy to recognize. The females, some time before spawning, show extended bellies and the anus is prominent, round and red, while for the male it is small, lengthened and pale. Also at this time the colour shades of the male become brighter. In rainbow trout, for example, the lateral, iridescent band is very marked, while the belly of the brown trout male turns blackish and the colouring of brook trout becomes very bright. The lower jaws of old males of all species turn up into a kind of hooked beak shape (Fig. 118).

B   Origin and rearing of brood fish

Brood fish can be taken from natural open waters or can be raised in ponds. However, Atlantic and Pacific salmon, Arctic char and American lake trout practically always come from open water. Rainbow trout are generally raised in ponds. Brown trout and brook trout come from both.

Brood fish coming from open waters such as lakes or from running water are caught when swimming upstream, just prior to spawning. The two sexes are separated and the fish are then put into long shaped ponds or in holding tanks crossed by a sufficiently abundant current. Females found to be ripe at the moment of capture are stripped immediately.

The fish are caught either with large scoop nets (Fig. 122), in permanent fish traps (Fig. 124) or electrically (Fig. 125).

In Europe the acclimatization in open water of rainbow trout and brook trout has given good results in a few places only. Brood fish ready for spawning caught in running waters as they swim upstream to the spawning grounds are nearly always brown trout. To obtain fry for repopulating running water these are preferred as parent fish.

In industrial fish farming the parents of rainbow trout and to a lesser extent of brown trout and brook trout are best raised in ponds. They can live off natural food only or be given good artificial food. This can be carried out in small ponds in which up to 200 young reproducers per are (one-fortieth of an acre) are kept (Fig. 121). They can also be raised in large ponds (Fig. 120) where from one to ten large brood fish per are are kept according to the biogenic capacity. In certain cases rainbow brood fish can be kept in large carp ponds provided the ponds are not too warm or too muddy.

If the brood fish are raised in ponds and are fed artificial food then this must be reduced 1 or 2 months before spawning and stopped just before this period, that is to say when the fish want to leave the ponds and swim upstream. At this time the ponds must be emptied, the brood fish sorted out and placed in holding tanks. Instead of emptying the pond, if the location permits, a fish trap (Fig. 124) can be installed upstream. The brood fish will swim into it and are caught and as they are near maturity a long holding period is avoided.

Certain writers (Buschkiel) believe that lack of sufficient space in small ponds and artificial food are responsible for the degeneration of the sexual product of brood fish. The failure to grow in a restrained space would be due to the disappearance of certain nutritive substances which only exist as a trace, to rejection through the metabolic mechanism of toxins remaining in the medium, or to insufficient space. All these can be avoided by assuring that the ponds are crossed by a strong current of water.

C   Factors influencing the potency of milt and eggs

1   Food

Food is of capital importance to the value or potency of the sexual products of farm bred salmonids. This is realized when raising trout for reproduction and raising trout for consumption are separated. Trout fed exclusively or almost exclusively on normal, artificial foods, give sexual products of only mediocre value. Eggs of trout given natural foods are rose-coloured and translucent. However, white eggs which are not translucent can give excellent descendants. It is by these that the value of the eggs should be judged and not by exterior appearance.

The use of natural reproducers or those fed exclusively with natural foods nevertheless produces difficult problems. It is difficult, for example, to obtain sufficient numbers of rapid growth reproducers. Further, it is difficult to appraise how they will react to the satisfactory use of artificial food as a basis for fattening, because they have not been fed this way. It is difficult to judge their resistance to diseases which might occur in farm breeding, and they can, even, introduce certain diseases as well as parasites.

Basic feeding with good quality artificial food accelerates growth and permits the production of young, large females. Nevertheless feeding must be in moderate quantities and the quality of food must be excellent. Artificial feed should be given at an early age, continuously and regularly, for sudden interruptions or considerable variations in feeding can provoke sterility. By selection among animals fed in this way it is possible to detect subjects capable of producing descendants able to make the best use of artificial food.

2 Age and size

Trout are ready to start spawning at the end of their second year. According to the species, the best male milt is furnished by males of from 2 to 4 years. Females are not used before their third or fourth year and should not be used after their sixth year. The older females, because of their weight, give a greater number of eggs but the loss is quite significant and a part of their descendants may be sterile. The number of sterile reproducers increases with age. Buschkiel estimates that it passes 15 per cent at 3 years to 50 to 60 per cent at 6 or 7 years for rainbow.

The current weight for farm bred genitors varies between 350 and 1,000 gm (12 oz and 2 lb or over). The quantity of eggs and milt increases with the size of the genitor. The diameter of the eggs increases with the size of the female; the larger the egg the larger and more resistant the young. The size of the male, however, does not influence the size of the alevin. There is an advantage in using males with precocious growth for this is a sign of a natural predisposition to profit from food and this faculty can be hereditary.

The sex of the descendants is influenced by the relative size of the genitors. Descendants of older females fertilized by young males is predominantly masculine.

D Improvement by selection

The aim of selection is to improve those strains of particular interest, that is to say, those of vigorous and rapid growth well adapted to their medium. A strain is characterized by a certain number of inherited morphological and physiological characteristics likely to be passed on to descendants living under the same conditions as the parents.

A general rule is that a region in which conditions are predetermined leads to the constitution of a strain best adapted to the conditions in which it must develop. However, this strain should possess rather strong faculties to enable it to adapt itself easily to different conditions.

When it is understood that it is not possible to obtain complete satisfactory adaptability under all possible circumstances, it is useless to try to improve the local strain by influencing all the hereditary characteristics by the introduction of fresh blood. It is far better to use, regularly, the best elements among local strains. These are chosen from among those with the best growth. Account must also be taken of the ability to make the most of artificial food, the faculty of adaptation to living in a pond and also resistance to disease. Taking the best growth as a criterion there is every possibility that at the same time subjects possessing a maximum of other qualities will be chosen.

The importance of selection is very great in the cultivation of salmonids because growth differences are considerable between trout of the same age. Fingerling trout of one summer, of the same origin and raised under the same conditions can measure between 5 and 15 cm (2 and 6 in). Differences in growth between fish of the same age can emanate from external causes, related to the medium being favourable or unfavourable, and which can, in all probability, be made uniform. Differences will also emanate from internal causes brought about by the appearance of recessive factors which can lead to unfavourable hereditary characteristics. These recessive factors are helped by inbreeding. However, this can be overcome by eliminating descendants of poor growth. But this should not have unfavourable economic consequences since the product is numerous and only a certain number need be eliminated. Being young their economic value is minimal.

At the time of choosing brood fish, all those with curvature of the vertebral column, atrophy of the gill cover or those which are manifestly diseased (blind, etc.) must be eliminated.

II Artificial Fertilization

1 Historical record

The discovery of artificial fertilization of trout eggs is supposed to be very old and the technique to have been practised in the Middle Ages by the monk Dom Pinchon. This, however, is not proved. It was certainly discovered in the eighteenth century and described by Jacobi in 1765 in the Hannoverschen Magazin, but no one paid attention and it was forgotten. It was then rediscovered by two Vosges anglers Gehin and Remy in 1842, and thanks to Prof. Coste of the College de France it was widely publicized. This was the start of the artificial cultivation of salmonids. In 1854, France created an important station at Huninge in Alsace. The discovery of the dry method of fertilization by the Russian Vrassky, and its application between 1856 and 1870, led to important technical progress. Fecundation by this method permits success often exceeding 95 per cent.

2   The period for reproduction

The period for reproduction of trout varies:

1.   According to the species: brown trout and brook trout spawn from October to January and rainbow trout normally from January to May. Brook trout, however, have a tendency to precede brown trout by several weeks.

2.   According to the strain and to the individual fish, some are precocious, and others late. It has been possible to obtain strains of rainbow which spawn from the start of the autumn.

3.   According to the regional climate, local circumstances and their influence on the temperature of the water. Brown trout spawn that much earlier as the water grows cold in the winter, that is earlier in the mountains than in the plains. Rainbow that much earlier as the water rewarms rapidly in the spring.

4.   According to the movement of the water: trout are ripe earlier in flowing water than in ponds.

5.   According to the health of the brood fish so that those in good health and which are well fed spawn first.

The age of the reproducers would have no influence on the precocity or tardiness of the spawning period.

Differences in maturity can be considerable. It is an advantage to have precocious strains at least for rainbow trout. This prolongs the length of time for raising during the good season and permits the production of fry already used to feeding from the start of the summer growth season. Besides, the more forward the strain the better it can resist attacks of whirling disease.

3   Characteristics of good sexual products

Brood fish living in open waters are caught as they swim upstream to the spawning grounds. At the time of capture these wild brood fish are already mature or near to it. They must not therefore be placed in holding tanks. If this is done, then it should only be for a short time.

If the brood fish are raised in ponds then the latter must be emptied as sexual maturity approaches, for the brood fish then have a tendency to group near the water inlet and try to escape from the ponds in order to swim upstream. One or 2 months before spawning the amount of artificial food fed to the brood fish must be reduced if this is the way they are fed. Brood fish continue to feed until spawning but the privation to which they are now subjected does not interfere with normal sexual maturity.

As complete sexual maturity approaches the brood fish are caught and placed in small ponds or in holding tanks where full maturity will follow. Good dimensions for holding tanks are 2 to 2·50 m (2 to 3 yds) long, 0·75 m wide (2½ ft) and 0·80 m (2 ft 8 in) deep. Trout withstand this holding very well but not so other species such as pike. If the brood fish are kept in ponds, they should not have gravel bottoms as this can incite the trout to spawn. The ponds must also be free of vegetation and of any obstacle hindering fishing-out by net. It is also useful to cover the holding tanks with netting. Tanks should have sufficiently high banks to stop the brood fish jumping out of the water in the effort to escape. The tanks or ponds should be crossed by an abundant flow of water which should have a favourable effect on maturity; but the current must not be excessive.

The sexes must be kept separated and once a week at least the state of sexual maturity must be checked and those selected found to be ripe and ready for stripping. If the abdominal vent of these is pressed slightly then the eggs or milt will emerge. Inspection must be more and more frequent as the temperature of the holding water grows. In a fixed medium the start of the reproduction period is nearly always on the same date. If the late season is mild then it will be a little more precocious.

If the degree of maturity is right then the milt will be white and creamy; poor milt is watery and curdled. Milt which is not ripe will demand strong pressure and will be mixed with blood. But the milt will be over-ripe if first of all a watery liquid runs out followed by an equally watery milt.

Ripe eggs are stripped with ease. The eggs only stay ripe during about 8 days. Over-ripe eggs give fry in which males are dominant and among which many are often malformed. The loss is higher than normal at the time of fertilization, incubation and hatching. It is the same when the eggs are not sufficiently ripe.

If the female does not spawn, which is also the case when the brood fish remain in ponds, then the eggs are reabsorbed progressively. This can lead eventually to the sterility of the fish.

4 The basis of artificial fertilization

At first the "wet method" of fertilization was practised by which the eggs were gathered in a pan half filled with water and the milt was then added. Carried out quickly this method gives good results and the percentage of fertilization is high.

But for a long time the "dry method" has been preferred and this permits almost total fertilization of the eggs. By this method the latter are gathered in a dry receptacle (a basin or pail strainer), the milt is added and only after careful mixing is the mixture poured into a pan half filled with water. This method of fertilization gives a maximum chance of success.

The reasons why the dry method has been adopted are as follows. When the milt is not diluted the spermatozoa of trout keep their fertilizing qualities several days. At a temperature of from 4 to 8°C this quality varies, according to observations, between 2 and 10 days. Even if taken from males which have died it is still usable during from 12 to 24 hours. In water, on the other hand, the spermatozoa retain their motility over a much shorter lapse of time; 90 seconds only, but great motility ceases after 30 seconds. Also, plunged into water the eggs absorb liquid by their micropyle. This causes the eggs to swell and the obstruction of the micropyle. Now, fertilization follows the penetration by the spermatozoa inside the egg by way of the micropyle. When the latter is closed then, in consequence, penetration and the fertilization are no longer possible. The fertilization must therefore take place beforehand. On carrying out the operation dry, the egg is coated with a film of milt and so there is a maximum chance that the fertilization will succeed.

Pressure on the abdomen of the female in full sexual maturity causes a light coloured coelomic fluid to flow at the same time as the eggs. When a strainer is used this fluid runs through the holes. Recent writers estimate that fertilization is better if the liquid is conserved and mixed with the eggs as it helps the motility of the spermatozoa, increases the amplitude of their movement and prolongs their life. Indeed the latter can be doubled (Dorier). Instead of gathering the eggs in a strainer, it is preferable therefore to do so in an unperforated receptacle. The part played by this coelomic fluid in fertilization seems to vary according to the species. It is more important for rainbow trout than for brown trout.

Dorier points out that the fluid, like other liquids containing electrolytes, can be used to dilute the milt and that this action makes for longer conservation at low temperatures. Treated in this way it can retain its power of fertilization for as long as 2 weeks.

5 The method in practice

The following implements are indispensable: two or three carefully cleaned pans from 25 to 50 cm (10 to 20 in), a hand towel, chicken or goose feathers, and tweezers for cleaning the eggs (Fig. 126). All these instruments are placed on a table before starting. The operator should wear a clean waterproof apron.

Before fertilization the brood fish are sorted out. The ripe ones are put to one side, the females in one basin and the males in another. Those which are not ripe are returned to the holding tanks. The whole operation of sorting and fertilizing must be carried out with care because violence or roughness is harmful.

The female is taken out of the basin with the hand. The fish is then wiped delicately to avoid any water on the skin from dripping into the fertilization pan. The head of the fish is held firmly but not tightly in one hand. An experienced operator can do this with his bare hands but a beginner had better use the hand towel. The other hand then takes hold of the back end of the fish which is permitted to struggle. But it will calm itself eventually and it is possible then to strip. This can be done single-handed by a skilful operator for fish weighing up to 1 kilo (2 lb) (Fig. 127), but for large fish it takes two, one holding the head and the other the tail (Fig. 129).

The fish is held in a steeply inclined position with its head up, its back towards the operator and the whole of its underside held over the pan. The hand, starting from below the head, presses with the thumb and index finger, asserting pressure on the abdomen as it descends over the lower end of body down to the genital vent. If the eggs are ripe then they should flow out under this pressure. This stripping operation is repeated several times until the female is completely emptied of eggs. The fish endures this quite well provided it is carried out correctly. As there is practically no appreciable difference in the maturity of the eggs they can all be stripped at once. First of all the eggs nearest the genital vent, which are the ripest, are stripped and then the hand moves up progressively. If the stripping is normal then the eggs should flow continuously under the slight pressure of the fingers from the start to finish of the operation. For the stripping of the last eggs care must be taken not to exert such pressure that blood might mix with the eggs. It is possible there may be a little blood which means that the pressure was too heavy and broke a few veins. Such small haemorrhages are often without harmful consequences, but it is necessary, as far as possible, to avoid them.

According to the size of the brood fish this operation is repeated with two or four females. After that the male is stripped of milt by the same manipulation. Before starting, the male is carefully and lightly wiped, especially the lower part to avoid water mixing in with the milt. The underside of the fish must be turned downwards and the sides (Fig. 128) pressed, whereas with the female the abdomen is pressed. All the eggs are sprayed with milt. The quantity of spermatozoa in the milt is considerable: 10,000 million in a cubic centimetre of trout milt (according to Schlenk and Kuhmann, quoted by Dorier). Only a few drops are necessary to fertilize a large quantity of eggs. Each spermatozoon is a microscopic, motile corpuscle of one-twentieth to one-fiftieth of a millimetre in length identified by a large head and a threadlike tail. As it is rare to find a male whose milt is sterile, one male should suffice. If one has enough stock, and for the sake of security, the milt of two males can be used to fertilize the eggs of two to four females.

The eggs and the milt are well mixed with the aid of a feather and then poured into a pan half filled with water (Fig. 130). Then they are mixed again quickly and carefully three or four more times with the feather. The fertilization takes place the instant after the mixed eggs and milt have been poured into the pan of water. They are then allowed to rest for 20 to 30 minutes. At first one side of the eggs stick to the pan. But they swell with water, the volume is increased and finally they stick to the pan only by a tiny point and can be detached easily. Meanwhile other females are stripped and their eggs are placed in other pans.

The eggs of the first stripping are again given attention. Because of the excessive amount of milt the water is cloudy and therefore is replaced several times until the eggs are seen to be very clean and the water clear (Fig. 131). This cleaning process permits the removal of bad eggs and empty shells as well as other impurities, notably excrement which will be mixed with the eggs. The eggs are then poured slowly on to incubation trays where they are uniformly spread out in one or several layers (Fig. 153), by means of a feather, according to the cleanliness of the water and the amount of space. From this point a new stage begins.

If the incubation troughs are not near to where the eggs were fertilized, they are carried to the troughs in a closed receptacle filled with one part eggs to three parts water. The eggs and milt mixture can be carried dry before water is added and the fertilization taking place on arrival at the incubation trays. Vouga recommends this system. In special cases the eggs and milt can also be carried dry in thermos flasks. They will retain the fertilizing faculty within the limited times given above.

After fertilization, the females can be removed or held for control from 3 to 8 days later. This is a stripping control in order to eliminate the few eggs which were not taken the first time.

After this control, the brood fish having spawned are replaced in ponds or in open water. On industrial-size farms they can be made to spawn over three or four consecutive years. On certain particularly favoured farms, they are used only once after which they are fattened and sold for the table.

III   Incubation and Hatching

Incubation is that period during which the fish develops inside the egg. After hatching, the yolk sac of the young fry is progressively absorbed. After that period rearing fry may be done for some weeks in the incubation troughs or in special rearing tanks.

A   Incubation, hatching and rearing fry installations

1   General note

The incubation and hatching hall and the rearing fry installations must occupy a single building designed for that use. It must fulfil the following conditions: easy accessibility; sufficient space to assure easy, clean and rapid working; good light by both day and night, protection from the direct rays of the sun. It must also afford effective protection against frost. It should be airy and have a good water supply with satisfactory distribution and evacuation systems.

The preparation hall is the place reserved for storing and preparing food for the fish and other operations. It must be right next to the incubation hall but distinct from it so that the fry are not constantly troubled by the coming and going of the personnel. All the material can be ranged in order here. The shed for the preparation of food need only be relatively small and easy to clean. It is important that the incubation and the handling sheds should always be kept clean and in perfect order.

These premises for handling and incubation must be in a separate building from that in which the fish farmer lives. But it must be close at hand in order to save loss of time and give necessary facilities for supervision, though still distinct for the sake of hygiene.

The floor of the incubation shed can be in concrete or stone and at a slight slope (1 cm per metre), to facilitate cleaning and the removal of water which, inevitably, spreads round the troughs. As the floor, including the aisles, will always be more or less wet, it can also be covered with wood slats to save the feet from constant and disagreeable contact with the wet concrete which is cold in the winter. The walls and the roof should be sufficiently thick to protect the hatching house against temperature variations and to prevent extremes of cold and heat. They should also be protected from humidity by an appropriate facing. The ceiling and beams of the roof must be carefully painted or whitewashed.

Natural light for the incubation hall should be from windows preferably facing north so that the sun's rays do not fall on the eggs. While the room should be well lit it should also be possible to reduce the amount of light by means of shutters or blinds or even whitewash on the window panes.

As for electric light, taking into account the humidity, reduced power should be installed: 24 or 12 volts. It is prudent to keep the wires of the different powers well apart. A satisfactory electrical installation should include an inspection lamp which can be moved about on a trolley.

2 Importance of the hatchery installation
The importance of the buildings required in salmonid cultivation depends on the type of cultivation and the anticipated production.

The different types of salmonid cultivation can be limited to incubation only or to incubation followed by rearing of the fry. Incubation may be at industrial level or it may be artisanal. If it is industrial the hatching sheds will be more or less vast, but in artisanal cultivation the troughs can be set out in the open or housed in barns or cellars, etc. Details on artisanal cultivation are given in the appendix.

As for the space required, if the number of fry is decided in advance it is easy enough to calculate the number of eggs needed, taking into account a normal loss of around 10 per cent during incubation. If the eggs are uniformly spread out in one layer on the meshed incubation trays, there should be between 400 to 600 eggs per square decimetre (16 sq in) for diameters varying between 5·0 and 4·0 mm. It is simple enough therefore to calculate the surface required for the meshed incubation trays and the troughs which are a little larger.

To get an idea of the surfaces necessary for artificial hatching and rearing, account must be taken of details to be given later when the rearing of fry in troughs is discussed. It is possible to feed 10,000 fry for 1 month, 3,000 for 2 months, 1,500 for 3 or 4 months per square metre provided there is sufficient fresh, well aerated water.

Taking into account the amount of space strictly necessary for incubation and rearing fry, it is also possible to calculate the size of the building required. Naturally the space between the troughs, the corridors, the space required for handling food etc. must also be taken into account so the building will, in fact, be larger than the original dimensions. Far too often in old hatcheries the troughs are too close together in the incubation shed and this makes the work difficult. The space taken up by the incubation and rearing fry installation should be approximately one-third of the size of the cultivation building.

If incubation is not followed by the feeding of fry in troughs then the space taken up by the incubation installation can be reduced considerably by using vertical incubators.

3 Water supply for the incubation hall
Filtering of the water for incubation
The water supply must be installed so that the water flows by gravity to the incubation hall. It should be provided with a screen and should be designed to work according to the constant flow water supply model already described in Chapter I.

If the water is supplied by conduits then the pipes should be buried about 50 or 60 cm ($1\frac{1}{2}$ to 2 ft) in order to avoid damage caused by freezing in winter. However, if the water comes from a nearby spring the channel can be left open to the air for this will help to cool and aerate it at the same time.

In order to help oxygenize the water when it reaches the hall, if the ground level permits, it is a good idea to install the channel with a series of miniature waterfalls either inside or outside the building. These will help to stir the water up and aerate it. Inside the building the water supply channel can also be perforated for then the water will pour, like rain, into the spout feeding the troughs (Fig. 132).

When the water for incubation is not sufficiently clean it must be filtered. Spring water should not be filtered however.

There is a variety of filters both for use in the open air or under pressure. However, very few give completely satisfactory results.

Baffle plate type filters, which are commonly used, operate by sedimentation. They include vertical plates enclosing small sections in which a filter substance is placed between horizontal, perforated plates. The water flows into the first space at the bottom, passes up through the filter substance, flows over into the second space where it is again filtered down and so on to the last space. Each space or compartment must be at least 1 m and there should be at least two of them.

As for the filters they can consist of charcoal, coke, peat, wood fibre, gravel or sponges. Wood fibre and sponges are not very hygienic. Peat is one of the best filters. Sand and various fabrics have been used from time to time but they are not to be recommended as they choke up too easily.

A by-pass should also be provided so that the water can flow direct to the incubation hall when the filters are being cleaned out. The filters must also be provided with emptying plugs or bungs installed in the bottom. They must be placed at the lowest point of each section or filtering space.

These filters should not be too small for the larger they are in size the more effective they will be.

If supplying the cultivation installations with water is difficult, defective or insufficient, then a closed circuit installation can be put in (Leitritz, 1962). This should overcome the problem but it will complicate the mechanical side of the installation.

4 Troughs used for incubation and hatching

All the classical type incubation and rearing installations amount to a trough in which there are one or several meshed trays on which the eggs are placed. The central section, and at the same time the largest of the troughs, is where the trays are placed. It is separated from the upper and lower sections by partially perforated partitions. These control the circulation of the water inside the troughs and avoid the escape of fry after hatching.

The incubation installation must be so planned that at no point should it be away from the flow of water necessary for the development of the eggs and the fry.

The water enters each incubation trough at one end and flows out of the other continuously and without interruption. It must also be seen that as well as flowing from one end to the other of the trough it also fills it from bottom to top.

Vertical incubators are described under point 6.

a Principal types of incubation and rearing troughs. There are many models all of different shapes but they can be classified into two main types. These are short or Californian incubators in which, generally, one or two incubation trays are placed per trough, or long hatchery troughs in which several trays per trough can be placed. The first generally are only used for incubation; the second permits, successively, hatching and artificial rearing.

Long shaped troughs are generally used for industrial cultivation in which a large number of eggs are hatched and artificial food is distributed to the fry. They are also installed where the difference in level is insufficient. The Californian type is used if the amount of water available is insufficient and therefore can be re-used if necessary.

1. Californian incubator (Fig. 138). These incubators are always short (Figs. 139 and 140). They include an exterior trough in zinc sheet, wood or Eternite, holding one or two hatchery trays.

The sizes of these troughs vary between the following: $0.50 \times 0.20 \times 0.15$ m ($20 \times 8 \times 6$ in) and $1.00 \times 0.50 \times 0.25$ m ($40 \times 20 \times 10$ in). The first holds only one hatchery tray and the second normally holds two.

Californian incubators are also so arranged that the current ascends across the incubation trays (Fig. 138). In this way the water is well oxygenized and there is a partial decantation of suspended matter which settles between the bottom of the trough and the incubation tray. To obtain a satisfactory water flow inside the trough it is necessary that the overflow covers most of the width or even the whole width.

2.   Long shaped trough (Fig. 141). Long shaped troughs are used first for incubation. After that they can also be used for rearing and feeding fry during 1 to 3 months. They always contain several hatching trays.

The normal dimensions of long shaped troughs are between 2 and 3 m (2 and 3 yds) in length, 0·50 to 0·80 m (20 to 32 in) in width and from 0·20 to 0·35 cm (8 to 14 in) in depth. In order to facilitate the work the top of the trough should be about 1 m 10 cm (3 ft 7 in) from the ground.

The sides of the troughs can be made in different materials such as concrete or cemented bricks (the bottom should be a concrete slab), varnished zinc, Eternite, wood, wood covered with zinc sheet and various sheet metals such as aluminium, zinc or steel (Figs. 144, 145 and 146).

The depth of the trough should be between 25 to 35 cm (10 to 14 in). It is not advisable to reduce the depth otherwise the volume of water necessary for rearing will be insufficient. But it is not necessary to increase the depth either beyond 35 cm (14 in) for then the sorting of both the eggs and the fry would be too difficult. The water level should be 10 cm (4 in) below the rim of the trough to avoid the fry jumping over.

The widths of the troughs vary depending on how they are placed. If they are single troughs then the inside width can be 80 cm (31 in). If they are arranged in pairs then the width of each should not exceed 60 cm (2 ft).

As for lengths these can go up to 3 m (9 ft) and over though this is not recommended, for then eggs placed downstream run the risk of a shortage of oxygen.

If the troughs are in cement then the interior surface must be carefully smoothed and the angles rounded. In order to protect the eggs from too bright a light the inside walls and the bottom of the troughs should receive a coat of dark coloured waterproof paint.

b   General placing of incubation and rearing troughs. The general layout of incubation and rearing troughs can be very varied, such as halls with wide shaped installations and those with long shaped installations.

For the wide shaped types the troughs are generally placed perpendicularly either side of a central aisle (Fig. 146). For the long shaped types the troughs are generally placed perpendicularly on one side of the principal aisle (Fig. 145).

It is advantageous to use double troughs rather than single, for then fewer aisles are necessary.

The space between two troughs or between two pairs of troughs, side by side and parallel, can vary between 60 and 80 cm (24 and 32 in). The central aisle should not be less than 1·50 m wide (5 ft) so that it is wide enough for carrying cans, etc.

The upstream ends of the troughs should not be placed directly against the walls otherwise they will be constantly wet. In any case between the upstream end of the troughs and the walls of the building, a space should be left for the supply conduits and taps.

Californian incubators are often installed in tiers (Fig. 140), but never more than four or five, so that the water can flow from one to the other. If there are too many the water in the last trough will be deficient in oxygen.

At one time Californian incubators for salmonid cultivation were most often set out in tiers. On modern industrial farms these have been replaced by long shaped troughs when, of course, the water supply is sufficient.

c   The incubation trays. Normally incubation trays are in perforated stainless steel, aluminium or zinc or sometimes in plastic. Formerly copper, brass mesh or even glass rods were used (Figs. 149, 150, 151 and 152).

The holes in the sheet metal are from 1·5 to 2·5 mm in diameter and the copper or brass mesh made of wires from 0·5 to 0·75 mm spaced 1·5 mm apart. The glass rods were from 4 to 5 mm and spaced 2 mm apart; the latter are mounted on wood or zinc frames.

The sizes of the trays vary according to the sizes of the troughs. The width of the trays should fit the interior width of the troughs easily. Lengths rarely exceed 50 cm (20 in) and generally sizes are between 50 × 50 cm and 20 × 50 cm. The number installed depends on the length of the troughs.

The trays are placed in the troughs in such a way that the eggs are covered with from 3 to 5 cm of water ($1\frac{1}{2}$ to 2 in). The water could be deeper but in order not to interfere with control and cleaning of the eggs it should not exceed 10 cm (4 in). The trays stand on legs resting on the bottom of the troughs. They are regulated so that the required height is right. Legs are not necessary however when the trays rest on ridges fixed to the inside walls of the troughs.

The space between the sides of the troughs and the trays should be as small as possible in order to ensure the circulation of water from bottom to top across each tray. To make certain of this the space can be filled in with foam rubber.

The sides of the perforated or mesh trays can rise above the water in the troughs. They need not, however, but it is better that they should because after hatching and until their yolk sacs are absorbed, the fry stay on the trays and do not fall to the bottom of the troughs as happens when glass rods are used. After hatching, and before the trays are removed, the troughs should be well cleaned, which is not possible if the fry fall to the bottom immediately after hatching.

Trays in zinc and mesh must be painted (with varnish, bitumen, or asphalt lacquer) and should receive a new coat each year. This is most important for zinc trays when the water is acid, for if they are not painted zinc oxide forms which is toxic for eggs and fry. In alkaline water this inconvenience disappears rapidly.

After calculating the sizes of the trays and the average diameter of the eggs it is easy to find the number of eggs incubating. Thus, if there is one layer of eggs with an average diameter of 5 mm there would be 400 eggs per square decimetre (16 sq in). A 50 × 50 cm tray (20 × 20 in) can support 10,000 eggs and if the eggs have an average diameter of 4 mm only then the number would be approximately 15,000. If the water is clean and clear enough several layers of eggs may be superimposed.

The trays are provided with wooden, metal or Eternite lids as hatching always takes place in the dark.

d The water supply for incubation and rearing troughs. The water distribution in the hatching hall can be pressurized in pipes or be left in the open and flowing through a gutter.

Taps, or other devices ensure water distribution (Figs. 132 to 137). In principle, there is an individual water supply for each long shaped trough or for each group of short troughs if they are set out in tiers.

When the canalization is fixed to the walls it must be installed at the same level as that at which the water will be used. When the supply cannot be fixed to the walls, which is the case when the troughs are in the centre of the hall with a working space round them, the conduits are installed under the tiles and the supply pipes rise up to the height of the troughs.

A gutter for open water distribution can be in cement, wood or metal. At the end of the gutter an overflow must be provided in order to ensure a predetermined and constant water level and also to run off excess water.

Taps are fixed to the waterfall system some 20 to 30 cm (8 to 12 in) above the water level of the troughs. A very practical tap has been designed in iron or bronze. It does not need rubber and its opening is regulated by a small lever ("Perfection" type tap, Fig. 134).

Besides taps there are other more or less satisfactory systems. For example, a siphon system exists consisting of curved tubes which hook on to the rim of the water gutter (Fig. 135). This is a simple system but if the level of the water is not always the same the siphons will run dry.

It is also simple to provide the water trough with a few small holes facing the hatching troughs. The number of open holes controls the water supply for each hatching trough. Or, instead of several small holes there need only be one which can be partly closed, when necessary, by a small board or by inserting a plug.

The tap can also be replaced by an elbow-angled tube, the long arm being sunk into the wall of the water trough with the short arm (Fig. 133) hanging over the incubation trough. When this arm of the tube is turned upwards the water will not flow. By turning it gradually the water will start flowing and if pointed downwards it will be fully open and the water will flow freely.

Aeration of the water supply can be aided by installing small cornet shaped funnels on the taps in which the air and water are mixed through suction caused by the running water (Fig. 136). It is also possible to install, under the tap of each hatching trough, a plate which will cause the water to fall in a sheet. This means even more and better contact with the air (Fig. 137).

e Evacuation of water from hatching troughs. The evacuation of water from the hatching troughs is through an overflow (Fig. 148). Several types exist. These can be normal overflows occupying all or part of the upper part of the downstream wall of the hatching troughs.

Often they are cylindrical tubes with a diameter greater than that of the supply taps.

The simplest type of cylindrical overflow is formed by a tube fitting tightly into the evacuation hole of the hatching trough. This hole is indispensable anyway in order to permit the cleaning of the trough, regardless of the kind of overflow.

A slightly more complicated model is formed by two sleeves or sockets, one slipping into the other. The interior sleeve is sunk into the hole made in the bottom of the hatching trough so that the sliding movement of the other sleeve regulates the height of the water in the trough.

To help the evacuation of water at the bottom of the trough the cylindrical overflow can be enlivened by a kind of cylindrical collar with one end overlapping the overflow. As the lower part of the collar is pierced with holes the water to be run off passes through them into the sleeve and then into the cylindrical overflow by which it is evacuated. A similar system is used for the evacuation of water from aquaria (Fig. 482).

It is also a good idea to have two water levels in the troughs – one for hatching and one for rearing. In this case a cylindrical zinc or copper overflow is adopted and separated in two unequal parts by a small copper collar welded to the tube.

The small collar can also be replaced by a double trunk cone. By placing a metallic ring round the evacuation hole of the trough, a watertight joint is created. The overflow and the ring can be in copper or bronze.

To bring the water overflow to the general water collector, tubes are fixed to the evacuation holes of the troughs overhanging the axis of the collector and extending directly to it.

The general evacuation of water from the incubation hall is by means of an open channel sunk in the tiled floor. It runs along the principal aisle.

To facilitate working in the room place a wooden lattice, an iron grill or a plaque made in some hard material and resting on two ridges over the water evacuation channel so that its upper part is at the level of the tiled floor. Openings must be so arranged to allow the passage of water from the evacuation pipes of the incubation troughs.

f Screens to avoid the escape of fry. To prevent the escape of fry a vertical screen is placed up-stream from the overflow and about 5 to 20 cm (2 to 8 in) from the downstream end of the trough, depending on its length (Fig. 148). Normally a similar screen is placed upstream to keep the fry from approaching the water inlet fall.

These screens, like the hatching trays, are in perforated sheet metal or in copper or brass mesh. To aid the water circulation from bottom to top at the centre of the trough, the upstream screen can have holes in its lower half only and none in its top half and vice versa for the downstream screen.

Normally these screens are welded to the walls of the trough. Sometimes they are detachable and slide in a vertical groove sunk into the sides of the troughs. In this case they can be mounted on a wooden frame. To ensure that the frame fits the bottom and the sides of the trough tightly and so avoids all possibility of the fry escaping, a thin strip of lead can be nailed to it which, because of its malleability can be pressed with the fingers on to the bottom of the trough and also to its sides.

To replace the vertical downstream screen the overflow can be fitted with a cylindrical screen with a small collar at its base. If the latter is made of lead it will fit the bottom of the troughs splendidly thanks to a little pressure with the fingers.

5 Round shaped circulation rearing troughs

Since the end of the second world war there has been a trend towards replacing classical type hatching troughs, i.e. long shaped troughs for rearing fry for a few weeks or months, by tangential round shaped circulation tanks (Figs. 114, 164, 165 and 166).

The merit of these lies in the regularity and uniformity of the current and in consequence the equally uniform distribution of the fry in the trough and the improved chances of each one benefitting from its food and maintaining normal growth. The fry face the current and spread regularly over the whole of the rearing tank. In long shaped troughs, the spread of the fry is not so regular for they have a tendency to concentrate upstream and in corners. The fact that there is a better spread of fry in the round shaped circulation troughs means they can take a greater number of fish than can the long shaped troughs.

These tanks can be circular, or square with rounded corners. The diameter or side of these tanks is between 1·50 to 2·50 m (5 to 8 ft). The

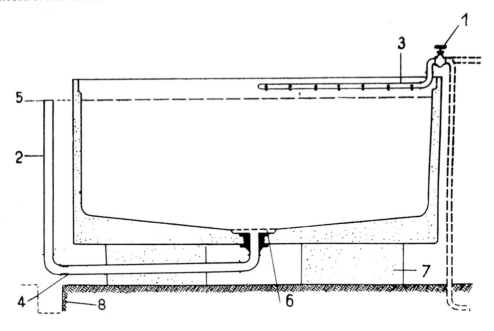

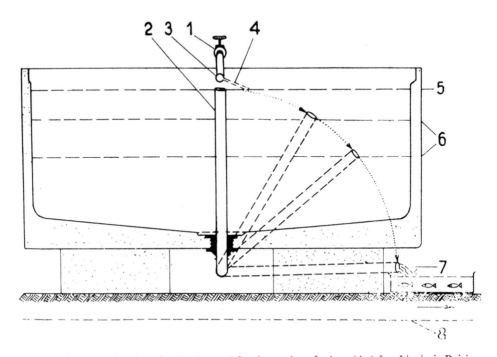

Fig. 114 Schematic drawing of a circular pool for the rearing of salmonids (after Livojevic-Bojcic, 1967).

tangential current is predetermined by the lateral and tangential supply of water under pressure and poured obliquely onto the surface. The out-flow of the water is from the centre of the tank.

Apart from the round or square shapes the characteristics of these tanks are very similar to those of the long shaped troughs. The materials most currently used are Eternite or plastic. The water supply is lateral and by an ordinary pipe or by a perforated water conduit fixed, perpendicularly, to the side of the tank. In any case the jet of water onto the surface is tangential.

The evacuation of the water is assured by a cylindrical overflow placed in the centre of the tank and consisting of a bronze or copper sleeve topped with a cylindrical mesh permitting the evacuation of the water but stopping the passage of the fry. In other cases evacuation is controlled by an elbow bent pipe the slant of which predetermines, at the same time, the depth of the water in the tank. When it is placed horizontally this device can also serve for emptying the tank (Fig. 114).

6 Vertical flow incubators

Classical shaped incubators are often criticized (whether they are short models placed in tiers, or long models) for giving only a small economic return after taking into account that they occupy rather a large part of the total area of a costly building in which the actual period of production is relatively short. This criticism or inconvenience is reduced by using long shaped troughs because they can be used for three supplementary months for rearing.

Over the past 20 years several vertical incubation models have been developed in the United States and in Sweden (Burrows and Palmer, 1955; Lindroth, 1956). The most popular in the United States is made up of one or two superimposed shelves each one enclosing eight incubation trays (Figs. 155 and 157). Each incubation tray is made up of an external basket in which, covered by water, there is an egg container tray with the eggs and the fry after hatching. This tray is covered with mesh. The whole device is made in plastic. The exterior dimensions of the single shelves are about 80 cm high × 62 cm wide (2 ft 8 in × 2 ft). The exterior dimensions of the water trays are 53 × 62 × 9 cm and those of the egg container trays 40 × 35 × 5 cm (21 × 25 × 3½ in and 16 × 14 × 2 in).

The water arrives above the case of shelves and is carried to the top tray. By rising inside the tray the water covers the eggs or fry and then flows over the top and falls onto the tray immediately below. From top to bottom it is used to serve, successively, the eight or 16 trays. After it has passed through the case it can be evacuated or recycled and used again.

Each tray can carry 10,000 salmon eggs or an even higher number of trout eggs which are smaller. If several double shelves were placed side by side the number of eggs incubated per square metre would be very high. A second advantage is the relatively small water flow required by the device which is no more than approximately 15 to 45 litres per minute (3 to 10 gal per minute).

This device can only be used for hatching and not for rearing. It needs very pure water and is rather costly to install.

Vibert (1959) tried out and described a vertical incubator in which trays were replaced by double bottom 12 litres capacity pails in plastic (2 gal 5 pt) and which used a rising current. Each pail when filled carried at least 80,000 trout eggs. The device was completed by a system permitting automatic disinfection by malachite green.

In Pennsylvania, Buss and Fox (1961) adopted jars for the incubation of salmonid eggs such as are used for pike and coregonids (Fig. 238). In order to stop the eggs moving a layer of gravel was placed in the lower part of the jar. The development of fungi (or water moulds) was avoided by the employment of formalin or malachite green baths. Large jars 35 cm in diameter and 80 cm high (about 13 in and 31 in) holding about 70 litres (123 pt) are used for incubation in Italy (Fig. 156). There are 50 litres of eggs (11 gal) or roughly 8,000 to 9,000 to the litre which means each jar holds at least 400,000 eggs. After the eggs are eyed they are transferred to the hatching and rearing hall. The hatching takes place in very pure water and at a relatively high temperature: 10 to 12°C.

7 Adjacent installations

The handling hall in which several operations take place such as artificial fertilization, the stocking and preparation of food, dispatch of the fish, are distinct from the incubation and rearing hall. It must be spacious, clean, light and airy and the floor should be either in cement or tiles. Sufficient space was not provided on many old farms for all the operations and the housing of material so that the work was not easily carried out.

At the time of artificial fertilization, holding

tanks in which the parents can be kept are necessary. These should be covered with a mesh lid. The fertilizing operation calls for a table, galvanized or plastic basins, in which to place the parents, as well as auxiliary tools such as small pans for the stripping, graduated glasses, formalin and feathers.

If the fish are given dry food then it must be stored in a dry place. If they are given fresh food it should be kept apart in cold storage.

The fish and meat are prepared in a kitchen provided with a butcher's block, a chopper or a large meat or fish mincer, and a sink.

Tools used in fish cultivation must be kept separate from the room in which handling operations are carried out or kept partly in the loft. These tools include different graduated glasses, carrying cans, weighing equipment, fish graders, basins, etc.

Appendix: Rural Salmonid Farms

The installations described above for the incubation of trout eggs are sometimes greatly simplified. This is the case of many amateur installations known as "rural fish cultivation" with the single aim of restocking angling waters. Such establishments are numerous in Switzerland and are not rare in France and other countries. Amateur fish farmers work for themselves or for local angling clubs.

The main installation is formed by a variable number of incubation troughs permitting the hatching of some 100,000 eggs but rarely more. The number can be reduced to 5,000 if the demand for restocking is limited.

The incubators are generally Californian type, it being understood that the troughs are rarely used for rearing.

Generally there is no incubation hall as such as the troughs, being in the open air, stand directly in the ground or on light scaffolding. If the troughs are under cover it is generally in a barn. Quite often, incubators are placed in a cellar.

In this case the water normally comes from the local distribution system but frequently it comes from a rheocrene spring in the neighbourhood. It is a good idea to place the troughs a few metres from the spring to permit sufficient oxygenation of the water but not too far so that the winter cold does not freeze it in the troughs.

A simple dam with a supply pipe for the incubation troughs and an overflow for the removal of excess water completes the installation which really is of artizan type (Fig. 142 and 143).

The incubation trays, the evacuation of the water, the screens for stopping the fish from escaping, are not special in any way.

The eggs to be hatched are either bought commercially or come from brood fish kept in ponds by the fish farmer or are caught by him in fishing waters with trap nets or other fishing devices.

The fry are liberated for restocking as fry with absorbed yolk sacs, or they are placed in ponds until they grow into fingerling trout to be harvested the following autumn and then liberated in angling waters.

B Care during incubation and hatching

1 Phases of incubation and hatching

There are three phases from the start of incubation till after the yolk sac is absorbed:

1st phase: from fertilization until the appearance of the eyes ('eyed eggs');

2nd phase: from the appearance of the eyes until the hatching;

3rd phase: from hatching until absorption of the yolk sac.

During the first days (10 to 15 for brown trout and 5 to 8 for rainbow) which follow fertilization, the eggs can be handled without fear. After that they become very sensitive to shock and must be left quiet at least until the eyes appear as two large black spots. They are then called "Eyed Eggs". This occurs about halfway through incubation.

Before the eyes appear it is already possible to see if the eggs have been fertilized or not. To do this it is only necessary, from the second day, to place the eggs in a mixture of three parts chromic acid (0.5 per cent), four parts sulphuric acid (10 per cent) and 30 parts alcohol (96 per cent). After a few minutes, if the eggs are fertilized, a two-part or four-part division appears, or an embryo if this is carried out later. After 8 days an embryo will also appear if an egg is placed in a solution of acetic acid or even vinegar.

During the second phase, eggs can be handled or transported provided humidity is maintained. An eyed egg is a commercial product.

Hatching lasts about one week. Normally, the tail emerges first. If on the other hand the head comes first then the fry will be in difficulty and might even die. The malformed embryos are generally the first to emerge.

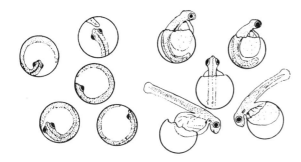

Fig. 115 Eyed eggs and newly-hatched trout fry (after Hey).

During the third phase, which is about half the length of the incubation period, the fry begin to look very different. This development is more a transformation of the organism than growth. At first, weighed down by their heavy yolk sacs the fry remain more or less inert at the bottom lying on their sides (Fig. 154) and only showing signs of life when disturbed. Little by little the yolk sacs are absorbed, the fry lengthen and also lighten in appearance. Immediately after hatching the average length of a trout fry is 15 mm, and by the time the yolk sac is absorbed it is 20 mm. At the same time it takes on colour, loses its transparency and grows more and more vigorous. Its fins can be discerned and it progressively becomes more and more mobile. It rises, moves around and finally starts swimming. The fry group together, pressing up against each other and congregating one on top of the other. When light is let in by taking the covers off the trough they look more like a lot of disturbed ants. At the end of the third phase their digestive tubes open and they can start to swallow small prey. At this moment the yolk sac is three-quarters absorbed and the fry can be released in open water if the intention is not to feed them. But also at this time they can be fed artificial food if they are going to be retained in a rearing trough.

The rapidity of development of the different phases of incubation depends on many factors which in turn depend, above all, on temperature. The development varies naturally according to the species. It is shorter for rainbow than for brown trout. But for any one species it is that much faster as the temperature becomes higher.

The period of development is given in degree days, that is, the number of days required for the incubation of eggs of any one species at 1°C.

The product of the incubation period expressed in days, by the average temperature of incubation shows a fairly constant number. For brown trout it is from 400 to 460 degree days, for rainbow trout from 290 to 330 degree days. Hatching extends over about 50 degree days for eggs of the same spawning.

The number of degree days being equal to the product of the average temperature of the water expressed in degrees centigrade, by the hatching time expressed in days, so it is possible, inversely, to determine the number of hatching days by dividing the known number of degree days by the average temperature of the incubation water. It is thus possible to draw up graphs which permit advance determination of the average time of incubation, through knowing the average temperature at which it is being carried out.

The absorption of the yolk sac takes about 220 degree days for brown trout and brook trout and 180 degree days for rainbow.

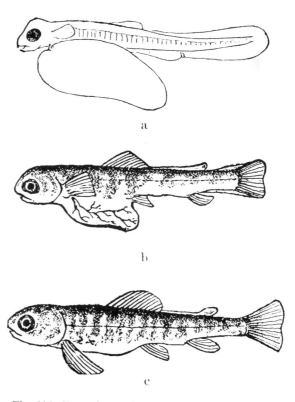

Fig. 116 Trout fry at three different stages: (a) immediately after hatching; (b) with yolk sac semi absorbed; (c) with yolk sac almost entirely absorbed (best time for releasing) (after Diessner-Arens).

2   Care to be given

During incubation and hatching, care must be taken of the eggs and, later, of the fry.

At regular intervals, in principle every 2 days dead eggs and fry must be removed.

It is easy to distinguish dead eggs from healthy ones from the fact that they turn white and are opaque. If they are not removed they will be attacked by Saprolegniaceae which will contaminate the healthy eggs. They are removed with forceps (either of wood or brass) or with a pipette comprising a glass tube 5 to 7 mm in diameter equipped with a rubber bulb. The tube is always heat softened at the tip in order to avoid injuring the healthy eggs (Fig. 153). More complicated devices are also used, some of them more or less automatic. Under good conditions the losses should not exceed from 3 to 5 per cent for brown trout (exceptionally 20 per cent) and for rainbow trout and brook trout between 10 and 20 per cent. If the percentage is higher, then the removal of the dead eggs takes a lot of time and this will reduce the profits of the farm. Generally, over the first days following fertilization, it is possible to detect the most important waste due to the death of defective eggs. The same applies to eyed eggs when they are washed. Unfertilized eggs can retain the normal external appearance of fertilized eggs for a long time. Information about the treatment of eggs during incubation is given in Chapter XV.

Immediately after hatching the empty shells are removed. One way to do this is to use a short, glass pipette but it is easier to siphon them off. If a few fry are siphoned off at the same time they can be recovered easily by pouring off the siphoned product into a receptacle. The fry will sink to the bottom whereas the shells sink very slowly and can be poured off simply by inclining the receptacle.

During the absorption of the yolk sac daily attention is necessary and fry which die during this period must be removed.

The eggs must be kept in darkness. As the absorption of the yolk sac comes to the end, so the fry must be helped, progressively, to get used to light. This is done by uncovering the troughs a little at a time. Also by darkening the windows of the incubation hall it is possible to filter strong light which will not upset the fry each time the covers of the troughs are removed for daily inspection.

The water used for incubation must be rich in oxygen. By placing small planks under the inlet of the water supply the flow will be uniformly spread.

In order to avoid accidents during incubation the metal or mesh trays used should be slightly convex in shape at the bottom so as to stop the formation of bubbles under the trays and so hinder the renewal of water for the eggs above. During the absorption of the yolk sac the water flow should be increased slightly to at least 1 litre per minute for 1,000 fry. However, the flow must not be too great otherwise it might provoke shock which would harm the eggs, help to accumulate bubbles under the trays or even crush the fry against the trough screens.

Mud deposits on the eggs must be avoided as they hinder breathing and favour the development of Saprolegniaceae. However, there is no need to worry about a very thin deposit. The eggs should not be moved before they are eyed. If the mud deposit is considerable it can be reduced by lightly lifting and shaking the tray and then lowering it again. After the eggs are eyed they can be easily washed either by removing the trays from the trough and spraying them with water or mixing them gently in a Zoug glass or MacDonald jar. The troughs and trays can then be cleaned at the same time.

During the whole of the incubation and especially before the eggs are eyed all shocks must be avoided. Eggs can only be transported therefore during the second hatching phase.

General hygiene must be carefully watched and the first signs of disease, particularly costiasis, must be looked for. If it appears it can be treated with formalin (Leger). This comprises 40 cm³ (2 in³) for 100 litres of water (22 gal). The fry should be dipped into the bath for 15 minutes. During this treatment the renewal of water is stopped. This treatment is easily supported by the healthy fry and those which are curable, but it hastens the deaths of those badly affected and which are condemned in any case.

In order to prevent external parasites Gerard (1974) recommends periodical treatment with formalin, every ten days from the fifth day after the yolk sac is absorbed. The bath contains 25 cm³ (1.5 in³) of formalin per 100 litres (22 gal) of water if the water temperature is below 10°C, 20 cm³ (1.2 in³) if the temperature is between 10 and 15°C, or 16 cm³ (1 in³) if the temperature is higher than 15°C The duration of the treatment is 15 minutes. Such treatment is applied in the rearing troughs themselves, after closing the water supply. This may be continued for the fry reared in the troughs, but feeding should stop 12 hours earlier.

C Counting and transport of eggs

1 Counting of eggs and fry

It is easier to count eggs than fry; indeed fry are counted only if it is absolutely necessary and then one of the three first systems described below can be used. The eggs are counted when eyed. At this time they can be handled without danger excepting during the few days preceding the actual hatching. It is possible to estimate the number of fry because the losses from now on until the actual hatching are insignificant.

One of the following methods can be employed:

First method. First 1,000 eggs are weighed on laboratory scales. On another set of scales the mass of eggs to be counted is weighed. As the weight of the eggs varies considerably the first weighing must be very precise for each lot.

Second method. 1,000 eggs are poured into a small graduated glass filled with water and the number of cubic centimetres it occupies are counted. Normally from 10 to 15 trout eggs occupy 1 cm³ but each lot must be controlled because variations in volume are important. In this way a large quantity of eggs can be measured in larger graduated glasses.

This is a quick method though the eggs must not be removed from water.

Another way is to use a 150 to 200 cm³ receptacle with a perforated bottom (9 in³ to 12 in³). This is filled with eggs which are carefully counted a first time. If the bottom is so built that it can be regulated (Fig. 159), then it can cube 1,000 eggs at a time.

Third method. A figured receptacle provided with a small curved pipe inserted halfway up is filled with water (Fig. 160). After having dried off 1,000 eggs or 1,000 fry on muslin they are poured into the receptacle. The water in it will, of course, be displaced and flow out of the tube into a graduated glass. The amount of water displaced can then be measured and the number of eggs or fry can be deduced by comparison.

The Schillinger apparatus (Fig. 161) is similar to the one described above. It comprises an Erlenmeyer flask mounted with a graduated glass. A funnel is placed in the top of the latter. The lower flask is filled with water up to the "O" mark at the base of the inserted glass. The eggs or fry are poured in up to a graduated line and the dis-

placed water then pours into the glass and is measured. Knowing the volume of a given number of eggs or fry it is possible to calculate the number of all the eggs or fry measured.

Fourth method. It is also possible to use a stiff rubber Brandstetter pad which includes a specified number of hemispheric cavities (200 to 500). An egg lodges in each of these (Fig. 158). This is a rapid system free from the problem of the relative sizes of the eggs.

Fifth method. The number of eggs occupying a known surface such as a frame used for the transport of eggs in small cases can be counted. By counting the number of frames it is then possible to determine the approximate total number of eggs.

2 Transport of eggs. Hatching boxes

This is easy enough and without danger for the eggs, provided it is carried out at the right time, that is when the eggs are definitely eyed and no fewer than 5 days before hatching. The transport of trout eggs is helped by their ability to remain alive out of water provided the temperature is not too high, the humidity is sufficient and they have all the oxygen they need. If the transport is carried out too early then the embryos will not support the shock. If it is too late then it will provoke premature hatching and death.

It is this relatively easy transport of eyed salmonid eggs which has permitted and facilitated the introduction and acclimatization of foreign species, the purchase of massive quantities of eggs from specialized or foreign farms, and the relatively easy method of restocking open water.

If distances are short then a thermos flask or damp moss is enough. But the eggs must not be packed tightly and no more than 2 kilos (4 lb) should be carried at a time.

For long distances up to two or even more days a model type of transport described below is used (Fig. 162).

Previously the essential elements for transporting eggs were square frames of wood (20 × 22 cm exterior dimensions) (8 × 9 in). They could also be rectangular (22 × 32 cm) (9 × 12 in) and made up of slats 20 mm wide and 6 mm thick. On the under surface a muslin or some other type of straining material is stretched. The frames are moistened and then filled with eggs carefully placed in layers so that they cannot roll about. Five frames are placed one on top of the other followed by a sixth, an empty frame. This can be repeated to a height of up to 20 or 30 cm (8 to 12 in). They are then placed on several frames, either left empty, or filled in with some kind of moss material. On top a frame with deep sides (3 to 5 cm) is placed. It has a slatted bottom (the slats are from 1·5 to 2 cm in width, placed 5 mm apart) on which ice cubes are placed. These frames are now being replaced by an insulating synthetic product (polyster) (Fig. 163). There are many models, some in sections and others not. The exterior dimensions of the frames are around 25 to 30 cm (10 to 12 in) but depth is increased to 5 cm (2 in). They take several layers of eggs.

Before being transported the eggs must be carefully washed. The frames are tied and can, if desired, be wrapped in impermeable paper but this is not necessary. They are then placed in a case leaving a space between the frames, on all sides, of several centimetres. This space is filled with any insulating material such as peat, wood shavings, moss, etc. The case is then covered by a lid. There is a handle on either side for carrying and the bottom includes a hole for the outflow of water. Formerly these cases were always made of wood but now plastic is used. It is best to place on top a large label inscribed "Live fish eggs, fragile, keep upright, handle with care, keep away from warmth, avoid freezing". The transport must be well organized and the addressee should be warned well in advance.

Under these conditions eggs stand up well to transport over long distances. Formerly, when the required isothermic precautions were taken, transport was possible over long distances even from Europe to America and New Zealand. Now, thanks to air transport, the time has been shortened and the transport of eggs greatly facilitated.

On arrival the eggs are unpacked carefully in the building where they will be hatched. They are sprayed with water for 15 or 30 minutes in order that they might get used to the water at local temperature, and are then placed on the incubation trays.

If the transport is carried out under good conditions losses should not exceed 1 to 3 per cent.

Fertilized trout eggs can also be used for restocking trout streams by means of hatching boxes (Vibert boxes) designed by Vibert (1975). The boxes are made of rigid, transparent plastic material. Holes are drilled in the walls, their diameter being smaller than the egg size but large enough for the fry to swim out. Vibert boxes are filled with 800 to 1,000 embryonated trout eggs and planted in areas showing the characteristics of natural spawning grounds.

D   Releasing fry with absorbed yolk sacs in open water

Salmonids for restocking can be raised up to fry, fingerlings, yearlings or older fish. If the aim is restocking with fry, cultivation ends when their yolk sac is absorbed. At this stage fry are released in the open waters.

1   The time to release

The time for releasing fry should not be too early, otherwise losses will be high. But it should not be too late either. The best time is when fry, freed from the weight of their yolk sacs, leave the bottom of the trough where they have remained so far. They rise to the surface, swim easily and behave like real fish. At this time the yolk sac is about a quarter or one-fifth of the size it was but the reserves it contains are still precious to the fry while they get used to their new surroundings and the food they will find. Fry which retain a part of the yolk sac are known as "fry with reabsorbed yolk sacs". The time for release is about 200 degree days after hatching. Fry which have been released too late or badly fed have large heads prolonged by elongated bodies (athrepsy).

The fry are siphoned from the hatching troughs to the transport cans by rubber or plastic tubes.

2   Distribution plan

Before the actual liberation of the fry it is necessary to plan their distribution. Cultivation of fish in cold water gives late fry whereas fish cultivated in higher temperature waters are more precocious. For this reason those waters which flood late are so restocked that the fry are already small fingerlings at the time of the spring floods and the melting of the snow. However, there is no necessity to exaggerate the dangers of flood for them as the fry take shelter behind rocks, stones, roots, submerged plants, etc. But in waters which lack these or have beds of moving pebbles there is grave danger.

The plan must also include careful distribution according to species and strains. Rainbow trout and brook trout are kept for carefully chosen waters where the fish have a real chance of acclimatizing themselves. In Central Europe lake trout fry are only released in tributaries of lakes.

The importance of distribution must depend,

evidently, on need; that is to say, the difference between productive capacity and real production.

In open waters, fry with absorbed yolk sacs are exposed to numerous enemies, starting with fish of the same species, fingerling trout and trout of 1 or 2 years or over.

3 Choice of place for releasing

The choice of place for releasing is a delicate problem on which the whole success of the operation depends. Too often the importance is underestimated and releasing is finished off in a hurry. Effective release is very important because all the care taken of the eggs at all stages of fertilizing, hatching, and rearing is finally at stake.

It is very important that the fry should be released so that protection is assured, that they have food and are able to escape from their enemies. Also that they can develop rapidly. As they grow step by step fry move around seeking out places which suit their size and their age. For example, for those lower trout water courses which are not suitable for the release of fry, it is best to liberate the fry upstream in zones favourable to natural hatching and from which they can find their own way to suitable, lower and deeper parts.

The best spots are those far removed from where large trout are found, where the current is not too strong and where there are aquatic plants either submerged or semi-submerged (water starwort, water-crowfoot and watercress), where the fry can find both food and shelter.

Areas near springs or the sources of water courses rich in aquatic plants offer the most suitable places for release. Indeed it is here that brood fish find natural spawning beds and where they spawn. Brooks and trickling brooks (Fig. 168) which are effluents of larger brooks and streams which it is proposed to restock, often offer better releasing spots than the water courses themselves. In any case fingerling trout will, eventually, reach and restock the more important courses as they grow and develop. In the large courses of the trout zone, as well as in the grayling zone and even more so in the barbel zone which includes too many enemies of fry (chub, bullheads, minnows, etc.),

fingerling or trout of two summers or even catchable fish are used for restocking.

Fry should not be released in polluted water nor in water near farms or mills where ducks abound.

4 Releasing the fry into the water

Liberation itself must be carried out with care, patience and attention. The degree of success depends also on the care taken over transport.

Gradually the fry must get used to the new water as it is different in composition, temperature and oxygen content from that in which they were bred. First of all getting accustomed depends largely on acceptance of the temperature. This is done in the following way. The receptacle is placed in the brook or stream and submerged to about three-quarters of its height. First, one-quarter of the water in the receptacle is replaced by the same quantity of water from the stream. After 10 minutes, the operation is repeated and half the amount of water is emptied and then replaced by stream water. Five or 10 minutes later the fry can be liberated with safety.

They should be released carefully in small numbers and at well chosen spots. As far as possible the lots should be from 50 to 100. The best course is to pour a few from the transporting can into a liberating pail (Fig. 486), after having equalized the temperature. The pail is then taken to a favourable spot where with the aid of a small muslin scoop-net fixed at the end of a cane (the length of which will depend on conditions) the fry are transferred from the pail to the water. From time to time the water in the pail must be renewed. This operation is quite simple as the top section of the pail is made up of close mesh permitting the water to be poured out without the fry escaping. The pail should be 18 cm high, 5 cm of which are mesh; it should be 15 cm in diameter. When the pail is empty a new lot of fry are collected from the can which has been placed meanwhile under a tap or submerged in water in the shade so that it will remain cool. In this way all the fry are liberated without any being touched – which is very important.

SECTION IV

PRODUCING TROUT FINGERLINGS AND TROUT FOR TABLE

I Rearing Trout Fingerlings

The raising of fingerling trout and other salmonids of less than 1 year is the most delicate part of salmonid cultivation. It must start in good time, that is, just before the yolk sac is entirely absorbed.

Young salmonids can be raised without artificial food (extensive farming) or with artificial food (intensive farming). Whichever method is used feeding generally starts in the rearing troughs and extends over a period of from 3 to 4 weeks.

The time it takes to raise young salmonids varies. Harvesting can start at 3 or 4 months but a common practice is to produce fish of one summer (fingerlings or summerlings) gathered in the autumn of the year they were hatched, or fish of 1 year (Jährlinge in German or yearlings in English) harvested in the spring of the following year. These different ages correspond to average variable sizes: 3 to 7 cm at 3 months (about 1 to 3 in); 7 to 12 cm at 6 to 9 months (3 to 5 in); 10 to 15 cm (4 to 6 in) at 1 year. However, the sizes of young salmonids vary considerably even at the same age. Independently of different individual growth, sizes differ from one species to another although this also depends on the food available and the intensity of artificial feeding.

If the fry are released in ponds then it is necessary to ascertain that they will not be endangered by enemies, and it is therefore a good idea to put the ponds under water only just before the fry are liberated. If, however, they will depend on natural food then this must be given time to develop before the fry are released. This can take several weeks. If feeding is with artificial food then, of course, this problem is of much less importance. Raising of fish for the table is carried over to the second year. But the same can apply for fish of two summers for restocking, which is regularly practised for Atlantic salmon and sometimes for trout.

A Feeding fry for 3 or 4 weeks in rearing troughs

The most difficult problem in raising fry is to get them used to eating. One proved method consists of giving the young salmonids artificial food for 3 or 4 weeks while still in the rearing troughs

and before liberating them in water courses or rearing ponds.

This practice was developed from efforts to get the young fish to take artificial food when released in ponds, because it had proved difficult. At this stage of rearing best results are obtained when the fry are concentrated in a limited space with very abundant food which is not the case when they are in ordinary rearing ponds. At the beginning, the fry will only take food which has not sunk to the bottom. If they are deprived of all natural food therefore they will soon become used to eating artificial food.

Feeding fry in rearing troughs results in rapid growth which is an important advantage. (1) The fry are stronger when liberated in open water and better able to resist such enemies as they may meet. (2) They are released at a stage when danger of infection by whirling disease is passed. This is primordial for rainbow and other species as well. (3) They are now accustomed to artificial food so if this feeding method is to be continued in ponds they do not have to get used to it. On the other hand, if the fry is no longer fed with artificial food then they have to get used to their new surroundings. This inconvenience is compensated for by other advantages. Total loss is less than when this method is not followed. On the other hand, feeding with artificial food over a period of 3 or 4 weeks exposes fry to certain alimentary sicknesses. However, taking into account the improvement in the quality of the foods used now, this problem is notably less during rearing.

This is the artificial food method to follow if the farm is equipped with more or less large ponds for rearing. On the other hand, if the fingerling trout are raised in small ponds and above all in spring or linear ponds then it is possible to achieve good results if feeding is exclusively natural.

The rearing of fry is carried out in hatching troughs from which the incubation trays have been removed, or in special tanks made in wood, concrete, asbestos cement or plastic, built along the same lines as hatching troughs or tangential current tanks (Figs. 164 to 167). A 2 mm mesh screen is placed downstream or in the middle if they are tangential current troughs. All must have their own independent water supply in order to avoid epizootics.

To succeed with this kind of rearing this method should be followed:

1. The surface of the rearing tanks must not be too small and should measure at least 15 dm² (90 in²). This means short incubators cannot be used.

2. The tanks must be crossed by an abundant flow increasing with the size of the fry but controlled to ensure that it does not carry or push them up against the evacuation screen. For 1,000 fish the flow should be around 1 to 2 litres per minute.

3. The inside of the tanks should be painted a light colour and exposed to a diffused light to secure full growth benefits.

4. Temperature must not be too low but about 12°C.

Under these conditions it should be possible to raise large quantities of fry within a small surface of water if sufficient food is given. One fry can be raised per square centimetre (10,000 to the square metre in a tank 2·50 × 0·40 m) (about 8 ft × 15 in) and even up to 30,000 per square metre for rainbow trout.

Organic waste, food remains and droppings fall to the bottom and must be removed regularly. So must dead fry in order that no accumulation should contaminate the water. Troughs must be well cleaned every day. The best and most effective way to do this is to use a siphon. Generally a supple rubber tube is used. It can have a metal, flattened out nozzle attached at the end so that fry cannot be sucked in. This nozzle should have a wooden handle to make manipulation easier. The nozzle is passed along the bottom to suck in waste like a vacuum cleaner. This water should be siphoned into a pail from which any fry accidentally removed may be recovered.

Formerly, when rearing started, it was recommended to mix artificial food with natural food when possible and if sufficient was available. But in practice it is only possible, really, to distribute Daphnia. Better growth of fry will result. Daphnia is particularly good for brown trout which, often, cannot get accustomed to artificial food. But finding natural food is difficult particularly in the spring. Further, it can be a danger, as natural food can carry disease. Before being distributed, the Daphnia must be carefully washed and, both

during and after distribution, the water flow must be reduced considerably.

After 3 or 4 weeks the fry are liberated in rearing ponds or in open water. Rearing can continue in troughs, however, but when it is, sorting out is necessary, otherwise cannibalism may follow. In any case the stronger fry will take all the food from the weaker fry and this could lead to some fry growing to twice the length of others.

Sorting is carried out more and more with large mesh strainers. A suitable size is about 0·40 × 0·20 × 0·15 m (16 × 8 × 6 in). The small fry will escape through the mesh and the large will remain in the strainer. Suitable strainers with appropriate meshes can be found for most sizes of fish. There are also automatic graders with adjustable openings varying between 5 and 30 mm (Fig. 469).

After sorting, the fry are distributed in the rearing tanks again but are now in smaller numbers. After 1 month, when the fry measure about 3 cm (1¼ in) the initial numbers are reduced to about one-third (3,000 per square metre); at 4 months (fry of 7 cm (3 in)), again by half (1,500 per square metre). These numbers can be doubled or tripled for tangential circulation tanks.

Instead of using troughs, the first rearing with artificial food can be practised in ditches or in small long ditch ponds not more than 20 to 30 m² (about 23 to 35 yd²) and 50 to 60 cm (20 to 24 in) deep. These ponds must be disinfected with quicklime 2 weeks before the fish are released. Start with about 1,000 fish per square metre or yard.

B   Methods of raising fingerling trout

General note

Raising fingerling trout and other salmonids of one summer and of one year can be carried out in many different ways either with or without artificial food, with or without manuring, in ponds with natural bottoms, or finally in ponds with built bottoms and banks.

Raising methods for fingerling trout can be grouped: (a) raising without artificial food: spring water ponds, linear ponds, ordinary ponds manured or not; (b) raising with artificial food: ponds with natural bottoms, artificial long shaped or circular ponds.

Ponds for fingerling trout should not be too deep; but sufficiently deep, however, to permit the fry to take shelter from the heat. Depths should be around 1 m at least at the monk.

Care must also be taken to ensure that fry

cannot escape. To this end screens with 2 mm mesh are used. These must be well adjusted and installed in order to prevent escape through any little aperture. In ponds these screens can easily become obstructed by leaves and algae. To avoid this a semicircular or a corner-shaped screen is placed in front of them; bars should be spaced 5 to 10 mm apart.

A screen with a 2 mm mesh should also be placed at the water inlet. But if possible a waterfall instead of a screen would be better since it would mean richer oxygenated water. But in any case it is necessary to install a device to keep large fish out, for if not they would play havoc with the fry.

Ponds should be stocked with fry with yolk sacs three-quarters absorbed or with fry of 1 month fed in troughs over a period of 3 or 4 weeks. In the latter case half the number of fry can be used as loss will then be much less.

The average stocking densities recommended per square metre are the following; obviously they are somewhat limited for the first two categories.

Type of food natural or artificial; Category of rearing pond	Fry yolk sac 3/4 absorbed	Fry fed in troughs 3 or 4 weeks
(1) No artificial food; ordinary ponds not manured	4–6	2–3
(2) No artificial food; ordinary manured ponds, spring ponds, linear ponds	6–12	3–6
(3) Artificial food more or less intense; ponds with natural bottoms	20–60	10–30
(4) Exclusive artificial food; intensive feeding; ponds with artificial walls and bottoms	50–100	25–50

The loss percentage depends on size; the smaller the fry the greater the loss. But it also depends, essentially, on health. The healthier the fry the better the results. In ponds natural enemies must also be taken into account. They are best dealt with by carefully drying out the ponds in winter and disinfecting them with quicklime.

Fingerling trout are harvested in the autumn. After being gathered they are taken to the shed and graded. Sizes should vary between 6 and 15 cm (2½ to 6 in) averaging from 8 to 12 cm (3 to 5 in).

The principal size categories are easily recognized. They are: large (10 to 12 cm), weight (9 to 15 gm), exceptional 12 to 15 cm; average (8 to 10 cm), (5 to 9 gm); small (6 to 8 cm), (1·5 to 5 gm).

After emptying, the rearing ponds are left dry throughout the winter for liming. They are placed under water again 15 days at least before the fish are liberated.

1 Natural raising without artificial food

This kind of farming, entirely natural, is current practice for salmonid cultivation for stocking and particularly for brown trout. It is not industrial cultivation. The advantage is fingerlings are only fed naturally and this avoids difficulties which artificial food presents. In all probability the method produces fish best adapted to stocking open water such as brooks, streams and natural lakes. Unfortunately the number of fish available is limited, so the method is being replaced more and more by the artificial food system even for such species as brown trout and salmon intended for restocking open water.

Natural raising means using only natural nutritive fauna which develops in ponds. Natural food for fry varies considerably according to the environment in which they are raised, and the choice they are offered obviously depends on the species present. In ponds, chironomid larvae and mayfly nymphs play a very important part. In spring ponds and linear ponds freshwater shrimps can be very abundant provided the water is sufficiently alkaline.

a. *Raising methods.* Rearing without artificial food is practised in ponds fed by springs, in linear ponds and in normal ponds, the latter being manured or not.

Spring Water Ponds. Ponds supplied by spring water are probably best for the production of fingerling trout not fed with artificial food. They are generally small ponds rarely larger than 10 ares (1 are = one-fortieth of an acre) and are fed with a spring in the pond or in the immediate vicinity. Sometimes they are open, sometimes shaded and sometimes situated in woods. In any case, the very fact that they are near springs means they have a regular temperature. In the summer the water remains cool. This is an advantage. What is even more important is that at the end of winter and the start of spring the temperature is relatively high, so that the fauna have developed abundantly by the time the fry are liberated.

Linear Ponds. Linear ponds which already exist or which can be arranged as such, are fairly numerous in mountain country. They are rarely built as such but they can be adapted from old mill-races, either saw-mills or channels once used for irrigation (Figs. 170 and 172). These are long narrow ponds often several hundred yards in length but rarely more than from 2 to 5 m (6 to 16 ft) wide. The longitudinal slope is generally slight and the bottom is gravelly. Banks are grass covered. However, in some cases the slope is steep and the bottom stony. These ponds should be divided up by small dams built of stones which provide hiding places and also ensure sufficient depth. Each time they are emptied the dams are pulled down for harvesting the fish and then rebuilt before being placed under water again.

From time to time, for the production of fingerling trout, natural brooks are used, 1 to 1·50 m wide (3 to 5 ft). These are specially arranged (Fig. 169). The fingerlings are harvested either in the autumn or the following spring (at 1 year) by various methods including electric fishing.

Ordinary Ponds. Ordinary ponds are used for the natural raising of fingerlings. They are small and of various shapes: circular, square, rectangular (Fig. 173). They are not supplied with water by springs but by a small water course such as a brook or a small stream. They cover only a small area generally not exceeding 50 ares ($1\frac{1}{4}$ acres). Larger ponds are difficult to fish. They are crossed by a regular but not excessively strong current from 2 to 5 litres per second per hectare.

The value of these ordinary ponds varies and depends on the richness of the nutrition they provide. If the water is alkaline then the nutrition should be rich and good and the ponds should be productive. On the other hand if the water is acid and poor in food then raising fingerlings will be difficult. They can be improved by the use of fertilizers.

Fertilized Ponds. The rearing of fingerling trout in ordinary, fertilized ponds (Fig. 171) is practised in the same way as rearing in ponds which are not manured but production is higher thanks to the more natural food thus stimulated. This shortens the critical period during which fry get used to feeding themselves.

The best fertilizer, superphosphate, is distributed at a ratio of 2 to 3 kilos (4 to 6 lb) per are (one-fortieth of an acre) either before or after the pond has been placed under water but anyway at least 15 days after liming. Eventually mineral fertilizers will continue to be used during growth but in lesser quantities. At the same time small quantities of liquid manure, hay or lucerne can be thrown onto the surface where they will decompose slowly thus stocking the pond with natural food.

The use of mineral and organic fertilizers should, as far as possible, be used in such a way as to avoid a strong growth of algae which can be harmful because it can obstruct the screens. The decomposition of masses of algae can also cause pollution and reduce the oxygen content considerably.

b. *Production* by natural means and without artificial food can be greater than five fingerlings per square metre (square yard), under the most favourable conditions. In Switzerland some ponds and rearing channels for fingerling trout produce 10 fish per square metre or more (Fig. 171).

In general, it can be taken that the production of fingerlings equals or exceeds half the biogenic capacity of the water. It has been shown (Ch. X) that the biogenic capacity is between I for poor water and X for richest water. A pond with a biogenic capacity of VI can produce at least three fingerlings per square metre and five with the same surface if the biogenic capacity is X. But production will fall to half a fingerling per square metre if the biogenic capacity is I.

The survival rate of fingerlings harvested in relation to the number of fry liberated varies considerably. For fry with absorbed yolk sacs it is considered mediocre if under 15 per cent, good if between 15 and 25 per cent, very good between 25 and 40 per cent, and excellent if over 40 per cent. It has been noticed that the return obtained from almost identical ponds apparently under identical conditions varies one from the other without any obvious reason. It has also been found that the survival rate in ponds stocked with fry fed for 3 or 4 weeks previously is generally much higher.

The *stocking rate* also varies and is linked to the production target. It depends on the biogenic capacity. In ordinary ponds which have not been fertilized, an average of from four to six yolk fry are liberated per square metre. In ponds with alkaline water, fertilized ponds, ponds with springs and linear ponds, six to 12 fry are liberated per square metre. If, however, fish which have been fed 3 or 4 weeks previously are used then the number is reduced by one-third or by one-half, i.e. to two or three or possibly six nourished fry. In Switzerland the numbers used are higher reaching 30 fry when conditions are most favourable. But only small fingerlings are then harvested. On the other hand, in water which is sufficiently rich in food under-stocking must be avoided for this will mean poor use of the available natural food present and the production of only a small number of large fingerlings.

To obtain good results with natural rearing of fingerling trout stocking must be adequate, the development of natural food must be helped, a strict watch must be kept during raising and harvesting must be carried out with care.

b   Raising by artificial feeding

This form of rearing depends on artificial food being used more or less exclusively for fingerling trout. It is generally used in farms which fatten trout for eating but is also finding favour more and more for restocking.

Formerly, when fish farmers depended on fresh food, the method succeeded admirably for rainbow trout. Now, following the generalization of feeding salmonids with powdered and pelleted food this method includes all salmonids, brown trout and salmon among them, despite the fact that it is difficult for these species to completely accept food in this form.

The food is adapted (fresh or dry) to the size of the fish. The method is given in detail in Chapter XIII.

Ponds must be crossed by a moderate current which must, nevertheless, be stronger for high density population. This becomes more and more important as rearing becomes intensive. It can only be practised when the flow is sufficient to keep the water cool in summer.

In September, ponds are emptied. If the fish are for restocking they are liberated. If, however, they are kept a second year they must be distributed according to size among other ponds in which they will pass the winter.

Fingerlings can be reared in natural ponds or in circular or long artificial ponds.

Rearing in ponds with natural bottoms. These ponds are generally small and never more than from 4 to 5 m (4 to 5 yds) wide so that the artificial food can be distributed evenly. They vary in length from 10 to 80 m (about 30 to 250 ft) (Fig. 175).

They also vary in depth between 0·75 and 1 m (2½ to 3 ft) and in any case should not be too shallow particularly for brown trout. In ponds which are too shallow much of the food will fall to the bottom before being snapped up by the fry and will be lost or only be useful as fertilizer. Also, when a pond is too shallow it heats up too quickly.

Stocking rate is between 20 to 60 fry with absorbed yolk sacs per square metre (or 10 to 30 fry of 1 month) according to the intensity of feeding. Losses vary between 20 and 40 per cent.

Rearing in artificial ponds. When such ponds are used, food is exclusively artificial. Natural food can be used as an extra, however, but in any case the amount will be negligible. This is intensive or very intensive farming.

Such fish cultivation is generally confined to large establishments and in any case can only be practised by experienced farmers and then only if they have a regular supply of the right food. Furthermore, the water must be fresh and have a sufficient flow, but only those with an important water source are really favoured (Fig. 176). This kind of farming is not recommended if the water heats up in summer. Feeding must be given the greatest care otherwise there will be a risk of rapidly spreading epizootics.

Farming of this kind is practised in long or circular ponds (or pools). They must be built of solid and durable materials. One advantage is the ease of handling, such as draining which can be done completely and easily and under hygienic contions thanks to easy cleaning and disinfecting (Fig. 181).

Dimensions vary considerably. Averages are given below. For *long shaped ponds* (raceway ponds) (Figs. 177 and 178): 30 m (100 ft) long, 1·20 m (4 ft) deep, 3 m (9 ft) wide at bottom (if the sides are inclined (Fig. 177) the width at the surface is greater and can be 9 m (30 ft)). The longitudinal slope is ½ per cent. *Round or circular ponds* (Figs. 185 and 186) are about 5 m in diameter and 0·75 m deep (15 ft and 2½ ft). Circular ponds can be built from 8 to 12 m in diameter (26 to 39 ft) with a surface area of from 50 to 100 m² (about 500 to 1,000 ft²). In recent years, prefabricated tanks of smaller size have often been used. They are circular, square or rectangular and are made of fibreglass, metal or other hard materials. They are often set up on the ground. It is even possible to find tanks made of a strong and waterproof fabric, suspended from a collapsible metallic frame. The advantage of prefabricated tanks is their relatively low cost and the ease with which they can be moved around.

The stocking ratio is from 50 to 100 yolk fry (or 25 to 50 fry of 1 month) per square metre. This number can even be increased to 1,000 fry per square metre if the water is fresh and if, after a few weeks, the fingerlings are sorted out and the redistribution is at a much lower stocking ratio. However, in circular pools with a tangential flow, stocking can still be high: as many as 1,000 summerlings per square metre provided the water is fresh and plentiful. The flow required is about 10 litres per minute (10 pt) for 1,000 summerlings.

At the end of the first year, the average weight of fry fed with artificial food is around 25 gm, but they can reach 100 gm (3½ oz) or even more.

Production can be speeded up and also regulated if there is a constant supply of water at stable and somewhat high temperature, around 13 to 15°C for example. This can be done in constructed ponds equipped with a closed water circulation circuit for reheating (Grunseld, 1967). Operating this way, it is possible to obtain, by the autumn of the first year, rainbow summerlings of from 20 to 22 cm (7 to 8 in) weighing over 80 gm (2 oz).

Feeding fry and trout fingerlings with natural food developed in plankton pools has also been tried. But in fact the production of large quantities of food in this way is both difficult and chancy, and has never given encouraging results. It can be satisfactory however in the Mediterranean region: in Slovenia, for example, where there is plenty of fresh water and a regular and sunny climate. But in any case the amount of food produced in this way is never very large.

II   Rearing Trout for the Table and other Second Year Salmonids

After the first winter, trout fingerlings are graded and redistributed according to size, in the fattening ponds. This is done to produce fish for eating, notably rainbow trout but also some brown and brook trout.

But it can also be done for restocking catchable trout. In this case all three species can be raised. But in each case the fish should measure 23 cm (9 in) or more and weigh around 125 gm (4 oz).

It is also possible to try and produce second year fish for restocking but smaller than catchable trout. This is often the case for salmon and sometimes for trout. The lengths vary from 15 to 20 cm (6 to 8 in) and weight is around 35 to 80 gm (1 to 3 oz).

From the spring, second year salmonids should not be too crowded nor kept in small ponds. Throughout the process of raising salmonids the amount of space given must be progressively greater giving the fish more room according to size.

Farms should aim at producing table trout weighing 150 gm (5 oz), or larger trout weighing from 200 to 500 gm (7 to 17 oz).

This kind of farming can be practised in three principal ways: extensive, semi-intensive and intensive.

The *extensive* method is risky and not so remunerative. It is based on the consumption of natural food for trout raised in ponds. In order to be worthwhile the ponds should extend to 1 hectare (2½ acres) provided the water remains cool in summer. Production can reach 100 kg (220 lb) per hectare in average ponds and 200 kg (440 lb) in good ponds.

Under the same conditions but with a more restricted surface (only a few ares, 10 at the most) it is possible to rear a greater number of fish, in the same space and during the same length of time, by distributing artificial food. This is called improved extensive or *semi-intensive*.

On *intensive* farms, the fish are kept in a small quantity of water which is frequently renewed. There is almost no natural food and rearing depends on distributed food. Condensed intensive farming sets out to produce the greatest number of fish in a minimum volume of water. This is possible if the flow is abundant and the water fresh and cool which is the case with spring water or streams with a regular and abundant flow generally found in calcareous regions.

In ponds with sufficient fresh water, between 25 and 50 individuals per square metre can be stocked which gives a production up to 15 kg (33 lb) per square metre or 1.500 kg (1 ton 9 cwt) per are. Such results cannot be achieved on any other type of farm. It is even possible to produce 30 kg (66 lb) per square metre if there is enough water and an abundance of good quality food.

Ponds intended for intensive rearing generally cover from 100 to 500 m² (120 to 600 yd²). They are mostly oblong in shape, say from 25 × 4 m (80 × 13 ft) to 30 to 50 m (100 to 160 ft) by 5 to 10 m (16 to 32 ft). This shape has two advantages; first, it assures a good and easy distribution of food and secondly it permits the current which traverses the pond to cover all of it. Artificial circular ponds are, naturally, smaller and are not used so often for the production of second year salmonids excepting when they are intended for restocking. The trout and salmon used must be rather small in size. Prefabricated tanks are similarly used, especially for rearing fingerlings. Trials have been carried out on table trout production in vertical units, a kind of silo, in which the stocking density is very high (Berger, 1977). This system may also be used for other species.

Fattening ponds need to be traversed by an abundant current. If they are natural ponds then their banks will have steep slopes but if they are artificial then they will be vertical. Fattening ponds should be rather deep: from 1·50 to 2 m (4 ft to 6 ft) as this permits the fish to find coolness in the depths and to catch their food as it sinks to the bottom. In principle such ponds should all have their own supply and draining systems which is a great aid to draining and can help to avoid extension of disease. The water supply is at one end and evacuation at the other (Fig. 179).

In sum, local conditions will decide the choice of shape and the disposition of the ponds. Above all, hygienic conditions must come first. Further, an industrial salmonid fish farm situated in a plain is entirely different from one installed in the mountains (Figs. 174, 175, 176 and 183).

Normally, thanks to artificial feeding, rainbow trout reach a suitable weight for serving during their second year and the best of them in 16 months. Generally it is more economical that a saleable size should be reached as soon as possible. In order to reduce the rearing time, food should be distributed without interruption at a temperature of from 12 to 16°C, which means the best use is then made of the food.

During the second year normal losses are between 5 and 10 per cent.

Fig. 117 Brown trout (*Salmo trutta fario* L.).

Fig. 118 Rainbow trout (*Salmo gairdneri* Richardson).

Fig. 119 Brook trout (*Salvelinus fontinalis* Mitch.).

Fig. 120

Fig. 121

Fig. 123

Fig. 122

Fig. 125

Fig. 124

Fig. 120 Large pond for unfed brood fish.   Parc de Vizille, Isère, France.

Fig. 121 Pond for artificially fed brood fish.   Fish farm at Etrun, Pas-de-Calais, France.

Fig. 122 Large dip-net for catching brood fish in running water. Motiers, Switzerland.

Fig. 123 Electric fishing apparatus. Schuld, Eifel, Germany.

Fig. 124 Trapping device for brood fish. On the left, the fish trap, on the right, the screen. Ruisseau de Martin Moulin, Achouffe, Belgian Luxembourg.

Fig. 125 Electric fishing of brood fish in running water. Reaction to direct current: the fish swims towards the positive electrode. Kriesbach, Zürich, Switzerland.

Fig. 126

Fig. 127

Fig. 1~8

Fig. 129

Fig. 130

Fig. 131

Fig. 126   Accessories necessary for artificial fertilization of salmonid eggs. Sieve or basin for gathering the eggs, basins for receiving the mixed eggs and milt in a little water, towel, feathers, graduated glass for measuring cubic content and counting eggs. Behind, storage tanks.

Fig. 127   Stripping the eggs from a small female brown trout by pressure on the abdomen.

Fig. 128   Pressing out the milt of a male brown trout by applying pressure to its sides.

Fig. 129   Stripping a large lake trout (weight: 11 lb; length 2 ft 7 in). Fish culture station at Boudry, Neuchâtel, Switzerland.

Fig. 130   Mixing, with a feather, eggs and milt poured into a pan containing a little water. In front: eggs at rest during 20 to 30 minutes after mixing.

Fig. 131   Washing the fertilized eggs in order to remove milt in excess and impurities.

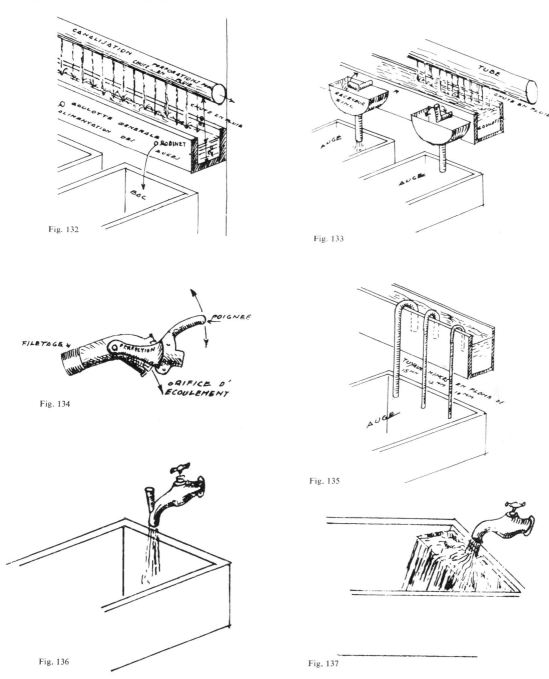

Water supply for hatching and rearing troughs, and water aeration equipment (after Charpy).

Fig. 132   Water supply by perforated pipes.
Fig. 133   Overflow water supply system by means of small elbow pipes.
Fig. 134   "Perfection" type taps.
Fig. 135   Water supply system by siphons.
Fig. 136   Tap with aeration cone.
Fig. 137   Tap with an inclined glass or wooden slat.

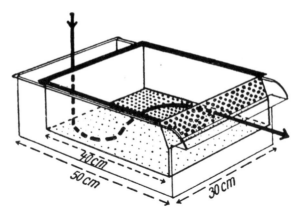

Fig. 138   Drawing of Californian incubator (after Schäperclaus, 1933).

Fig. 139   Small Californian incubators installed in parallel.
Fish hatchery of Vizille, Isère, France.

Fig. 140   Industrial or large scale fish hatchery with Californian incubators arranged in tiers. Old model (after
Buschkiel, 1931).

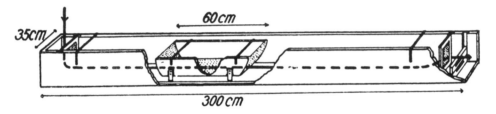

Fig. 141   Incubation and rearing trough for salmonids. Drawing shows only one of the four
incubation trays (after Schäperclaus, 1933).

Fig. 142 Rural fish hatchery supplied by water from a rheocrene spring, situated on the flank of wooded valley. Two troughs in sheet zinc; each contains two incubation trays. Achouffe, Belgian Luxembourg.

Fig. 143 Rural fish hatchery. A small dam situated near a rheocrene spring; overflow and water supply pipe of a sheet-zinc trough including two incubation trays; suitable for incubation of 10,000 eggs. Ry Pirot, Brabant, Belgium.

Fig. 144 Large-scale salmonid fish farm. Hatchery shed with long, wooden incubation troughs, in pairs, for hatching and rearing fry. Fish hatchery at Colverton, United Kingdom.

Fig. 145

Fig. 146

Fig. 147

Fig. 148

Fig. 145   Industrial salmonid hatchery. One series of paired troughs in concrete. Water supply gutter with "Perfection" type taps. Fish farm at Maulévrier, Seine Maritime, France.

Fig. 146   Industrial fish hatchery with two series (only one visible in the photo) of long, paired, metal troughs for incubation and rearing. Nimbus Hatchery, Sacramento, U.S.A.

Fig. 147   Long incubation troughs in Eternit. The incubation trays are in perforated aluminium. Fanure Fish Farm, Roscrea, Ireland.

Fig. 148   Screen and overflow system for incubation troughs. Fish hatchery at Leatown, West Virginia, U.S.A.

Fig. 149

Fig. 150

Fig. 151

Fig. 152

Fig. 153

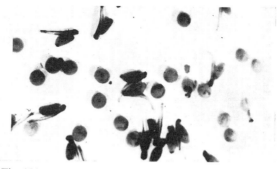

Fig. 154

Fig. 149 Incubation tray in perforated zinc sheet and glass rods incubation tray. Jonkershoek, South Africa.
Fig. 150 Square perforated incubation tray in zinc sheet. Water is forced to flow up through each tray. Stäfa, Switzerland.
Fig. 151 Copper trellis incubation trays coated with ashphalt. Boudry, Neuchâtel, Switzerland.
Fig. 152 Incubation basket on right and superimposed incubation trays on left, installed in a long metal trough. Nimbus Hatchery, Sacramento, California, U.S.A.
Fig. 153 Egg cleaning with a pipette by picking up dead eggs.
Fig. 154 Trout eggs at the point of hatching; devins with vitelline sac or unreabsorbed yolk sac. (Photo Hey.)

Fig. 155   Fig. 156

Fig. 157

Fig. 155   Vertical flow incubator with two series of eight superimposed trays. Humbold State College, Arcata, California, U.S.A.

Fig. 156   Hopper-shaped incubator for salmonid eggs; capacity 50 litres (11 gal). Fish farm Ospedaletta, Venezia, Italy.

Fig. 157   Vertical flow incubator showing details of two incubation trays.

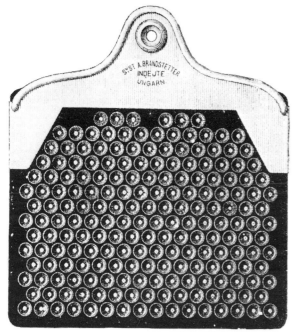

Fig. 159

Fig. 158

Fig. 160

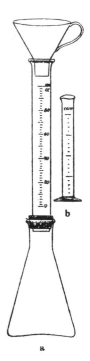

Fig. 161

Fig. 158   Brandstätter egg counting board (after Diessner-Arens).
Fig. 159   Regulable measure for cubage counting of the eggs. Height 80 mm, diameter 54 mm; 1,000 eggs can be counted at a time. Fish hatchery at Jonkershoek, South Africa.
Fig. 160   Equipment for counting eggs and fry. Riedwil, Canton of Berne, Switzerland.
Fig. 161   Schillinger apparatus for counting eggs and fry (after Diessner-Arens).

Fig. 162 Shipping container for air transport of eyed salmonid eggs.

Fig. 163 Small divided shipping case in insulated synthetic material for transporting eyed salmonid eggs.

Fig. 164 Circular tanks in Eternit, diameter 2 m (7 ft), for rearing trout fry over a period of 6 weeks. Capacity 20,000 fry. Fish farm of Poschiavo, Grisons, Switzerland.

Fig. 165 Elements of equipment for the central evacuation of water in circular tanks.

Fig. 166 Small concrete rearing tanks with tangential water circulation. Fanure Fish Farm, Roscrea, Ireland.

Fig. 167 Long rearing troughs in Eternit for feeding fry.

Fig. 168 Brook of Bellefontaine, a tributary of the Semois in the Belgian Ardennes. Prairie streamlet, very fresh, thanks to neighbouring springs. The bed is stony and gravelly. There is abundant nutritive fauna. Excellent for restocking fry with absorbed yolk sac.

Fig. 169 Brooklet used for growing trout fingerlings. Tipping fry with absorbed yolk sac in the spring and harvesting fingerlings in the autumn. Miniature falls built to increase depth of water, to decrease speed of the current and provide alevins nooks and crannies. Specially suitable for brown trout. Canal du Vivier, Vallorbe, Switzerland.

Fig. 170. Winding channel produces trout fingerlings with or without artificial feeding. Weber Canal at La Heutte, Canton of Berne, Switzerland.

Fig 171   Rectangular ponds for brown or rainbow trout fingerlings. Fish farm of Motiers, Swiss Jura. Ponds with high natural production (alkaline water) increased by the use of organic manure. Production rate as much as 10 trout fingerlings per square metre (1·20 yd²).

Rearing trout fingerlings without artificial feeding. This method is particularly suitable for brown trout. Under good conditions, production can reach five trout fingerlings per square metre; elsewhere it can fall below one trout fingerling per square metre (Figs. 172 and 173).

Fig. 172   Use of long narrow ponds, formerly millraces or races of saw-mills, or irrigation canals. Vallée du Ruisseau de Valire Chevral, Achouffe-Wibrin, Belgian Luxembourg.

Fig. 173   Harvesting trout fingerlings in normally shaped ponds. Domaine de Mirwart, Belgian Luxembourg.

Fig. 174 Intensive salmonid fish farm in the mountains. Rectangular ponds in tiers. Fish farm of Zeiningen, Bâle-Campagne, Switzerland.

Fig. 175 Salmonid fish farm in low hills region of the Baltic plain. Danish type of fish farm.

Fig. 176 Example of intensive salmonid culture: long ponds with beds and banks in concrete; very abundant water output. Fish culture of Etrun, Pas-de-Calais, France.

Fig. 177   Raceway fattening ponds with bottom and banks entirely in concrete. Alsea Hatchery, Corvallis, Oregon, U.S.A.

Fig. 178   Rearing salmonid of 1 summer in concrete raceways. Feeding exclusively artificial. Green River Hatchery, Seattle, Washington, U.S.A.

Fig. 179   Rectangular, concrete fattening raceways. Leaburg State Hatchery, Oregon, U.S.A.

Fig. 180   Long type salmonid rearing basins with banks and beds in concrete. Grayling State Fish Hatchery, Grayling, Michigan, U.S.A.

Fig. 181 Concrete raceways for fattening salmonid. Note lower end which is deeper in order to facilitate hygienic treatment. Provincial fish farm of Laurentides, St-Faustin, Quebec, Canada.

Fig. 182 Circular concrete ponds. Radial water supply, oblique to the surface, which assures favourable circulation of water and the regular dispersion of the fish. Fish Culture Station, Leatown, West Virginia, U.S.A.

Fig. 183 Intensive industrial fish farm. Rectangular concrete ponds fed by abundant calcareous springs. Total surface under water: 2·50 ha (6 acres). Production per square metre (1·20 yd²): 10 to 15 kg (22 to 33 lb). Fish farm of Pederobba, Valley of Piave, Italy.

Fig. 184 Industrial salmonid culture can be very intensive and production can reach and even pass 10 to 15 kg (22 to 33 lb) per square metre (1·20 yd²), if the water supply and the food are very abundant.

Fig. 185 Rearing salmonid in concrete circular ponds. Feeding exclusively artificial. Fish culture station of Jonkershoek, South Africa.

Fig. 186 Regular distribution of fattened trout in circular ponds. Fanure Fish Farm, Roscrea, Ireland.

Chapter IV

BREEDING AND CULTIVATION OF CYPRINIDS

THE cultivation of cyprinids has been practised for centuries on the continents of Europe and Asia. Its extension to other continents is much less. It is probably the oldest form of more or less intensive fish cultivation in fresh water yet at the same time it covers the very widest areas.

The principal cyprinid, carp, is popular in both Europe and Asia; there are also associated species. In Europe, until now the first of these is tench. In Asia, especially in the Far East, carp are associated with other species and often supplanted by other cyprinids under the group name of "Chinese carps" and "Indian carps". It is also associated with different Anabantids and species of other families.

The cultivation of cyprinids is principally concerned with the table, but in certain regions they are also raised for restocking.

Until fairly recent times the evolution of the cultivation of cyprinids was slow. Over centuries it was extensive. In the nineteenth century progress was made, starting in Central Europe thanks to controlled reproduction in Dubisch-type spawning ponds. Little by little the practice of feeding and fertilization became established and intensified the production. Thanks to the use of dry concentrated food even more progress and better results can be expected. Other progress which will contribute to the spread of this type of farming will include the relatively recent technique of artificial fertilization. This may be expected to have a profound influence on the cultivation of carp and will permit the spread of cyprinids of Asiatic origin which are of great interest.

SECTION I

GENERAL CONDITIONS FOR INSTALLING A CYPRINID FISH FARM

1 General Arrangement of a Cyprinid Fish Farm

Formerly, in Europe, cyprinids in general and carp in particular were cultivated by very simple methods (wild carp cultivation, Femelbetrieb), in which fish of all ages including parent fish were kept together. They were fished out once every 2 to 4 years. When the time came for emptying, the best fish were chosen either for sale and eating and the rest were put back in the ponds. These included young fish but also fish of poor growth. Finally the latter reached spawning age and the resultant spawning of that poor stock led to a decline of the farming.

This kind of wild cultivation gave mediocre results. Nearly all the ponds included, in different proportions to the carp, other species such as roach, tench, pike, and perch, as well as some undesirable species such as catfish which may form an important part of the harvest. Often enough the ponds were unhealthy. This kind of primitive cultivation is no longer practised by modern farms of course, but there are still far too many who still practise it as it was done of old or with only slight and insufficient improvements.

The rational rearing of carp is practised under different age classifications and in such a way that it is possible to control growth at all ages. According to the region, the production of saleable carp takes from 2 to 3 years. Normally, in western Europe, it takes 3 years but only 2 years in central and southern Europe because the climatic conditions are more favourable.

Three-year production passes through the following stages of harvesting:

First year: the production of young or fingerling carp which is practised in different ways, calls for the use of several categories of ponds known as spawning ponds, first rearing ponds and ponds of second rearing.

Second year: carp of two summers.

Third year: production of carp for sale or eating.

Apart from the ponds mentioned above, wintering ponds are necessary for second year fish as well as ponds for holding and stocking fish before sale for the table. Considering the large number of fish stocked, these two last categories should be located near the house so that they can be kept under observation. The ponds stocked with young carp should be supplied with very pure water to avoid risk of infestation from parasites carried by older fish.

As the cultivation of carp is always more or less extensive such a variety of ponds means that for a farm to be complete it should cover a sufficiently large area: 30 to 50 hectares (75 to 125 acres). The way the area is divided proportionally between fish of different ages will vary according to the farmer's plan: namely whether fry are to be produced for final utilisation by the farm itself or whether it will plan a surplus for sale to other farms which practise only partial cultivation.

It is rare for a farm to produce the exact number of fry and second year fish to meet its planned production requirements of carp for the table. Most large farms have a surplus of small carp which they sell to small farms which raise only partially and only for fattening.

Here, according to Schäperclaus (1933), is a division of areas for a complete farm in Europe.

	Types of ponds		
Years	Ages of fish (1)		Percentages
I	Spawning ponds	0·25 %	
	Ponds for first rearing (K_{o-v})	2·75 %	} 13 %
	Ponds for second rearing (K_{v-1})	10 %	
II	Two summers carp ponds (K_{1-2})	23 %	
	Wintering ponds	3 %	} 26 %
III	Fattening ponds (K_{2-3})	60 %	
	Holding ponds	1 %	} 61 %

Note

(1) In Europe the growing season for carp corresponds with the summer and is referred to as carp of one, two or three summers. These are designated in Germany as K_1, K_2, K_3. After wintering during which time there is no growth, the fish have, respectively, 1, 2 or 3 years. They have not changed practically either in size or weight and are designated by the same symbols. Carp of 6 weeks are shown as K_v and fry with absorbed yolk sacs as K_o. Similar symbols can be used in which the capital letter differs according to the species.

The cultivation of carp, however, might well become intensive and even very intensive in the same way as in salmonid cultivation. Today such techniques are used in Indonesia and Japan, but are not very common. Intensive farming has been carried out in Germany though only experimentally (v. Sengbusch *et al*, 1967).

II   General Characteristics of Water

The characteristics of water which have been mentioned and discussed for salmonid cultivation also apply to cyprinid cultivation. There is therefore no point in returning to them. Nevertheless, fundamental differences do exist between the two types of cultivation concerning the quantity of water required and also its temperature.

Cyprinids raised on farms are warm water fish which means that they not only support but need minimum temperatures of 18°C and even up to 30°C. For many raised cyprinids the optimum of growth is in water between 20 and 28°C.

To ensure and maintain such temperatures in raising ponds it is indispensable, at least in temperate climates, that there should be only a small flow of water into the ponds. This flow compensates for evaporation and seepage, and also maintains the temperature of the water within the limits given above. Minimum quantities of water have already been given in Chapter I. In Europe it is estimated that in general the need is for 1 litre of water per second per hectare (2½ acres) for cyprinid ponds during the growing season – the summer.

In order that the ponds should warm up enough they should not be too deep. A depth of about 1 m is recommended. Given that farming is generally extensive, ponds for this type of farm will be large and flat, shallow and sunny as opposed to ponds used for the intensive raising of salmonids which are small and deep and crossed by a strong current amounting to hundreds of litres per second per hectare.

SECTION II
PRINCIPAL REARED CYPRINIDS IN EUROPE

Fig. 187   Improved common carp (*Cyprinus carpio* L.), mirror-carp type.

I Carp—the most important Cyprinid

1 General Characteristics of Carp

Carp (*Cyprinus carpio* L.) (Fig. 187) is the principal farmed cyprinid in Europe. Its dorsal fin is long and comprises three or four single rays of which the last is thick and spinous and has 17 to 22 small soft rays. The mouth is terminal, the lips thick and they can also be projected forward. Carp cannot be confused with any other cyprinids. It can be differentiated from the crucian carp – which it resembles – by its four barbels on the upper lip. The two anterior barbels are short and thin and the two posterior are long and thick. Its back is greenish-brown and its belly yellowish-white. In the Far East, orange, yellow and white carp are bred. Carp can reach 80 cm (2 ft 8 in) in size and can weigh from 10 to 15 kilos (20 to 30 lb).

It is a warm water fish. It can only be raised in water which warms sufficiently during the growing season. In Europe, carp spawn in water at least 18 to 20°C at the end of the spring. Its growth optimum corresponds to summer temperatures of between 20 and 28°C. Development diminishes as the temperature falls. Below 13°C its growth will be greatly reduced and it will stop eating when it falls below 5°C (Fig. 382). In ponds with an average summer temperature between 15 and 18°C they can live and even grow sufficiently but they will not reproduce. When it is very cold carp take refuge on the bottom of the deepest part and become lethargic. Their growth rate can be followed more or less clearly from their scales which also assist in determining age. This is done by a scalimetric method.

Carp originated from eastern Europe (basins of the Black Sea, the Azov Sea, and the Caspian Sea). They are also found in Asia from the basin of the Aral Sea as far as China and the Amour Basin. They are ideal for raising in warm temperature and shallow ponds. In temperate climates such ponds are only found in the plains. Carp like water rich in grass, calm and warm in summer. They live in ponds and shallow lakes, canals, and in the water courses of the bream zone.

The introduction of carp raising to central and western Europe goes back to the Middle Ages. Their farming is particularly widespread in Germany, Poland, Czechoslovakia, Hungary and Yugoslavia. In France the important cyprinid pond regions are the Dombes, Bresse, Sologne, Limousin, Brenne, Forez and Lorraine.

The raising of carp is widespread in the Far East and in Israel. Apart from Asia and Europe, carp were introduced before the end of the nineteenth century into the United States, South Africa and Madagascar where they were appreciated in different ways. In the United States and Canada carp are generally considered a nuisance for they have invaded the warm, turbid and open waters where they compete with the indigenous species. Also in the absence of appropriate voracious fish in sufficient numbers, their expansion gets out of control.

Their eating habits are omnivorous. Carp eat planktonic organisms as well as animalcules living near the banks and on the bottom. They can be fattened with seeds of leguminous and cereals or with dry concentrated food.

Carp flesh is esteemed in many countries but it easily takes on a taste of mud due to the *Oscillatoria* algae which get fixed in its muscular system. This taste can be removed by soaking the fish in clean, constantly changing water.

Carp are greatly appreciated by anglers. They are wary, crafty, very resistant and difficult to catch. They can live a long time out of water.

Carp are raised for eating and restocking. In Europe, those for restocking are generally of one or two summers and fish for eating two or three summers old. Average weights in autumn for carp are: 1st year, 35 to 50 gm (1 to 2 oz) and they are from 9 to 12 cm (3½ to 5 in) long; in their 2nd year they average 350 gm (250 to 500 gm), 12 oz (8 to 16 oz); 3rd year, 1,250 gm (1,000 to 1,500 gm), 2½ lb (2 to 3 lb).

Fish for the table must meet certain requirements:

1. They must meet the weight requirements of the local market: 1,000 to 1,500 gm in Central Europe (2 to 3 lb), 500 to 750 gm in Israel (1 to 1½ lb), and 75 to 100 gm (2½ to 3½ oz) in Indonesia.

2. The flesh must be firm and not too fat, the head should be small and there should be only a few bones. The gonads should be only slightly developed (that is less than 10 per cent of the body). Fleshy and thick set fish with wide, high backs are in demand. This is measured by an index in an H/L form giving the ratio between the height of the body and its length but not taking into account the caudal fin (Fig. 188). For those carp intended for raising, the index varies between 1/2 and 1/3. For wild carp taken from running water (which are much longer), the ratio varies between 1/3 and 1/4.

3. Whether the fish is preferred with many or few scales depends upon the region.

The following must be taken into account when raising carp:

(a) Rapid growth which reduces to a minimum necessary rations for maintenance and produces

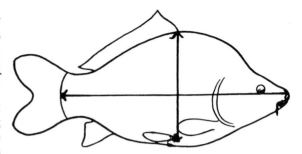

Fig. 188   Drawing determining the body ratio H/L of a carp (after Maier-Hofmann).

saleable fish before reaching sexual maturity. Rapid growth results, notably from the good use made of natural food and from artificial food. This tends to reduce the inedible parts such as the head and the skeleton while producing short fine bones.

(b) Resistance to disease and other causes which weaken the fish (pollution, transport and wintering).

(c) If good parents are produced they give resistant fry of rapid growth which perpetuate the characteristic qualities of the brood fish.

2 Improving carp by selection

Selected wild carp from the Black Sea and the Caspian Sea area have provided the stock for raising these fish in Europe. They are long in shape, have rounded bodies and are completely covered with scales. The head is relatively large. Mass selection carried out in Europe over many years has permitted the breeding of an improved carp, more thick set and of which there are now many different races and lines.

By "race" is meant fish of the same species having certain similar ties and generally originating from the same region. These racial characteristics are hereditary as in the Bohemian race for example.

By "line" is meant fish of the same species having between them certain filial ties. They are the issue of one or more couples. A line differs genetically from other lines of the same race.

In Europe the best known races are those of Aischgrund, Galicia, Lausitz, Franconia, Bohemia or Wittingau and Hungary. They are different, one from the other, by their form indices, their scales, the relative thickness of the head, the presence of bumps on the neck, more or less developed (this is a mass of thick flesh behind the head). As for form and shape the Aischgrund and Galicia races are thick set while the Lausitz, Bohemia and Franconia races are longer in shape.

As for the exterior appearance there are four principal phenotypes with different scales cover (Fig. 189): carp with scales (not to be confused with wild carp), mirror carp, carp with one row of scales, naked carp or leather carp, according to where the carp is completely covered with scales, or covered with a more or less reduced number of scales or without scales at all. The scale carp (Schuppenkarpfen) is completely covered with scales. The mirror carp (Spiegelkarpfen) has scales of different sizes on its sides and generally one row of scales high on its back. Carp with a single row of scales (Zeilkarpfen) have the row on their sides. It may be complete or incomplete but the scales are of identical size. They also have a few scales at the base of the fins and on top of the back. Leather carp (Nacktkarpfen) have very few scales – found generally at the base of the fins.

The aim of mass selection has been essentially to produce a type of carp suitable for eating. It is of rapid growth, has few scales (except in certain regions) and is thick set. In general, different regions breed their own line of carps but these different lines possess, probably, a rather great genetic uniformity.

In selected carp there is a correlation between the scales on the one hand and the physiological characteristics on the other: speed of growth, possibilities of survival, resistance to disease. Carp with scales and mirror carp have a more rapid growth and greater possibilities of survival than carp with one row of scales or naked carp. The last two are also less resistant to disease and often present deformities of the fins. The growth of carp with scales is slightly higher than that of mirror carp but the latter is preferred for eating.

In the light of their genetic constitution it is not possible to raise a pure type of carp with one row of scales, or naked carp (heterozygotes) while it is possible to obtain pure lines of carp with scales and also mirror carp (homozygotes). Among the four different types, only carp with scales and mirror carp seem to be suitable for raising. Schäperclaus (1961) recommends, for eating, the rearing of mirror carp with a complete row of scales below the dorsal fin as well as a few irregular scales at the root of the tail.

In several countries carp are also raised for restocking of waters for angling so it would also be of interest to select carp for their angling value. Experiments in ponds (Beukema, 1967) show that under the same conditions wild carp are much more suspicious and 10 times more difficult to catch than improved carp (carp with scales and

Fig. 189   From top to bottom: carp with scales, mirror carp, carp with single row of scales, leather carp. (Document Jaarverslag 1960-1961, O.V.B., Low Countries).

mirror carp). Crossing between improved carp and wild carp has produced fish with intermediate characteristics.

It is important that efforts should continue to improve the hereditary value of bred fish in such a way that the qualities looked for are also found in their descendants. Among them rapid growth and resistance to disease are the most important. Other factors must also be taken into consideration: scales, shape, the number of bones, adaptability to new conditions such as salt water and their sports value.

Methods used in carp selection

The selection of carp can be carried out in two ways: progressive mass selection which was practised in the past with very useful results and individual selection with control of the descendants.

A combined method, that is, mass selection followed by individual selection is recommended by Kirpitschnikov (1966) and Schäperclaus (1961) who describes the process as follows. Brood fish are chosen, 10 females and 20 males, according to their exterior appearance – shape, scales, absence of malformations, sufficient but not excessive gonads. They are placed together in a sufficiently large spawning pond. Theoretically this gives 200 crossing possibilities. The fry from these brood fish are reared in one large pond. The spring of the following year the K_1 are sorted out according to their individual growth and exterior appearance. Only a maximum of 5 per cent is kept for raising through the second year. They are then placed in wintering ponds on their own and not with other fish. During the third year rearing continues separately or mixed with other fish eventually after marking. On farms which are subject to serious infectious abdominal dropsy, the fish should be injected in the spring with living organisms which will provoke the disease in order to obtain resistant individuals. In the autumn there is another selection of K_3 and then again in the spring and so on until they can be used as brood fish aged from 5 to 7 years.

From this time individual selection can be practised. The best brood fish are placed in small spawning ponds, one male and one female in each one. The descendants of each couple are reared separately in identical ponds and the best K_2 from the best descent are chosen as future brood fish.

Individual selection calls for several ponds or the marking of the fish if several descents are to be raised in the same pond. But practically, possible combinations for individual selections are few.

On any one farm it is possible to obtain, after a certain time, a pure line adapted to the conditions of that environment. Crossing with lines from other stations is not practised unless inbred degeneration shows itself or unless it is considered desirable to produce a new characteristic in the cultivated line. Future brood fish can be fed artificial food throughout the first and second and even the third year. This also tests the ability to assimilate artificial food. This is very important.

II Other Cultivated European Cyprinids

Carp are rarely reared alone. They are generally accompanied by other cyprinids and other fish. The principal cyprinids raised in Europe at the same time as carp are given below. For Asia the list and the most usual associations are given in Chapter IX.

Tench (*Tinca tinca* L.) (Fig. 215). In cyprinid cultivation tench are nearly always associated with carp. However, the favour this fish enjoyed formerly is now diminishing notably for economic reasons. It now seems more advantageous to substitute for tench different age groups of carps or Asiatic cyprinids.

Tench flesh is finer than that of carp but growth is much slower. They support colder water than carp and can get used to water with only a slight oxygen content which factor is valuable when the fish have to winter in badly arranged ponds as far as the renewal of water is concerned. They like flat ponds which heat up well. There should be a deep ditch in which they can shelter when the weather is very cold. The ponds should be rich in natural food. Tench are well suited for stocking in ponds with plenty of grass as well as in ponds which cannot be drained. Kept in water which suits them tench reproduce too abundantly and the surplus fry simply aggravates the inconvenience of their slow growth.

The raising of tench for stocking is of interest for repopulating lakes and running water in which the fish grow, but do not reproduce well.

Roach (*Leuciscus rutilus* L.) (Fig. 216). Roach exist and multiply abundantly in rather warm waters in the summer which are not drained regularly. Roach have the same disadvantages as tench: their growth is slow and their reproduction is too abundant. Commercial production of this fish is of no interest for the table. On the other hand it is advantageous to produce it at times for restocking canals or those waters which do not restock well by natural means. The fish is bred to this end both in France and Belgium.

Rudd (*Scardinius erythrophthalmus* L.), closely resembles the roach and is found in its company though it is less abundant. It is the most herbivorous fish in temperate western Europe.

Crucian Carp (*Carassius carassius* L.) (Fig. 191). The cultivation of this fish has only slight interest. Its main characteristic is that it can survive in water very poor in oxygen. It is of slow growth and its flesh is scarcely appreciated.

Goldfish (*Carassius auratus* L.). Cultivation is not to be recommended unless sales conditions are satisfactory. This fish is only cultivated for ornamental purposes, aquaria and ponds in parks or public gardens. Goldfish need warm water and develop particularly well in temperatures exceeding 22°C.

Golden Ide (*Idus idus* L.) (Fig. 217). Golden ide is an orange coloured variety of the common ide. It has no interest for the table but is good for the decoration of aquaria, ornamental ponds in parks and public places. Indeed this is its particular interest since it is the only decorative species which lives on the surface. Other species of "gold" fish (goldfish, gold tench, gold rudd) remain more or less in the depths. Golden ide have another advantage since they grow quickly and reach 50 cm (about 20 in). To this end golden ide are cultivated but only in small numbers considering the limited demand.

III   Asiatic Cyprinids

For centuries, if not for millennia, the people of the Far East, the Chinese especially, have raised fish in ponds. Apart from carp (cultivated in most places) there are two other principal groups of cyprinids which are cultivated at an important level. They are, first, "Indian carps" which group several genera such as *Catla*, *Cirrhina* and *Labeo*, largely cultivated in India, Pakistan and Burma, and also "Chinese carps" under which name are assembled different species with very different feeding habits, and which are raised and mixed together in China, Thailand, Malaysia, Vietnam and Japan. Among the best known species are *Ctenopharyngodon idella*, which is herbivorous: *Hypophthalmichthys molitrix*, which is plankton eating (phytophagous); *Aristichthys nobilis*, which is planktophagous (phyto- and zooplankton); *Mylopharyngodon piceus*, malacophagous (molluscs); *Cirrhina molitorella*, which live off fauna and detritus at the bottom of ponds.

Of these species the first four were introduced into the Soviet Union around 1950, from China, and also into Rumania in 1959. After 1960 the first three were introduced into Hungary as well as into other eastern European countries. Since then successful artificial fertilization of these species in Russia, Rumania and Hungary has contributed to its wide dispersal not only in central and eastern Europe but also in western Europe. They have also been introduced into the United States.

The interest these have for Europe is that they grow quickly, their flesh is appreciated and above all they live off vegetation or plankton not eaten or insufficiently eaten by carp and other European cyprinids.

The three principal Asiatic cyprinids used in Europe are the following: grass carp, silver carp and big head. They originated in China where they live and reproduce in large rivers with strong currents. They spawn during the floods and summer rain periods and in troubled and fast flowing warm water (25 to 30°C). For centuries these species, as well as others have been cultivated, mixed together by fish farmers and Chinese peasantry. As fry they were exported regularly to Formosa, Malaysia and Japan as well as to other eastern Asiatic countries, to be raised for eating. The growth of these fish is closely tied to temperature.

Grass carp (*Ctenopharyngodon idella* Val.) (Fig. 226) is a cyprinid with a long body covered with large scales. Its fins are short and powerful and it closely resembles the chub (*Squalius cephalus*). This fish originated in the Amur river and the great rivers of China from the north down to Canton in the south. In open water it can reach 50 kg (110 lb). It is a typical and exclusive grass eater living off the higher water vegetation as well as filamentous algae.

Silver carp (*Hypophthalmichthys molitrix* Val.) (Fig. 227). The body of this fish is covered with small, silver scales. The head is round and broad and its eyes are well forward though turned down. The branchiospines are long and provide a fine filter capable of retaining planktonic organisms.

This species can reach 20 kg (44 lb). It lives off green and blue planktonic algae.

Big head (*Hypophthalmichthys* (*Aristichthys*) *nobilis* Richard.) (Fig. 228) resembles the previously named species. The head is larger and the belly rounder. It is darker in colour, somewhat yellowish, and has darker mottled patches. The branchiospines filter is larger than those of the preceding species and is adapted to filtering zooplankton on which it lives principally, as well as on phytoplankton. If the food it requires is abundant the species grows quickly.

SECTION III
REPRODUCTION OF CULTIVATED CYPRINIDS

I Natural Reproduction

A Semi-controlled or uncontrolled natural reproduction

1 Semi-controlled reproduction of European cyprinids in ponds

a Cultivation in one phase. This method is very old and simple and is currently followed for European cyprinids including carp. Only one kind of pond is used in which the brood fish are placed and where the fry remain from birth to the following autumn or spring – that is up to 6 months of age or a year. The ponds must be relatively large, from $\frac{1}{2}$ to 2 hectares (1 to 5 acres). They should be shallow, from 1 to 1·25 m (3 ft to 4 ft) at the monk, and sheltered from cold winds. All this will facilitate the re-warming of the water necessary for spawning and growing the fry. Naturally the level of the water must be constant. The parent fish are immersed in the spring before the normal spawning period. The Bokrijk, Campine pond 1·50 hectares (4 acres) (Fig. 212) is the kind of pond that can be used.

This system has several advantages. It avoids all handling of the fish other than for releasing the spawners, and thanks to the considerable amount of water in the ponds the effects of cold spells are attenuated and this regulates spawning.

It has the inconvenience, however, of uncertain results which vary in a proportion as wide as 1 to 20. These variations in reproduction – only more or less satisfactory – are due to many causes, notably atmospheric conditions favourable or unfavourable, and this means insufficient precise information is available as to the causes of success or failure of the reproduction. An unspecified number of fry remain in the pond throughout their growth. They are mixed with the brood fish which cannot usually be recaptured but which can destroy both eggs and fry. To sum up, it is a primitive method.

During the course of the summer it is possible to try and establish probable results by observing the fry in those sunny parts of the ponds which are more or less free from vegetation.

As for breeding carp, this method can be used on large farms with plenty of space. It is also used if there is a shortage of experienced personnel to carry out rearing according to the techniques of controlled reproduction. When this method is used then there should be five "sets" per hectare. Each "set" should hold two males and one female (for more detail, see the section on controlled natural reproduction).

This method is generally employed for the reproduction of tench which is difficult. The fry are very small and catching them is very difficult. Therefore it is not possible to use, successively, the system of spawning ponds and rearing ponds as used for carp. Beyond this, in water which suits them, that is, warm in summer, tench multiply abundantly.

For reproduction, tench need warmer water than carp, from 20 to 24°C. The spawners are placed in special ponds or in second year carp ponds, or in fattening ponds used as such. Recognition of the sexes is easy with tench.

If only one pond is used for the reproduction of tench and raising yearlings, then from 10 to 20 females per hectare and at least the same number of males can be accommodated. If mixed with carp in ponds of second or third year growth then from two to six female tench per hectare are used with at least an equal number of males. When mixed conditions are employed it is better to avoid too many tench fry which would, because they are so small, lead to harvesting difficulties.

The same method is used for roach as for tench. About 15 couples of fish weighing from 200 to 500 gm (7 to 17 oz) are placed per hectare. The same method is used for rudd and other European cyprinids.

b Raising in two phases. Semi-controlled reproduction by the two-phase method is also, but not very often, used for European cyprinids. When this is so a spawning pond and a fingerling pond are necessary.

The brood fish are placed in the spawning pond (Fig. 193) with average dimensions of from 5 to 25 ares. After hatching the fry are left in the pond for several days or weeks according to the species of fish and the size of the pond. The latter is dried out more or less early in the summer and the fry transferred to a fingerling pond or ponds where they will remain until the autumn or the following spring.

This method is sometimes used for carp, more rarely for tench, but can also be used for roach and other cyprinids. Relative details are given in the section IV: Rearing cyprinids.

2 Uncontrolled breeding of Asiatic cyprinids in open water

This, in fact, is not really a cultivation technique as the farmer does not intervene. He waits until after the spawning to harvest the fry, grade them and transport them over more or less long distances in order to liberate them in ponds where they will be reared. Here, according to Chimits (1958) are details on catching and grading the fry.

For centuries there existed in China an important fishing industry for fry of Chinese carps, and in India and Pakistan for Indian carps. In China, Chinese carps reproduce mainly in the Yang-tse-Kiang, swimming upstream to spawn.

The catching of very large numbers of fry is carried out in suitable places by experienced fishermen. They use very fine nets shaped like a pocket. The very small fry, transparent and scarcely visible, are placed in linen receptacles floating on the water. From this moment multiple grading operations are carried out, as well as counting and transporting by carriers using their backs or some sort of vehicle. This lasts sometimes as long as 2 weeks and is done practically without loss, which is very surprising and shows that fish farmers in Asia are really masters of their craft. Similar methods are used in several Far Eastern countries for catching by the sea and counting and transporting *Chanos* (milkfish) fry (Ch. VIII, section III).

One of the most delicate operations is sorting fry of several species; normally about five. First the fry of species other than carp are removed. This is easy for the predatory fry, which are larger, occupy the top part of the receptacles. The separation of five different species of carps is more delicate. The fry are poured into large jars and allowed to rest for a few minutes. It will be found that they will group and stratify by species in the receptacle either through physico-chemical and biological tropism, or because of simple gregarious instinct. They will range thus: silver carp at the top, big head carp second, and grass carp third, followed by mollusc-eating carp and then mud-eating carp. With the use of very fine nets and skilful manipulation it is possible to capture the fry layer by layer. This operation is repeated two or three times and in this way excellent sorting is obtained.

The fry are carefully counted. They are then transported over long distances, sometimes very long distances, to various farms in China and even beyond to countries of east and south-east Asia.

B Controlled natural reproduction

Controlled natural reproduction of cyprinids is carried out in small spawning ponds. The spawners are placed at the right time and if atmospheric conditions are favourable, they will then spawn rapidly. The fry hatch out after a short time and are gathered, counted and transferred to rearing ponds. This method is particularly good for carp but is different according to whether the Dubisch spawning system is used or artificial spawning nests or kakabans.

1 System of Dubisch spawning ponds
(Laichteiche, Dubischteiche, Streichteiche)

Origin
These ponds were created in the nineteenth century by an Upper Silesian fish farmer named Dubisch (1813-1888); they were later modified somewhat by Hofer. Both types, Dubisch and Hofer, are used exclusively for spawning; the Dubisch type in particular.

Shape and lay-out (Figs. 195 and 197)
Spawning ponds need to be small, sunny and sheltered from wind. For this reason they should be surrounded by a fence (2 m high) of reeds (*Phragmites communis*) or something similar (Fig. 197). The spawning ponds are square in shape (8 × 8 to

10 × 10 m) (24 × 24 ft or 30 × 30 ft) but should not exceed one-fortieth of an acre. They can also be rectangular (6 × 15 m) (18 by 45 ft). They must be shallow, about 20 to 30 cm (8 to 12 in) of water. Dubisch ponds are surrounded by a peripheral ditch about 40 to 50 cm (16 to 20 in) deep which flanks the banks. In the centre there is a grass covered spawning board with a 1/15 slope. The ditch is used to hold the parents. It also helps to gather them as well as the fry and allows the drying up of the centre section on which the eggs are laid. Hofer type ponds do not have a peripheral ditch and the slope of the grass spawning board in the centre is regular and corresponds to the progressive deepening upstream to downstream. A harvesting ditch is dug along one side, ending at the monk.

The peripheral ditch (Fig. 198) must be without vegetation completely and must be carefully cleaned before being put under water. The centre spawning board should be covered with meadow grass of good quality neither too short nor too long, to which the eggs will stick after spawning. Rye grass (*Lolium perenne*) is recommended. If the grass is too high then it should be cut and removed a short time before the pond is put under water. Removal is necessary to avoid rotting.

Water supply
Water for spawning ponds must be very pure and should be able to warm up and stay sufficiently warm. This can be assured by various methods.

The water should be renewed as little as possible, just sufficient to recompense loss through seepage and evaporation. The bottom of the spawning pond must be watertight and the outflow arranged to avoid leakage. This is done by packing clay between the two series of stowboards of the monk.

To help maintain the rewarmed water of the spawning ponds special warming ponds can be installed (in German a Vorwärmer). These are flat and their size can vary from 10 to 25 ares, for example. They hold no fish in order to eliminate parasites and they are put under water only a short time before spawning is due.

How the ponds work (Fig. 199)
There should be from two to 10 spawning ponds according to the importance of the farm. One pond is never sufficient because of possible failure. A number of ponds is not difficult to set out because they are small and not costly. They should always be grouped together.

Ponds are put under water at a time when it is possible to ascertain that the water will be sufficiently warm, will remain above 18°C, that there will be no cold spells or that temperatures will not fall steeply at night. The brood fish and the eggs are very sensitive to low temperatures and also to sharp changes in temperature. In central and western Europe spawning ponds are not generally put under water before May 15 but rarely after the start of June.

A fall in temperature at night and the return of cold which lowers the temperature of the water below 16°C will stop the brood fish reaching maturity and could even lead to the females failing to ripen. Length of hatching will also be prolonged or even stopped, and if this goes on for too long then the eggs will be killed. Indeed this is the reason why spawning at the end of April or at the beginning of May is often fruitless.

One way to avoid night cold is to cover the spawning ponds at night with panels fitted with glass panes or some plastic material. The water can also be warmed directly (coal, oil, propane or butane gas). But if such apparatus may be effective they also complicate the running of the farm.

In open water (natural ponds or calm water courses) spawning can be early or late according to whether the warm season is early or late. In western Europe carp have been known to spawn in April and in other years spawning is recorded as late as May-June or even later. Often, because of seasonal temperature variations, spawning is divided and does not take place all at once. For this reason spawning periods differ, some being early and others late. This leads to fry of different sizes, the early developers being the stronger.

Until spawning and from the time of their leaving the wintering ponds, the brood fish are kept in holding ponds. These are rather large but the water must not be too warm in order to avoid premature spawning; nor too cold either as this would slow down sexual maturity. A good temperature for these holding ponds is from 15 to 16°C. The sexes can be separated.

One to three "sets" can be installed for each spawning pond – a "set" comprising two males and one female. Brood fish are selected from those between the ages of 5 to 8 and 10 years weighing 3 to 7 kilos (7 to 15 lb) (Fig. 192). The males are recognized from the females by their milt, and only slight pressure on the belly is necessary to cause it to run. The females have swollen bellies and the anal vent is prominent. Males are generally ripe between the ages of 3 and 4 years while females ripen between 4 and 5 years. Too young parents should not be used (at least 4 years for males and 5 years for females) nor should they be too old (over 8 years for males and 10 for females) although older females may still be apt for reproduction.

Brood fish are placed in spawning ponds immediately after they have been put under water. Before this the parents should be given a 15-minute bath in a solution of 2·5 per cent sodium chloride in order to destroy parasites on the skin and gills. If the temperature remains right, which is not always the case, spawning, habitually, takes place within 48 hours after the ponds have been stocked and sometimes only 24 hours later. Carp spawn near the surface. They make a lot of noise ridding themselves of their eggs by means of agitated movements. A female gives about 100,000 eggs per kilogramme (2 lb) of weight. The eggs are transparent and they stick to the submerged vegetation (Fig. 196). After swelling they are from 1 to 1·5 mm in diameter.

Under favourable circumstances about 50 per cent hatch, but sometimes this amounts to no more than 15 per cent. After spawning and before hatching the brood fish are removed in order to stop them destroying part of the fry or passing parasites on to them. Parasites can indeed destroy a whole nursery in this way. Removal is also necessary to avoid poaching. Besides this the parents also dig into the peripheral ditch, stir up the water and upset the development of the eggs. Brood fish are removed with scoop-nets. Sometimes it is necessary to lower the water level and dry part of the spawning pond for a few hours. This operation is not dangerous for the eggs if it is carried out in the fresh and early hours of the morning or in the evening. It can also be done if the weather is overcast. The brood fish are taken to large ponds, where they will find good natural food. They stay there until the following spring and are not fed artificial food.

When the temperature is between 18 to 20°C carp eggs take from 4 to 6 days to hatch. The eyes are noticeable after 2 or 3 days. On hatching, fry measure 5 to 6 mm. They have very small yolk sacs which are absorbed rapidly. Two or 3 days after hatching, or 1 week at the most, the fry must be caught and placed in nursery ponds. Average length will then be 9 mm.

Immediately after this the spawning ponds must be dried out and disinfected with quicklime. They will remain dry until the following year.

If, because of unfavourable weather conditions the females fail to ripen, or if the first spawning is destroyed, then spawning should be tried again with reserve brood fish. Different spawning ponds are used from those which served for the first spawning. It is also a good idea to keep more reserve fish than would, ordinarily, be needed just in case of shortage.

Capture and counting of fry

Fry are caught without the ponds being dried out. Normally the water is lowered slowly and in such a way that the fry abandon the grassy parts, which dry up, and take refuge in the ditches under water. The fish are caught with shallow scoop-nets 40 to 60 cm (16 to 24 in) in diameter (Fig. 200) or with wooden rim devices lined at the bottom with gauze or very fine muslin (sifting silk).

The fry should be caught on a sunny day when they will be on the surface. They should be placed immediately in floating cases with wood frames and very fine mesh, or in white painted tanks filled with water from the spawning pond.

Naturally it is not possible to make an exact count of the fry caught. This is done therefore by approximation. Having dispersed the fry uniformly in the holding tank, a certain number are fished out by means of a receptacle the volume of which is known, ½ litre for example (1 pt). The fry are poured into an enamel pan and then poured slowly into another pan and counted as they pass from one pan to the other. As the volume of the tank is known it is not difficult to calculate the number of fry it contains. As many fry as desired can be caught using the same graded receptacle.

This method of calculation is very approximate. A much better way is to use two cases each with perfectly watertight movable divisions (Timmermans, 1956) (Fig. 207). The larger of the two cases should be divided into six sections for example and the smaller one into four. Each case has installed, low down at its extremities, a wide tap permitting quick and complete drying of the end divisions. Having removed the dividing panels from the large case an important number of fry are placed in it. This mass is spread out equally, as far as possible. and then, as quickly as possible, the panels are slipped into place. The tap at one extremity is then opened. Water and fry pour out into the second case the divisions of which have also been removed.

The fry are again spread out as equally as possible and the panels are replaced. The fry contained in one terminal section are poured out into a receptacle and counted. This means a 24th part of the large case is counted and the figure being known, it is not difficult to obtain quantities corresponding to a part or a multiple of the 24th part. This system works well and resembles the divisional system used by the Chinese to count Asiatic carp (Chimits, 1958).

Appreciation
In Central Europe, where the spring is late but regular, the spawning pond system gives good results and is advantageous for all. In western Europe, however, where the spring is more uncertain and there are frequent cold spells even as late as June, the system can be disappointing. For that reason the system of spawning and rearing in the same pond is fairly often preferred.

2 Artificial spawning nests system
While in the Far East, and most particularly in Indonesia, the reproduction of carp is sometimes carried out with Dubisch ponds, more often small ponds installed with artificial spawning nests made of vegetable fibre are used. These are called "kakabans". Spawning and hatching ponds are used successively (Huet, 1956).

Spawning ponds
These are small oblong spawning ponds about 25 to 30 m² (30 to 35 yd²), completely free of vegetation, the bottoms of which have been dried out for several days before being put under water. This latter operation must be done in the morning with clear, well aerated water. In the afternoon of the day the ponds are put under water, the kakabans are put in place and the brood fish released.

Kakabans are made (Fig. 201) of fibres taken from *Arenga pinata* palms. The fibres are set out in layers from 1·20 to 1·50 m (4 to 5 ft) long and 40 cm wide (15 in) and pressed down lengthwise by two bamboo slats so that they are about 4 to 5 cm thick ($1\frac{1}{2}$ to 2 in). The kakabans should be placed side by side and attached perpendicularly to a long bamboo pole which supports the ensemble and is held in place by four posts driven in pairs at each end. The weight of the kakabans keeps them just below the surface of the water.

About five to seven kakabans to 5 m² (6 yd²) of surface are required for each kilo of female. Normally the carp spawn the same evening and the following morning the spawning should be finished. It is a good idea, during the night, to turn the kakabans upside down to help the uniform distribution of the eggs which stick to the vegetable fibres. The following morning the kakabans are transferred to the hatching ponds.

Hatching ponds
These are 20 times larger than the spawning ponds. Before being put under water they are kept dry for several days to help mineralize the mud.

When the kakabans are transferred they should be shaken slightly in the water in order to remove the mud which might cover them. The kakabans are then laid out in the same way as in the spawning ponds but two long bamboo poles are placed over them crossed by several banana tree trunks so that the kakabans are at least from 8 to 10 cm (3 to 4 in) under water. A space of from 5 to 8 cm (2 to 3 in) is left between each kakaban.

Five days after placing, the hatching should be over. The kakabans are removed, washed and dried. The fry remain in the hatching pond for 3 weeks. One week after hatching, rice bran is distributed. On emptying, normally, 15 to 20,000 fry are harvested per kilo of female. Under the best conditions it is possible to obtain up to 30,000. Three-week-old fry measure about 2 cm.

As described above, the kakaban system is not used in Europe. Nevertheless the artificial spawning method has been and is still used for certain species. Formerly it was the practice to put solid vegetation which would not rot into the spawning ponds of cyprinids. These were generally branches of briar (Guenaux, 1923) (Fig. 194). Artificial spawning nests were also placed in large ponds used as spawning and rearing ponds. The method can still be used for carp and other cyprinids such as roach. It is also used regularly for pike-perch (Ch. VI). After spawning the nests are transferred to hatching ponds.

II Artificial Reproduction

1 History
Although the artificial reproduction of salmonids has progressively spread throughout the whole world for over a century, the artificial reproduction of cyprinids – a recent discovery – is still difficult and often remains uncertain.

The difference between the possibility of and

the ability to apply artificial reproduction both to salmonids and to cyprinids, springs from the different genital structure of the fish of these two families and, above all, of the female (Wurtz-Arlet, 1967). The ova of female salmonids at maturity fall directly into the abdominal cavity and only normal pressure need be applied to the abdomen for the eggs to emerge. Only slight pressure on the flanks of the males is needed to obtain the milt. There should be no problems. It is not the same with cyprinids, particularly for females whose eggs cannot be liberated until complete maturity has been reached as far as that is possible with captive fish. After spawning, the eggs of many cyprinids, notably carp, stick to the substrate or to one another. To make mass incubation possible this must be eliminated.

Experiments in the artificial fertilization of cyprinid eggs were tried in 1937 by Probst in Germany and at about the same time by Steinmann and Surbeck in Switzerland. The results were not very encouraging. Between 1950 and 1960 experiments which produced better results were tried in the U.S.S.R. by Gerbilski, in India by Chaudhuri and Alikunhi, in Formosa by Tang and in Hungary by Woynarovich. These experiments were carried out with grass carp, Indian carps or common carp. For the latter, as for Asiatic carps, the major problem was to achieve maturity. This was done by hypophysation or an injection with a pituitary extract. For common carp maturation was not a major problem but the adhesion of the eggs was extremely difficult to avoid. For Asiatic carps as for common carp, techniques are now perfected so that artificial reproduction gives large numbers of fry both for Asiatic carp and common carp. The important progress in Europe in this field is principally due to the work of Woynarovich (1964). The techniques used for the artificial propagation of warm-water fish (common carp, channel catfish, danubian wels, tench) have been well described by Woynarovich and Horvath (1980).

The methods described below refer mainly to common carp.

2 Maturing spawners

First, only carp in excellent health should be chosen and they must be handled with the greatest care. The females should not be too large and should weigh about 3 to 6 kilos (7 to 13 lb).

Pre-maturation extends over several weeks or even several months and corresponds to the period when winter ends and the warmer spring weather begins.

The temperature rises gradually. When it reaches 18°C and stays there – which can also be assured by artificially heating the holding pond – the brood fish are placed in the spawning or maturing ponds which should be at a stable temperature between 17 and 19°C. In central Europe the fish are left from 10 to 8 days between March 15 and 31, from 8 to 5 days from April 1 to 15, from 5 to 3 days from April 15 to 30 and from 3 to 1 day from May 1 to 15.

3 Hypophysation of spawners

If carried out at the right time, i.e. when maturity is sufficiently advanced, an injection of pituitary extract will help towards complete maturity of the spawning fish and the emergence of eggs and milt which permits natural or artificial fertilization of the eggs.

Hypophysation is an intramuscular injection towards the top of the body near the first ray of the dorsal fin halfway between that fin and the lateral line. It is between 2 and 3 cm deep ($\frac{3}{4}$ to 1 in).

For the female the practice is one hypophysis per kilo, using 1 cm³ of the solution for two hypophyses. For the male one injection of one hypophysis in 1 cm³ of the solution is sufficient.

Before injection, the hypophysis is reduced to powder in a mortar. The hypophysis of carp, small glands 2 to 3 mm in diameter, are extracted by trepanisation, dried with acetone and conserved dry. A 10 cm³ solution is made up of 3 cm³ of glycerine and 7 cm³ of a physiological solution (6 gm of sodium chloride per litre).

After injection, the fish are left in calm water at around 17 to 20°C. If the operation is a success then complete maturity will be reached in about 18 hours. Why this operation sometimes succeeds or sometimes fails is not completely known as yet. Generally, if for 50 per cent of the females complete maturity is achieved, it is considered satisfactory.

Actually, for carp, these injections are not indispensable and artificial fertilization can be practised without them.

4 Artificial fertilization of eggs

The eggs and milt are gathered from perfectly ripe fish (Figs. 202 and 203) in the spawning ponds or maturation tanks in which the first active signs of natural spawning will be noticed. The brood fish must be handled with great care and the eggs and milt gathered completely dry. For a start the volume of the milt should be about 5 to 10 per cent of the total volume of the eggs. The milt and eggs must be well mixed, quite dry, per portion of 150 to 200 cm³. The mixture is then poured in a fertilization solution. Afterwards the eggs are treated with a second solution to avoid sticking.

The fertilization solution comprises for 10 litres (17 pt) of very pure water 30 gm of urea and 40 gm of sodium chloride. For a given volume of eggs at the start (generally 150 to 200 cm³), two volumes of solution are used at most. First, half a volume of the solution is added and mixed without stopping during 3 to 5 minutes. After that the remaining solution ($1\frac{1}{2}$ volume) is added in small quantities every 5 minutes over a period of $1\frac{1}{2}$ hours. During mixing the eggs absorb the solution and grow considerably in volume.

The fertilized eggs are then treated against sticking. Woynarovich recommends the use of tannin but for some specialists this is not free from danger for the eggs. 15 gm of tannin is added to 10 litres of pure water. One and a half to 2 litres are then poured into a 5 litre receptacle and mixed by hand with a portion of the fertilized eggs for 10 seconds. The solution is then rapidly removed. The operation must be carried out twice more but each time with a weaker solution. Washing must be with pure water between each operation.

5   Incubation of eggs

Incubation is carried out in Zoug jars (Fig. 204). Details are given below when the incubation of pike is described (Ch. V). There are about 120,000 common carp eggs per litre. At the beginning, as the eggs are very fragile, the water flow must be very slow, no more than 1 to 2 litres per minute. This is then increased gradually. Incubation is carried out at a temperature of 20 to 24°C which necessitates warming the water. At these temperatures incubation takes only 4 or 5 days. Hatching can be in jars or in the trays on which the eggs were laid at the start of the hatching (Figs. 205 and 206). It is on the very fine cloth

covering these trays (perlon stitch 0·2 to 0·4 mm) that the yolk sacs are absorbed. The time required is very short – no more than 2 to 4 days.

6   Advantage of this method

The advantages of this method are considerable. In the case of Asiatic carps artificial fertilization was made possible thanks to hypophysation which permits multiplication outside the natural habitat. This is an enormous advantage and the most is made of it both in Europe and Asia.

There are other important advantages for European and Asiatic cyprinids. First, it is possible to obtain a high number of fry of the same age with greater certainty. This is a great help in high production countries. Next, notably for carp, it is possible to obtain spawning several weeks in advance which is helpful in countries with relatively warm and regular climates such as in Eastern and Central Europe. It is also an advantage, though to a lesser but nevertheless real extent, in Western Europe where cold spells often come late, are frequent and can be mortal for young fry liberated too early in the rearing ponds. In these regions it is best to be wary of spawning which is too precocious. This can be partially remedied, however, by feeding the fry in rearing troughs or in small well sheltered ponds, but it is a delicate and uncertain operation.

Finally, artificial fertilization leaves possibilities open for improving selection. Numerous crosses, both intra- and interspecific, become possible and it is also possible to make more trustworthy morphological tests, as well as testing rapidity of growth and resistance to disease of the experimented strains.

Naturally this requires proper housing and installations, holding ponds or tanks, incubation jars and hatching troughs all supplied with heated water the temperature of which can be regulated. But the installation can be used for the artificial reproduction of other fish such as pike and coregonids which reproduce at a different time of the year.

SECTION IV

REARING CYPRINIDS OF ONE OR TWO SUMMERS AND CYPRINIDS FOR CONSUMPTION

I   Rearing Carp

A   **Rearing fingerling carp**   (The first year)
Methods of producing fingerling carp differ widely.

Some are quite simple and include only one or two categories of ponds while others are more complex and call for three sorts of ponds. One or other of these is used according to circumstances, local conditions, local climate in particular, topographical conditions and the personnel available. The major difference between the two groups lies in whether the fry are the result of controlled reproduction (natural or artificial) or whether the reproduction was natural and semi-controlled.

1 Rearing by means of natural or artificial controlled reproduction

The principal characteristic of these methods is that they rely on the controlled reproduction of carp. This is achieved by either natural reproduction in spawning ponds of the Dubisch type or on artificial nests, or by artificial reproduction. When they are a few days old, that is after the yolk sacs are absorbed, the fry are transferred in specific numbers to rearing ponds. According to each case only one category of rearing pond is used: rearing fingerling carp in one phase or, successively, two categories of different rearing ponds. These are called first and second rearing ponds (rearing fingerling carp in two phases).

a   Rearing fingerling carp in two phases

During the first stage, 4 to 8 week fry are produced in the first rearing ponds. At the end of this stage, that is at the start of the summer, the young fish are transferred to second rearing ponds where they stay until the following autumn or spring.

1   *First rearing ponds* (Vorstreckteiche, Brutvorstreckteiche)

Characteristics of the ponds. First rearing ponds (Fig. 209) are shallow, about 50 to 70 cm deep (20 to 28 in). They are also relatively small. The carp fry are placed in these ponds one week after birth at the latest, that is to say, the beginning of May to the beginning of June. They remain for 4 to 8 weeks.

These ponds have one main purpose, that is to see the fry through their most dangerous period as quickly as possible, i.e. until they have reached about 5 to 6 cm (2 in) in length. After that they will not be subject to skin and gill parasites such as *Dactylogyrus, Costia, Chilodon, Ichthyophthirius*. In order to aid rapid growth at this stage and to eliminate risks from parasites the following method is recommended:

1.   If possible, water used for first rearing ponds should be absolutely pure, that is it should not have traversed a pond stocked with fish but should, if possible, have traversed a warming pond. Where the water passes into the pond, a screen should be placed to keep enemies out, particularly pike fry. This screen should be horizontal and should have a fine mesh.

2.   First rearing ponds should be fairly shallow for, if the depth is too great it stops the water from heating. On the other hand, if it is too shallow changes in temperature will be felt too easily.

3.   The size of the pond will depend on the length of time it will be occupied by the fry. If the period is short (4 weeks) the pond could be between 0·25 to 1 hectare ($\frac{1}{2}$ to 2 acres). If it is long (8 weeks) then the pond can cover as much as 2 hectares and even more (Fig. 212). In any case the size of a pond must not be too small bearing in mind that living space counts, but neither must it be too large, otherwise draining will not be easy or complete.

4.   The actual stocking of the pond will be proportionate to its productivity, the length of time it will be used and the desired size of the fish. Experience helps to predetermine the relative value of a pond. On average, 50,000 fry are released per hectare and under the most favourable conditions 200,000 or from five to 20 fry per square metre.

5.   The ponds are put under water some time between the start of the whole operation and the hatching of the fry. Eight days or so between the pond being placed under water and the release of fry in the first rearing ponds should be sufficient to ensure the development of the infusoria and the crustaceans which will feed the carp fry while their enemies will not have had time to develop.

6.   Efforts must be made to increase this natural food as much as possible so that the fry is literally plunged into the food. To this end Schäperclaus suggests:

(*a*) In the spring, before being put under water, cereals or graminaceae should be cultivated.

(*b*) Between filling with water and releasing the fry, the grass should be cut, piled in heaps and left in the pond.

(*c*) When the fry are about 3 cm (1 in) long the grass should be spread out. Its decomposition will help to develop and multiply the food. The oxygen content must however be carefully watched.

(*d*) Mineral and organic fertilizers must be applied carefully.

(*e*) Moderate distribution of liquid manure and blood should be regular and repeated as well as meat and fish meal which will also help to multiply the natural food of the fry.

During the first stages of development it is not possible to give the young fish artificial food. Their growth can be controlled regularly by placing white painted planks on the bottom of the pond. These will permit the growth of the fry to be watched.

After draining, the pond can then be left dry and treated as a meadow. The principal and secondary drainage ditches must be treated with quicklime. If the fry are found to be contaminated with parasites then the pond must be treated accordingly. The pond can also be put under water again and used for the second stages.

Value of the method. From mid-June to the beginning of August this method produces fry of from 2 to 3 cm ($\frac{3}{4}$ to 1 in) if captured early and from 8 to 9 cm (3 to $3\frac{1}{2}$ in) if caught late. Their average weight is below 5 gm (Fig. 218). Loss can be as high as 75 per cent.

Controlled stocking of the first rearing ponds is the principal factor on which the advantages of the method depend, for it avoids over-stocking the pond, poor growth and the risk of epizootics which can occur if all the fry of the spawning ponds are released into the first rearing pond. This is sometimes done. On the other hand the method calls for careful handling and experienced personnel.

2 Second rearing or fingerling ponds (Brutstreckteiche)

Young carp are liberated between mid-June and the beginning of August after having been removed from the first rearing ponds. After the growing period, i.e., in the summer and autumn, they pass the winter here and stay until the following spring when they are transferred to the fattening ponds.

Characteristics of the ponds. The ponds are put under water 8 or 15 days before stocking if there is sufficient water to fill them rapidly. If not then this must start earlier.

After emptying, normally at the end of March or the beginning of April, the ponds stay dry until the next stocking in June or July. While the ponds are dry the soil may be turned over and fertilized in order to improve its productivity. They can also be sown but this must not be done with a single species but with a mixture of 50 per cent small and large leguminous plants such as lupine and serradilla and 50 per cent graminaceous growths such as rye and oats (Fig. 379). Half the pond only is sown, in rows, in order to avoid too much rot

when they are put under water. This also permits the fry to swim around easily between the submerged stalks. Under no circumstances should more than one-third of the seed customarily used over the same area in agriculture be employed.

Fingerling carp ponds (Fig. 210) can be any size. They should not be too large, however, so as to avoid drainage complications. Mehring recommends between 2 and 5 hectares (5 to 12 acres) although on some farms carp fingerling ponds cover 15 to 25 hectares (35 to 60 acres). As the fingerlings will pass the winter in the ponds they should not be too shallow and must be traversed by an adequate but not strong current. Depth should be at least 1 m in order to avoid invading vegetation. It is advantageous to have ponds around 1·50 to 2 m ($1\frac{1}{2}$ to 2 yds) near the monk. This will help in winter.

Stocking rate. This depends on natural productivity, fertilization and the artificial food distributed. The latter can commence and be carried on regularly.

Stocking also depends on the size of the fish required. If the average size is from 9 to 12 cm ($3\frac{1}{2}$ to 5 in) then an average of 5,000 Kv per hectare can be stocked (50 to the are). If larger fingerlings are desired (12 to 15 cm) (5 to 6 in) then only 1,000 to 1,500 are stocked. If, on the other hand, the aim is fingerlings of less than 10 cm (6 to 9 cm) ($2\frac{1}{2}$ to $3\frac{1}{2}$ in) then 10,000 per hectare can be stocked and even more. The recommended length is 9 to 12 cm. They should weigh around 50 gm (around 2 oz) (Fig. 219).

In good ponds losses are generally limited to 10 to 15 per cent but can be higher: 20 to 35 per cent and even 50 per cent. On the whole, from fry with absorbed yolk sacs to fingerling carp, a 75 per cent loss can be considered normal.

Value of the method. The principal advantages of using second rearing ponds are as follows:

1. As it is possible to control the stocking, it is also possible to estimate in advance with a sufficient degree of approximation the number of fingerling carp which will be available the following spring. Further, control by fishing some specimens in summer will enable growth to be followed.

2. As the fingerling carp are left to winter in these ponds, autumn fishing is suppressed and wintering in small ponds which are poor in nutritive fauna is also avoided—thus two operations which could favour sickness and epizootics of the fry are eliminated.

b   Rearing fingerling carp in one phase

This method obviates first rearing ponds. It developed out of problems set at one time of finding sufficient water to supply ponds in summer and above all the failure of the two-phased method as a result of negligence and clumsiness.

To start with by using one stage only, draining in mid-summer is avoided and this helps a lot. On the other hand, by this method, it is not possible to follow the progress of the cultivation nor to estimate its eventual success. As the loss of small fry of 8 days is very high, 40,000 per hectare must be released.

Improved method. If the first rearing stage is practised without the use of a special pond then it is possible to use the same pond successively for the first and second rearing stages. At the start the pond is only partially filled. After 4 or 8 weeks it is fished out, then put under water completely and this time stocked adequately as for the two phases method. If enough water is not available to fill it completely, it is not dried out and the first stage is controlled by trial fishings. If the population is too great then an effort must be made to reduce it. Then the pond is completely refilled for the second stage of rearing.

c   General appreciation of rearing with controlled reproduction

This method is currently used in Central Europe and it has made a great contribution to development and progress in the cultivation of cyprinids. Its principal advantages are that the stocking rate is successfully controlled, it avoids over-stocking and aids rapid growth of the fry while protecting them against epizootics and disease.

But it also has certain inconveniences, above all for small farms because it is necessary to have several categories of ponds (spawning, first and second rearing ponds). On the whole it is rather a complicated method calling for many ponds, lots of work, lots of experience and also sufficient water and ground which can be laid out for all the ponds. Consequently the semi-controlled reproduction method is often used since it is apparently simpler. But the results it gives are that much more uncertain and chancy as this farming breaks away from classical method.

2   Rearing with semi-controlled natural reproduction

This method of rearing can be carried out in two ways, one very different from the other. For the first there are two stages using large spawning ponds which also serve as first-stage rearing ponds from where the fry are transferred to fingerling carp ponds or second stage rearing ponds. The second system includes only one stage. All the fingerling are reared in a single pond. This is the system of spawning and rearing in the same pond.

a   Use of large spawning ponds and fingerling carp ponds

Natural reproduction and the first rearing stage takes place in the same pond (Fig. 193). This is fairly large, from 5 to 25 ares (five to twenty-five fortieths of an acre). A rather high number of brood fish are released using 5 to 7 females and 10 to 14 males per 10 ares. Reproduction is supervised but it cannot be influenced. Success depends on atmospheric conditions.

After hatching, the fry are left in the ponds for 2 or 3 weeks. After that, the ponds having been used for hatching and first rearing are dried out and the fry transferred to carp fingerling ponds. They measure about 2 cm ($\frac{3}{4}$ in) and can be handled without too much difficulty. Further, as loss will be pretty constant and can be estimated in advance, the young fish can be left in the ponds until after the winter. This method is to be recommended if there is a lack of spawning ponds and if the water supply is a bit short.

The method has the advantage of avoiding the difficulties of controlled reproduction. On the other hand it has the inconvenience of calling for a high number of brood fish and, above all, does not permit the control of reproduction. This means it is not possible to estimate the number of fry until after the pond has been emptied. So if the reproduction is faulty nothing can be done to put it right. But on the whole it is a relatively sure method and nearly always gives positive if varied results.

The second stage comes after emptying the reproduction and first stage rearing pond. This method is practically identical with the second stage of rearing with controlled reproduction and, from that moment, it offers all its advantages notably those resulting from controlled stocking.

b   System of spawning and rearing in the same pond

This method has been described in section III under the heading "Semi-controlled reproduction: rearing in one stage". A single category of ponds is used. The parent fish are placed in the ponds in the spring. They spawn and stay there with the fry until the following spring.

The value of this system will already have been appreciated. It is easy but the results it gives each year are very variable and this is a great inconvenience.

B   Rearing carp for consumption
(Second and third year rearing)

1   General characteristics of cultivation
Carp for sale for the table must meet the requirements of the local market. These differ considerably just as the time taken to satisfy them differs.

The first requirement is the weight of the fish. There are others such as quality of flesh and the number of scales. These factors have been dealt with above. In Central Europe the demand is for carp weighing from 1,000 to 1,500 gm (2 to 3 lb). Sometimes carp of 1,000 gm (2 lb) meet requirements but at other times carp weighing from 1,800 to 2,000 gm ($3\frac{3}{4}$ to 4 lb) should be produced. In Eastern Europe carp weighing 500 gm (1 lb) are acceptable. In Israel fish weighing from 500 to 750 gm (1 to $1\frac{1}{2}$ lb) are produced and in Indonesia marketable fish weigh from 75 to 100 gm (3 to 4 oz).

The time it takes to grow a fish varies with the climate. If the aim is a fish between 2 and 3 lb then in Europe the normal time is around 3 years. In regions where the climate is more favourable, with high summer temperatures and longer growing time, the same sized fish can be produced in 2 years.

In European carp cultivation the fish are referred to as being of a certain number of summers – that is 1, 2 or 3 summers (ein-, zwei-, dreisommerige Karpfen) which corresponds to 1, 2 or 3 periods (or years) of growth. Referring to carp in this way also covers, very judiciously, the character of the warm water needed by these fish which only grow during the summer or about 4 months of the year in western Europe. The production factor which corresponds with the productivity factor K_1 (temperature) is examined below (Ch. X). In those regions where, by reason of a more favourable climate, the annual growth period is longer (6, 9 or 12 months) fish for eating of the same weight are obtained over a period in inverse proportion to the growing time. This means that in tropical regions carp of 2 or 3 lb can be produced in 12 months of consecutive rearing.

In referring particularly to carp of from 2 to 3 lb for eating, which are normally produced in three summers in central and western Europe, it is possible to produce them in two summers in southern Europe; this applies particularly in Yugoslavia, but only on condition that the stocking has been properly calculated and production intensified by fertilization and that artificial food has been distributed.

According to the policy of each enterprise European carp for consumption are produced in three or two summers by the successive use of second year ponds (Streckteiche), then the third year ponds or fattening ponds (Abwachsteiche); or in two years by using only the fattening ponds. If the farm is based on 3 years the 2 summers carp will weigh about 250 gm ($\frac{1}{2}$ lb) and measure 20 to 25 cm (8 to 10 in) (Figs. 222 and 223).

Fattening ponds, for second and third year fish, are put under water in the spring after having been dried out in the autumn. After emptying, the carp are kept in holding ponds until sold. If the farm is a 3 years establishment then carp of 2 summers are transferred to ponds where they pass the winter.

Maintenance (in particular fertilization for the fattening ponds), artificial food for the fish and stocking must be carefully controlled so that production is regular and meets the requirements of the programme. These three very important questions are treated separately (Chapters XII, XIII and XIV).

Specialized productions. 1. If the production of fingerlings was too great the first year, then so as not to lose this surplus, the fish can be placed in storage ponds where they are tightly packed. In the second year they will scarcely grow at all and the following year they can be considered and treated as fingerlings of 1 year.

2. Carp are sometimes only saleable at the end of their fourth summer if they have been held packed together during the third year and in consequence have made only slight growth during that year.

3. Some farmers leave their ponds under water for years at a time (*emptying every several years*). This is bad policy, for then global production over the period will, in fact, be less than all the total annual production obtained by other methods. If the ponds are kept under water constantly they do not gain the advantages of winter drying; they become acid and grassy and it is difficult to stock

them adequately. When such ponds are used for growing carp it is necessary to add pike.

2 Traditional rearing: extensive, semi-intensive and intensive methods

The method of raising carp for the table, like the raising of trout, can be carried out in several ways. There is extensive raising without artificial food being distributed, there is semi-intensive in which only a small quantity of artificial food is distributed and there is intensive raising in which an abundance of artificial food is distributed (mostly vegetable cereals and legumes). Ponds can be fertilized or not.

In all cases, whether artificial food is used or not, it is nearly always deficient in proteins and a degree of intensity comparable with that of salmonid cultivation is never reached. Under these conditions the natural food found by the fish in the ponds always plays an important part in the feeding of reared fish. Further, the production of large quantities of fish for the table can only be achieved in large size ponds. They can cover, sometimes, tens and even hundreds of hectares (Fig. 211).

3 Intensive condensed rearing

Although still rare, the production of fattened cyprinids under intensive and condensed conditions (maximum of fish in a minimum of water) has been tried. In Europe, only the experimental stage has been reached (v. Sengbusch, Meske, Szablewski and Luehr, 1967). In Indonesia the practice is current though limited (Huet, 1956). This is also true in Thailand and Japan (Chiba, 1965).

Farming is practised in a very limited space. The water must flow and be renewed, and the food distributed must be abundant and frequent. In Japan, where some ponds are from 20 to 100 m² (25 to 120 yd²) average production is 100 kg (220 lb) per square metre (square yard). Weights of 220 kg (about 500 lb) per square metre have even been attained in one pond. In some parts of Indonesia farming is carried out in bamboo cages with the fish closely packed (Figs. 224 and 225).

For intensive condensed farming to succeed, it must fulfil the following conditions: (1) temperature must be high, around 25°C; (2) the water must flow, be well oxygenated and also constantly changed in order to eliminate harmful physiological excretions which accumulate in static water; (3) artificial food must be abundant and distributed

frequently, though never in excess, and it must be rich in proteins.

C Wintering and holding ponds

In regions where the winters are long and hard, adequate measures must be taken to protect the carp during that period. In Europe the wintering of carp differs according to age. Generally 1 summer carp (K_1) pass the winter in the rearing ponds in which they passed the summer (Fig. 212). These ponds must be suitable for wintering but if they are not the carp must be placed in wintering ponds. Wintering ponds are used regularly for carp of 2 summers (K_2). As for carp of 3 summers (K_3) destined for the table, they are not wintered in the proper sense of the word but are placed in storage ponds until delivered to their final destination.

Wintering and storage ponds must be located near the farm building to ensure that they can be properly watched. This is necessary for two reasons – wellbeing, and protection from thieves.

1 Wintering ponds

1. Utility. Wintering has two aims:

(a) To permit the drying out of the growing ponds. This has great advantages.

(b) To protect the fish against those winter dangers likely to affect the flat shallow ponds used for summer growth. These can freeze over in winter.

These special dangers are: (a) "frostbite" by contact with ice, (b) asphyxiation under the ice, (c) danger of increased acidity in the water of regions with low alkalinity through winter rains and melting snow. Such dangers are less threatening, or, in any case, are easier to fight in wintering ponds than in the large growing ponds.

2. General characteristics. Wintering ponds should be deep and the greater the threat of icing, the deeper they should be. Depth near the monk should be at least between 1·50 and 2 m (5 to 7 ft) in Central Europe; this means that an important part of the total surface of the water will exceed 1·20 m (4 ft) in depth. In Western Europe it can be less.

The sizes of wintering ponds can be relatively limited from 0·50 to 3 hectares (1 to 6 acres) (Fig. 212). It is better to have small wintering ponds in good condition than large ponds which are more or less silted up and filled with grass. The size depends on the degree of stocking. This is given below. It is best to have a wintering pond for each species of fish (carp and tench) and even for each age group (fish of 1 and 2 summers).

3. Hygiene. Good wintering ponds enable the fish to be kept under as good sanitary conditions as possible; that means also that they are completely rested and are not hungry.

Throughout the summer these ponds should be kept dry. The soil should be turned over if necessary, limed from the start of spring and excessive mud should be removed. The bottoms of wintering ponds should be rather hard but not excessively so. Acid beds should be avoided as well as ponds with peaty soil which decomposes and emits toxic gas. Wintering ponds should not be used as summer growing ponds.

Fish placed in wintering ponds must be healthy. Those which are not should be removed or at least separated. Skin parasites must be destroyed and handling must be gentle in order to avoid injury to the fish.

During the winter the fish should be left as quiet as possible – the ponds should not be located near frequented paths. Skating on the ice must be forbidden as indeed must any other activity involving the iced-up ponds.

4. Oxygenation and temperature. The water should be cold and well oxygenated. To ensure this, care should be taken to see that the supply is not too rich in organic matter and that the bottom does not have too much mud which is liable to rot. Precautions should also be taken against an insufficient water supply. The pond must not be too shallow and the density of the population should not be too high.

To ensure sufficient oxygenation the slow renewal of the water must be maintained in such a way that the whole pond is renewed with water. This means a change every 10 or 12 days or even more. The flow should be enough to oxygenate the water though sufficiently slow, at the same time, to avoid the fish losing weight which would result from excessive movement in struggling against a current. The water supply must be pure and clear and not too acid. Precautions must be taken against rainwater and, above all, water from melting snow, which must be evacuated if possible by a by-pass. In order to ensure that the water flow extends over the whole body of the pond it is a good idea to evacuate it from the bottom (Fig. 25).

Apart from renewing the water there is really no other way of ensuring sufficient oxygenation. Making holes in the ice is ineffective because they are too small and also because they are only temporary – in fact this practice can be harmful since it will disturb the resting fish.

Removing snow from the ice will permit the light to penetrate the water and promote photosynthetic action by the green plants of the pond. Artificial aeration is sometimes possible by using a windwheel engine, as is done in Czechoslovakia.

In eastern Europe, where the winters are very hard, special measures, such as the installation of "wattle fencing", are taken to protect the fish against harm by freezing (Fig. 190). Horizontal beams are placed on vertical posts making 2 metre squares (6 ft). Inside these squares, by means of barbed wire, 1 metre squares are made. The horizontal level of this construction is arranged so that it can be submerged a few centimetres. By the time the hard winter arrives the beams are well embedded in the ice. When the ice reaches a thickness of about 20 cm (8 in) the water level of the pond is lowered. There will then be a cushion of air between the water level and the ice which will stay in place due to the framework holding it. In this way further icing is stopped and the aeration of the pond is assured. It is even possible to drain the pond if so desired. This interesting method is expensive but very effective.

The temperature of the water should be neither too low nor too high: not less than 1 °C nor more than 4 to 5 °C. The average should be around 2 to 3 °C (Kostomarov). If the temperature is too high the fish will need to eat more, but as the amount of food available will not be sufficient they will lose weight. This loss of weight becomes more evident as the water warms. A good water supply should come from a spring some distance away so that it can be sufficiently oxygenated and chilled.

5. Feeding the fish. The fish must be kept in excellent health and be provided with abundant food reserves which will enable them to resist the winter fast.

Young fish (K_1) and small fish (S_1 and S_2) reach their winter rest period later than large carp, lose weight more quickly and suffer more through loss of weight. This can be remedied by helping natural and providing artificial food for the fish.

Aiding the development of natural food means putting the pond under water at the start of October so that the different insects, notably the

Fig. 190 Wintering pond with wattle fence device as protection against freezing. Fish farm of Zator, Cracow, Poland.

chironomids which are in abundance at this time, can lay their eggs in the ponds under water.

On the other hand, as long as the ponds are not too cold, that is the water is over 4°C, wintering cyprinids, and above all the young fish, can still feed satisfactorily. Under these conditions the light distribution of artificial food is recommended.

6. Categories of wintering fish. Cyprinids placed in wintering ponds are generally second year carp. Third year carp are delivered for eating during the winter. As for first year carp it is better, if possible, to let them winter in second rearing ponds in which they passed the summer growing, for their winter rest is not so important to them and they would lose too much weight in wintering ponds.

7. Stocking. Loss. Stocking is generally around 50 kg (110 lb) per are, but this can be doubled and reach 10 tons or more per hectare.

Losses during the winter vary. They are higher when the winter is long and rigorous. In weight, loss can reach 10 or 15 per cent. Losses in numbers can reach 10 per cent and even more for first year carp. For second year carp loss is around 5 to 10 per cent and for older fish less than 5 per cent.

2 Storage and holding ponds

Storage and holding ponds (Figs. 213 and 214) are intended to hold carp for eating from the time the growing pond is emptied in the autumn until the fish are sold. A large part of the fish is sold for Christmas and the New Year but sometimes storage can go on until Easter.

In most respects, storage ponds are more or less the same as wintering ponds, notably concerning oxygen and the temperature of the water as well as the nature of the bottoms. This does not apply to size. Storage ponds are small and built of earth sometimes with reinforced banks (stone, wood or concrete). Generally they comprise fewer than 10 ares but they can go up to 30 and even more. The best sizes are between 5 and 15 ares.

They should be easy to watch and easy to get at particularly for trucks which will take the fish away.

The density and stocking is important and can be as much as 500 to 1,500 kg per are (9 cwt to 1 ton 9 cwt). In this case the flow of water must be sufficient to renew the pond once or twice daily.

The enforced fasting will improve the quality of the flesh but there will be a slight loss of weight, up to 2·5 to 3 per cent (and even up to 10 per cent over a period of several months). Care must be taken to ensure that there are no phenols in the water. Even the smallest amount will impregnate the flesh, though it will not upset the fish.

II Rearing Tench and other European Cyprinids

A Tench breeding and cultivation

1 Aims and general characteristics

Normally, tench (Fig. 215) are associated with carp. Production is for the table or for restocking.

Certain aspects characterize the rearing of tench. First the fish grow much more slowly than carp particularly over the first year. Reproduction takes place in high temperatures and late in the year. One year tench fry are small and difficult to catch. In water which suits them, tench reproduce abundantly and this excessive spawning aggravates all the problems of slow growth.

If tench are produced for the table the fish will be ready at the end of their third year and will weigh around 250 gm (8 oz). Tench intended for repopulating open waters or lakes and ponds are generally second year fish. First year fish are far too small and too likely to be eaten by voracious fish.

2 Growth of tench

Tench grow slowly, particularly when young. One year tench (preferably one summer tench) are small. Following the successful reproduction, tench of 1 summer measure from 3 to 8 or 10 cm (1 to 3 or 4 in) and weigh from 0·5 to 10 gm. Two year tench measure from 12 to 20 cm (5 to 8 in) and weigh, on the average, between 30 and 80 gm (up to 1 to 2½ oz) (Fig. 221). Third year tench measure 22 to 30 cm (8 to 12 in), weigh 200 to 350 gm (7 to 12 oz) and, exceptionally, up to 500 gm (1 lb).

There is little interest in raising tench beyond this for they will reproduce in excess in the ponds. If they are kept a fourth year then it is best to separate the males from the females. There is, however, one interesting factor and that is to obtain races (or strains) of rapid growth. To this end the same selection method is used as for carp.

From the start of the second year the growth of females is well above those of the males. Indeed the growth of males is 30 per cent less.

3 Reproduction of tench

Reproduction is simple but existing methods are not very advanced. Until now no one has succeeded or even tried to reproduce tench artificially. Spawning is natural and semi-controlled in one or two stages. These methods have already been described for carp. Small ponds are used for spawning (a few ares to 1 hectare). They are used exclusively for this. Otherwise the spawning fish are placed in second or third year carp ponds.

In Czechoslovakia (Kostomarov, 1961) artificial spawning ponds (Fig. 194) are used. Artificial nests from branches of coniferous trees are placed in the spawning ponds and after spawning the nests are transferred to hatching ponds which serve, at the same time, as rearing ponds.

Tench need high temperatures, 20°C and even more to reproduce. This means they spawn late, in June or even July. Spawning can be spread out which explains why in the breeding ponds the fry harvested are of clearly different lengths.

The parent fish chosen are from 4 to 6 years old and weigh from 0·5 to 1·5 kg (up to 3 lb). The differences between the sexes are easily recognized even out of the spawning season. The male's ventral fins are far more developed than those of females. There are a large number of eggs but they are very small. A female weighing 500 gm (1 lb) can yield 200,000 to 300,000 eggs. They measure less than 1 mm in diameter. They also stick. The hatching period is around 120 degree days or about 5 days. With natural, semi-controlled reproduction it is possible to harvest between 20,000 and 200,000 tench of 1 summer per hectare. The individual weight is between 10 and 8 gm and 1·0 to 0·5 gm.

4 Rearing methods for tench

a Production of first year tench. This production can be carried out in one or two stages.

1. Production of fry in one stage. For this, either special ponds for the purpose or second or third year carp ponds are used.

Harvesting young tench of 1 summer is difficult for the fry are extremely small and hide in the mud easily. Best results are obtained when the water level is slowly reduced at night. The smaller the tench the less they can support handling.

Spawning in special ponds is best, at least in so far as it facilitates harvesting. However, it has the inconvenience of not giving much weight, for in reality production only starts in July. This inconvenience is not very important as the area given over to production is not extensive. It is also compensated for by the high economic value of the production.

If spawning takes place in second or third year

carp ponds, then the value of the tench fry is added to that of the carp. On the other hand this complicates the harvesting as the carp will disturb the water when the pond is being emptied and this will lead to a more or less high loss of fry.

2. Production of fry in two stages. Tench parent stock are placed in special ponds, four to six females to eight or a dozen males per $\frac{1}{4}$ hectare ($\frac{1}{2}$ acre) for example. The ponds are emptied at the end of August and the fry, which measure about 3 cm (1 in) are released in young tench ponds for rearing. They stay there until the following spring. This method has a double advantage. First it avoids the necessity of autumn emptying which is pre-judicial to the young fish and secondly it permits controlled stocking of the rearing ponds which will ensure a good growth of fry. However, raising tench in two stages is rare.

b *Producing second and third year tench.* Tench are not reared alone but with carp and possibly with roach.

If the aim is tench for the table then the fish are mixed with carp as the two species feed, practically, in the same way. But this will be detrimental to the carp particularly if the percentage of tench is too high. Generally tench of the same age are mixed with second or third year carp but never at a ratio of more than 20 or 30 per cent tench of the same age. Wintering of tench of 2 summers is carried out under the same conditions and over the same time as carp of the same age, or in separate ponds. During the second or third year loss of tench can amount to 10 or 20 per cent and sometimes higher than that for second year tench.

If the aim is production for restocking second year tench (Fig. 221) then carp, tench and roach of 2 years are reared at the same time. But some-times only two species are used. The ratio of mix-ing is left to the farmer. He can, for example, stock one-third of each or decide to produce no more than one-fifth carp, two-fifths tench and two-fifths roach, or of course in any other ratio.

B Rearing roach and rudd

Roach (*Leuciscus rutilus* L.) (Fig. 216) is only reared in a few western European countries. If it is practised then it is not for the table but only for restocking water courses or angling ponds for coarse fish anglers.

Roach for stocking are fish of 1, 2 or 3 summers. Their growth is slow, slower even than that of tench excepting during the first year. Roach of 1 summer are larger than tench of the same age. This is because they reproduce at the start of the growing season, in May, while the temperature of the water is around 15°C, while tench reproduce 1 to 2 months later. This puts the roach in front.

The same methods of reproduction and rearing can be used for roach as for tench.

According to the success of the spawning, roach of 1 summer (Fig. 220) can measure around 6 to 10 cm ($2\frac{1}{2}$ to 4 in) and can weigh from 2 to 8 gm. If they are not too densely packed it is possible to obtain 1 summer roach of 10 to 12 cm (4 to 5 in) weighing up to 15 gm ($\frac{1}{2}$ oz). On the other hand if they are packed together then the size can fall as low as 4 cm ($1\frac{1}{2}$ in) and weigh around 1 gm.

Roach of 2 and 3 summers are produced by con-trolled stocking methods. Roach of 2 summers measure 14 to 18 cm (5 to 7 in) and weigh between 30 and 60 gm (1 and 2 oz). At 3 summers they can be longer than 20 cm (8 in) and can weigh 150 gm (5 oz). During the second and third year, losses can be between 5 and 30 per cent.

Methods used for roach are also suitable for rudd (*Scardinius erythrophthalmus* L.) but they are rarely raised.

C Rearing golden ide (orfe)

The raising of golden ide or orfe (Fig. 217) (*Idus idus* L.) is limited to the production of fish intended for decoration. They are well suited for this since they swim near the surface of the water and grow rapidly to 30 or 40 cm (12 to 15 in).

The same methods can be used for ide as are used for the reproduction and raising of carp.

Golden ide is a cyprinid which spawns early. This takes place at the end of April as soon as the temperature of the water reaches 14°C. The same reproduction methods are applicable as are used for carp, notably the use of Dubisch type spawning ponds as well as artificial fertilization. Whatever the method used the number of fry is always small as a lot of eggs are not fertilized. For brood fish, males of 3 or 4 years are used and females from 4 to 6 years.

After only a few days, that is as soon as the yolk sacs are absorbed, the fry are placed in rearing ponds. Generally, rearing is in one stage (only one category of pond) during the first year. The best way is to reserve one small pond for the young ide but it must be rich in natural nutritive fauna. Over following years, similar methods

Fig. 191   Crucian carp (*Carassius carassius* L.)

are used as for carp. The ide can be mixed with other cyprinids of the same age. Growth depends, as for all species, on the amount of food available and the temperature of the water. At one year, ide can weigh 5 to 20 gm and even 35 gm (over 1 oz). At 2 years they can weigh 50 to 100 gm (1½ to 3 oz). At 3 years from 150 to 350 gm (5 to 12 oz) and during the fourth and fifth years they can grow to between 200 and 500 gm (7 oz to 1 lb). Under the best conditions these weights and sizes can be surpassed.

D   Rearing species of the genus Carassius

Species of the *Carassius* genus are to be found among European and Asiatic fauna.

Crucian carp (*Carassius carassius* L.) (Fig. 191) are indigenous and common to warm European waters principally in Central and Eastern Europe. They resemble carp but, having no barbels, the difference between the two is easily recognized. Their growth is much slower than that of carp but, on the other hand, they withstand high temperatures and sparsely oxygenated water. They are resistant to disease and, in particular, to infectious abdominal dropsy. They can be raised in the same way as carp but because of their slow growth they are not welcome and are always eliminated.

Crucians interbreed with carp. The hybrids are sterile but present certain advantages. Their growth is better than that of pure crucians and they are hardier than carp. They could be used in Rumania.

Goldfish (*Carassius auratus* L.) are of Asiatic origin. They differ from crucians since they have fewer scales on the side (29 to 30 instead of 32 to 33 on the average). Being highly polymorphic, by means of selection it is possible to obtain certain deformations, notably of the eyes and the fins. They also offer considerable colour variations: red, yellow, white-rose with or without black patches.

Biologically, goldfish are very close to carp and crucians. They are of slow growth. They spawn in warm water around 25°C and their growth as well as their colouring development are particularly marked in water temperatures around 30°C. In Europe they are cultivated in Italy for ornamental value. Methods of raising are similar to those of carp but they will not be discussed in this work.

Gibele (*Carassius gibelio* Bloch) are Asiatic and seem to be identical with a wild form of *Carassius auratus*. Gibele were introduced into Eastern Europe, notably in the Soviet Union. They do not seem to show better growth nor are they generally of greater interest than the crucian. Gibele females can be fertilized by carp, crucians and other cyprinids; should the hybrids be females only they evoke no particular interest.

III   Rearing Asiatic Cyprinids in Europe

1. The advantages of this cultivation. The principal advantages of the cultivation of Asiatic and European cyprinids together are as follows: (1)

fish·production is increased by transforming into fish for eating an abundance of organic vegetable matter which would not, otherwise, be used; (2) this supplementary production is obtained without competing with or reducing the normal production of edible fish; (3) on the contrary from the fact that this leads to the disappearance of algae, reeds and other water plants which would otherwise be unusable or even harmful, more space for normally cultivated fish is provided and the possibilities of production are increased; (4) production is even intensified by the fact that the abundant droppings from grass eating fish have good manuring action and intensify and accelerate fish growth; (5) finally the flesh of these fish is good to eat.

Until now Asiatic cyprinids have been introduced only into ponds. They are not liberated in open water and it seems there is no doubt that the effects on water plants in this habitat would be less marked if they were.

2. Reproduction. Before 1960 these fish were only known to reproduce naturally in the water courses of their origin in China. Since then, the discovery of artificial spawning has permitted the fertilization of their eggs after hypophysation of the parents. This technique is used in Europe and notably in the Soviet Union, Rumania and Hungary.

Asiatic cyprinids reproduce in the summer in waters reaching 25 to 30°C. The semi-pelagic eggs, measuring from 2 to 2·5 mm in diameter, swell after spawning to 5 to 6 mm. They hatch after 32 to 40 hours and incubation is carried out in Zoug jars.

Sexual maturity of the parent fish is rather late and the slower the growth the later the maturity. This should be reached in from 4 to 5 years in China, 6 to 7 years in Rumania, 8 to 9 years in the Kiev region. Parent fish weigh 5 kilos (10 lb) or more.

3. Growth. Asiatic cyprinids are typical warm water fish and their growth is closely dependent on temperature. Thus in summer, with atmospheric temperatures in July, as given below, grass carp at the end of their third year would weigh in Turkmenistan and in China 2,500 gm (5 lb) (24 to 32°C), 1,300 gm (2½ lb) in Rumania (23 to 26°C), in the region of Moscow 850 gm (29 oz) (18 to 20°C), and in the region of the Amur River (16 to 20°C) 250 gm. In Hungary, according to Antalfy, cited by Bank, 1967, grass carp reach the

following weights: 1st summer 50 gm (2 oz); 2nd summer, 800 gm (28 oz); 3rd summer 2,000 gm (4 lb). Until now little information on the subject has been available from Western Europe. In Belgium during a year with a favourable summer and a warm autumn (1969) grass carp grew in one season from 1,475 gm (3 lb) to 2,350 gm (4 lb 12 oz) – this by the end of October of their fourth year.

4. Raising methods. The introduction of Asiatic cyprinids in Europe is too recent to know what rearing methods are best for them. In fact the methods used are the same as for carp.

For the first year Asiatic carps are raised alone but later, in second and third years of rearing, carp are mixed with them.

To succeed with Asiatic carps the following conditions should be achieved:

1. The temperature of the water should be high. It is true that the fish can withstand cold water but they will only start growing satisfactorily in water with temperatures about 18°C. Their optimum growth is in water with temperatures around 25°C or even higher.

2. Food must be plentiful. When young, all species need water rich in plankton. Later on feeding regimes will differ. For grass carp the water must be rich in vegetation. The most suitable are given in the chapter devoted to artificial food (Ch. XIII). Silver and big head carp also need water rich in plankton, for if it is not both warm and rich in plankton the species will not succeed.

3. Finally the fish must be well protected from their enemies – above all, when they are young but also during the second year. All fish, voracious or not, must be removed from the ponds holding the young. Even coarse fish and frogs must also be eliminated. As a protection, wire netting should be placed round the young fish rearing ponds.

During the first year, rearing takes place in one or two stages and in ponds specially devoted to this. Stocking depends on the amount of food available. If one stage operation is followed, with feeding (the young fry receive a finely minced mixture) and mineral and organic manuring, then stocking is around 5 to 15 fry per square metre. If it is a two stage operation, feeding the fry from the start, the first stage lasts only a few weeks and during that stage stocking can be 10 times as great; say 50 to 150 fry per square metre. If the farm is run well then from 20 to 60 per cent of 1 summer fish of the fry released can be harvested.

During the second and third years Asiatic carps are added to other raised carp at a ratio of from 40 to 200 per hectare.

In order that the raising of Asiatic carps might give good results some precautions should be taken, apart from the conditions already presented. Control devices both for the supply and evacuation of the water must be in good condition for the fish have a tendency to escape easily. Ponds must have an easy emptying system always in perfect condition otherwise the fish risk death. Emptying must not be too late, for if it is carried out during low temperatures, the healing of wounds which they might receive at this time could be difficult.

If the necessary conditions for raising Asiatic carps exist then they can be mixed with carp and production might well increase by 20 to 40 per cent in weight and even 80 to 100 per cent, if circumstances are favourable for the simultaneous addition of the three species of Asiatic carps.

IV Accompanying Fish in European Cyprinid Cultivation

Cyprinid cultivation often calls for mixing several species of European and Asiatic cyprinids. It is even opportune, under certain conditions, to mix in small voracious fish. This is not recommended for cyprinids of 1 year but it may be for older fish. In any case despite the farmer's efforts, undesirable fish, sometimes in appreciable numbers, are bound to enter the rearing ponds.

Pike and other species are sometimes released into cyprinid ponds voluntarily and fish of various sizes, the names of which are given below, are liberated. On the other hand this is not always desired but it happens anyhow when the ponds are supplied with water from streams in which pike and other species live and reproduce. The fry of these fish are carried by the current through the very fine mesh of the screens. They can also be introduced, as eggs carried by aquatic birds.

Pike (*Esox lucius* L.) (Fig. 229). These fish are introduced into cyprinid ponds for two reasons:

1. Pike play the role of predators. They can destroy carp fry and other cyprinids fry which might have hatched in the fattening ponds and get rid of coarse fish accidentally introduced into the ponds where they compete for food with the carp and other fish being raised. Pike destroy fry and other undesirables, and turn them into good quality flesh. They can consume prey weighing over 30 per cent of their own weight.

For this police role small pike of 1 year are used but they must only be introduced into second or third year carp ponds. About 10 to 50 small pike of 1 year are introduced per hectare. They must be small and should be no longer than 20 cm (8 in).

The wintering of young pike is difficult because of their very pronounced cannibalism. This is even more pronounced when the young pike find themselves in a confined space. For the winter, only those which will be used the following spring are kept. They are held in ponds with carp of 2 summers and as far as possible they should be about the same size.

2. Pike can also be mixed with cyprinids of 1 year and over for the production of pike of 1 summer, or 1 year, which will be used for repopulating both natural and artificial lakes and running waters. In 1 year young pike can reach 20 to 35 cm (8 to 14 in) and even more. With the development of spinning by anglers the demand for young pike for repopulating waters is considerable.

It is also possible to stock pike in the form of fry with absorbed yolk sacs, but the results are uncertain and variable. It is far better to use 6 week fingerling pike. They can be released at a ratio of 100 to 200 per hectare and at the end of the year 50 per cent of that number might be harvested.

Pike-Perch (*Lucioperca lucioperca* L.) (Fig. 260). Just as for pike, pike-perch can be associated with cyprinids. They also play a police role in the fattening ponds by destroying fry and valueless fish which might hatch and develop there. Pike-perch flesh is of excellent quality and association with second and third year carp in limited quantities not only means the destruction of undesirable fish, but at the same time produces fish with a high commercial value.

Pike-perch have a smaller mouth than pike and at the same size eat only smaller fish. Further, their growth is less rapid. They are very sensitive to handling and if draining is not carried out with care then it can lead to considerable loss, particularly among first year pike-perch. Young pike-perch are produced in special ponds (see Ch. VI) not mixed with carp but with other cyprinids such as roach, the fry of which serve as their food.

Largemouth Bass (*Micropterus salmoides* Lac.) (Fig. 266). Largemouth bass can be mixed with cyprinids practically in the same way and under the same conditions as pike-perch. The raising of bass of 1 year is carried out in special ponds (Ch. VI). Later, also as police fish, they can be mixed in with carp and other cyprinids. Naturally they should be chosen at a size which will not destroy the principal fish being raised but only the fry and undesirable intruders. Bass can also destroy tadpoles and frogs. As with pike-perch, black bass has flesh of excellent quality. They are warm water fish and support high temperature water if it is very well oxygenated.

Danubian wels (*Silurus glanis* L.) (Fig. 299). In central Europe, notably in Yugoslavia and Hungary, Danubian wels are often associated with carp and other cultivated cyprinids in the same way as pike and pike-perch. Young Danubian wels are produced in large growing ponds used for cyprinids, but generally special ponds are used (Ch. VII). During their second year they are placed as secondary fish in the carp fattening ponds. They also play a police role and at the same time produce a fast growing much appreciated fish for the table. In Yugoslavia they are released in carp fattening ponds at a ratio of 100 of 1 year per hectare, weighing from 100 to 250 gm (3 to 8 oz) each. By the time autumn draining comes round they should weigh about 1 kilo (2 lb).

Rainbow trout (*Salmo gairdneri* Richardson) (Fig. 118). Sometimes rainbow trout are placed in cyprinid raising ponds as accompanying fish. They can withstand temperatures as high as 25°C, at the surface. On condition that it is possible to ensure a sufficient oxygen content and drying out under good conditions, it is a good policy to grow trout spawners in cyprinid fattening ponds where they can find good natural food. It is also possible to place 1 year trout in the ponds to be harvested as edible fish weighing 150 to 200 gm (5 to 7 oz). However, this association is only possible in rather small ponds (1 hectare for example). The ponds should be somewhat deep so that they do not heat up too much, and emptying should be carried out without difficulty.

Generally the association between rainbow trout and cyprinids is favoured in angling ponds in which the trout intended for catching within a short time are liberated outside the summer season.

Undesirable fish. It would be exceptional not to harvest variable quantities of unwanted fish when cyprinid ponds are emptied. Generally they belong to the Cyprinidae, Cobitidae, Siluridae, Percidae, Centrarchidae and Esocidae families.

Their abundance is that much greater if the ponds are badly arranged and badly run. The fish interfere with cultivation in many ways: (1) they compete for food with the basic fish and this reduces production of the latter more or less considerably; (2) the intruders are sometimes voracious and can do a lot of serious damage in the cyprinid nursery ponds; (3) they might also be carriers of some parasites for the basic fish.

Among the undesirable fish Cyprinidae come first and principally crucian carp (*Carassius carassius* L.), roach (*Leuciscus rutilus* L.), rudd (*Scardinius erythrophthalmus* L.), bleak (*Alburnus alburnus* L.), able (*Leucaspius delineatus* Heckel), all high multipliers of slow growth and above all harmful, principally to basic cyprinids, as competitors for food. Certain Cobitidae such as the pond loach (*Misgurnus fossilis* L.) present the same inconveniences.

Then come members of the Percidae, Centrarchidae and Esocidae families. First of these is the perch (*Perca fluviatilis* L.) and the pumpkinseed (*Eupomotis gibbosus* L.), the latter imported from North America. While young they compete for food and later, as they are voracious, can do a lot of damage in the rearing ponds. When pike-perch (*Lucioperca lucioperca* L.) and, even more so, pike (*Esox lucius* L.) introduce themselves into ponds they can do a great deal of harm. The latter can wipe out all of the young carp or young tench.

Siluridae must also be mentioned and among the most harmful the catfish (*Ictalurus melas* Raf, incorrectly identified as *Ameiurus nebulosus* Lesueur) imported from North America around 1880 and which have become acclimatized in some warm waters of Europe. Omnivorous, voracious and of slow growth they are the very image of undesirable fish in ponds. On farms where the water inlet is badly controlled, in which ponds are neglected and where emptying takes place only after several years, catfish can represent, in weight, as much as half the harvest.

Although it is never possible to ensure complete protection from the harm which can be done by undesirable fish, above all the voracious such as perch and pike, it is possible to reduce it by arranging and running the ponds correctly. Good arrangement means, notably, that ponds can be emptied completely by a good network of ditches. The entry of the supply water must be well controlled preferably by a horizontally adjusted screen. Rational running of ponds requires, among other things, periodical and really effective drying out.

Fig. 192   Brood carp known as "royale". Etang du Der, Haute-Marne, France.

Fig. 193   Semi-controlled natural reproduction of carp. Rearing in two stages. Spawning pond around 25 ares ($\frac{1}{2}$ acre). Neerijse, Belgium.

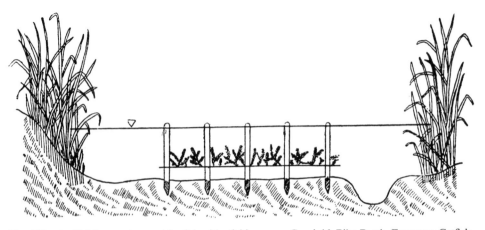

Fig. 194   Artificial spawning nest for fish with sticking eggs: Cyprinid, Pike-Perch, European Catfish (Document Jeleonski, in Kostomarov.)

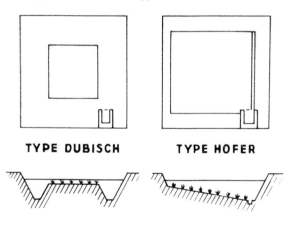

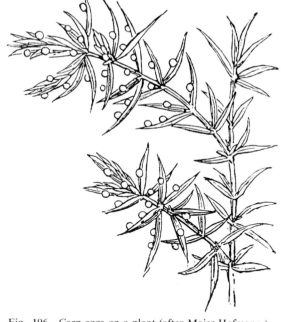

Fig. 195 Schematic drawing of Dubisch and Hofer spawning ponds.

Fig. 196   Carp eggs on a plant (after Maier-Hofmann.)

Fig. 197 Dubisch spawning ponds before filling with water. Fence made of reeds placed round the ponds for protection against wind and cold. Milicz, Wroclaw, Poland.

Fig. 198   Dubisch spawning pond; part of the peripheral ditch. Valkenswaard, the Netherlands.

Fig. 198

Fig. 199   Filled Dubisch ponds. Linkebeek, Belgium.

Fig. 200   Dubisch spawning pond. Catching carp fry. Linkebeek, Belgium.

Fig. 201   Kakabans or artificial spawning nests for carp. The kakabans are made of palm tree fibres arranged in layers pressed down by two bamboo laths. Tjisäät, Soekabumi, Indonesia.

Fig. 202 Artificial spawning of a female carp. Fish hatchery of Dinnyes, Hungary.

Fig. 203 In the background: milt ejection on eggs. Foreground: eggs and milt. On left: mixed eggs and milt.

Fig. 204 Carp eggs incubating in Zoug jars. Fish hatchery of Dinnyes, Hungary.

Fig. 205   Hatching trays for carp fry. Stretched perlon; mesh of 0·2 to 0·4 mm. Fish hatchery of Szeged, Hungary.

Fig. 206   Carp fry on trays after hatching. Linkebeek, Belgium.

Fig. 207   Counting boxes for cyprinid fry. Groenendaal, Belgium.

Fig. 208   A general view of an important cyprinid farm in Central Europe. Fish farm of Poljana, Croatia, Yugoslavia; total area 900 hectares (2,200 acres). In the foreground, holding ponds. In the background, fattening ponds from 14 to 160 ha. (from 35 to 400 acres).

Fig. 209   Nursery ponds; manuring with abundant organic manure. Fish farm of Der, Haute-Marne, France.

Fig. 210   Second rearing pond before being filled with water. Fish farm of Golysz, Cracow, Poland.

Fig. 211   Large ponds for rearing third year carp. 100 hectares (250 acres) in area. Fish farm of Poljana, Croatia, Yugoslavia.

Fig. 212   Wintering pond for first and second year cyprinids. This pond is also used as a nursery pond. Area: 1·50 hectares (4 acres). Fish farm of Bokrijk, Limburg, Belgium.

Fig. 213   Holding ponds for consumption carp. Area: 5 ares (5/40ths of an acre); depth 1·25 m (4 ft). Fish farm of Milicz, Wroclaw, Poland.

Fig. 214   Holding ponds (Halterteiche). Fish farm of Draganici, Croatia, Yugoslavia.

Fig. 215   Tench (*Tinca tinca* L.).

Fig. 216   Roach (*Leuciscus rutilus* L.).

Fig. 217   Golden ide (*Idus idus* L.).

Fig. 218 Six-week-old carp (K$_v$). Average weight: 4·0 gm (2¼ dr). Average size: 6·0 cm (2½ in).

Fig. 219 Carp of 1 summer (K$_1$) Average weight: 72 gm (2½ oz). Average size: 14·5 cm (5½ in).

Fig. 220 Roach of 2 summers. Average weight: 22·0 gm (about ¾ oz). Average size: 12·5 cm (5 in).

Fig. 221 Tench of 2 summers. Average weight: 65·2 gm (2 oz 4 dr). Average size: 13 to 22 cm (5 to 9 in).

Fig. 222   Two summers carp (K_2). Average weight: 259·0 gm (9 oz). Average size: 23·5 cm (9 in).

Fig. 223   Carp of 3 summers (K_3). Weight: 1,468 gm ($3\frac{1}{4}$ lb). Size: 42 cm (1 ft $4\frac{1}{2}$ in).

Fig. 224   Intensive rearing of carp in bamboo cages installed in a fast running stream well aerated and very rich in organic matter from sewage water. Tjibunut, Bandung, Indonesia.

Fig. 225   Carp reared tightly in bamboo cages placed in water rich in organic matter. The carp are fed with rice bran. Tjibunut, Bandung, Indonesia.

Fig. 226   Grass carp (*Ctenopharyngodon idella* Val.).

Fig. 227   Silver carp (*Hypophthalmichthys molitrix* Val.).

Fig. 228   Big head carp (*Hypophthalmichthys (Aristichthys) nobilis* Richard.).

Chapter V

SPECIAL TYPES OF FISH CULTIVATION FOR RESTOCKING

THE two preceding chapters, devoted to the cultivation of salmonids and cyprinids, dealt with species cultivated both for repopulation and eating.

The present chapter deals with the cultivation of fish belonging to families each of which is very different from the others. The principal link between them is that they are all produced for restocking: eggs, fry and summerlings. Cultivation of all the species concerned rests on artificial fertilization and incubation. In general, fertilization is the same as for salmonids. Incubation, on the other hand, is normally carried out in receptacles known as MacDonald jars or Zoug jars.

The first section describes the cultivation of pike

spread throughout Europe and North America. Very similar methods are used for the production of coregonids cultivated for restocking certain lakes in temperate regions of the northern hemisphere. In these regions the shad is also sometimes cultivated. The section which follows is devoted to a relatively limited form of fish cultivation; that of grayling destined to restock certain temperate running waters in Europe and North America. The cultivation of sturgeon which follows, is of importance, above all, to the Caspian Sea region but is less in evidence in the United States and Europe. The last type of cultivation described concerns the pejerrey or silverside. This is practised in the Argentine and in some other countries of Latin America.

SECTION I

CULTIVATION OF PIKE

The development of esociculture, or the raising of pike, is relatively recent.

While it is realized that an equilibrium between voracious and non-voracious fish must be balanced, it is also understood that the capture of voracious fish has often been excessive and that it has now become necessary to reintroduce them by artificial cultivation. In Europe, common pike being the most characteristic fresh water voracious species, it comes to mind first. Common perch, on the other hand, often possess the faculty of reproducing themselves sufficiently to maintain their own balance and sometimes become overcrowded.

In some respects the raising of pike is tied to salmonid cultivation for restocking. However, it presents different peculiarities relative to the catching of the brood fish, hatching and the raising of small pike. On the whole, raising is more delicate than for salmonids.

Esociculture is essentially cultivation for repopulation. It is difficult to imagine production of pike in ponds for eating. They feed only on living

fish and it would be difficult to obtain these economically, and next, if pike are stocked too densely they destroy themselves by cannibalism. This characteristic dominates the whole of pike cultivation.

Esociculture concerns two species: the common or Northern pike (*Esox lucius* L.) (Fig. 229) and the muskellunge (*Esox masquinongy* Mitch.) (Fig. 230). Common pike are found in Europe, Asia, and North America and are typically voracious. The muskellunge is an American carnivore which can grow to a much greater size than the common pike. They are found in the United States and Canada in the regions of the great lakes and also in Ohio, Wisconsin and Minnesota.

I   Ecological Role of Pike

After having been accused, unjustly and over decades, of being insatiable carnivores and destroyers of other fish, pike have now been rehabilitated and are also being reintroduced into waters from

151

Fig. 229   Northern pike (*Esox lucius* L.).

which over-fishing more or less eliminated them. Highly esteemed for the excellent quality of their flesh, and having few bones, they play an important economic role. They are also great favourites with anglers. Their correct place is in all still and running waters inhabited by cyprinids. They are not found in the salmonid lakes in the mountains or in the running waters of the trout and grayling zones. Living on the surface, they like calm, shallow, grassy waters where they can hide among the plants, specially among the reeds. They are found in still, warm waters in summer and in water with a slight current, particularly in lateral swamps of the barbel and bream zones.

In open cyprinid water, pike are both useful and indispensable for they transform the mass of coarse fish of only mediocre economic or sports value into good quality food, and at the same time assure better growth of the remaining coarse fish. As Walter so rightly pointed out, the presence of pike as an equilibrating factor is just that much more necessary where the population is made up of mixed species, for that complexity brings about a decline in individual growth and, thereby, the economic value of the species. Further, this complexity limits the effective control of the natural population.

In fish hatcheries, pike must be kept away from trout and also from the rearing ponds of young and small cyprinids. They can be released as fry or small pike of one year in carp fattening ponds. Here they give a supplementary production by consuming hatched unwanted carp fry.

Pike fry eat only prey which moves: first, small crustaceans: young *Cyclops*, and cladocerans and later large cladocerans, adult *Cyclops*, chironomid larvae and ephemeropteran nymphs. Before attacking, the young pike remain perfectly still for an instant; then the body curves into an S shape and the fish, in a flash, attack their prey.

After having fed on plankton, from the time their size exceeds 10 cm (4 in) the small pike become voracious and live off perch, roach and other fry of coarse fish which spawn only a short while after they spawn. They are of rapid growth and at 1 year measure from 15 to 20 cm (6 to 8 in). They can, under favourable conditions, reach 30 to 40 cm (12 to 15 in). Later they need about 5 kilos of fish to give 1 kilo of growth. Males do not grow as large in size as females.

II Capture or Rearing of Brood Fish

Pike spawn at the start of the spring, at the end of March or beginning of April, but more generally from the end of February to the end of May or as soon as the snow and ice melt and the waters start to warm up to 8 to 10°C.

They spawn in calm, shallow and grassy waters. In still waters they spawn that much earlier if the lake is small and shallow and the water heats that much more rapidly. Pike are sedentary fish which spawn where they live. They scarcely migrate. They spawn in pairs and their eggs stick to vegetation.

Brood fish are caught on the spawning grounds in open water at the proper season; but they can also come from special ponds arranged for the purpose.

In open water, the brood fish are most easily caught in the belt or on the borders of *Scirpeto-Phragmitetum*. In lakes at average or low altitudes, where the green belt is well developed, bow nets are laid. The nets are drawn daily (Figs. 231 and 232). The pike swim around the reeds, among which wattle fences direct them into the trap nets. Main hatching stations should be planned for high production with an important supply source behind them which will ensure a sufficient number of brood fish.

Parent fish find it difficult to withstand being held in artificially built tanks. Males which reach sexual maturity can be kept for a few days, but females will not generally reach their sexual maturity in captivity even if only recently caught. However, some reach maturity provided it is only a question of 2 or 3 days. The temperature of the holding ponds must not be too low and never, in any case, lower than the water in which they are intended to spawn.

It is possible to use pike brood fish from artificial ponds. They are held in ponds specially devoted to this end having passed the growth season in the ponds with small coarse fish for food. The coarse fish should be about three times the weight of the pike. At the end of winter, generally about the start of March but before the waters start warming up, the brood fish are transferred to small ponds which are well sheltered. They should be supplied with spring water of an almost constant temperature. These are the maturing ponds. The fish must be closely watched and must also be controlled each time one or other of the pair shows signs suggesting the start of natural spawning. In order that spawning should be regular it is essential that the temperature of the pond should be both sufficiently high and constant. Cold spells or sudden heat can be very harmful and create the risk of stopping or definitely compromising the maturity of the female.

This method has been followed for more than 10 years at the experimental fish station of Linkebeek (Brussels) but only on a small scale. The raising pond covers about 7 ares and each year 18 females of 3 years are liberated. They should weigh around 18 kilos (40 lb) at the end of the year. Near March 1 they are transferred to the maturity pond which covers about 6 ares. Generally about 12 to 15 females spawn, that is two-thirds, and they give, on average, a total of 250,000 eggs. Some females do not reach maturity at all or spawn in the pond before they have been controlled. Females which spawn normally give from 20,000 to 25,000 eggs per kilo (2 lb).

III   Artificial Incubation of Pike Eggs

1   The shed and the fish cultivation apparatus

If the equipment generally used for the hatching of salmonids is completed with a few extras, it can also be used for esociculture.

Incubation does not take place on trays, on which the eggs do not move in the same way as for trout, but in jars in which the eggs are constantly slowly moved round and round (Figs. 235 to 238).

Most often Zoug jars are used. These are large inverted bottles which were given their name because they were first used on the shores of Lake Zoug (Switzerland). The jars, open at the top, terminate at the bottom in a narrow neck. They are set up vertically, the neck turned downwards, linked to a pressurized water pipe and separated from it by a tap topped by a nozzle. This nozzle supports the jar and retains its equilibrium. In some installations the jars are linked to the water pipe by a rubber pipe and held in position by means of a collar. Each is provided at the bottom end with a fine 1 mm mesh screen which stops the eggs from descending into the pipe. The normal height of the jars is around 60 to 70 cm (24 to 27 in) and the diameter is from 15 to 20 cm (6 to 8 in). The jars hold from 6 to 8 litres (12 to 17 pt) and take about 1 to 5 litres of eggs. The average flow is around 4 litres per minute (7 pt) but it can vary between 2 and 6 litres per minute (3 to 10 pt). The water flows over the top of each jar and runs down the outside.

Other models exist. One variation is a jar provided with an outlet at the top (Fig. 235). Kannegieter glass or plastic vases, which are based on and are an improvement of the old von dem Borne incubators which were made in metal, are also used (Figs. 239 and 240).

Instead of Zoug jars, MacDonald or Chase jars (Fig. 265) can be employed. They are in current usage in America for the incubation of pike eggs as well as walleye and coregonids. They are cylindrical glass vases, provided with feet, between 40 and 50 cm high (15 to 20 in) and 15 to 20 cm (6 to 8 in) in diameter. The water is carried inside to the bottom by a pipe or tube and the flow keeps the eggs moving slowly and continuously just as in the Zoug jars. The eggs occupy only one half of the jar.

Certain pike farms also rear salmonids and this can be their principal cultivation. Others work with pike only for restocking. In this case the water used is often taken from the lake to be restocked. The farm is installed on the lakeside and the water is pumped or held in a reservoir before being supplied to the hatching jars. This lake water must be well filtered.

The hatching shed (Fig. 237) should not be too dark. Screened daylight is not harmful to the eggs but they must be protected from the direct rays of the sun.

The water used for incubation must be clean, well oxygenated and the temperature must be adequate. The best temperatures are between 8 and 15°C for the hatching of pike eggs. On those farms using lake water the temperature is normally satisfactory. The water heats up gradually passing from 8 to 9°C at the start and then from 12 to 14°C at the termination of the hatching. Spring or brook water may be too cold and in that case it should be artificially heated.

2   Artificial fertilization

The artificial fertilization of pike is carried out in the same ways as dry fertilization for salmonids. Although females are sometimes ripe at 2 years of age they are not normally stripped until they are 3 or more. Males, on the other hand, are smaller and are ripe 1 year earlier. They can be used from the second year.

Pike eggs are much smaller than salmonid eggs, no more than from 2·5 to 3 mm in diameter. A litre (pt) of eggs generally totals between 50,000 to 70,000 at the time of spawning but they swell quickly and, from the day after spawning, 1 litre content will fall to between 35,000 to 45,000 eggs. Females can measure 1 m (3 ft) and even more and at spawning time their bellies distend considerably. The weight of the ovaries at this period is almost one-third of the total weight of the whole fish.

Females must be really ripe so that when they are held vertically with their heads pointing up, the eggs should flow almost by gravity from their bodies (Fig. 233).

Males reach maturity earlier than females and often towards the termination of the spawning period it is impossible to run milt out of them. They may have very little milt and for this reason artificial fertilization must be carried out very carefully (Fig. 234). But during the spawning period they can give milt more than once. If there is a shortage of males and little milt, it is possible to draw it off with a pipette.

The eggs and milt should be covered with 2 to 3 cm (from ¾ to 1 in) of water in the pan in which they are placed after mixing. The spermatozoa live from 60 to 120 seconds at temperatures between 15 to 5°C. Everything must be carried out with the greatest care. Above all, neither the eggs nor the milt should make contact with the water before mixing. They must be quite dry. To this end, the back half of the females and above all that of the male must be carefully wiped before stripping and fertilizing. Under these conditions only a small amount of milt is necessary. According to Lindroth, 0·02 cm³ of milt only is necessary for 1 litre of eggs. This small quantity nevertheless contains 400 million spermatozoa. For trout there would be only half the amount but that would still be enormous.

After fertilization the eggs are left to rest for 20 or 30 minutes. They are then carefully washed. As pike eggs are far more delicate than those of trout they must be handled with the greatest care. When washed they are poured, very gently, into the incubation jars. Those which stick to the sides should be detached very carefully with the help of a feather attached to a stick.

3   Incubation

During incubation the flow of water in the jars must be regulated in such a way as to ensure that the eggs are stirred gently without being spilt. They are carried up by the flow of water and they tend to descend because their density is greater than that of the water. The flow must be regulated to ensure that when the eggs ascend to half or two-thirds of the height of the jar, they fall to the bottom where they are carried up again by the current and the circuit re-starts. From the second day after fertilization and for 2 or 3 days the eggs will stick a little one to the other or to the sides of the jars. At this stage they should be left alone and not be separated forcibly at all as this might kill them. They will come unstuck on their own. They may be helped very gently with a feather, however.

The incubation period for pike eggs is much shorter than that of salmonids: an average of 120 degree days. After 30 degree days the embryo can, already, be seen clearly.

The number of unfertilized eggs is higher than for salmonids. It can indeed be 10 or 20 per cent and even more than 30 per cent. Regular control of incubation is possible by means of samples taken and examined in a glass tube or placed in a plankton chamber. Because of the great number of eggs and the relatively high loss level there can be no question, as in the cultivation of salmonids, of removing the bad eggs individually. However, they can become covered with fungi and so contaminate eggs in their vicinity. But as they form patches which tend to rise to the surface from the mass they sometimes eliminate themselves and in any case can normally be siphoned off (Fig. 236). This is done with a rubber tube provided with two glass tubes, one at each end. Siphoning starts by filling the tube with water after which one end is held with one hand for seeking out the spoiled eggs while with the other hand the tube is pressed, more or less, to remove the other eggs by suction. If loss is as high as 30 per cent then the incubation has been carried out under bad conditions and it is best to eliminate the whole spawning.

When the eggs reach 70 degree days it is time to despatch them if they are to be disposed of as eyed eggs. This phase must be respected if failure is to be avoided.

The eggs do not remain in the Zoug jars right up to the completion of the hatching. Just before that, when the eggs are about 110 to 120 degree days and first hatchings are noted, they are siphoned out of the jars and placed on hatching trays similar to those for salmonid cultivation (Fig. 237) except that the holes do not exceed 2 mm in diameter. The fry are supported on the trays at the moment of hatching which would not be the case if they remained in the jars in which they move constantly. Besides, after hatching, the fry have a tendency to fall to the bottom of the jars where they could not absorb their yolk sacs. It is possible to lay out at least 10 eggs per square centimetre, that is some 20,000 eggs on a tray 50 × 40 cm (20 × 16 in).

However, before taking them from the jars and placing them on the trays they must be counted. The water flow is stopped and the eggs fall in heaps. After a few moments it is easy to read the volume they occupy because the jars are graded. A sample is taken as a base, that is an average of from 35 to 45 eggs per cubic centimetre. This shows they are far larger than they were at spawning when there were from 50 to 70 per cubic centimetre. After that, the number of bad eggs is established in the following manner. A glass tube about 3 mm in diameter is plunged into the heaped eggs while one end is stopped with the thumb. When the thumb is removed the eggs will enter the tube. This is carried out three or four times and each time the number of bad eggs per 100 is counted (Fig. 242). The hatching loss being normally slight it is possible to know the number of fry each tray will produce from the volume of eggs placed on the trays. Hatching is rapid. When it is over, the empty shells are siphoned off.

At hatching, the pike fry or larvae (Fig. 241)

do not resemble adult fish at all. They are grey-yellow in colour and measure about 8 to 9 mm in length. With the exception of the pectoral fins, the other fins are not shaped. Neither the mouth nor the opercula are pierced and the gills are unformed. On the anterior part of the head, in front of the eyes, the fry have adhesive papillae, thanks to which they can stick vertically (Fig. 243) and remain suspended.

The absorption of the yolk sac takes about 160 to 180 degree days which means it takes longer than hatching which is around 120 degree days. A few hours after birth the fry leave the bottom of the trays, where they fell immediately after hatching, and attach themselves vertically by their papillae to the sides of the troughs, In order to provide a greater surface to which the fry can suspend themselves, a few plants are submerged in the troughs: starwort, horn-wort, and Canadian pondweed.

The fry do not remain perfectly still and can be seen swimming for an instant and attaching themselves again or resting on the bottom of the trough. Their internal structure gradually modifies. Colours deepen and the fins develop. Then, suddenly, the fry abandon the vertical position and take up a normal swimming position. The rest of the remaining yolk sac is rapidly absorbed and it is time to release the fry. From fertilization to release, the time taken for the development of the egg and the fry is around 300 degree days.

When fry take up a normal swimming position it is time to change their environment. This will often be in open water for repopulation: a lake or a water course. Sometimes the fry are taken to rearing tanks or ponds in order to produce pike fingerlings for restocking.

IV Liberating Fry in Open Water

Pike fry are transferred to their new habitat when their yolk sacs are almost absorbed. The alevin must be able to swim freely in the water and it is not necessary to wait until the yolk sac is totally absorbed. Provided that absorption is quick then the optimum releasing stage is short.

The fact that the fry are released in open water with their yolk sac absorbed and that they are already agile, means their chances of survival are good. In effect, it is during incubation and the absorption of the yolk sac, that is when they are passive, immobile and suspended from reeds, etc, that eggs and fry risk destruction by their natural enemies such as small perch and other fish as well as carnivorous insect larvae. It is best to liberate fry with their yolk sac absorbed and after they have developed their reflexes and have enough speed to escape, than when the eggs are eyed as has been recommended at times.

Fry, carried in cans or in plastic bags, are distributed regularly in small lots and in shallow water which should be calm and grassy (Fig. 244). The best way to release fry is to use a boat, to follow the lines of reeds and to employ a siphon or a small net about the size of a soup ladle with stretched muslin at the bottom. A mug can also be used. It is absolutely fundamental to success that the fry should be carefully distributed in grassy spots which are rich in plankton. The results will not be satisfactory in spots devoid of plants (such as canals or rectified water ways). This is also true of those spots where the water is turbulent. Before liberation, the temperature of the water must be equalized carefully by mixing. For this reason, releasing the fry takes time.

The exact number of fry which should be liberated along a given length of bank has not been determined. In running water, according to Prud'homme, if 10 per cent of the fry released become adults and give one pike per 10 m (32 ft) for a river with average flow, then one fry per metre of bank (1,000 per kilometre) ($\frac{5}{8}$ of a mile) seems to be the logical maximum.

For repopulating fairly large lakes the best thing is to have a stocking station situated on the banks of the lake. For repopulating small lakes and running waters of the bream zone, it is not possible to have a large number of hatching stations, in so far as the latter can only be established near to the centre where the brood fish are caught, and such centres need not be very numerous. The best policy is to establish several principal hatching stations near important brood fish capturing centres. From these stations, if the distances are not too great, the fry with absorbed yolk sacs can be transported direct at the time of release.

It is also possible to carry this out in two stages. The first is in the principal incubation station up to the time the eggs are eyed and can be transported. At this stage, they are taken to secondary hatching stations in which only hatching trays are installed. This station can be supplied with spring water, brook water, lake water or by water from the course to be repopulated. The clarity of the water is not of the same importance as for the hatching station.

Hatching eyed eggs received by the station just previously, holding the fry during the absorption of the yolk sac and releasing them, form a series of simple operations easily carried out by a secondary station.

V Production of Fingerling Pike

Instead of liberating fry with absorbed yolk sacs, which in any case is risky, it is better to liberate fingerling pike. Losses are normally small.

The first efforts to produce fingerling pike were inspired by salmonid cultivation methods which included harvesting fry at the "summerling" stage; that is at least at 6 months. It did not give encouraging results. This was due to the feeding habits of pike, which are carnivorous from an early age.

The production of fingerlings depends on the following essential principles. First, pike, no matter what their age, feed off living prey only and from the moment they reach from 6 to 10 cm ($2\frac{1}{2}$ to 4 in) they are carnivorous. Because of this it is impossible to feed young pike with non-living food whether it be meat or fish. They can only be given living prey. Nor can they be fed dry concentrates. Nevertheless, experiments with dry concentrates have been carried out successfully with pike fingerlings in the United States (Graff, 1968).

Till they are from 6 to 10 cm, according to the density of the raising, the young fish, fed first on plankton and minute aquatic fauna, do not need to reserve for themselves such vital space which would lead them to destroy any neighbour considered too close. Consequently, it is possible to raise young pike in crowded conditions until they reach a size of 6-10 cm. However, beyond that size they must be thinned out and left with their own living space.

Taking this into account, three methods are used for producing young pike. First, rearing is from 3 to 4 weeks in rearing troughs. They can be well packed. Then comes rearing in ponds without feeding, fish of from 6 to 8 weeks, and finally, subsidiary extensive raising of one-summer pike in ponds.

1 Raising fingerling pike of 3 or 4 weeks in rearing troughs

Until they are 3 or 4 weeks of age, which equals 3 to 5 cm (1 to 2 in), young pike can live packed

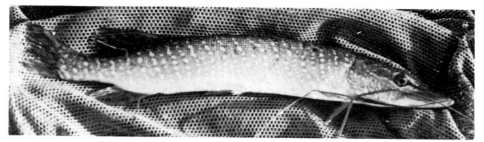

Fig 230   Muskellunge (*Esox masquinongy* Mitch.)

Fig. 231   Fig. 232

Bow-nets used for catching brood fish pike in the reeds border of a lake. Fish hatchery of Stäfa, Zurich, Switzerland. (Fig. 232: Photo Hey.)

Fig. 233   Artificial spawning of a female pike.   Fig. 234   The pike eggs are sprinkled with milt.

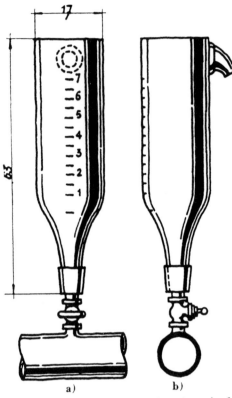

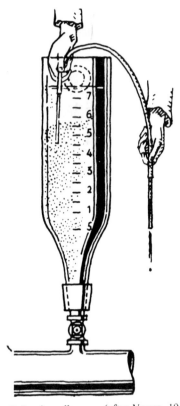

Fig. 235 Zoug glass incubator with a lateral nozzle: front and lateral views (after Vouga, 1938).

Fig. 236 Siphoning pike eggs (after Vouga, 1938).

Fig. 237

Fig. 238

Incubation shed for pike eggs using Zoug glass jars. On the left (Fig. 237): a general view; in the foreground: hatching troughs; in the background: Zoug jars. On the right (Fig. 238): incubating eggs in Zoug jars. Fish hatchery of Limal, Belgium.

Fig. 239

Fig. 240

Fig. 242 Egg control.

Plastic incubation funnels (Kannegieter type) for the incubation of pike eggs and eventually of whitefish or coregonids. Top left (Fig. 239): a double series of funnels. Top right (Fig. 240): a close view of an incubation funnel. Fish hatchery of Nieuw Vennep, the Netherlands.

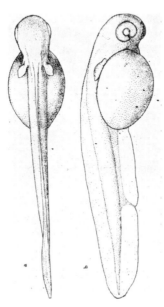

Fig. 241 Day old pike fry (after Vouga, 1938).

Fig. 243 Pike fry on plant fragments and also sticking to the sides of a glass pan.

Fig. 244   Tipping pike alevius after reabsorption of the yolk sac, into a pond, following homogeniz-
ation of the water in which they were transported and the water to be restocked.

Fig. 245   Long, concrete feeding tanks used for the intensive production of small pike fingerlings of
3 or 4 weeks from 3 to 5 cm (1½ to 2 in) long. Fish hatchery of Nieuw Vennep, the Netherlands.

Fig. 246 Long feeding tanks for small pike fingerlings from 3 to 4 weeks. Scharfling, Austria.

Fig. 247 Gathering plankton with which to feed fry. Lake of Scharfling, Austria.

Fig. 248 Flat, shallow, grassy pond for the production of pike fingerlings from 6 to 8 weeks. Valkenswaard, the Netherlands.

Fig. 249 Pike fingerlings of 6 weeks: 83 per cent from 7 to 10 cm ($2\frac{3}{4}$ to 4 in) and 4·2 gm ($2\frac{1}{4}$ dr); 13 per cent from 11 to 13 cm ($4\frac{3}{8}$ to $5\frac{1}{2}$ in) and 7·7 gm ($4\frac{1}{2}$ dr); 4 per cent from 14 to 19 cm ($5\frac{1}{2}$ to $7\frac{1}{2}$ in) and 20·0 gm ($11\frac{1}{4}$ dr).

together without showing important signs of cannibalism. In order to keep them under these conditions and also avoid cannibalism they must be fed abundant, appropriate food. They are given zoo-plankton taken from rich waters. Under these conditions the station must be situated near a lake where abundant quantities of plankton can be harvested (Fig. 247). A pond rich in plankton can also be satisfactory since it can be helped by appropriate manuring.

The rearing troughs, built in durable materials, are either long in shape (Figs. 245 and 246) and have a longitudinal current, or they are circular with a tangential current. These troughs can be stocked up to 2,500 per square metre and can give a 60 per cent survival rate which corresponds to a production of 1,500 fingerling pike of from 3 to 4 weeks per square metre. This kind of installation exists notably in Austria, Holland, and Switzerland. In Holland, concrete troughs are used from 375 × 175 × 45 cm (12 ft × 6 ft × 1 ft 6 in) (van Drimmelen, 1968). As the raising period is limited to a few weeks the same troughs can be used twice during the course of the season – that is, for early rearing and for late rearing successively.

The rearing troughs must be cleaned carefully every day. It is recommended to disinfect regularly the equipment used, with quaternary ammonium compounds. A bath of malachite green is sometimes given preventively. In order to treat infections, Steffens (1976) recommends the following:

Trichodina: bath of table salt (0·4%) for 10 hours.

Costia and *Chilodonella:* bath of formalin (200 ml or 31 in³/m³) for 30 minutes.

Ichthyophthirius: bath of malachite green (0·15 g/m³) for 2 hours.

Myxobacteria: three successive 10 hour baths of trypaflavine (3 g/m³).

In order to prevent the viral disease caused by the *Rhabdovirus* group, eggs can be treated with iodophores. Gerard (1974) recommends a 10 minute bath at the ratio of 50 mg of active ingredient per litre of water.

2 Raising fingerling pike of from 6 to 8 weeks in ponds

The fry are normally released in the ponds at the end of April and the fingerlings are harvested in mid-June. The average duration of rearing

is 6 weeks but in effect it can last from 5 to 8 weeks.

Ponds used for the production of fingerling pike must be flat, rich in plankton and easily warmed (Fig. 248). Cyprinid second rearing ponds serve very well and can, in this way, give the fish farmer a supplementary and appreciable income. In fact successively, and in the same ponds, first fingerling pike of high commercial value are raised and then young carp of 1 summer.

Naturally, draining must be complete for if any young pike were left they would, subsequently, cause havoc among the carp fry released a little later. After harvesting the pike, the ponds are left dry for several days and quicklime is spread over such patches of water as may still exist. The multiplication of natural food in the pike ponds is helped by spreading manure such as compost and dung. Good results can also be obtained if grass is sown in the pond before it is placed under water. When it rots it will provide abundant plankton.

Under normal conditions between 10,000 and 25,000 fry are stocked per hectare in the ponds and the normal harvest gives between 2,000 and 5,000 young pike per hectare or an average survival rate of 20 per cent. The operation is, however, chancy since the return can vary from between 5 and 50 per cent. It depends largely on atmospheric conditions which can favour or harm the production of natural food indispensable to the fry. For example, cold spells after liberation are very harmful. In order to avoid this risk it is best to liberate the fry late, that is during the second half of April. Average size of 6- to 8-week-old pike is from 9 to 10 cm (3½ to 4 in), but the largest can reach 15 cm (6 in) and even more. This indicates that they have eaten the smaller fish, a fact which is found regularly at each draining. Average weight is 5 gm and outer limits lie between 1 and 20 gm.

3 Raising fingerling pike of 1 summer or of 1 year in ponds

This is a supplementary production, for after reaching 2 months pike become carnivorous and cannot be packed together. But it is possible to raise pike up to 1 summer or 1 year in ponds in which they are not a risk for the principal fish being raised. This is practised with carp of 3 years. In this way a secondary production of important economic value is realized and at the same time provides a means of controlling unwanted fry such as perch which may have hatched or which have managed to enter the ponds.

If carp are the principal fish being raised and if the water inlet is protected by an appropriate device such as a good, horizontal sunken screen, the young pike will not find small fish in the pond on which to feed and consequently at the end of their growing stage they will be rather small themselves. But it is not the same if the operation is mixed and includes cyprinids, roach and tench for example, which can reproduce in these ponds.

Young pike can be liberated, as fry with absorbed yolk sacs, at a ratio of about 500 to 1,000 per hectare (Anwand-Grohmann), but the operation will be more certain of success if the pike are a few weeks old and are placed at a ratio of about 10 to 50 per hectare. In the autumn they will measure from 20 to 40 cm (8 to 16 in) and can weigh up to 200 gm (7 oz) and more. Losses are slight.

VI   Transport of Eggs and Fry

It is possible to transport pike eggs before or just after fertilization. Eyed eggs and fry with absorbed yolk sacs can also be transported.

1. Transport of unfertilized eggs. Sometimes ripe females and males are captured on the spawning beds. The female is stripped of her eggs by the dry method. These are sprayed with milt and then carried mixed, and without water, in jars to the station where fertilization is completed. If the time taken to transport the eggs does not exceed 1 hour then fertilization should normally result.

2. Transport of fertilized eggs. Immediately after fertilization the eggs, having been carefully washed, can be transported. They are carried in receptacles filled with water. The eggs are poured in very gently and then one-third of the water is removed. The ratio for transporting is one part eggs to three parts water. Transport should never take longer than 8 hours.

3. Transport of eyed eggs. This is done immediately after the eggs have reached 70 degree days. As with recently fertilized eggs they are carried in receptacles one part eggs to three parts water and temperature is maintained from 8 to 4°C by means of moss and ice.

Eyed eggs can also be transported on trays in the same way as trout eggs, using packing cases which have been carefully insulated and lagged. They can be carried over long distances. In this way it has been possible to transport eggs from Switzerland to Morocco.

4. Transport of fry. They should only be transported when they can swim freely in the water. The fry are transferred with great care in transport cans or plastic bags. The water must be kept fresh. From 300 to 600 fry can be transported per litre of water.

SECTION II
CULTIVATION OF COREGONIDS

Coregonids live in the temperate fresh waters of the northern hemisphere. They are lake dwellers. In Europe they are found in the great alpine oligotrophic lakes (lakes of Geneva, Constance, Neuchâtel, etc.), as well as in some lakes of the Baltic plain. They need fresh and pure water. Under similar conditions coregonids are also found in Asia and North America.

Coregonids are related taxonomically to salmonids but differ from them. The mouth is small and the colour is different; light and brilliant, silver on their sides and bellies. They also have relatively large scales. The flesh of this fish is greatly appreciated.

They belong to two principal groups. One is plankton eating (*Coregonus albula* L.) while the other, at the bottom, lives off the nutritive fauna

(*Coregonus lavaretus* L.) (Fig. 250). Each lake has its own special type of coregonid. There are a great many different types but they are considered as purely local since they can change if transported to different water.

The cultivation of coregonids is very similar to that of pike. They are exclusively cultivated for repopulation and have developed very well for restocking certain lakes, where they are the most important species as far as the fishing industry is concerned.

I   Capturing the Spawning Fish

Coregonids spawn during almost the same period as brown trout, that is from the end of October

Fig. 250   Whitefish (*Coregonus lavaretus*) L., "Fera" type from Lake of Geneva.

to the end of December. The principal period is between November 15 to December 15.

They differ from one lake to another being subject to regional climate and the shape and conditions of the lake. In general, the earlier the lake cools in the autumn and rapidly reaches a homothermy corresponding to the completion of the full turnover, so coregonids will spawn early. But the species differ; *C. lavaretus* spawn somewhat later than the others.

Coregonids are caught by means of gillnets or trail nets, but preferably by the former. The nets are laid out in the evening on the spawning grounds and the living fish are harvested in the morning.

In a given lake the spawning period rarely lasts longer than 10 days. At the start more males than females are captured, but this is reversed at the end and sometimes leads to problems for the good fertilization of the eggs.

The weight and size of the brood fish vary according to the species. *Coregonus albula* can weigh between 30 and 200 gm (1 to 7 oz) while *Coregonus lavaretus* can vary from 1 to 3 kilos (2 to 6 lb). As for age, this is generally 3 years and more.

The diameter of the eggs also varies according to the species and is between 1·6 to 2·4 mm (*C. albula* L.) and 3·1 to 3·7 mm (*C. lavaretus* L.). The number of eggs per litre is from 180,000 to 80,000 for the first and 31,000 to 27,000 for the second. According to Schäperclaus, a female of the first species gives from 2,000 to 10,000 eggs, and the second 10,000 to 19,000 per kilogramme of weight. In any case at fertilization the female never gives all her eggs.

II   Artificial Fertilization and Incubation of Coregonid Eggs

It is essential that the eggs should be quite ripe at the time, if artificial fertilization is to succeed. Indeed, for that reason, the parent fish must be captured on the spawning grounds at the time of reproduction. It is easy to recognize the ripe eggs for they strip easily from the body of the female. Also they are light and transparent. Eggs which are not sufficiently ripe are white and not transparent.

If the brood fish are not quite ready at the time of capture they can be held in storage, but only for a few days. It is also a good thing to hold some males in reserve to meet a possible, subsequent deficit.

The artificial fertilization of coregonids is carried out in exactly the same way as for salmonids and pike, that is to say by the dry method. As for incubation this is the same as for pike, that is in Zoug carafes (Fig. 251) or MacDonald jars. The water is regulated in such a way as to ensure a slow and regular mixing of the mass of eggs. Only exceptionally are coregonid eggs hatched on trays in the same way as salmonid eggs.

The percentage of unfertilized eggs is rather high. Quartier estimates it as around 25 per cent, but this can be lowered if the water is filtered and is left untroubled. Loss can also be higher. But even in this case it is possible and advantageous to make use of the fertilized eggs. When the unfertilized eggs are clearly visible, which is possible 2 or 4 weeks after fertilization, it is essential to seek them out and remove them, for if they are not they will be attacked by *Saprolegnia* and this will be a

Fig. 251 Zoug glass jars for incubation of whitefish eggs, and hatching tanks. Fish hatchery of Stäfa, Zurich, Switzerland. (Photo Hey.)

risk for the healthy eggs which could be contaminated.

The way to eliminate the eggs more or less automatically is to reduce the current in the jars so that the mass of eggs continues to move. As the dead eggs are somewhat lighter they will tend to gather on the surface of the mass. They can then be siphoned off and, if worth it, the healthy eggs, also siphoned, can be put back in the jars or transferred to other jars to be hatched separately.

As opposed to eyed pike eggs it is possible to leave coregonid eggs in the jars until hatching. This takes a rather long time, between 300 and 360 degree days. The eggs become eyed after 160 degree days. According to Ammann and Steinmann the relatively low hatching temperature of from 3 to 5°C is not only an advantage but is practically indispensable. Many early failures, with coregonid fry were caused, without doubt, by hatching in water of too high a temperature: spring water for example. Such water speeds up hatching and the fry, less vigorous than those hatched in colder water, are liberated too early in water without the natural food necessary for the young coregonids.

At the time of hatching the current lifts the fry out of the jars. They fall into rearing troughs under the jars and are held by a very fine mesh screen. They measure only 10 mm in length and are less than 1 mm in diameter. Several jars can be placed together so that all the fry fall into the same trough.

III Liberating Coregonid Fry

Coregonid fry have very small yolk sacs which are rapidly absorbed, so from 3 to 5 days after hatching they should be released in the water to be restocked.

They are liberated in lots and in shallow water well removed from the normal habitat of pike and perch. According to Schäperclaus, normally success is only 10 per cent after releasing and sometimes as low as 1 per cent. This is a high loss and there is still plenty of room for progress in the techniques employed in liberating coregonid fry. The ideal spot for liberation, either the littoral zone above the limnetic zone, or at the limit between the two should be more precisely known for the different species of coregonids.

Instead of liberating fry with absorbed yolk sacs, fish a few weeks old, and which have been fed, can be liberated. But it is difficult to raise such fry for they live off natural, living food exclusively, that is, plankton. It is possible to concentrate the latter in small ponds or even in nursery troughs by pumping or by fishing out of the lakes to be repopulated, as is done for the production of young pike.

IV Transport of Eggs and Fry

It is possible to transport the eggs at the time of fertilization or when they are eyed. The fry are transported at the time of restocking.

Before fertilization, the eggs and milt can be

transported separately, always dry, to the incubation shed. However, it is easier to carry eggs which have just been fertilized, in a receptacle containing water and the eggs. Before transporting, the eggs must be carefully washed in order to avoid fermentation during transportation. In a container filled with 15 litres (26 pt) it is possible to carry from 3 to 4 litres (5 to 7 pt) for a period of from 6 to 8 hours. During transportation the can must be protected against wide and sudden changes of temperature. On arrival at the hatching shed the eggs must accustom themselves slowly to their new environment and this is done by replacing, gradually, the water in the container with incubation water. This operation can take from 30 to 60 minutes. Finally the eggs are poured very gently into the incubation jars.

When the eggs are eyed they can be carried on frames in the same way as trout eggs. The transported eggs are counted volumetrically. The volume occupied by 1,000 eggs in a graduated glass filled with water is measured. Otherwise the total quantity to be dispatched is measured in a larger graduated glass.

Finally the fry can be transported at the time of release. Receptacles with a capacity of 15 to 20 litres (26 to 35 pt) can carry 1,000 fry per 1 or 2 litres. But before liberating them in the water to be repopulated the temperature must be equalized.

V   Cultivation of Coregonids in Czechoslovakia

In Czechoslovakia the cultivation of coregonids has peculiarities of its own. The fish are not indigenous.

In 1882, Susta, a well-known fish scientist, introduced *Coregonus lavaretus maraena* Bloch (Grosse Maräne) fry which originated in Madüsee, a Pomeranian lake, into ponds in the region of Trebon in Bohemia. The result was a scarcely-hoped-for success. Since then, coregonids have been reared in lakes in the region. At 4 years they reach 1 to 1·5 kg (2 to 3 lb) and sometimes even more.

The species does not reproduce in ponds and has to be bred artificially. The parents are chosen from fish for eating aged 4 years and, exceptionally, 3 years. Spawning takes place in the middle of November and a female weighing 1·5 kg can give 20,000 to 40,000 eggs with a diameter between 2·6 and 3·2 mm. The eggs are hatched in Californian incubators with water at low temperature around 3 to 4°C. Hatching starts about the beginning of March. Eight days later the fry are released in the ponds.

Fish of 1 summer which feed off plankton are harvested in the autumn. They measure from 12 to 14 cm (about 5 to 6 in) and weigh around 10 to 15 gm. They are wintered in special ponds and are separated from other species. In the spring they are placed in ponds where they stay, normally, for 2 years, living off plankton and insect larvae. At the end of this period, if their weight is insufficient, they are kept until they are 4 years old. They are sold for eating at weights varying from 0·7 kg (1 lb 8 oz) to 1·5 kg (3 lb). Generally coregonids are smoked. At the time of draining, coregonids must be handled with care for, like pike-perch, they cannot withstand rough handling or turbid waters.

SECTION III

CULTIVATION OF SHAD

Shad are clupeids and are anadromous migrants like salmon. After spawning in fresh water the young shad stay to complete their early growth and then return to the sea to continue growing, which can last for from 2 to 4 years for males and from 3 to 6 years for females.

They swim upstream in the spring, from March to May, and spawn in June in waters with a relatively high temperature around 17 to 18°C. Shad spawn in noisy shoals at night and in the

principal rivers or their tributaries. They spawn not far from the banks and in the shallower depths, and this allows massive captures.

As they frequent the lower courses of the great rivers, shad have been greatly reduced by pollution.

In cleaner waters, efforts have been made to remedy this depopulation by the use of artificial fertilization. This is rarely practised in Europe now but in the United States it is more current.

The principal species in Europe is *Alosa alosa* L.

In North America several species are found principally in the area of the Atlantic slopes. The principal one is *Alosa sapidissima* Wilson. It was introduced into the region of the Pacific slopes.

Cultivation of shad is specifically for restocking like that of coregonids.

The brood fish are captured on the spawning. grounds. Artificial fertilization follows immediately Shad eggs are small, no more than 1·5 to 2 mm in diameter. A female can give 40,000 to 50,000 eggs per kilogramme (2 lb) of weight.

After fertilization the eggs are incubated in Zoug or MacDonald jars. When the temperature is around 18 to 20°C incubation is rapid and takes no more than 6 to 8 days. After hatching, the fry have only very small yolk sacs which are absorbed very rapidly. The fry must be released in the water for repopulation, immediately.

SECTION IV
CULTIVATION OF GRAYLING

1 Graylings

Graylings, like coregonids, belong to the small-mouthed group of salmonids and are part of a special family, the Thymallidae. The genus *Thymallus* is, among others, characterized by a large and high dorsal fin and also by its relatively large scales.

Grayling are found in Europe, Asia and in North America. The species in Europe is known as common grayling (*Thymallus thymallus* L.) (Fig. 252); but it is not found in the Spanish peninsula nor in Ireland. While grayling are found in certain lakes of Scandinavia in which they can even preponderate, they are primarily river fish. They are characteristic of what is called the grayling zone which belongs to rheophile running waters between the trout zone and the barbel zone. The grayling zone (Fig. 253) is characterized by alternating between rapid currents and sections with more moderate currents. The beds of the water courses are stony and gravelly. In the typical water of the grayling zone the most abundant species are brown trout, common grayling and running water cyprinids – chub, dace, barbel and "hotu" (*Chondrostoma nasus*); the cyprinids preponderate in the lower part of the zone. In North America *Thymallus signifer* Richardson is the species living in certain water courses of British Columbia, Alaska and the upper Missouri.

2 Artificial reproduction of grayling

Grayling spawn in the spring from March to May but above all in April. The reproduction period is tied to the temperature of the water which should be around 10°C. Grayling spawn in pairs on the high gravel and shallow beds.

The cultivation of grayling resembles that of properly - called salmonids. Nevertheless, they present certain peculiarities. The first is that the brood fish must be caught in open water, for grayling do not support captivity well, at least when they are adult. The best way is to catch them

Fig. 252   Common grayling (*Thymallus thymallus* L.).

Fig. 253   View of the river Ourthe occidentale between Ortho and Bertogne (Belgian Ardennes). Example of a small river in the grayling zone.

with electrical devices on the spawning beds at the time of reproduction. On the other hand, in Yugoslavia (Svetina, 1958) it has been possible to bring grayling to sexual maturity by keeping them for 2 months after capture in a by-pass channel installed parallel to a water course of the grayling zone.

Spawning and fertilization are the same as for salmonids. Sexual maturity is generally reached at 2 years for males and 3 years for females. Grayling eggs are rather small, measuring from 2·5 to 3·5 mm in diameter and there are from 3,000 to 6,000 eggs per kilogramme (2 lb) of female. Hatching can be carried out on trays as for trout, or in Zoug glass jars as for coregonids. As grayling eggs are somewhat delicate they cannot be kept close together on the trays. As their density is slightly greater than that of water, hatching in jars gives good results. When the eggs are eyed they are transferred to trays for hatching. The perforations in the trays must be small because grayling fry are very small. The period between fertilization and hatching is between 180 and 200 degree days.

3   Production of grayling fingerlings
Young grayling of 1 summer or grayling fingerlings can be raised in the same way as trout fingerlings. Success is, however, more difficult and to be sure of it the water must not be too cold though it must be rich in natural food.

A good temperature for raising is around 15°C. Raising with success has been tried in troughs with liver and spleen being distributed as food as well as *Daphnia*, produced in special tanks, and plankton fished out of lakes and ponds. The simplest way is natural raising in rearing ponds which are very rich in natural food; in this case particularly well in calcareous water. From 5 to 10 fry can be released per square metre. If the food is plentiful – the young grayling grow rapidly and reach from 10 to 12 cm (4 to 5 in) in the autumn – then they are harvested at 1 to 2 per square metre.

SECTION V
CULTIVATION OF STURGEON

I Sturgeons

Sturgeons are found in the northern hemisphere, in Europe, Asia and in North America. The family includes four genera, the principals of which are *Huso* (two species) and *Acipenser* (18 species).

Once important in numbers but now disappearing, sturgeons are thinly spread out over an immense area. Most of the species are anadromous migrants but some live only in fresh water such as the sterlet: *Acipenser ruthenus* L. – relatively small and no longer than 1 m. The most abundant migrating species at present are found in the brackish water of the Caspian Sea, and also in the Black and Azov Seas. In Western Europe, *Acipenser sturio* L. (Fig. 254) was widespread at one time but is only relatively abundant now in the Gironde in France and in the Guadalquivir in Spain.

Sturgeons have disappeared from many rivers in which they were found previously. Among the causes for their disappearance is the overfishing of species reproducing at an advanced age, that is, over 15 years. Then there are such obstacles as dams which hinder spawning migration, and the pollution of estuaries where the young sturgeons have to live for several years.

Sturgeons have long lives and some live to over 50 years. They can grow to 4 metres (13 ft) in length and weigh over 200 kg (440 lb). They live off benthic organisms: crustaceans, molluscs, worms, insect larvae. The mouth of the sturgeon is "infer" (situated underneath the snout) receding backwards, tubular and protractile. It is preceded by four barbels. The caudal fin is heterocercal and the upper lobe is more developed.

To reproduce, sturgeons swim upstream until they find adequate beds in the gravel of the barbel zone. Some species have seasonal strains which swim up rivers and spawn at different periods. Sturgeon eggs are very plentiful and give greatly appreciated caviare. With the exception of fresh water species, young sturgeon return to brackish water at one year and after a more or less long stay in this transitory habitat make for saltier waters.

II Sturgeon Cultivation

Sturgeon cultivation is carried out for restocking, particularly in Russia. It is relatively old as the first artificial fertilization goes back to 1868. Fertilization worked well but incubation was difficult because of the adhesiveness of the eggs. This problem was resolved in 1914. Two problems which have only been partially resolved remain however. The first is finding enough ripe females, and the other, rearing fry to a sufficiently advanced stage to ensure the success of the restocking.

On the whole, the different phases of sturgeon cultivation are similar to those of other restocking cultivations based on the capture of brood fish from open water, followed by artificial fertilization and hatching.

1 Capture and maturation of parent fish

A ripe female must have a sufficiently soft belly so that slight pressure of the hand on the abdomen causes the eggs to run but not to stick. A similar pressure on the belly of a ripe male will cause the milt to run.

Formerly, females were simply brought to maturity by holding them in a small elongated pond near to the river of their origin and close to the place of incubation. It was traversed by a somewhat strong current. The ponds were long channels, from 100 to 150 m (110 to 170 yds). The bottom of the downstream section was of earth and was also the longest. The bottom of the upstream section was comprised of pebbles.

To help maturity, pituitary gland extract can be injected as the normal spawning time approaches. This method has advanced since 1937 and today is in current usage. Guerbilski, according to Rostami, recommends a dose of extract from 30 to 60 mg, dry weight, for females, according to the species. Larger doses have also been recommended. For males only half this dose is used. Between 24 and 36 hours after injection the brood fish should be mature. This method ensures maturity for about 80 per cent of the fish injected. To aid control after injection the brood fish are placed in holding tanks a few metres long situated upstream from the holding canal. The bottoms of these tanks are in concrete and covered with pebbles.

To ensure maturity it is necessary to know the biological groups of each species of sturgeon. This allows determination of the time to capture the brood fish of each group, how long they should be held in ponds, and the exact temperature necessary in order to obtain eggs by hypophysation. According to each group, sturgeons reproduce either in the spring, at the end of that season or in the autumn, but in general between April and August and sometimes up to October.

2   Artificial fertilization

Artificial fertilization differs slightly from the method generally used for other fish. Particulars of this are given below according to Rostami. The number of eggs given by a female can amount to several millions because of the size of the animal and in relation to the species and the region. But practically the number of eggs generally obtained by artificial fertilization is between 100,000 and 300,000, for the eggs do not ripen simultaneously and a great many are no good for fertilization. Diameter is from 2·5 to 3·0 mm and for the *Huso* 4·0 mm.

Because handling is so difficult and because the fish are so large they have to be killed. A ripe female is stunned and suspended by the operculum from a hook. The gauze pad inserted in the genital vent to avoid premature loss of eggs is removed and the abdomen is then gently pressed. The eggs are gathered in a pan placed under the fish. They are given a first washing which lasts a maximum of 5 minutes. This removes the blood and the mucus. A little water is then poured on the eggs which are sprinkled immediately by the milt from two or three males. Mixing is done with the hands and the eggs are then allowed to rest for 3 to 4 minutes until fertilization is terminated. This coincides with the hardening of the shell which will be both resistant and elastic.

Then comes a second washing to remove excessive sperm and above all the sticky coating covering the eggs. In this way the eggs will neither stick together nor to the bottom of the hatching apparatus. The water used for washing should contain 10 per cent finely powdered clay or chalk. Running water should be used or renewed several times. Gradually the viscosity will disappear and after 20 to 30 minutes the second washing should be terminated.

3   Artificial hatching and releasing of fry

The hatching of sturgeon eggs is carried out in relatively simple incubators of various models though generally of Californian type such as those used for the hatching of salmonids. The dimensions of the incubator vary between 60 to 75 cm (2 to 2½ ft) long, 40 to 65 cm (15 to 25 in) and 25 to 40 cm (10 to 15 in) deep. Some models are simply cases with perforated bottoms and others have the bottom, sides and covers perforated. Other perfected models are real incubators installed with a single hatching tray in a hatching trough which is slightly larger in order to ensure a satisfactory circulation of the water (Fig. 255).

Incubation is carried out directly in the river water. The incubators are placed in lines or in series across the river (Fig. 256). Care must be taken to clean off the mud from the eggs and dead eggs must be removed. The cleaning of the eggs is helped by washing them. During washing, 15 gm of clay per litre of water is added twice, with a 3 minute interval. Because of the current, the dead eggs will rise to the surface and will then be siphoned off. Wastage during incubation amounts to 20 or 30 per cent.

The incubation time taken depends on the temperature and this can vary between 15 and 20°C. In any case it is short, no more than 90 degree days (6 days at 15°C).

The yolk sac is absorbed rapidly – in 5 or 10 days. If the fry are not intended for raising then they should be released at this stage in places known to be free from predators. Despite the enormous quantity of fry liberated at this stage – tens of millions in the Caspian basin – the results are very poor and it is estimated that they give no more than 0·3 per thousand adults.

4   Raising fry of 2 to 3 weeks or 4 to 6 weeks

As the liberation of fry with absorbed yolk sacs give such poor results efforts are being made to produce fry better able to defend themselves – that is, fry aged from 2 to 3 weeks or 4 to 6 weeks.

1. *Raising fry of 2 to 3 weeks.* This process is short and lasts no more than 20 days, of which 5 to 10 are for absorption of the yolk sac. The fry are raised in round concrete tanks of 2 to 4 m (6 to 12 ft) in diameter or in rectangular tanks of 2 × 1·5 m (6 × 5 ft) which float in the growing ponds. The depth of the tanks is slight, from 20 to 40 cm (8 to 15 in). Stocking rate is high, from 40 to 60 fry per square decimetre. After the absorption of the yolk sac, the fry are fed with daphniae, chironomids and oligochaetes which are grown in special ponds. The raising system is good (waste is from 20 to 30 per cent) but the methods

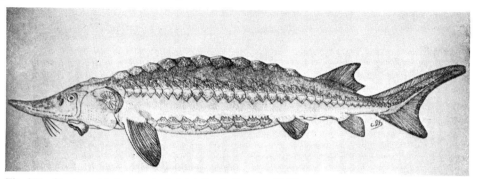

Fig. 254   Sturgeon (*Acipenser sturio* L.) (after Spillmann, 1961).

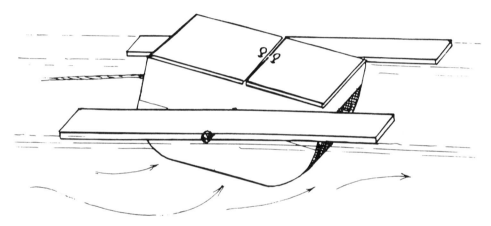

Fig. 255   Perfected Seth-Green incubator (after Roussow, 1955).

Fig. 256   Seth-Green incubators on a river (after Rostami, 1961).

used for growing the food to be distributed produces problems. For this reason it is often preferable to raise young sturgeons of from 4 to 6 weeks in one stage.

2. *Raising young sturgeons of 4 to 6 weeks.* This farming is practised in ponds from 0·5 to 2·0 hectares (1 to 5 acres) and 0·5 to 2·0 m (2 to 7 ft) deep. The ponds are stocked at a ratio of from 4 to 6 fry per square metre, either with absorbed yolk sac fry or with 2- to 3-week-old fry already fed in tanks.

To increase their productivity the ponds are fertilized with mineral fertilizer (superphosphate or ammonium sulphate) distributed at a ratio of 200 kg per hectare (220 lb per acre) over the whole surface, or organic fertilizers such as hay spread out along the banks at a ratio of 1,000 kg per hectare (8 cwt per acre).

Raising in ponds takes from 4 to 6 weeks. The young sturgeons live off plankton (cladocerans, copepods), and food on the bottom of the pond (chironomid larvae). Production is from 75 to 125 kg per hectare (85 to 140 lb per acre). At the end of raising, the young sturgeons weigh from 1·5 to 3·5 gm. Loss is estimated around 20 to 30 per cent.

5 Raising sturgeons of 2 years and over

This raising of sturgeons of saleable size is not practised very much. It has been tried in Hungary with sterlet (*Acipenser ruthenus*). The young fish are caught in the water courses such as the Danube, the Theiss, and are released in cyprinid growing ponds. Sterlet take a long time to grow: about 175 to 325 grm (6 to 11 oz) per year. Unger estimates that it takes about 5 or 6 years to reach a kilo (2 lb). The flesh of sterlet raised in ponds is fat.

SECTION VI

CULTIVATION OF PEJERREY OR SILVERSIDES

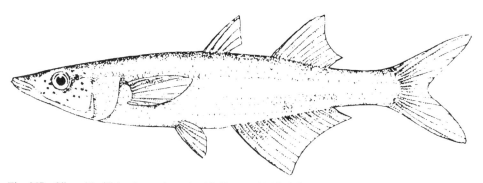

Fig. 257   Silverside (*Odonthestes bonariensis*) Cuv. and Val. (after Conroy, 1967; after Sterba, 1962).

Pejerrey, or silversides, are atherinids. Several kinds are found in South America, the majority being small fish living in the brackish waters of estuaries.

Among pejerrey the most important species for cultivation is *Odonthestes bonariensis* Cuv. and Val. (Fig. 257), cultivated in the Argentine since the start of the century. They develop well in rivers, lakes and brackish water lagoons. They are used for repopulating inland waters and reservoirs. They are suitable for both fresh and also slightly

brackish water. They are also used for the same purpose throughout Bolivia, Brazil, Chile and Uruguay.

Pejerrey are omnivorous. They can grow to between 35 and 40 cm (14 to 16 in) and even to 50 cm (20 in) and weight up to 3 kg (6 lb).

They do not reach sexual maturity before one year. At a first spawning a female can give 2,000 to 3,000 eggs. Subsequently they can give 30,000. The eggs measure 1·6 mm and adhere to aquatic plants and other submerged objects. Reproduction takes place in two periods: March and April and also in September and October, but the second is the most important.

Pejerrey are raised and cultivated easily. They are used for restocking only – not for eating.

The brood fish are difficult to keep in captivity. They are caught in open waters and this presents no problems. Spawning and artificial fertilization is the same as that of salmonids, pike and coregonids. When they are stripped the brood fish easily lose their scales and must be killed. After spawning, the fertilized eggs are carefully washed in order to clean off the waste scales and other impurities. The eggs are hatched either in Zoug or MacDonald jars. 40,000 to 50,000 eggs are placed in jars of 6 litres (1 gal 2 pt). The optimum hatching temperature is from 18 to 20°C and they take 180 to 220 degree days to hatch approximately in 10 days.

Pejerrey fry are difficult to handle and also to transport. They are grown in various sizes for repopulation, in small ponds 15 × 3 m (50 × 10 ft) in which they receive abundant food such as plankton (the growth of which is helped by manuring), daphniae, *Tubifex* and other natural food. Finely minced liver gradually replaced by bran is also used. Other artificial foods are also probably utilizable.

Rearing continues for a maximum of 6 months and at that age the fish can measure 20 cm (8 in) and weigh 100 gm (3 oz). No real attempt has been made to raise the fish to edible size. During raising the pejerrey tolerate the pond temperatures to be around 10 to 28°C, but the best temperature is around 18°C. Production varies per year between 30 and 200 kg per hectare (33 to 220 lb per acre).

The introduction of pejerrey into Europe and Israel has been considered.

Another atherinid, *Chirostoma estor* Jordan, is raised in Mexico under the same conditions.

Chapter VI

BREEDING AND CULTIVATION OF PERCIFORMES

THE cultivation of perciformes is widely spread throughout the entire world in both temperate and tropical regions. In the former, in Europe, are found pike-perch and in the United States walleye. In the warm temperate regions of North America and other continents black bass is also included. In the intertropical regions the cultivation of *Tilapia*, which are cichlids of African origin, started to develop on an important scale in the decade 1940 to 1950.

SECTION 1

CULTIVATION OF PERCH

Cultivation of the common perch (*Perca fluviatilis* L.) (Fig. 258) is only occasionally practised. In fact the general tendency is to eliminate them from ponds rather than to introduce and raise them.

Perch reproduce easily in a wild state and, understandingly enough, are considered undesirable in cyprinid cultivation. They get into ponds very easily either as eggs or as small fry, and develop and compete with the basic fish being raised.

Nevertheless, perch flesh is excellent eating and the fish is also greatly appreciated by anglers as a sports fish. They are voracious and fulfill an important equalizing or balancing role by helping to secure and maintain the desired balance between voracious and non-voracious fish in cyprinid angling waters. However, they are difficult to transport over long distances, although this position has improved since the use of plastic sacks containing an oxygenated atmosphere has become possible.

Under special circumstances it is sometimes necessary to consider perch for repopulation purposes, particularly when spawning grounds and favourable conditions for natural reproduction are lacking. This can happen notably in canals and canalized rivers and also in the running waters of the barbel zone where perch spawning grounds are rare or where running water cyprinids, which are destroyers of perch eggs, are numerous.

Repopulating with perch is carried out with fish of 1 summer. These measure at least from 7 to 10 cm (3 to 4 in) in length while they weigh around 3·5 to 5·0 gm. When the population in the rearing

Fig. 258   Common perch (*Perca fluviatilis* L.).

174

ponds is not very dense these sizes can be increased and perch of 1 summer can reach 15 cm (6 in).

The multiplication of perch presents no problems. It can be carried out either by natural or artificial fertilization. Perch spawn in the spring about April, shortly after pike do, and when the temperature reaches 12°C. Perch eggs stick together

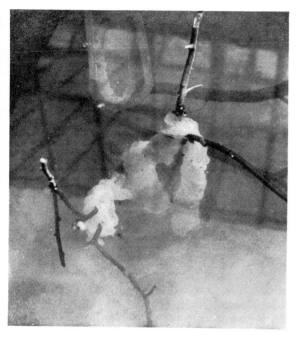

Fig. 259   A string of perch eggs on a sunken branch.

and form ribbons which are rather like flattened tubes about 6 or 7 cm (2⅓ or 2⅔ in) wide. They can reach 1 m (1 yd) in length (Fig. 259). These ribbons are pale yellow and are also semi-transparent.

When natural spawning is used, a few brood fish from 20 to 30 cm (8 to 12 in) long, both male and female and in equal numbers, are taken. These are easily recognized, because just before spawning time the bellies of the females balloon out and the milt of the males is plentiful. These brood fish can be placed, at a ratio of two pairs per hectare, in second or third year carp ponds. Operating under these conditions, the fish culture station of Bokrijk (Huet and Timmermans, 1966) obtained from 1,000 to 5,000 fingerling perch per female.

Stripping can also be used. If the female is ripe there is no difficulty in obtaining a ribbon of eggs which are then fertilized by spraying with milt. These can be incubated in Zoug jars. Just before hatching the ribbons are removed from the jars and are placed in hatching troughs. The troughs must have very fine mesh and models used for carp are suitable. Perch fry are so small that they are scarcely visible. It is also possible to transfer the ribbons of eggs, when they are ready to hatch, directly to the ponds. Here they are placed on the submerged plants and branches to avoid their falling into the mud. If this is done, however, the number of fry likely to hatch from the ribbons can only be approximated.

The ecology of American perch (*Perca flavescens* Mitchill) (Yellow perch) is practically the same as that of European perch.

SECTION II
CULTIVATION OF PIKE-PERCH

I   General Characteristics of Pike-Perch

Belonging to the family of Percidae, pike-perch (*Lucioperca lucioperca* L.) (Fig. 260) are, after pike, the most typical voracious fish living in the cyprinids waters of Europe. Except when they are very young and then live off plankton and small fauna, they are exclusively voracious. However, at sizes equal to those of pike they are only able to swallow smaller prey because the mouth is small. Pike can consume prey almost 50 per cent of their own weight whereas pike-perch are limited to

between 10 and 15 per cent of their own weight and 50 per cent of their own length (Steffens, 1960a).

Pike-perch are typical warm and calm water fish. They do not live in rheophile waters. They are found in the water courses of the bream zone and in lakes with warm water in summer. For example, they are abundant in Lake Balaton (Hungary) and are also well developed in the Ijselmeer (Holland) where they have been introduced. They prefer sandy or sandy-muddy bottoms rather than muddy-clay bottoms. Pike-perch accommodate themselves better in turbid waters than do pike, But they need water with a normal oxygen content. They cannot withstand polluted waters poor in oxygen; shallow, grassy and muddy waters are not suitable either. Further, they do not endure handling, particularly when young, and this is one of the principal difficulties met with in rearing them. The flesh of pike-perch is excellent and is much appreciated for the table.

Pike-perch grow rapidly. According to Tölg and Penzis (1966), under average conditions in Hungary they reach at 1 year 50 gm (1⅔ oz) and 13 cm (5 in); at 2 years 100 gm (3⅓ oz) and 25 cm (10 in); at 3 years 350 gm (12 oz) and 32 cm (1 ft); at 4 years 700 gm (24 oz) and 40 cm (15 in); and at 5 years 1,200 gm (2 lb 10 oz) and 47 cm (18 in). If feeding conditions are good then these weights and sizes can be surpassed easily and at the end of the third year, pike-perch can weigh 500 to 1,000 gm (1 to 2 lb) and measure 40 to 50 cm (15 to 20 in). The females grow more quickly than the males. Pike-perch feed on different cyprinids and other fish such as roach, rudd, bleak, bream, loach, gudgeon and also perch and pope.

They spawn in the spring soon after perch when the temperature of the water reaches 12 to 16°C. This generally happens at the end of April or the beginning of May. Pike-perch spawn in couples. They choose water at least 1 metre in depth (3 ft) with pretty hard, sandy or gravel beds. The male digs a kind of nest about its own length. Here the eggs are laid, preferably on aquatic roots, at a ratio of 200,000 per kilo (2 lb) of female. They are 1·0 to 1·5 mm in diameter and stick to the substratum. They are watched over by the brood fish which also fan them with light movements of their fins. At temperatures of 11 to 14°C, they hatch after 8 to 10 days.

Originating in central Europe and studied, above all, in Hungary, the cultivation of pike-perch is now developing throughout temperate Europe. From the end of the nineteenth century pike-perch were introduced into many waters of western Europe believed to be suitable. The natural limit in the west was then the Elbe though now it is the Atlantic Ocean. Pike-perch are cultivated for restocking purposes and the aim is to produce young fish of a few weeks or of 1 summer. But it is also (though this is secondary) cultivated for the table and then is grown to saleable sizes in cyprinid ponds. The cultivation aims of pike-perch are as follows: (1) The production of young fish for restocking; (2) production for the table though this

is secondary; (3) to play a police role by eliminating small undesirable fish without value that are found in cyprinid ponds.

II Semi-controlled Reproduction of Pike-Perch

In view of the existing interest in the species and the growing demand for fry, equally for repopulating free waters such as lakes and slow flowing streams as for the restocking of farm ponds, efforts are naturally afoot to increase the supply and meet that demand.

Artificial fertilization methods are not used. Reproduction is natural, either semi-controlled or controlled on artificial nests. Females of 4 years and over and males of 3 years and over are used as parents.

If *natural semi-controlled reproduction of pike-perch* is the chosen method then the brood fish are liberated at the start of the spring – end of March or the beginning of April – in spawning and rearing ponds. The ponds must always be rather large, ½ hectare (about 1 acre) at least, and the method employed is similar to that used for the reproduction of European cyprinids, namely, spawning and rearing in the same pond.

According to the circumstances of each case the ponds used are either exclusively reserved for raising pike-perch or they are used principally for the raising of carp for the table. In the first case, density can be relatively high and five or six pairs of brood fish can be placed per hectare (2½ acres). To these, an equal number at least of cyprinid pairs can be added such as roach or rudd, for example. The fry of these will be forage fish for the young pike-perch. In the second case fewer reproducers are liberated (one pair per hectare) and only a few forage fish are added as the fry will live off plankton and small aquatic fauna.

The *natural and controlled reproduction of pike-perch* is also carried out in ponds by means of artificial nests. In effect this is really the "kakaban" system used at times for carp. There are several different methods of using this system but the following is the most generally practised.

In the spring (the season of reproduction is at the end of April or the beginning of May), the brood fish are placed in spawning ponds where artificial nests have been installed. In Hungary the spawning ponds used are in fact holding ponds for marketable carp. They are rather deep (depth

Fig. 260   A 4-year-old pike-perch brood fish. Weight: 1,230 gm (2 lb 11 oz); size: 52 cm (1 ft 8 in).

Fig. 261   Pike and pike-perch of 1 summer grown under the same conditions in a pond rich in forage fish. Pike: size 35 to 41 cm (1 ft 2 in to 1⅓ ft); pike-perch: size: 16 to 18 cm (6½ to 7½ in).

Fig. 262 Artificial pike-perch nest covered with eggs. (Photo Woynarovich.)

over 1 m) (over 3 ft) with hard and clean bottoms, and rather small, no more than 10 ares. A large number of brood fish are liberated: from 20 up to 100 pairs. According to Woynarovich, 5 to

10 m² (50 to 100 ft²) is sufficient for a couple of brood fish. In these ponds, completely free of vegetation, an equal number of artificial nests is also placed (Fig. 262). These nests are square from 50 to 60 cm (20 to 24 in) and are formed by a wire netting on a wooden frame. On the upper side of the wire netting rigid fibres are fixed to which the eggs will stick after spawning. These fibres can be the aquatic roots of water plants, willow or alder, or branches of briar or resinous plants. Sometimes bits of old fishing nets are used or very fine artificial leaves made in plastic. The underside of the netting is weighted with a brick or tile and all rest on the bottom of the pond. Pieces of wire about 75 cm (30 in) long are attached to the corners of the nest and the upper ends are bound into a kind of ring which can be easily hooked when the nests are withdrawn from the water (Fig. 263).

At the spawning season, that is, normally during the first 15 days of May, the nests must be inspected regularly. Those covered with eggs must be withdrawn from the spawning pond and, using the same methods as for kakabans, transferred to the hatching ponds which are also rearing ponds. This can be carried out directly when it is seen that the nests are filled with eggs. It is also possible

Fig. 263   Lifting a pike-perch artificial nest from a pond for control. (Photo Woynarovich.)

to introduce a transitory incubation phase in an aspersion chamber.

The aspersion chamber technique has been developed by Woynarovich (1960b). The fibres covered with eggs are suspended in a small chamber where the temperature can be maintained at 14 to 15°C and where the eggs, sticking to the fibres of the nests, are kept for several days under a very fine spray of pulverized water from an atomizer. The water spray can be continuous or alternate. This method has the advantage of eliminating any risk of the eggs being choked by mud. It also stops their being covered by fungus growth. Just before hatching the eggs are transferred to the hatching and rearing ponds.

When pike-perch eggs become eyed they can be transported even over long distances. The eggs still stick to the fibres on which they were laid. During transport the fibres and the eggs should not be pressed together. This can be avoided by placing damp moss between the fibres. The eggs are transported dry and in a pack in which the temperature remains fresh and the atmosphere damp and oxygenated. By placing the eggs in a plastic sack, which is sealed and provided with oxygen, this proves quite possible.

Fertilized eggs of pike-perch can also be used for restocking open waters by means of hatching boxes (Corchus boxes). These are made of rigid, transparent material whose walls are perforated in such a manner that fry can escape after hatching. The Corchus boxes are filled with part (15 × 15 cm or 6 × 6 in) of an artificial nest covered with eggs and immersed in the waters for restocking.

III Raising Young Pike-Perch for Restocking

In many ways the production of young pike-perch resembles that of pike fingerlings. Among the similarities is the necessity for the young fry to be plunged literally into an environment very rich in natural food. These include, progressively, rotifers, *Cyclops* larvae, *Cyclops* and other forms of zooplankton. After having reached 15 mm, pike-perch fry also eat insect larvae, plankton and the bottom fauna (chironomids, *Corethra*, *Cloëon*). In an environment which suits them the pike-perch grow rapidly and turn voracious very quickly (Woynarovich, 1960a). If there are no forage fish fry available then cannibalism develops as soon as the fish reach 4 cm (1½ in).

If pike-perch of a few weeks or of 1 summer are being produced then it should be done in ponds. Intensive rearing of young pike-perch has been tried in tanks but the method has never been generalized because of the difficulty of finding appropriate food for very young fry which scarcely measure more than 5 mm.

Pike-perch of a few weeks (6 to 8 weeks) can only be produced under the controlled reproduction method. One or several nests filled with eggs ready to hatch are placed in the hatching and rearing ponds which should not be too large (about 10 ares) in order to avoid draining problems. The environment must be very rich in natural food and to this end should be fertilized. It is possible to arrange for ponds containing up to 200,000 fry per hectare (about 20 per square yard).

Pike-perch of 1 summer (Fig. 261) can be produced either by the semi-controlled or controlled methods of reproduction. In the first case the brood fish and in the second case the nests with eyed eggs are placed in the ponds. In both cases the production of young pike-perch can be a principal or a secondary aim.

If the production of young pike-perch is the principal objective then forage fish (roach and rudd) must also be placed in the ponds. But the forage fish parents must not be permitted to eat the pike-perch fry immediately after hatching. The nests should be protected by wire netting or a lattice built round the nests. With this type of farming production in weight is small but economically important because of the high commercial value of young pike-perch. The results of this type of farming are very variable. According to Steffens (1960b) and Willemsen (1961) the most one can hope for is a harvest of 15,000 pike-perch summerlings per hectare, but under very good conditions this may reach to 30,000 summerlings. Average size is around 8 to 10 cm (3 to 4 in) and weight around 3 to 5 gm. According to the number of young pike-perch of 1 summer harvested, sizes can vary between 8 and 20 cm (3 and 8 in) and the weight between 3 to 5 and 40 gm (up to 1 oz).

If, however, the production of young pike-perch is secondary, it will be ancillary to the principal production which is generally carp for eating or second year carp. But in any case even if the production of the young pike-perch measured by weight is far below that of the carp, the commercial value will be equal to it (Steffens, 1960b).

When production includes liberating the pike-perch brood fish into ponds (semi-controlled reproduction) they should be third year carp ponds. If artificial nests are used (controlled reproduction) then the ponds can be either second or third year carp ponds. But precautions must be taken to protect the nests by a lattice or suitable branches to avoid the young pike-perch being exterminated by the carp. This method can give very good results and produce pike-perch of 1 summer and of good size.

IV Rearing Pike-Perch of 2 Years and Over

This is always a secondary production but it is still quite possible. It has a double objective. The first is to eliminate undesirable fish in cyprinid cultivation and to transform them, at the same time, into good quality flesh; and secondly to produce young pike-perch brood fish. The fact that pike-perch are less voracious than pike means they can become larger when kept in carp ponds. But they cannot be kept in first year ponds without risking the destruction of the whole carp population.

The production of second year and older pike-perch is always ancillary and is limited to a few dozen fish per hectare (40 for example) for second year pike-perch, and only a few over that age. Pike-perch of 2 summers can weigh around 150 gm (5 oz) and measure 20 to 30 cm (8 to 12 in) and those of 3 summers 300 gm (10 oz) and more, and measure 30 to 45 cm (12 to 18 in).

Pike-perch do not fast during the winter. For this reason forage fish must be placed in the winter ponds for food. The total weight of the forage fish should be around once to twice the weight of the pike-perch brood fish.

SECTION III
CULTIVATION OF WALLEYE

I General Characteristics of Walleye

Walleye (*Stizostedion vitreum vitreum* Mitch) (Fig. 264), often referred to as "yellow pike-perch", are American perciforms largely spread over the eastern regions of the United States and southern Canada. They are a characteristic species of the great lakes region where they are most abundant. They are also abundant in Wisconsin and Minnesota, and down towards the south they are found in north Georgia, Alabama and Arkansas.

These fish prefer lakes and great water courses with fresh clear water with gravel or sand beds. They do not like water with muddy bottoms. The species was introduced successfully in a number of lakes which were too shallow and consequently too warm for American lake trout (*Salvelinus namaycush*). In warm waters, walleye compete with black bass.

Fig. 264 Walleye (*Stizostedion vitreum vitreum* Mitch.). (Document Wisconsin Conservation Department.)

Externally, walleye strongly resemble European pike-perch. Their general conformation and colouring are the same. The eyes are glassy. The lower lobe of the caudal fin is white and the posterior end of the spinous dorsal fin has black patches.

Biologically, the two fish are also close together since they are both voracious. Walleye live in small groups preferably in poorly grassy water. They prefer the depths if the water is clear, coming to the surface to hunt preferably at night. They grow quickly if the food at their disposal is adequate and abundant. In Minnesota, average size is 12 to 13 cm (5 in) at 1 year of age, 30 cm (12 in) at 3 years and 45 cm (18 in) at 5 years. Females grow faster than males.

Walleye spawn early in the spring, that is as soon as the water warms and only a short time after pike when the water is around 10°C. Spawning can take place from the beginning of April to the middle of May according to the different regions. Males are sexually mature at 2 or 3 years and females at 4 or 5. At the time of reproduction some spawn where they are, that is, on the gravel bottoms of the lakes, while others migrate, swimming up the tributaries of the lakes to find currents with gravel beds on which to spawn. Walleye do not make nests but spawn on gravel bottoms of lakes and running waters.

Just like pike-perch, walleye are enjoyed for both fishing and eating and have a considerable commercial and sports value.

Walleye are cultivated only for restocking and the objective is always to use fry with absorbed yolk sacs, or 1 summer fish.

II Artificial Reproduction of Walleye

Artificial reproduction of walleye is possible and is carried out by methods very similar to those used for pike.

Brood fish are not raised in ponds. They are caught in open water during the spawning migration. This is done by means of trap-nets or gill-nets which are laid at night on the spawning beds, but the nets must be handled with great care as the fish are delicate. Eggs of ripe brood fish are fertilized at once but if some parents are not ripe they can be kept in storage or holding tanks for a short time or in ponds should they need to be held for a longer period.

The spawning and fertilization of walleye eggs is carried out according to the same techniques

as for pike. The eggs are free. They are about 2 mm in diameter and the harvest varies from 15,000 to 45,000 per pound of female. At the time of spawning the eggs are very soft and fragile and must be handled with the greatest care. They are also slightly sticky for a few hours following spawning. After spawning, but before incubation in jars, the envelopes must be hardened. This is done by plunging the eggs into a bath containing starch or clay for 1 or 2 hours. Fine particles of either will stick to the shell. After this the eggs must be carefully washed before being placed in the jars. To complete the hardening and before tipping the eggs into the jars some practitioners spread the eggs out on elongated trays covered with stretched, fine linen and leave them for from 8 to 10 hours.

In the United States jars (Fig. 265) replace the bottles used in Europe for pike and coregonids. Their function is identical. During incubation the eggs, like those of pike, are kept moving very slowly, but because walleye eggs are so fragile the jars are filled up to only one-quarter or one-fifth.

Fig. 265 Glass incubation jars for pike and walleye eggs. State Fish Hatchery, Clear Lake, Iowa, U.S.A.

The ratio is 130,000 eggs per litre. Loss is from 20 to 25 per cent. The dead eggs, which are lighter, have a tendency to become covered with fungus. They rise to the surface where they can be siphoned off.

The time necessary for incubation depends on temperature. At 10°C it is about 20 days. Fry hatch in the jars and start swimming immediately. At this time they measure about 12 to 13 mm. They are carried out from the jars by the current to a channel which in turn takes them to a receiving pond in which the yolk sacs are absorbed within a few days. When this is finished the fry are released in open water or in ponds where they will be raised to the summerling stage.

As walleye eggs are so delicate they can only be carried over short distances. This is done either in receptacles or on trays which, when used, must be covered with moss and kept damp and fresh.

III Raising Young Walleye for Restocking

Just as for the incubation stage, so the techniques used for the production of walleye for restocking resemble those used for pike. But these techniques need to be improved.

Feeding walleye resembles feeding pike. First they feed on rotifers and nauplii, then on cladocerans and copepods, followed by insect larvae such as chironomids and mayfly nymphs, but by the time they reach 8 to 10 cm (3 to 4 in) they become voracious.

1. *Production of young fingerlings*. The production goal is fish from 7 to 10 cm (3 to 4 in) long and an average of 7 to 8 gm in weight. This can be achieved in a few weeks just as for pike. The same techniques can be used also when the goal is to place walleye fry in an environment which is very rich in natural food. This is possible if mineral fertilizers and manure are used but excessive manuring must be avoided.

Some writers have recommended stocking 25,000 to 50,000 fry per acre; that is 60,000 to 120,000 per hectare. But the figure seems somewhat high and stocking from 25,000 to 40,000 per hectare should perhaps be sufficient.

2. *Production of summerlings*. This method has been used. Fry with absorbed yolk sacs are released and at the start of the autumn the summerlings are harvested. The harvesting, however, rarely accounts for more than 10 per cent and gives from 6,000 to 12,000 fish per hectare measuring from 12 to 15 cm (5 to 6 in). Average production in terms of weight per hectare is 50 kilos (110 lb) which is relatively small.

Such results will be understood when the way walleye feed is understood – that is, like pike, they become carnivorous as soon as they have passed 10 cm (4 in) in length. In order that young walleye should not devour each other it is necessary to liberate forage fish in the ponds at the end of the spring. These will reproduce and their fry will serve as food. But the system is risky. As for pike it seems that the production of young fingerlings, the value of which as fish for restocking is as high as large fish, is preferable. Further, production costs for raising young fingerlings is very much lower.

SECTION IV

CULTIVATION OF BLACK BASS AND OTHER CENTRARCHIDS

The Centrarchidae are North American fish. Their cultivation, especially that of black bass, is well developed in the United States of America.

Since the end of the last century these fish have been introduced into several western European countries as well as into North Africa and South Africa. Success is limited but it can be considered viable in regions with warm water in the summer such as in France south of the Loire. This result is normal when it is remembered that black bass are typically warm water fish. In South Africa their acclimatization is a success and they are considered satisfactory.

I Cultivated Black Bass

Here are the principal species of cultivated black bass:

1. **Largemouth Black bass** (*Micropterus salmoides* Lac.) (Fig. 266) are considered to be one of the great freshwater carnivorous fish. They are calm water fish suited for lakes and ponds situated in the plains as well as in waters with feeble currents, muddy beds and rich in plants. They like water which warms up even to 32°C; but, though the water can be warm, it must not be polluted nor poor in oxygen. In North America they are found

in fresh waters from Canada to the Gulf of Mexico and from the Atlantic coast to the Rocky Mountains.

2. Smallmouth Black bass (*Micropterus dolomieu* Lac.) (Fig. 267), are voracious fish originating from that part of the United States north of a line crossing southern Ohio and the centre of Missouri. They are found in moderate running water courses which include both calm and fast currents as well as in clear relatively cool-water lakes. They appear to be particularly suited to the running water courses of the barbel zone. In Belgium they became acclimatized in the waters of the lower Semois, a large river belonging to the upper type of the barbel zone. They lived there quite naturally for many years but now the population is diminishing because of insufficient reproduction.

3. Spotted Black bass (*Micropterus punctulatus* Raf.) originated in the waters south of the Ohio and the centre of the Missouri. In running water they appear to be suited to the inferior type of the barbel zone and to the bream zone, that is, to those streams which are warm in summer.

Figure 268 shows the principal distinctive characteristics of the three species.

Black bass are carnivorous except when young. The young first feed principally off rotifers and small crustaceans: cladocerans, copepods and later off water insects. As soon as they reach from 6 to 8 cm (2 to 3 in) they live off larger and larger fish, frogs and crayfish indifferently. The essential is that they obtain the prey they need. Largemouth black bass are the most voracious; the other two are happy enough with smaller prey.

Growth of black bass is rapid but closely tied to the food available and to the temperature. They stop growing when the temperature falls below 10°C. Largemouth black bass can weigh over 10 lb but in the United States average weight for adults is around 2 to 4 lb. In France, Wurtz-Arlet (1952) gives the following weights: 1 year, 10 to 50 gm (up to 1½ oz), at 2 years, 125 to 200 gm (4 to 7 oz), at 3 years, 350 to 450 gm (12 to 15 oz), at 4 years, 500 to 850 gm (17 to 29 oz), and at 5 years, 750 to 1,200 gm (26 to 38 oz).

II Methods of Cultivating Black Bass

Raising black bass is similar in many ways to that of cyprinids. It rests on natural, controlled reproduction in ponds, the first goal being the production of summerlings or fingerlings at the end of the summer of the first year.

For smallmouth bass and spotted bass rearing terminates after the first summer except for the brood fish. They are produced only for restocking running and still waters. For largemouth bass, for which the production goal is fish for angling aged 2 years and over, raising is continued after the first year in fishing ponds.

For the raising of black bass the following categories of ponds are necessary: breeding ponds, rearing ponds, growing ponds (fishing ponds).

As the raising of the three species differs only in minor details they will be treated together. In the United States the three species are raised in the following order of importance: largemouth bass, smallmouth bass and spotted bass. In Europe, only the first of the three is raised in practice.

1 Breeding ponds

Breeding and cultivation of black bass is carried out in ponds by controlled natural reproduction. They are not fertilized artificially.

Black bass spawn at the end of the spring; that is the end of April to the start of July according to the region and in temperatures which must be relatively high. The best temperature is between 20 and 21°C. The brood fish reach maturity slowly over a period of from 2 to 3 weeks in temperatures around 18°C which favour the success of the spawning, but if the temperature does not reach 17 to 20°C they will not spawn at all. Before the spawning season starts the brood fish are kept in spring water with regular and cool temperatures around 10°C.

Like all centrarchids, male black bass build nests and guard both eggs and fry. The nests are quite elaborate according to the species though the best are built by smallmouth black bass. In a wild state the nests are built on a firm bottom: stone, gravel or sand or among the roots of aquatic plants cleaned of detritus and mud. Measuring from 2 to 3 ft in diameter, according to the size of the male, the nests are not found in deep water and are generally no more than a few feet deep and in a spot protected from the current. When the nest is finished it is an excavation of an almost circular shape (Fig. 271) and free of all fine matter leaving only the larger matter in the centre. At spawning, the eggs are laid in small quantities at a time and are fertilized immediately. They are around 1·5 to 2·5 mm in diameter, according to the species, and they stick to the stones, gravel or roots. The moment spawning is terminated the female is pushed aside by the male who may, sometimes, call a second female to spawn in the same nest where he will also fertilize her eggs. Females yield from 2,000 to 10,000 eggs according to the size.

The dimensions of the spawning ponds vary in size and generally, in the United States, cover between half to 2 acres. Smaller ponds do just as well.

The bottoms must be specially arranged for spawning. To this end, before the pond is filled, small heaps of gravel must be laid out. They are placed near the banks of the pond in such a way that after the pond is filled they will be covered by about 2 to 3 ft of water. They should also be 3 to 5 m apart (9 to 15 ft) to avoid competition between the males.

If dense spawning is desired within a relatively limited surface, then spawning stalls placed side by side can be set up by the bank on one side of the spawning pond (Fig. 270). These stalls are about 1 m (1 yd) long, closed on three sides by either concrete or wood. Gravel is placed on the bottom on which the spawning fish will place their eggs. On the open side pointing towards the middle of the pond, a screen can be slipped in at the right time to permit the capture of the fry after they are hatched and before they are dispersed. However, this stall spawning system is not used much in the United States.

The brood fish weigh from half to 2 lb, are aged from 3 to 4 years and measure 25 cm (10 in) and more. Davis (1956) recommends more females than males at a ratio of three to two.

The number of brood fish placed in a spawning pond varies considerably according to the way in which the pond will be used. If the fry are to be taken away as soon as they are ready to leave the nest, then 40 or 50 couples to the acre can be placed – that is, practically one couple per are (one-fortieth of an acre). If spawning stalls are used the density can be higher. But if, after spawning and hatching, the fry are to remain in the pond until the end of the summer, then far fewer brood fish should be used. According to different writers 10 females and five males per acre or only three females and two males are recommended.

If climatic conditions are regular and favourable then the spawning season will be brief, but if they are irregular then the season can go on for over

one month and the result will be risky and irregular.

Hatching time is linked to temperature. At 18°C it extends for about 10 days but is shorter at higher temperatures. During hatching and immediately afterwards the young fry are guarded by the males (Fig. 271).

2   Production of black bass of 1 summer

The production of black bass of 1 summer (fingerlings or summerlings) can be carried out in different ways although, principally, in one or two phases. If raising is in two phases it can be carried out as extensive, semi-intensive or intensive. The essential factor, as far as the last methods are concerned, is not the question of size but of density. As for the surface area this can cover a few ares or even several hectares.

The fingerling ponds are treated in the normal way, that is, they must pass the winter dry, be placed under water a few weeks before planting the fish and they must be regularly fertilized.

a   Rearing fingerlings in one phase
The raising of fingerlings in one phase corresponds to the method of spawning and raising in the same pond as described for the cultivation of cyprinids. In spring a certain number of brood fish are placed in the spawning pond. They spawn and, in the autumn, the pond is drained and the fingerlings are harvested. In between, if possible, the brood fish are removed. With this method one can hope to secure 5,000 fingerlings from 8 to 10 cm (3½ to 4 in) in the autumn.

But the method can only be practised in relatively large-sized ponds: 10 ares (¼ acre) at least (Fig. 269). The small number of brood fish mentioned represents ten to three females for five to two males per acre. The difference between the sexes of bass are difficult to recognize, nevertheless it is unwise to place only a few fish in the pond for it would not then be certain that the two sexes were present.

Although apparently simple, this kind of rearing is not the most highly recommended. It does not permit control of the volume of fry stocking, and because of this the return varies considerably. Also, as the fry are not all of the same age, there is a risk of cannibalism.

Variation.   These inconveniences, particularly cannibalism, can be reduced if the rearing period is shortened. It can be limited to from 4 to 6 weeks after which the black bass fingerlings will measure

1 or 2 in. in ponds rich in natural food. Harvesting will take place during the month of July. The 4- to 6-week-old fry can be released in the water for restocking, especially in farm ponds. With this system the density of the brood fish can be higher than if draining were carried out only in the autumn.

b Rearing fingerlings in two phases

This method includes two phases. Reproduction is carried out in spawning ponds from where the fry are transferred to nursery ponds where they will be reared extensively or more or less intensively.

The harvesting of the fry is carried out by methods which differ slightly according to the species. Directly after hatching the fry remain in groups protected by the male, As they grow, so, progressively, they will leave the nest.

Smallmouth bass fry do not remain grouped for long so they must be caught before they start to disperse. And to be sure of capture in time it is necessary to isolate the nests. This is easy when spawning stalls are used (Fig. 270) for then it is only necessary to slip the screen into the open end. If spawning stalls are not used then the nests should be surrounded with fine wire screens, circular in shape and fixed to a rigid armature. The fry are then captured with a very fine scoop net.

Capturing spotted fry is more difficult as they are very small and slip between the finest mesh. On the other hand, largemouth fry can be captured later, when they measure 1·5 to 2 cm (up to $\frac{3}{4}$ in). Until they reach that size they remain grouped and swim together along the side of the pond banks. If the density of the brood fish is high then the fry of several spawnings will mix and form schools. They are captured with landing nets, seine nets with very fine mesh, trap nets or specially designed two-winged traps (Hesen, 1929).

From the spawning ponds the fry are taken to the rearing ponds. In order to avoid dangers of cannibalism it is important that the fry should all be the same size and be released in the pond in one operation.

1. *Extensive rearing.* This method (Fig. 273) of production rests entirely on the natural food available in the ponds. The latter must be increased to the highest possible density by good maintenance and rational fertilization, that is, intense but not excessive application. This fertilization is also necessary to develop plankton on which the young

black bass will feed. Fertilizers, both mineral and organic, should be applied in small but repeated quantities.

In well fertilized ponds, extensive production can reach 120 to 150 kilos per hectare (120 to 150 lb per acre). Stocking rate is from four to five fry per square metre (about 10 ft²); 20,000 per acre or 50,000 per hectare. Survival rate is around 25 to 50 per cent which means a production of between 4,000 to 10,000 fingerlings per acre or 1 to 2½ per square metre. The size of fingerlings produced in this way is about 7 to 8 cm (3 in) (Fig. 274). If the ponds are relatively poor then stocking must be reduced proportionately.

2. *Semi-intensive rearing.* If the aim is to produce large fingerlings (6 to 8 in = 15 to 20 cm) then the young fish must be able to eat small fish or forage fish. But this kind of rearing is more complicated for there are no ideal forage fish.

For first year black bass, bluegill (*Lepomis macrochirus*) are direct competitors for food. In the United States the best forage fish for the young black bass are golden shiners (*Notemigonus crysoleucas*) though other small species are used such as the bluntnose minnow (*Pimephales notatus*).

In this kind of farming it is necessary to release the parents of the forage fish only after the black bass fry have reached a certain size, otherwise the fry risk being devoured by the forage fish. The number of forage fish released should be between 10 to 20 per are (400 to 800 per acre).

3. *Intensive rearing.* Bass fingerlings can also be raised intensively and by similar methods to those used for pike fingerlings and salmonid fingerlings.

This intensive rearing is carried out in oblong ponds or in artificial troughs. In Pennsylvania long troughs are used measuring 2·50 × 0·60 × 0·60 m (8 × 2 × 2 ft) (Fig. 272).

While still young it appears to be necessary, as for pike, to feed black bass with young living prey. It is also necessary to have an important source of plankton which can be fished from lakes or, as for pike, can be produced under closed conditions while growing daphniae. The latter need water which is rich in organic matter, and has a favourable constant temperature.

When black bass measure 2 cm they can be fed with non-living food. The same food as for trout is used: meat, fresh fish or pelleted feed ground according to the size of the bass. The bass are grown to about 2½ in.

Fig. 266   Largemouth black bass (*Micropterus salmoides* Lac.).

Fig. 267   Smallmouth black bass (*Micropterus dolomieu* Lac.).

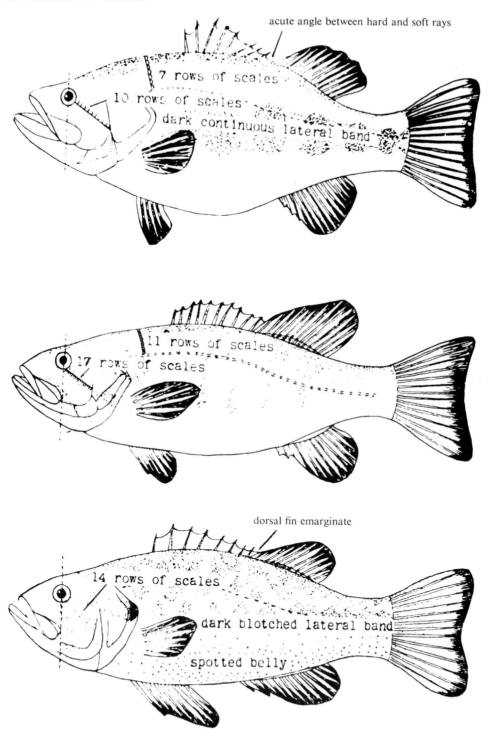

Fig. 268 Schematic drawing of reared black bass. From top to bottom: largemouth black bass, smallmouth black bass, spotted bass (after Hey.).

Fig. 269 Shallow pond, hot in summer, covering 1 acre and suitable for raising largemouth black bass and bluegill. U.S. Fish Cultural Station, Lamar, Pennsylvania.

Fig. 270 Spawning pond with cages or boxes for breeding black bass. Jonkershoek Fish culture station, South Africa.

Fig. 271 Largemouth black bass nest. The gravelly bed is cleaned by the male. (after Mraz et al., 1961).

Fig. 272   Long metal troughs for feeding and storing black bass summerlings ($1\frac{1}{2}$ to $2\frac{1}{2}$ in). Pleasant Gap State Fish Hatchery, Bellefonte, Pennsylvania, U.S.A.

Fig. 273   Rectangular pond, about 1 acre in size, for rearing black bass and bluegill. U.S. Fish Cultural Station, Leetown, West Virginia, U.S.A.

Fig. 274   Black bass summerlings of approximately $2\frac{1}{2}$ in. Upper Spring Creek Hatchery, Bellefonte, Pennsylvania, U.S.A.

Fig. 275   Bluegill sunfish (*Lepomis macrochirus* Raf.). This is the best forage fish in the United States for largemouth black bass.

Fig. 276   Fishing with a seine net for controlling a mixed population of largemouth black bass, bluegill and channel catfish. Auburn Experimental Fish Farm, Auburn, Alabama, U.S.A.

3 Rearing black bass in angling ponds

The further cultivation of black bass is carried out principally with the largemouth. This cultivation is somewhat special since it is only practised in ponds used for angling and not for the production of fish for the table.

The goal is to obtain a population able to maintain itself in equilibrium over several years, on the condition that relatively intensive fishing is undertaken regularly to remove the annual production. This kind of rearing is practised in angling ponds often situated near a farm where the pond not only serves for farming fish but primarily provides the water needs of the farm itself. The Americans call these ponds "farm ponds" and they have many uses (Fig. 353). They should be sufficient in size, at least 10 ares ($\frac{1}{4}$ acre), but it is better that they should not exceed 20 hectares (50 acres).

As largemouth bass are exclusively voracious they cannot be raised alone. Stocking should be mixed, generally with a vegetarian or omnivorous fish capable of high reproduction in order to provide sufficient prey for the predators. This is the role of the forage fish. In the United States "bluegill sunfish" (*Lepomis macrochirus* Raf.) (Fig. 275) usually play this part.

Other non-voracious fish capable of reproducing at a high level and not measuring more than 20 cm (8 in), can very well play a similar role. In tropical and sub-tropical countries *Tilapia sparrmani* can be used and in the temperate countries of Europe, common roach (*Leuciscus rutilus*) and rudd (*Scardinius erythrophthalmus*). In South Africa, tench (*Tinca tinca*) give excellent results as forage fish in the Paardevlei lake, Somerset West.

Ideal forage fish must grow rapidly, be very prolific and have a long spawning season, but they should not compete for food with the voracious fry. They must also spawn early, that is at an age of one year and quickly reach an adult size sufficient to ensure they will not be eaten by the carnivorous fish with which they are associated.

American writers, especially Swingle (1950), have tried to determine the ratio between associated voracious and non-voracious fish in order to maintain a balanced population.

A balanced population is one which gives, over several consecutive years, a satisfactory annual harvest of fish for angling. Its characteristics should include, at the same time, an adequate proportion between voracious and non-voracious fish and a

sufficient weight for those fish caught by anglers.

As far as the weight ratio between the voracious and non-voracious fish of all types is concerned, it should be between 3·0 to 6·0. As for the ratio for angling fish, including both voracious and non-voracious fish, this should be over 33 per cent of total weight. The most satisfactory is between 60 and 85 per cent. Swingle suggests for catchable bass at least 6 oz each and for bluegill $1\frac{1}{2}$ oz. Other writers consider a bass weighing 4 oz as being already adequate.

In order to obtain these ratios, American writers suggest the following quantities of fingerlings should be liberated. The figures refer to fish per acre which, multiplied by 2·5 gives the numbers per hectare.

In principle, one bass fingerling is released for 10 bluegill fingerlings (sometimes two bass for 15 bluegills) and the following proportions are advised: 100 to 150 bass for 1,000 to 1,500 bluegill to the acre of fertilized pond, and 50 bass to 500 bluegill per acre of unfertilized pond.

Such a stocking rate should ensure a balance for at least 3 years after which the whole population must be eliminated and a new population released.

Normally both bass and bluegill should spawn the year following restocking. Fishing starts from 12 to 18 months after release and the weight harvest should be at a ratio of 1 to 3 or 4. A fertilized pond can produce 200 to 250 lb per acre but for this the angling must be intense. Indeed, this level of angling must be maintained to keep the pond in balance.

III Rearing other Centrarchids

Apart from basses other centrarchids are raised in the United States, notably bluegill (*Lepomis macrochirus*), redear sunfish (*Lepomis microlophus*), green sunfish (*Lepomis cyanellus*), black crappie (*Pomoxis nigro maculatus*), white crappie (*Pomoxis annularis*) and rock bass (*Ambloplites rupestris*).

The most important of these are the bluegill (Fig. 275) generally added to largemouth black bass in the angling ponds.

Bluegill spawn over a longer period than bass. Their spawning season extends, normally, over several months from the end of May to the beginning of August. For forage fish this is an advantage. Spawning starts at a temperature around 20°C. As with other centrarchids, the male bluegill arranges a nest which he defends against intruders but the nests themselves can be close to each other. Hatching takes place after less than 5 days, after which the male protects the young fry.

On the whole, rearing bluegill hardly differs from that of black bass. In fact only one phase rearing is used. The brood fish are released in the spawning ponds and a few months later the fry are harvested. Bearing in mind that bluegill spawn rather late, the same ponds can be used for the reproduction of black bass at the start of the season, and after the harvest of fry the ponds can be dried out and used again for the reproduction

of bluegill. The latter are harvested in the autumn or in the following spring and are then used for stocking the farm ponds.

If a great number of fish are required, say up to 40 fry per square metre, then two or three pairs of bluegill can be placed per are. These brood fish measure from 12 to 18 cm (5 to 7 in) and weigh from 100 to 150 gm (3 to 5 oz). If a smaller number of larger fish are desired then a basis of half to one couple per are is sufficient and about 10 fry per square metre will be harvested at an average weight of 1·5 gm measuring from 1 to 2 in.

SECTION V
CULTIVATION OF TILAPIAS AND OTHER AFRICAN CICHLIDS

Introduction

The rearing of African cichlids is a new branch of fish farming. In Central Africa, apart from a few cold water trials, fish cultivation only really started after the Second World War, although experiments had been carried out previously. It is warm water cultivation, principally, up to now of tilapias. Having started well it ceased to make progress after 15 years and even began to decline. The reasons for this and the prospects for the future cultivation of tilapias will be discussed in this section.

Since 1924 *Tilapia nigra* have been kept in ponds in Kenya and since 1937 some experiments in rearing tilapias have been undertaken in the Zaire. These were intensified and organized on a scientific basis from 1946. At that time efforts were devoted principally to *T. melanopleura* and *T. macrochir*. Fish culture research stations were set up in several African countries. Conferences on African fish cultivation were held, the first at Elisabethville in 1949, the second at Entebbe in 1952 and the third at Brazzaville in 1956.

Simultaneously, the cultivation of *T. mossambica* started developing in Indonesia in 1939 and from there extended over the whole of the Far East. Starting from that region or from Africa (via Europe) rearing trials with different species of tilapias were tried in Central America, South America and in the southern states of the United States. Cultivation also developed in the Near East.

The particular interest of *Tilapia* to equatorial

and tropical countries is that they develop well in fresh water just as in brackish water. Interest, unhappily, is notably reduced because of excessive reproduction and all its resultant problems.

I   Cichlids suitable for Cultivation

A   Tilapias

Belonging to the Perciformes and to the Cichlidae family, *Tilapia* form a genus of fish found in the intertropical waters of Africa.

Tilapia are robust fish. They withstand high water temperature well and their respiratory demands are slight. They are easy fish to transport which, with easy reproduction, accounts for the success of their dispersion outside their natural area.

Tilapia are warm water fish. Their optimum development occurs at temperatures above 20°C and even up to 30°C and more. The lowest temperature they can withstand is between 12 and 13°C except for certain species (8°C for *T. sparrmani*). Many species are euryhaline and can withstand brackish waters, *T. mossambica* and *T. nilotica* for example.

The classification of *Tilapia* is difficult and not yet definitely fixed. Some writers have called for a revision of the classification of this genus. There has been frequent confusion in nomenclature of the species and some names are synonymous.

Cultivation trials have been carried out with 15 or 20 different species. Some gave positive results. These are given below in two groups:

1. One herbivorous and macrophytophagous group of the species, the principal of which are *T. rendalli* (= *T. melanopleura*), *T. zillii* and *T. tholloni* have high reproduction potential and do not practise mouth incubation. These species have only few gill rakers: eight to a dozen, on the first branchial arch. The food requirements of the herbivorous species make them of particular interest, for their artificial food can comprise varied or different leaves and this makes feeding generally easy.

2. A microphagous and omnivorous group of the species with lower reproductive potential, practises mouth incubation. The principal members of this group are *T. macrochir*, *T. mossambica*, *T. nilotica* and *T. nigra*. The number of gill rakers on the first branchial arch is higher: 14 to 20 for *T. mossambica*, 20 to 26 for *T. macrochir*.

Other species, among many more, are *T. sparrmani*, *T. galilaea*, *T. heudeloti*, *T. andersonii*, *T. esculenta*, and *T. leucosticta*, which are of less interest for cultivation. Some could be just geographical forms of certain of the preceding species.

T. mossambica (Fig. 280) originated in the east and the south of Africa down to Natal. The species was discovered in Asia in 1939 in the east of Java and its introduction into that region has never been satisfactorily explained. During the Second World War and later this species became spread over Java in many ponds with brackish water and then in fresh water, in open water as well as in artificial ponds. The Japanese introduced them into most of the Indonesian islands, in Malaysia and Taiwan (Formosa). After the war they were also introduced into Thailand, the Antilles and the south of the United States.

T. rendalli (= *T. melanopleura*) (Fig 279) originated in the west of intertropical Africa. It spread through the farms of Zaire from 1947 and has been introduced into numerous ponds in equatorial Africa. Efforts to acclimatize it have been made in Thailand and Brazil.

T. macrochir (Fig. 278) originated in the southern part of Central Africa. Since 1947 it has been cultivated in Zaire and elsewhere in Africa where it is associated with *T. melanopleura*.

T. nigra is used in Kenya particularly. *T. nilotica* appears to grow well in West Africa.

B   Other cichlids

Among cichlids other than tilapias with which experiments in cultivation have been carried out,

the following genera and species must be mentioned. Just as with tilapias these cichlids protect their young at the time of reproduction. Their eating habits differ widely one from the other.

Hemichromis fasciatus Peters (Fig. 281). This is a small voracious cichlid which can be used to reduce the spawning of tilapias in climates in which the latter are precocious and too prolific. In balanced cultivation it is necessary to reduce the tilapia offspring though not to eliminate them. Gruber and Mathieu (1959) estimate that an addition of 2 per cent of *Hemichromis* is enough to keep the balance right.

Hemichromis fry must be produced in separate ponds. The spawning ponds should be small, about 1 are (one-fortieth of an acre) in which one or more pairs of *Hemichromis* are placed. These fish do not make nests but they protect their young until they reach 20 mm. As soon as they reach 5 cm (2 in) they feed off small fish, tilapias among others.

Other cichlids, notably two *Serranochromis*, have been the subject of experiments in Katanga. The aim was to test them as controlling predators.

Haplochromis mellandi Blgr. This species was found to be interesting for Katanga from the fact that it feeds abundantly off molluscs if given the chance and is therefore a means of keeping bilharziasis down. In several African countries *Astatoreochromis alluaudi* have also given encouraging results to the same end.

II   Reproduction of Tilapias

African cichlids and tilapias especially, make their nests on the bottom of the water in which they live. The female spawns in the nest made by the male. The microphagous species such as *T. mossambica* and *macrochir* practise mouth incubation, that is to say, after hatching the female holds the young in her mouth as a protection against danger. *T. melanopleura* and *sparrmani* belong to a different group which does not do this.

The external differences between the sexes are not very clear. During spawning the colouring of the male is more marked than for the female.

1   Sexual characteristics

The external differences between the sexes are based on the fact that the male has two orifices under its belly, one is the anus and the other a urogenital aperture. The female has three: the anus, the genital and urinary apertures (Fig. 277).

The anus is easily recognized. It is a round hole.

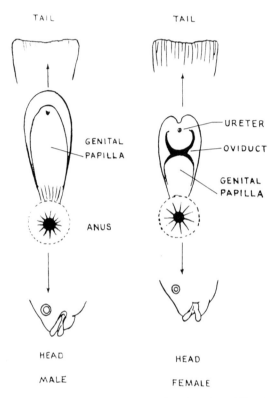

Fig. 277   Diagram of the genitalia of adult male and female *Tilapia mossambica* (after Maar *et al.*, 1966).

The urogenital aperture of the male is a small point. In some species (*T. macrochir*), the genital papilla is well developed and is divided into long whitish filaments which can measure several centimetres at the time of spawning. This papilla is also present in *T. mossambica* but it is not filamentoid and is less developed.

The urinary orifice of the female is microscopic and is scarcely visible to the naked eye, while the genital orifice is an opening in a line perpendicular with the axis of the body. This transversal opening situated between the anus and the urinary orifice is more or less easy to recognize in *T. mossambica* and *T. macrochir* but more difficult for *T. melanopleura* because of the darker colouring of the belly.

2   Size of the brood fish. Time and frequency of reproduction

Tilapias become adult during the second 6 months of their existence. At this age the size varies according to the species, the quantity of natural food found in the environment and also the living space at the disposition of the fish. In most regions tilapias start spawning when the brood fish measure no more than 15 cm (6 in) and even less.

As for most fish, temperature plays an important part in reproduction. The average daily temperature during the spawning period should be at least 20 to 21 °C, which implies that temperatures during the daytime will be higher. In equatorial, low altitude climates where the water is constantly warm, spawning goes on throughout the year. It is not the same in climates with more accentuated variations of temperatures including a hot and rainy season and a dry season which is generally colder. Under these conditions tilapias do not spawn during the cold season.

In regions with constantly warm temperatures they spawn throughout the year at 5- or 7-week intervals, or an average of eight times per year. In other regions spawning is similar but less frequent in accordance with the length of the hot season.

3   Nest building and spawning

1. *Construction of the nest.* The male constructs the nests, the shape of which differs for each species.

T. macrochir (Fig. 282) constructs its nest on the bottom of the pond at a depth of between 30 and 150 cm (1 to 5 ft). The diameter of the nest varies between 50 and 200 cm (about 2 to 8 ft) according to the size of the male. The nest is circular and complex, including an outer section with rays which extend from the central section which is in the shape of the trunk of a cone, the upper face of which is concave.

T. mossambica constructs a nest similar in shape to that of *T. macrochir* but it is architecturally simpler. It is, in fact, a simple bowl. The diameter of the depression varies between 30 and 90 cm (1 to 3 ft). The nest of *T. nilotica* is very similar to the latter (Fig. 284).

T. melanopleura spawns in shallow water either on a rather steep bank of 45 to 60° for example, or on the bottom at a spot which is raised. The nest is composed of from five to 10 hemispheric holes of 8 to 12 cm (3 to 5 in) in diameter in a collection of about 50 to 75 cm (20 to 30 in) in diameter

(Fig. 283). Normally *T. melanopleura* prepares its nest in the sides of the banks of the ponds while *T. macrochir* spawns on the bottom.

2. *Spawning and incubation.* When the male has dug the nest the female lays the eggs in it and they are fertilized. After spawning, tilapias do not abandon their eggs but watch over them or retain them in the mouth until hatched (mouth breeding).

The eggs of *T. melanopleura* measure around 2·5 mm in diameter. They are laid in one of the holes of the nest, generally the largest, and stick one to the other on the upper side of the hole. After spawning the female stands guard over the nest to ward off predators. Normally she takes up a position in the hole itself with her head sticking out.

After spawning and fertilization those eggs belonging to the mouth breeding species (*T. macrochir*, *T. mossambica*, etc.) are held in the mouth by the female and kept there until hatching. During this period the female's mouth is distended. Incubation does not take long.

The number of eggs varies according to the species and the size of the brood fish. There are, for each spawning, several hundred at least and several thousand at most. Finally it is possible to count on several hundred fry (300 was the average during Ineac trials at Yangambi in 1956 and 500 at other trials).

3. *Hatching.* The eggs are hatched about 6 days after spawning. After hatching the fry are still protected by their parents. As the fry grow so they disperse. Gradually the school of fry breaks up and at the end of 10 days it is practically disintegrated. The fry now swim around in shoals.

III Feeding and Growth of Tilapias

1 Feeding

a Natural food

Tilapias show marked preferences for certain food. Some feed off micro-organisms predominantly while others are herbivorous. *T. mossambica* are omnivorous and above all plankton eaters. *T. melanopleura* are phytophagous, feeding off filamentous algae and aquatic plants. They prefer submerged, semi-submerged or floating vegetation. They also go for emergent plants provided the stalks are not too ligneous. *T. macrochir* is a fish which feeds off small micro-organisms and essentially, at all ages, plankton and biological cover. Most tilapias also feed, quite abundantly, off insect larvae, crustaceans and detritus.

b Artificial food

In fish cultivation in ponds tilapias do well by feeding off the most varied artificial food.

However, during the early age, tilapias do not do this and the amount of natural food should be increased by fertilization. By the time they reach 4 to 5 cm (2 in) they start taking artificial feed.

Tilapias can easily be fed with numerous plants, farinaceous vegetation and different kinds of waste (Fig. 228). Plants are particularly suitable for the herbivorous, while farinaceous food is satisfactory for all species (Figs. 289 and 290).

The principal plants used are the leaves of cassava, sweet potatoes, eddoes, banana trees, pawpaw, maize, canna and several plants and legumes. The tubers and peelings of cassava are not very suitable.

Among the farinaceous food are generally found meal waste and more particularly cassava bran and flour, chips and balls of rice, corn flour, cotton and groundnut oil cakes.

Industrial and domestic waste, such as decomposed fruit, brewery draff, coffee pulp and wastes from local beverages are usable. Finally, all tilapias like termites.

In reality, the production of tilapias is only viable if they are fed intensively. Tilapias can be considered as remarkable transformers of waste and by-products. They are able to use and transform vegetable waste in the same way as trout transform animal waste.

If the fish are fed regularly, production in ponds can be doubled, tripled and even increased 10 times and over.

The conversion rate of distributed food is not known as yet with sufficient precision. Certain concentrated feed such as oil cake should have a conversion rate of from 3 to 5, while that of leaves should be from 15 to 20. Rice chips should have a conversion rate around 8.

Feed which is not consumed decomposes, and turns into organic manure which favours the development of abundant plankton.

2 Growth

Growth of tilapias varies considerably; first it varies according to the species and secondly according to the individual fish. It also depends on the feed available, both natural and artificial. In rich waters growth is much faster and as with all species of fish it is also tied to the temperature.

In open water the large species reach a length

of 40 cm (15 in) and weigh around 1,200 to 1,300 gm (40 to 45 oz). For cultivation it is best to produce, in certain regions, a saleable fish around 20 cm (8 in) which corresponds to a weight of 150 gm (5 oz). In certain rich waters *T. macrochir* and *melanopleura* reach these sizes after 10 or 11 months but this is not the case when the water is poor. Here, the same species will not even reach 100 gm (3 oz) in one year.

The male grows faster generally than the female because, where there is mouth incubation, during the incubation period the female does not eat at all.

IV   Methods of Rearing Tilapias

The principal rearing methods or groups of methods are given below. But no matter which method is used for the cultivation of tilapias, a micro-organism eater and a phytophagous tilapia are generally associated.

A   Rearing by mixed age groups

1   Permanent mixing of age groups

Rearing young and older fish is carried on continuously in the same pond. The latter is stocked with fish of all ages and sizes from fry to brood fish. Quantitatively, stocking varies according to the intensity of the production of the farm but it is generally high. This method, known as the "mixed method", corresponds to the old "Femelbetrieb" method utilized formerly for the cultivation of carp. The aim is not to produce size-graded fish but fish of all sizes in as great a quantity as possible.

After 3 or 4 months the stock will almost be up to the capacity of the pond. This level is maintained through intermediary fishing, but if it is not enough then there will be over population and dwarfing. After 8 to 12 months the pond is drained and the fish harvested. A quantity of fish of all sizes is selected for restocking while the rest are sold for eating.

This is a simple method which can be used by all. If the fish are very well fed artificially they will give a high production amounting to several tons per hectare per year. But the precocious reproduction of tilapias, above all in regions which have no cold season, soon leads to over population of young fish (Fig. 285). This means that, practically, the major part of the harvest will be made up of small fish and the proportion of large fish will be very small (Lemasson and Bard, 1966).

Considering that ponds are stocked again with fish of different sizes including those of poor growth, this method hinders genetic improvement. In order to remedy this it is necessary to renew the source of fish for stocking periodically.

2   Temporary mixing of age groups

Rearing young and older fish takes place in the same pond though successively at first. The pond is stocked with young fish of more or less the same size. They are reared and allowed to fatten and to reproduce once only. The pond is drained when the largest fry, already hatched in the pond, can be used for a new stocking.

After draining, the adult fish are sold for eating but weight per fish is small, around 60 to 100 gm (2½ to 3½ oz). Fry are used either totally, or in part, for further stocking.

This method has the advantage of giving a fairly large proportion of more or less size graded fish (Fig. 287). But it has the inconvenience of giving a smaller weight production than the preceding method. This happens with all types of farming which sets out to produce fish graded for size.

The method, as such, is rarely practised for long. Often it changes rapidly to one of permanently mixing ages, for the first draining is followed by stocking with fish which are of more or less mixed sizes. This is the result of fry not of the same age being hatched from the first stocking. The following stocking basis comprises, therefore, fish of different sizes and ages.

B   Rearing by separated age groups

Rearing young and older fish are two separate operations that are carried out in different ponds. This method is practised to produce fish graded in size for the table (Fig. 286).

Spawning and rearing young fish are carried out in spawning ponds (Fig. 291) used only for the production of fry which are sufficiently large as to be released in fattening ponds. In these only fish of the same size are liberated.

The *reproduction ponds* serve as spawning ponds and also for first rearing. They are small ponds though no smaller than 1 are (one-fortieth of an acre) nor larger than from 5 to 10 ares at the most, and in which from one to five pairs are placed per are. Only one species should be raised in each reproduction pond. The fry are harvested at around 4 cm (1⅝ in), a size which they can reach in 2 months according to the water quality and the quantity of food they find in it.

In the *growing ponds*, the aim is to produce, as rapidly as possible, fish weighing 100 gm (about $3\frac{1}{2}$ oz) for eating. This is often difficult, bearing in mind the premature reproduction of tilapias which reproduce sometimes when as large as only 10 cm (4 in) in water poor in natural food. When this is the case the harvest is mixed as a matter of course and the method becomes the same as the preceding one (temporary mixing of ages).

With a view to helping the rapid growth of these fish for the table and holding back premature reproduction, they must be abundantly fed, for the faster their growth the larger their size at the time of the first reproduction.

Variation. Another method is to rear, not in one but in two phases, by means of an intermediate stage in a second rearing pond. Fry from the reproduction pond should be liberated in those ponds, fed for 2 or 3 months and then harvested at sizes from 8 to 10 cm (3 to 4 in). They should be transferred to the properly called growing ponds. This method, which is not used, would, so it seems, reduce the partial non-usage of the growing ponds at the start of stocking, and should also permit them to be better controlled.

C Rearing with controlled reproduction

To avoid and remedy more or less the mixing of ages which results either voluntarily or involuntarily, as in the preceding methods, the following systems can be used.

1. *Association of tilapias with a predator.* To control excessive or undesired reproduction of tilapias reared according to methods just described, a predator can be added. This has been tried, notably with *Hemichromis fasciatus* (Fig. 281) (Zaire and the Cameroun), *Lates niloticus* (Nigeria) and *Micropterus salmoides* (Madagascar).

The method is a delicate one to apply and the results are uncertain. Sometimes the activity of the predator is too brutal and there are not enough fry available for restocking. Sometimes it is not brutal enough and therefore it does not prove a remedy at all so that the result is over population and dwarfing of the size of the stock.

2. *Monosex cultivation.* The inconveniences resulting from premature reproduction of tilapias can be avoided if the ponds are stocked with fry of one sex only. Male tilapias grow more rapidly than females and two methods are possible.

a. *Sorting of sexes.* It is possible to identify the sex of most tilapias at a rather early age. After this is done, the growing ponds are stocked with fish of one sex. But the method is complicated because it requires several categories of ponds. It is also very delicate, needing great care and accuracy. Mistakes are almost inevitable and it takes the presence of only a small proportion of the sex unselected to fault the operation and so upset it. The method can only be used by experienced fish farmers, aided by very conscientious personnel. But in any case it seems very difficult to follow.

b. *Hybridization.* Following the interesting results obtained in Malaya with hybridization of *T. mossambica* and *T. hornorum* from Zanzibar, hybridization trials were undertaken in certain African countries (Meschkat, 1966). A fair number of trials produced hybrids but rarely 100 per cent of one sex which was the aim. However, a satisfactory result was obtained in Uganda in 1965 by crossing *T. mossambica* males and *T. nilotica* females. Other hybrids, all males, were obtained in Uganda and the Ivory Coast with *T. nilotica* females and males of other species.

V Tilapias Rearing Systems in Africa

By the term "system of fish cultivation" (Lemasson and Bard, 1966) is meant the practice of a method of fish cultivation under determined ecological and socio-economic conditions.

A Types of raising ponds

There are two principal categories of ponds, large and small, the difference between them being not one of size only.

1. *Large ponds* (Fig. 292). These are ponds with variable dimensions, in general at least 10 ares but which can cover several hectares. They are installed with classic supply, control, and water evacuation devices.

Barrage ponds should only be built when there is no fear of an influx of water likely to provoke a rupture of the dike. This can be the case in forest regions. In savanna country and high altitude regions it is recommended to lay diversion ponds and the by-pass channels must be adequately large. Barrage ponds should be provided with an appropriate overflow.

Wild fish (*Haplochromis*, *Hemichromis*) must be carefully controlled at the entrance of the pond.

To this end, at Rwanda, a double horizontal screen system was installed.

2. *Small ponds* (Fig. 294). These are used in family fish cultivation in which the construction of individual ponds is encouraged. Such ponds should have only a small surface, though without being too small. One are is considered to be a minimum and 5 ares may be a maximum. The ponds are grouped in complexes of varying sizes. The emplacement of the complex is of capital importance. Intensive cultivation will be practised so the distance between the complex and the village where the owners live should not be too great. Many such groups leave much to be desired in this respect.

The ponds should be of as simple construction as possible, which means it is not enough just to dig a hole to be filled with water. The water supply and drainage capacity must be ensured without difficulty as well as the overflow. The use of durable materials should be avoided and the water be supplied and evacuated by simple devices such as bamboo pipes. It must be certain that the ponds shall receive sufficient water which calls for a normal, working supply channel. It is also necessary to be sure that the ponds can be dried out. This can be done by simply making a hole in the dike. Nevertheless the latter must be robust though not excessively so. Dikes should be grass covered in order to avoid gullying. The depth of the water must be sufficient (0·50 m at least) (20 in).

B Rearing systems (Lemasson and Bard, 1966; Huet, 1957)

1. *Communal fish cultivation* (Fig. 293). This is subsistance cultivation. Large ponds are constructed with public funds and local labour. Cultivation is carried out by the local community (village, school or mission) which uses the production for its own subsistance.

Farming conditions are generally mediocre. Stocking, intermediate fishing and draining are generally carried out at random. Often no one takes the trouble to feed the fish. Under these conditions production is poor and never reaches more than a few hundred kilos per hectare (2½ acres).

This was the first system of fish cultivation to take root in Africa but, considering the generally poor results, notably because of neglect, its development was limited. However, in certain regions such as Maniema (Zaire), positive results were undoubtedly obtained.

One particular case is that of communal cultivation in water reserves laid down for a variety of reasons (domestic needs, conservation of the soil, irrigation) in regions with a long dry season. If these reserves can be drained regularly or dried naturally they can be stocked when the reservoir is put under water, and harvested after drying. Such trials were undertaken in Kabylia (Algeria) in 1961.

2. *Family cultivation* (Figs. 294 to 296). This is also subsistance cultivation though different from that just described. It is practised in such a way that the peasant can produce in his own pond, constructed and exploited by himself without expense and demanding only a little work, at least part of the protein necessary to feed his family.

This is carried out in family ponds. Too often, however, the ponds are badly planned, laid out and maintained. Stocking is generally variable and vague. Artificial feeding which should be intensive in order to compensate for the small size of the pond, is neither regular nor abundant. Furthermore production is very variable and often there is none at all, though it can reach 25 kg (55 lb) per are per year. Generally it is made up of small fish. Harvesting starts with intermediate fishing, generally with lines and the ponds are drained after several months but sometimes only after 2 or 3 years.

This system of fish cultivation developed considerably in inter-tropical Africa between 1953 and 1958. Since then it has declined for a variety of reasons. Among them, difficulty of exploiting a badly constructed and laid out pond, failure of the farmer to persevere, insufficient help and advice and discouragement because of poor results which in any case were due to poor farming.

3. *Artisanal cultivation* (Fig. 297). This is not subsistance farming, for production is intended to be sold on the market, totally or partially, or to meet the food requirements of an agricultural exploitation.

Artisanal cultivation and exploitation are carried out more or less rationally. Stocking, feeding, intermediate fishing and draining are relatively well regulated so that in consequence production is rather high.

If such a system of cultivation could be integrated with an agricultural farm (Fig. 298) then the manner in which it is run and its production could be greatly improved.

Until now, artisanal cultivation has been principally aimed at the production of fish for eating. There are no specialized farms to meet the demand, nor transport arrangements of fry such as exists in south-east Asia.

VI Future Prospects of the Cultivation of Tilapias

1 Species suitable for cultivation

Tilapias are without doubt suitable fish for viable cultivation in inter-tropical regions but because of their excessive reproduction they are not ideal for producing graded fish for consumption. Happily, thanks to their adaptability, they offer precious material for hybridization and the selection of varieties and interesting strains. Hybrids permitting monosex cultivation have already been obtained. Until now it has been a question of microphagous species. To produce species of herbivorous hybrids would be of great interest, taking into account the ease with which they can be fed artificial food.

Tilapias can be mixed not only among themselves, e.g. phytophagous and microphagous species, but it is also interesting to mix them with fish of other genera such as *Heterotis* in Africa, common carp and grass carp (*Ctenopharyngodon*) in Asia. In Israel, tilapias (*T. nilotica* principally) are mixed with carp, the latter being the principal species; tilapias grow, in the summer, to 180 gm (approximately 6 oz). In Malaya *T. mossambica* is mixed with *Puntius javanicus* and *Ctenopharyngodon idella*.

2 A good cultivation policy is necessary

For tilapias, as for all other cultivated fish, the raising of fry and the production of fish for the table are separate operations. In other words the aim must be to replace the mixed age method by a separated age method.

Also such an orientation means setting up a service to make it known, as well as an advisory service with specialized rearing stations at its disposal capable of providing farmers with fry. It is also necessary to avoid wasting effort, as interest should be focused particularly on a specially favoured environment from physiographical and socio-economic angles (Lemasson and Bard, 1966). The need for a sufficiently developed and rationally organized fish cultivation policy has been outlined by Huet (1957).

It is also necessary to encourage the creation of artisanal fish cultivation by artisans who are specialized either in the production of fry for restocking, or in the production of fish for the table, or for both.

3 Intensive artificial feeding

All production of fish for the table should be as intensive as possible. Intensive production is particularly suitable for family cultivation where it is necessary to compensate for lack of natural production which is very limited in small sized ponds. Fortunately, tilapias benefit greatly from different food of poor economic value.

Under rural conditions the goal should be to produce 25 kg (55 lb) per are (one-fortieth of an acre). If it is accepted that natural production would reach 5 kg (11 lb), then production with feeding should be 20 kg (44 lb). In taking a conversion rate from 12 to 15 then 240 to 300 kg (approximately 500 to 660 lb) of food should be distributed in the year in a pond of 1 are which, more or less is the same as distributing 1 kilo ($2\frac{1}{4}$ lb.) of food per are per day.

Fig. 278 *Tilapia macrochir* Blgr. Typical example of the microphagous species.

Fig. 279 *Tilapia rendalli* Blgr (*melanopleura* Dum.). The best known of the phytophagous species.

Fig. 280 *Tilapia mossambica* Peters. Successfully introduced in fresh water ponds, in brackish water and in rice swamps as well as in certain open waters. First cultivated in Indonesia in 1939.

Fig. 281 *Hemichromis fasciatus* Peters. A small voracious fish of the cichlid family more or less capable of controlling excessive reproduction of *Tilapia* in ponds.

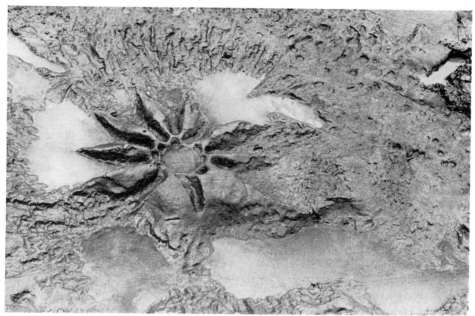

Fig. 282   Nest of *Tilapia macrochir*. Tilapia make breeding nests. Their construction differs from one species to another. *T. macrochir* arranges its nest at the bottom of the pond. Its construction is somewhat complicated.

Fig. 283   The nest of *T. melanopleura* is a group of shallow juxtaposed holes dug into the bank of the pond.

Fig. 284   The laying nests of *T. nilotica* resemble those of *T. macrochir*, but are of a much simpler shape.

Fig. 285 The breeding of tilapias in ponds presents no problems. Raising in mixed age groups produces a population comprised mostly of small size fish.

Fig. 286 (left) The production of graded tilapias of a good size is a desirable objective, but which is difficult to achieve.

Fig. 287 (above) Different methods can produce graded fish: abundant feeding, association with a predator, monosex culture.

Fig. 288 The culture of tilapias must be as intensive as possible by artificial feeding based on the use of vegetable products of little value. Above can be seen leaves of banana trees and cassava, rice bran, palm and ground nuts cakes, sweet cassava and cotton seed.

Fig. 289 Cassava leaves are greatly appreciated by the herbivorous tilapias. Distribution from the bank of a pond.

Fig. 290 Bits of rice and chaff are suitable for feeding all tilapias. They can be distributed from a small boat.

Fig. 291   View of the principal breeding centre at Gandajika, in the province of Kasai, Zaire. It includes a number of breeding, holding, production and experimental ponds.

Fig. 292   In 1948 an important fish culture station was set up near Lubumbashi (Katanga), Zaire, in the Wangermee valley. The fish were abundantly fed with flour-mill waste. Production frequently reached 3 to 5 tons per hectare (2½ acres) per year.

Fig. 293   The cropping of fish in communal ponds is carried out collectively. After draining, the pond maintenance work and improvements are carried out, for example green manuring; this is also carried out collectively. Kakoto pond, Kasongo, Zaire.

Fig. 294   Family fish culture is generally practised in small diversion ponds of around 1 to 2 ares (1 are = one-fortieth of an acre) grouped together, and in which intensive fish culture is tried out. Kinkala, Congo-Brazzaville.

Fig. 295   A group of individual ponds at Tshiamala-Panga in Dimbelenge territory, Zaire. Each pond covers around 1·50 ares (1 are = one-fortieth of an acre). Joined to the group are a few nursery ponds (in the foreground) covering a slightly greater surface.

Fig. 296   A group of individual ponds in the Iloma sector, Kabare territory, Kivu, Zaire. On the flanks of the valley one can see banana plantations in concentric rows to avoid erosion. The leaves of the banana trees and banana waste can be used for food.

Fig. 297 An example of artisanal and native fish culture. In this type of fish culture one peasant owns six ponds covering a total of 18 ares. Production is semi-intensive and the small fingerlings are sold to individual owners of ponds. Kamukunga, Kabinda, Zaire.

Fig. 298 Certain agricultural enterprises try to produce fish with which to feed their workers. If the feeding of the fish is abundant, production is high and can reach and even exceed 2 tons per hectare per year. Walthoff pond, Kasongo, Maniema, Zaire. The fish are fed with waste from rice factory.

Chapter VII
BREEDING AND CULTIVATION OF CATFISH

CATFISH are a sub-order of fish grouping several families, all of which are more or less similar in appearance. They are found almost throughout the world; there are also marine species.

Their bodies may be either long or short but are generally cylindrical in cross-section and thick. Some are greatly elongated and anguilliform (or eel-like) in shape. The skin is generally naked – i.e. without ordinary scales, but the body may be covered with bony plates which could be considered as being a secondary squamation. The fins are soft-rayed but many species have fins with one to three hardened bony rays which are used for defence. According to the different species, the latter are either in front of the dorsal fin and the pectoral fins, or in front of the two pectoral fins. Exceptionally, they may be in front of the adipose fin only, which is either more or less developed, though sometimes absent altogether. One to four pairs of barbels, which sometimes are very long, are characteristic of catfish, and it is because of these that the species are referred to by that name. These fish have many small teeth planted in rows but they are not powerful. The behaviour and feeding habits of catfish vary. They are fish which live on the bottom.

Although they are widespread throughout the world their cultivation is not very developed. In Europe it is limited to the cultivation of the Danubian wels. In North America cultivation is relatively important, primarily for different species of *Ictalurus*. Certain catfish are raised in the Far East. In Africa *Clarias* are found after cultivation ponds have been drained; in fact, however, the rearing of these catfish has not been well studied.

I Breeding and Cultivation of Danubian Wels

1 General notes

Danubian wels (*Silurus glanis* L.) (Fig. 299) are the only European catfish with the exception of a second species found in the Ionian Sea area. Danubian wels live in rivers and lakes of Central Europe and Asia and more particularly in the basin of the Danube and in the Caspian Sea area. They are not found in southern Europe and are only present, almost in isolation, in a few western European countries from Denmark to eastern France (Doubs) and Switzerland.

Their dark skin is without scales, and the anal fin, which is very long, is separated from the caudal fin by a slit. The head is big, even massive, and the mouth has six barbels of which the upper two are very long.

Except when young, when they live off plankton and, later, small aquatic fauna, Danubian wels are voracious. They live off fish, batracians and other aquatic animals and are of rapid growth. At 1 year they can weigh more than 100 gm (about $3\frac{1}{2}$ oz) and at 2 years can exceed 1 kilo ($2\frac{1}{4}$ lb). They can live to a very old age and even attain a length of more than 2 m (about 7 ft).

The flesh of Danubian wels is tasty and appreciated, particularly from those weighing between 2 to 4 kilos (4 to 8 lb). In Central Europe, where there are no eels, Danubian wels replace them for eating as first quality fish.

The cultivation of Danubian wels is well developed in Central Europe, particularly in Hungary and Yugoslavia. Rearing aims and objectives are many and include production of fish of 1 summer intended for restocking ponds, water courses and lakes. Danubian wels of 2 summers can also be grown in carp fattening ponds where they play the same role as pike or pike-perch: they control undesirable fish. Finally they can be raised to over 2 years as fish for the table. The first two rearing systems are the most currently used.

Around 1885 a small American catfish (*Ictalurus melas* Rafin. and not *Ameiurus nebulosus* Lesueur) was introduced into Europe. This species is of slow growth. It destroys fry and competes with indigenous fish. It is therefore considered undesirable.

2 Rearing Danubian wels of 1 summer

a Rearing in several phases

It is now recommended to use, successively, spawning ponds, hatching basins and ponds of first and second rearing.

1. *Spawning ponds.* The reproduction of Danubian wels can be practised as a method of natural, controlled spawning on artificial nests, such as kakabans used for carp, or nests for pike-perch.

Danubian wels are warm water fish which reproduce at the end of spring, during the month of May, when the waters of Central Europe reach and stay at a temperature of from 20 to 22°C.

Until this time the brood fish are kept in holding ponds used in carp cultivation. They winter and feed off forage fish. When the time comes they are transferred to the spawning ponds. These are small ponds covering from 1 to 2 ares and at least 1 m (3 ft) deep into which one or two couples are placed. Identification of the sexes is difficult except at the spawning period when the belly of the female is distended. The brood fish used weigh from 4 to 12 kg (9 to 27 lb). They are 4 years old at least. The nests, placed in the ponds previously (Fig. 302), are three-footed trestles covered with submerged roots of trees, generally willows or alder. The eggs measure 3 mm in diameter and stick to these roots very well.

2. *Hatching basins.* After spawning, the eggs on the roots are transported to hatching basins in the same way as with kakabans. The basins used at the fish cultivation station at Szarvas (Hungary) measure 3 × 3 × 0·50 m (about 10 × 10 × 2 ft). These basins are fed with running water and are covered. The young catfish are averse to strong light which in any case must be filtered. At a temperature of 20 to 24°C hatching takes place within 3 days. The yolk sac is absorbed in approximately 8 days. Fry start feeding off plankton and about the 12th day they are caught at night and transported to the first rearing ponds. It is possible to harvest from 20,000 to 100,000 fry per female.

3. *First rearing ponds.* These are small ponds, no more than a few ares in size but rich in plankton and natural food. Ten to 15 fry are released per square metre (one square yard) to be harvested 3 to 5 weeks later. These young fish measure from 5 to 7 cm (2 to 3 in) and weigh from 1 to 5 gm. Survival rate can amount to from 30 to 50 per cent.

4. *Second rearing ponds.* Catfish of 1 summer are produced in these ponds. To achieve this, only small ponds are used, from 10 to 20 ares maximum, in which 10 fry are placed per square metre. The fry are fed artificially. As with trout they can be fed with fresh meat or fish with a blender added, or with pelleted food. They must be fed several times a day and are harvested in the autumn.

b Other rearing methods

A two-phase method can also be used with spawning ponds and rearing ponds. The spawning ponds can cover 10 ares ($\frac{1}{4}$ acre) and at this size can take five or six couples at least. The same number of nests as there are couples must be installed. After about 5 weeks, when the fry measure 5 cm (2 in), they are harvested after which 1,000 to 2,000 per hectare ($2\frac{1}{2}$ acres) are released in second or third summer carp ponds. In the autumn the fish of 1 summer weighing from 40 to 100 gm (about $1\frac{1}{3}$ to 3 oz) are harvested. Survival rate is around 10 to 20 per cent. If stocking is less dense then growth can reach 100 to 200 gm (about 3 to 7 oz) for the first summer.

Another simpler though less certain method can also be used. From one to three couples can be liberated per hectare in carp fattening ponds. Harvest is done in the autumn. The results vary considerably and can be either poor or excellent. However, instead of stocking with brood fish it is possible to use egg-covered nests transferred from the spawning ponds. They should be transferred dry the day after spawning, but only in a damp atmosphere. Moss should be placed between the roots to which the eggs stick. This transport operation can take several hours.

3 Rearing Danubian wels of 2 summers

The rearing of Danubian wels of 2 summers is carried out in carp fattening ponds. Like pike or pike-perch, Danubian wels play the normal role of predator. At the same time they produce flesh of excellent quality.

One hundred Danubian wels of 1 year can be liberated per hectare. In the autumn they will weigh 1 kg ($2\frac{1}{4}$ lb).

II Breeding and Cultivation of American Catfish

1 General notes

The cultivation of American catfish is widespread throughout the United States and, because it is a warm water fish, is expanding greatly in the southern states.

Catfish belong to the family Ictaluridae. The most interesting species for cultivation is the "Channel catfish" (*Ictalurus punctatus* Rafinesque) (= *I. lacustris* Walbaum) (Fig. 300). This fish is easily identified for it has a bi-lobed caudal fin and the anal fin has from 24 to 29 rays. The species is indigenous to the great rapid water courses of the Mississippi basin from the region of the great lakes to the Gulf of Mexico. They can grow to 25 lb and are well suited to turbid water.

The farming of this species is suitable for the production of fish for restocking lakes, streams or fishing ponds as well as for the table. The species is of interest to anglers for it grows rapidly in warm water and can be caught with a variety of baits. It is a good fighter and its flesh is of excellent quality. For commercial production the species has many advantages, for a great number of fry can be obtained with only a few brood fish and the fish respond well to artificial feeding. As they do not reproduce too easily in ponds, there is no need to fear over-population. It is an omnivorous species which feeds off all kinds of animal or vegetable food. When the weather is warm, catfish feed principally at night, and large specimens also feed off small fish. Channel catfish, however, are not resistant to a lack of oxygen. Also they must be handled with care.

Some fish farmers specialize in the production of fingerlings for restocking, while others produce fish for eating or catchable fish for the stocking of fishing ponds.

In the latter event they can be associated with other fish, for instance, at a ratio of 50 catfish per acre for 100 black bass and 50 redear sunfish. In small ponds of 10 ares, however, only catfish are released. On some farms, small cyprinids (*Pimephales promelas*) are associated with catfish.

Several other catfish are also cultivated. For angling, besides channel catfish, white catfish (*Ictalurus catus* L.) and brown bullhead (*Ictalurus nebulosus* Lesueur) are used. In commercial cultivation, channel catfish, white catfish and blue catfish (*Ictalurus furcatus* Lesueur) are found. Under special conditions other species of catfish such as bigmouth buffalo (*Ictiobus cyprinellus* Val.) are reared, mixed with channel catfish, in the rice fields in the south-east of the United States. Bigmouth buffalo are catastomids. Although the methods given below concern channel catfish, they are also generally suitable for all the other species.

2 Reproduction of channel catfish

Under natural conditions channel catfish lay their eggs in holes dug in banks or any other obscure spots formed by trunks or submerged roots and a variety of different debris.

Reproduction takes place in specially arranged spawning ponds using a natural, controlled reproduction method, although methods differ in different states and even on different farms. Hypophysation has been tried but is not generally practised.

a Spawning

The species reproduces in the spring when the temperature of the water is sufficiently high around 70 to 75°F (21 to 24°C). This takes place generally in April in the southern states of the United States and in May or June in the northern states. If the aim is to obtain successive, staggered spawning by prolonging the spawning period, then the brood fish should be held in spring water.

Brood fish should weigh at least 2 lb or more (3 to 10 lb) and should be at least 3 years of age. During the year they are divided up between different ponds of the fish farm. They can feed off natural food or be given artificial food. In winter they are concentrated in ponds, a few dozen ares in size, at a ratio of from 50 to 100 kg (110 to 220 lb) per hectare. They are fed moderately.

At spawning periods the brood fish are placed in a pond covering from 10 ares to 1 hectare and the couples are isolated in spawning enclosures (Fig. 303) at a ratio of one couple per enclosure. The latter should measure 3 × 1·50 m (about 10 × 5 ft). Once they were constructed in wood but now they are wire-meshed. They must be driven well into the ground in order to prevent the brood fish escaping after making holes in the bottom of the pond.

At the time of spawning, identification of the sexes is quite possible though delicate. The female is not as coloured as the male, which has a larger head. The genital aperture of the male is small and slightly prominent while that of the female is depressed and more or less swollen. When the couples are placed together the male should be slightly larger than the female.

Spawning takes place in an obscure cavity and therefore it is also necessary to place, in the enclosure, an adequate spawning receptacle. Wooden kegs (1 m long × 0·25 m sides) (3 ft long × 10 in sides) can be used Fig. 304), as well as 10 gal milk urns, earthenware pots or metal barrels. When kegs are used one side should be open whereas when an urn is used it should be placed on its side. The opening should point towards the centre of the pond or in a different direction. An aperture on the upper side will allow spawning to be observed and also the fry. The spawning receptacle must be covered with 2 ft of water and held in place with pegs driven into the bottom of the pond. Before spawning the male cleans out the receptacle. Spawning takes place inside and lasts several hours. The eggs are laid at intervals and each partial spawning is fertilized. Finally they form a glutinous mass in the shape of a flattened sphere. Eggs are about 2·5 mm in diameter.

Until spawning takes place the receptacle should be visited daily in order to ascertain the exact date of the spawning. After spawning the male becomes aggressive and reacts sharply when disturbed. The female is chased off by the male if the latter is the larger of the two; if not, there is a risk the female might devour the eggs after having attacked and killed the male. The female should, therefore, be removed from the enclosure.

When hatching is completed and the fry released into the rearing ponds or in troughs, the enclosure is prepared for a second spawning. Males can be used again, and over the spawning period it is possible to make use of them two or three times. They should be fed between each spawning.

One simplified method of reproduction is to place the spawning receptacles in the rearing ponds which at the same time serve for spawning and rearing. There are no enclosures and the fry disperse in the ponds after the absorption of their yolk sacs. Some twenty couples are placed per acre.

b Incubation
Several techniques can be used. In the first the male looks after the incubation in the spawning receptacle. He guards and protects the eggs and fans them with a continual movement of his fins. Hatching takes place 8 or 10 days later at a temperature of 75°F (24°C). During hatching the male must not be disturbed for this could lead to his leaving the nest.

Another method is to remove the eggs after spawning. The gelatinous mass is placed in incubation jars similar to those used for pike, or in oblong incubation troughs used for salmonids. The number of eggs, 3,000 to 20,000, depends on the size of the female. Water for hatching must be

sufficiently warm. It must also be well aerated. This can be assured by using a paddle to stir the water at the side of the eggs which are held in place in the trough by a mesh screen.

The absorption of the yolk sac takes about 1 week. If this happens in the hatching troughs the fry can be kept in for several days. They start eating from the fifth day.

3 Production of channel catfish fingerlings
Fry are transferred to the rearing ponds either when the yolk sacs are absorbed or after pre-feeding in the troughs. If the pre-feeding method is used, techniques similar to those for rearing salmonids are employed but, of course, in water which is sufficiently warm. The fry can be reared in troughs during a first phase lasting about 6 weeks. Then, in larger basins, for a second phase lasting 4 weeks.

But whichever method is used, the rearing of fingerlings in ponds is always intensive. This has been helped since the spread of rearing by using concentrated feed in the form of meal or pellets.

Rearing ponds are varied in size and are generally between 1 and 5 acres.

In order to avoid the destruction of the fry by enemies such as large insect larvae, the ponds are put under water only a short time before stocking. If the fry are transferred directly from the spawning receptacle to the rearing pond they can first be placed in mesh screen enclosures (four to the acre) for protection from their natural enemies. A shelter is placed inside each enclosure (a barrel lying on its side for example) in which the young fry are fed for 2 weeks. The shelter is then removed and the fry disperse. Feeding continues, always at the same spots.

The quantity of fingerlings produced per acre varies according to the amount of food distributed and the sizes required. The following ratios are mentioned: 100,000 3 in fry, 75,000 4 in fry, 50,000 5 in fry and 30,000 6 in fry. Other methods include releasing 100,000 fry per acre at the end of the spring for harvesting at 2 or 3 in. in the autumn. Survival rate varies between 50 and 85 per cent. It is important to distribute quantities of food in the form of meal or pellets. Much study is done to determine the best food for this cultivation. Appropriate food should contain, according to the Fish Farming Experimental Station at Stuttgart, Arkansas, U.S.A.: 32 per cent protein (half animal and half vegetable), 15 per cent fish meal (which also includes protein and other nutrients), 5 per cent fat, 6 per cent cellulose, 1 per cent calcium, 1 per cent phosphorus and $\frac{1}{2}$ per cent different vitamins. Feeding is never carried out below 60°F (15·6°C) and is reduced or suppressed if the water is too warm and exceeds 30°C.

These fingerlings, which are to be kept throughout the winter, should be concentrated at a ratio of 100,000 per acre. Before being released they are disinfected in a formalin, permanganate or acriflavine bath. Throughout the winter the fish are fed moderately; one feeding per day equal to 1 per cent of their weight. They are fed 6 days per week.

4 Production of second and third year channel catfish

The production of second year channel catfish can be carried out in ponds. The stocking rate is around 1,000 to 2,000 fish per acre or, exceptionally, 5,000. The fish are fed daily (six times per week) during the growing season (210 days) and the daily ration is 3 per cent of weight. It is necessary to ascertain that the food is consumed regularly and this can be controlled by submerging feeding trays. The individual growth of second year fish can reach or exceed 1 lb. Production is between 1,000 and 2,000 lb per acre.

Rearing channel catfish can be continued into the third year and stocking is at a ratio of 1,000 fish per acre. Normal individual growth reaches 2 lb during the third year and the final weight is 3 lb. The daily ration is 2 per cent of the weight of the fish.

III Breeding and Cultivation of Clarias in the Far East

1 General notes
The cultivation of clarias (family Clariidae) is very popular in the Far East. They are indigenous to India and to the Philippines and are raised in India, Pakistan, Malaya and Vietnam. Highest production is in Thailand. Two species are raised: *Clarias batrachus* L. (Fig. 301) and *C. macrocephalus* Gunter.

These fish are popular for cultivation because they grow rapidly and also because they give a high production and are sturdy and resistant. They can withstand transport out of the water over long distances, and can also survive in water which is poor in oxygen and which is not suitable for other species. Clarias can live under all conditions and

can be raised in a restricted environment. The quality of their flesh ensures a high price.

Their feeding habits are very varied. They eat insects, worms, crustaceans, decomposing organic matter and also fish meal, rice, and rice bran as well as chopped vegetables. They can grow to 40 cm (about 16 in) in size.

2 Reproduction of clarias
Clarias reproduce during the rainy season (May to October) and, in ponds, perhaps all the year round.

The sexes can be recognized at the period of reproduction because the genital aperture is different and the belly of the female is swollen.

Clarias make their nests in the shape of holes of from 20 to 30 cm (8 to 12 in) in diameter and of variable depths. The holes are generally from 20 to 30 cm below the surface of the water, on the bottom of a rice swamp or in a bank with weeds and plants. The eggs are laid in the nest and stick to the roots of the plants or to the bottom of the nest. After spawning, the male watches over the eggs which are from 1·3 to 1·6 mm in diameter. They hatch out after 20 hours in temperatures from 25 to 32°C. There can be as many as 2,000 to 5,000 (15,000) fry to a nest.

Until now most fry (millions) have been gathered in their natural habitat (rice swamps, irrigation canals, marshes), but they can also be bred artificially. This can be done in special laying ponds arranged in the same way as for channel catfish, but more simply, in which a certain number of couples are placed. Before doing this, nests 20 to 35 cm (8 to 14 in) in diameter are installed about 20 to 25 cm (8 to 10 in) below the surface of the water. Aquatic plants are also put in place in the nearness and on which the spawning fish can lay their eggs. The nests are examined every 3 days and the final result is around 80 per cent fish spawning.

Reproduction after an injection of pituitary extract has also been tried with some success. It is possible to incubate clarias eggs either in jars or in hatching troughs. The water must be changed regularly and the unfertilized eggs must be removed. Yolk sacs are absorbed after 5 days.

3 Production of clarias fingerlings
The production of clarias fingerlings offers no problems and can be carried out intensively within a limited space. It is possible to raise almost 10,000 fry per square metre in a space between 2 and 3 m²

($2\frac{1}{2}$ to $3\frac{1}{2}$ yd²). The water should be shallow and in any case less than 1 m in depth. Artificial feeding includes zooplankton, fish flesh and ground nut oil cake. Within 2 weeks the fry reach 4 cm ($1\frac{1}{2}$ in). Rearing can go on for 3 to 4 weeks (7 to 10 cm) (3 to 4 in).

4   Rearing clarias for consumption

The aim generally is to produce fish 25 cm (10 in) long weighing 200 gm (about 7 oz). This size is reached in from 4 to 6 months in fattening ponds and permits two or three harvests per year.

Rearing is carried out in small ponds from 100 to 1,000 m² (120 to 1,200 yd²) with a depth from 1·5 to 3 m (5 to 10 ft). The initial depth of 1 m is gradually increased to 2 or 3 m. The bottoms and the banks of the ponds must be hard in order to avoid the fish digging into them. Ponds must also be surrounded by lattice fences made of bamboo or metal in order to stop the fish from escaping. It is also possible to raise fish in floating cages in conjunction with poultry.

Stocking is 50 fry of 7 to 10 cm (3 to 4 in) per square metre. The fish are fed abundantly with meat or fish offal and also with the waste from canning factories, as well as cooked rice mixed with bits of vegetables and oil cake. They are hand fed to the point of saturation, once at night and twice daily. The conversion rate is taken as 6. Heavy production can be secured – as much as 1,000 kg (1 ton) per are, or 100,000 kg per hectare (100 tons per hectare or $2\frac{1}{2}$ acres).

Fig. 299 European catfish or "wels" (*Silurus glanis* L.) Four-year-old fish; length: 1 m (3⅓ ft); weight: 5 kilos (11 lb).

Fig. 300 Channel catfish (*Ictalurus punctatus* Rafin.).

Fig. 301 *Clarias batrachus* L. (Thailand).

Fig. 302   Artificial spawning nests for European catfish. Szarvas fish culture station, Hungary.

Fig. 303   Spawning enclosure for channel catfish. Auburn, Alabama, U.S.A.

Fig. 304   Spawning cages for channel catfish. Auburn, Alabama, U.S.A.

Chapter VIII

SPECIAL TYPES OF FISH BREEDING AND CULTIVATION

SECTION I

CULTIVATION OF EELS

THE cultivation of eels is carried out for stocking and for eating purposes.

Eels are catadromous migrants. They spawn at sea and pass their principal growing phase and life in inland waters. Their habits are well known so there is no need to describe them in detail here. It is necessary to recall, however, that the growth of females in inland waters is far superior to that of males. Among the different species of eels spread throughout the world, three are of economic importance. They are the European eel (*Anguilla vulgaris* L.) (Fig. 305), the American eel (*Anguilla rostrata* Lesueur) and the Japanese eel (*Anguilla japonica* Temminck and Schlegel). All three are very similar in shape, growth and behaviour. European eels pass up to 2 or 3 years at sea in larval form while the two other species stay at sea only 1 year. Japanese male eels can weigh up to 250 gm (9 oz) and measure 60 cm (2 ft) in length.

I Cultivation of European Eels for Restocking

1 The need for restocking eels

The cultivation of European eels for restocking is limited to the harvesting and transport of elvers and young eels.

European eels pass their growing life in all the inland waters, both fresh and brackish, and in streams and rivers ranging from the trout zone to the bream zone. They are also found in still waters such as lakes, ponds, lagoons and marshes. Eels are fished industrially on an important scale in large rivers such as the Rhine, the Loire, and the Po, etc. The most important fishing occurs when the eels approach sexual maturity and swim down to the estuaries of large rivers to the sea. They are caught by fixed fishing installations (Fig. 308) or with the aid of large nets installed on the sides of special boats, called "schokkers" by the Dutch. In their earlier stage, when the elvers swim upstream in fresh water, they are captured in great quantities both for eating and for restocking inland waters.

The large seaward running eels are declining in numbers in many water courses and, in correlation with this, the number of elvers swimming upstream is also diminishing. This may be the result of the overfishing of elvers in the first place but it must also be due to pollution which is more and more evident in estuaries and the lower reaches of large rivers and other important streams. It is also doubtless due to dams which stop the elvers swimming upstream. One remedy being tried is to catch elvers for restocking in the lower reaches of

Fig. 305   European eel (*Anguilla vulgaris* L.).

217

streams. Such fishing installations exist notably in France, Great Britain and Italy.

Certain inland waters achieve very high eel production. For example, Lough Neagh in Ireland, which covers 35,000 hectares (87,500 acres) produces 700 tons per year (about 20 kg per hectare or 20 lb per acre). The introduction of elvers in Lake Balaton in Hungary in 1961 has given good results. After 4 years some eels measure 90 cm (3 ft) and weigh 1,000 gm ($2\frac{1}{4}$ lb); the production expected yearly is 4 kg hectare (4 lb to the acre) from this source alone. Good results were also obtained when elvers were released in Lake Einsidel in Austria.

In Spain and Portugal elvers are caught and eaten as such.

2   Capture of elvers for restocking

Two periods, for the run of elvers, have been observed at the mouths of the great European rivers. The first takes place during the winter and spring and the second at the start of the summer in June and July.

In the Loire during these periods "strings" of elvers, one after the other, mount the river at each favourable tide. They are composed of fry which are translucid at first and then pigmented. They are about 7 cm (3 in) in length and as thick as a matchstick. In that form they do not customarily travel more than about 100 km (60 miles) upstream from the mouth.

After the first warm weather, however, there is an important upstream migration by the larger elvers, or rather small eels, of which those between 15 and 20 cm in length (6 to 8 in) and the thickness of a pencil, dominate. These fry at one time formed easily recognized "strings" seen several hundred kilometres from the sea. The stronger fish probably comprise elvers of the preceding year which remained in the lower reaches of the river until they developed and had sufficient strength to swim over longer distances.

The capture of elvers for restocking is always made from the first upstream migration. Fishing starts in December, when the autumn is mild, but normally the best results are obtained between February and May.

The elvers are captured either with a wire mesh sieve or with a plankton net drawn by a motorboat. The fish are placed in tanks supplied with oxygen which enables them to be brought in perfect condition to the holding station where they are placed in floating boxes. They then await transport to their ultimate destination.

Elver capturing stations are installed in the lower Loire where the rise of the tide carries the fish into the river. They are carried upstream in dispersed form by the tide, but when the tide begins to fall they tend to struggle and form columns along the banks and the quayside where the current is less. This then is the "string" and it thence moves only at night.

Further upstream where there is no tide the "string" can be seen swimming every night in good weather making a few kilometres progress. During the day the eels remain on the bed in the bottom mud or sand showing their heads only.

Curiously enough these fry which fear daylight are, nevertheless, attracted by artificial light. Fishermen know all about this peculiarity and take care to provide themselves with lanterns when they go fishing. The nets used are round, have an 80 cm ($2\frac{1}{2}$ ft) opening and are approximately 35 cm deep (14 in).

Breathing is very important for elvers because they die rapidly if the receptacles in which they are placed after being caught have insufficient oxygen.

They are usually stocked on boats containing several tanks which are placed on either side with a passage in the centre; or they are placed in the middle with a passage running round. Each tank is 1·50 m (5 ft) long, and 0·75 m ($2\frac{1}{2}$ ft) wide and deep. The tank has holes in the bottom and its cover can be padlocked. Handling is by means of a pulley which permits the tanks to be plunged in and pulled out of the water.

Each tank can hold about 100 kg (220 lb) of elvers of which there are 2,000 per kilo. The boat must be moored in a spot where the water is clean so that dirt does not choke the holes, for movement or current must be maintained in the tanks. This can be ensured by current or tidal movements. The boat may also be equipped with a motor to stir up the water during calm periods.

3   Transporting elvers

Elvers can remain 1 or 2 days out of water either in moss or wet straw which should not be heaped. When being transported they are placed in small, wooden frames (with linen at the bottom) measuring 40 to 80 × 40 × 5 cm (16 to 32 in × 16 × 2 in). Five to ten frames are placed one on top of the other and the whole held by a hoop. About 1,000 to 5,000 eels are placed per frame according to the dimensions of the latter, the temperature and the length of time to be occupied in the transport. Each case should have one or several layers of moss or straw alternating with a layer of elvers (Figs. 306 and 307).

II   Production of Eels for Eating

The production of eels for eating is carried out either by extensive or intensive methods. They are slow growing fish which do best in sufficiently oxygenated warm water at a temperature over 18°C.

1   Extensive rearing methods

These methods are used, above all, in Europe. The production in ponds of eels for eating is often a secondary production, the eels being associated with cyprinids in growing or fattening ponds.

In some cases, however, eels are the principal fish stocked in the ponds. The latter are coastal brackish water ponds or lagoons such as are developed in Italy where they are known as Valli and the farming as valli-cultivation.

The principal operations practised in coastal ponds are the catching of elvers in the spring and the taking of large eels in the autumn or the beginning of winter.

The brackish water ponds communicate directly with the sea. The elvers enter the ponds themselves but it is best that the channel be well arranged, that is, equipped with sluices which permit the flow of fresh or sea water, at will. The system must work well, above all at the end of the winter and in the spring which is the principal migration period.

Eel production in the Valli reaches 30 to 40 kg per hectare (30 to 40 lb per acre) and can be increased to a maximum of 90 kg (200 lb) per hectare. Eels live off a variety of food. As they swim downstream the adults are caught in fixed fishing installations built across the channels (Fig. 308).

2.   Intensive rearing methods

1. *In Europe.* Over the past few years the intensive rearing of eels has been tried in Germany. The results were diversely received (Koops, 1967; Deelder, 1968). The trials were undertaken in ponds a few ares in size (1 to 7 ares) where the eels were fed with finely chopped fresh fish of slight value or specially prepared pelleted food. The food was distributed at the water inlet of the pond. It was found necessary to stock the young eels densely,

150 kg (330 lb) for example from 5 to 35 gm (up to 1 oz) per are. In this way densities can reach at harvest time from 500 to 750 kg (1,100 to 1,650 lb) per are. With fish weighing an average 90 gm (3 oz) at the start the harvest has given males weighing 150 gm (5 oz) and females weighing 500 to 600 gm (16 to 21 oz). The daily ration was around 10 per cent of the weight of the fish. Among the difficulties encountered there must be mentioned the escape of eels. Also they are difficult to harvest by draining the ponds for they dig themselves into the mud and can only partly be extracted by an electric fishing device.

2. *In the Far East.* The intensive cultivation of eels in Japan has been practised for 100 years. The cultivation of Japanese eels is also practised in Taiwan. In 1962 (Koops, 1967) there were 2,000 small artisanal farms (800 to 5,000 m²) (950 to 6,000 yd²) each run by the owner alone. Total production reached 10,000 tons. Since then these farms have increased in number and in 1967 production reached 20,000 tons (Fuji, 1968), and later figures show an increase to over 30,000 tons. Such an important development in the cultivation of eels is due to the manner in which techniques have been perfected and also the use of concentrated dry food for fattening.

The cultivation of eels was, originally, concentrated in central Japan where fresh and adequate quantities of food were to be found. But as the use of dry food became more abundant the cultivation of eels spread all over Japan, especially in the south. The extension of farming has, however, been limited by the problem of finding sites with sufficient supplies of good quality water.

According to Fuji (1968) eel farms can be divided into two categories, those producing eels of 1 year and those producing eels for eating.

1. Production of eels of one year. Elvers are captured in winter at the mouths of water courses. They measure from 6 to 7 cm (about $2\frac{1}{2}$ in) in length and weigh 0·16 to 0·2 gm. They are stocked in ponds until the autumn when they reach 15 to 20 cm (6 to 8 in) in length. At the harvest, weight is between 30 and 40 times what it was when they were first stocked, without counting the loss of fish during the rearing. Stocking can reach 300,000 fish per hectare (30 per square metre) (25 per square yard) (Hora and Pillay, 1962).

As cannibalism is common among young eels it is necessary to sort them out five or six times before they reach 10 to 15 cm (4 to 6 in). They are

Fig. 306   Fig. 307

Wooden cases with frames for the transport of elvers. These cases permit the transport of 20 kilos (44 lb) of elvers. (40,000 to 50,000).

Fig. 306   Top frame: ice; lower frame: moss or other water absorbent substance.

Fig. 307   The tipping of elvers into a transport frame. Utrecht, the Netherlands.

Fig. 308   Lago di Paola lagoon, Sabaudia, Italy. Catching device for the capture of eels migrating seawards.

then restocked again according to size but in different ponds.

First year eel farmers have two categories of ponds. The first are from 1 to 3 ares in which the population is dense and growth is slow, and the others from 10 to 20 ares which are the real growing ponds. The depths of these ponds is no more than 40 cm (about 16 in).

The food distributed to the young eels includes worms (*Tubifex*) at first, followed by finely chopped sea fish. But this fresh food is being gradually replaced by dry food.

2. Production of eels for eating. The eels are reared in ponds from 40 to 100 ares (1 to 2½ acres) and from 1 to 1·5 m (3 to 5 ft) in depth. The water flow must be slight, so much so that taking into account the high density of population this could lead to a shortage of oxygen which could be mortal. However, the phenomenon can be avoided if the water retains its oxygen through an abundance of phytoplankton specially rich in *Microcystis*. However, phytoplankton also uses up oxygen at night and is a danger if it dies suddenly.

Rearing takes place between April and November, and the fish are fed only if the temperature is over 15°C. Only dry food is distributed

Fig. 309   Eels for restocking or eating (small size). Length: average ± 40 cm (16 in); weight: 100 gm (about 3 oz 8½ dr).

Fig. 310   Fattened eels in Japan intensively fed with dry concentrated food (document: Dr Fuji).

and this can have a conversion rate of 1·5. The eels are fed, preferably, early in the morning. In Japan the aim is to produce eels from 100 to 120 gm (3 to 4 oz) at this stage.

Stocking is from 20 to 40 eels of 10 to 20 gm per square metre; stocking of 20 eels of 15 gm per square metre is considered normal (3,000 kg per hectare) (3,000 lb per acre). Density can be, finally, 1·2 to 2 kg (3 to 4½ lb) per square metre or 12 to 20 tons per hectare (Fig. 310). Loss fluctuates between 10 and 40 per cent. From time to time farms are affected by outbreaks of disease.

3. *In heated water.* The high yields obtained by eel farmers in the Far East are mainly due to the temperature of the water in which the eels are reared. The Japanese eel grows best when the water temperature varies around 23°C. The climatic conditions and the eventual heating of the water make it possible to reach such warm temperatures in these regions. For the European eel, it is considered that 20°C is an excellent temperature for its growth. Below 12°C eels eat little and grow very slowly. In temperate or cold climates, where intensive rearing in non-heated water only gives poor results, the utilization of industrial heated effluents is tried for the intensive production of table-eels which are of considerable economic value in many countries.   Problems connected with the utilization of such waters are discussed elsewhere in this book.

Table-eels can be reared either from elvers or from small eels weighing 20 to 40 g each. The rearing of elvers is the most critical and it is carried out in small-size tanks. The elvers are reared there

until they are small eels weighing about 20 g, which normally takes nine to ten months in 18–22°C water. Thermic shocks should be avoided and a progressive thermal adaptation is necessary. Water aeration and cleanliness are very important. Normally, the dissolved oxygen concentration should not fall below 5 mg/l. Under very good conditions, up to 5,000 elvers can be stocked per cubic metre of water. First, fresh food is used (liver, spleen, finely minced fish), and later, dry concentrated food mixed with water to form a paste, which can be enriched with codliver oil. Feeds are distributed on a feeding trough made of screening material and placed on the water's surface. Generally, food is distributed once a day and six times a week. As individual growth varies greatly, it is necessary to sort the fish at least once and to restock according to size.

For young eels (about 35 g or 1 oz), initial stocking density should not be less than 1 to 2 kg (2 to 4 lb)/m² (Koops and Kuhlmann, 1980). They recommend an initial stocking of 5 kg (11 lb)/m² in tanks with a 60 cm (2 ft) water depth. After two months, the density may reach 5 to 10 kg/m² (11 to 22 lb/m²). Grading is regularly carried out, for instance every two months, because eel growth is very heterogeneous. The biggest eels are sold for consumption and the others are graded according to size. Eels can be graded with a rigid-wire netting attached to a wooden frame. Normally, the eel density should not exceed 15 to 20 kg/m² (33 to 44 lb/m²). Starting from young eels of 20 to 40 g, it is possible to produce eels of 150 to 200 g (5 to 7 oz) within 12 to 18 months.

SECTION II
FISH CULTIVATION IN RICE FIELDS

I Interest of Fish Cultivation in Rice Fields

In appropriate regions the raising of fish in rice fields is considered one of the best and most rational means of using agricultural land.

It has been practised in the Far East for centuries and in that vast agricultural economy region it has reached a high degree of technical perfection. It is also practised in Madagascar and is now spreading in the south-east of the United States. In Europe it is not common, and in continental Africa is still in the experimental stage.

The advantages of rearing fish in rice paddies are many and can be considered of great importance to the rural economy of the regions in which it is practised. In the first place it contributes, at low cost, to the production of animal protein which in many countries are more or less chronically short. Taking into account the immense areas covered by rice fields, the fish cultivation potential they represent, even for extensive farming, is considerable. This is true, above all, for rice fields far from the sea, large lakes and other important fishing centres.

Besides, the methods of growing rice and cultivating fish are well matched which means that rice and fish can be raised simultaneously. The fish help to control weeds, molluscs and mosquitoes and so improve the health of the crops.

Controlling weeds is an important problem in rice cultivation. Herbivorous and algae-eating fish such as *Tilapia melanopleura*, *T. zillii*, *T. mossambica*, *Puntius javanicus*, and *Trichogaster pectoralis*, cultivated in Africa or in Asia, can be very helpful in this respect. At the same time the fish help to destroy the beds of culicids of which some (*Anopheles*, *Stegomya*) spread malaria and yellow fever.

The control of water molluscs is important in tropical rice swamps and in the fight against bilharzia. Malacophagous fish can also be used in biological control. *Haplochromis mellandi* have given positive results in Katanga.

The relatively deep water of fish-stocked rice swamps is also considered helpful in the fight against rats.

In the United States the extension of rice fields is being justified by the fact that it aids the conservation of soil fertility achieved by the rotation of crops.

II Principal Types of Fish Cultivation in Rice Fields

The different techniques used in the cultivation of fish in rice fields differ considerably from region to region. There are many reasons for this. Methods depend on local conditions such as climate, etc.; on the different species of fish being reared; on the varieties of rice cultivated; on different practices used in the cultivation of rice and the fish being reared, as well as manuring and artificial feeding.

In the first place there are important differences in capturing and rearing methods. In one the rice fields are not stocked but simply attract wild fish which are then captured. These wild fish populate the fields freely when in flood. When fish are farmed, the swamps are stocked in the same way as ponds.

It is necessary to differentiate between simultaneous and alternate production when referring to harvesting rice and fish. In the first case the rice and fish are grown together. This is real rice-fish cultivation. When production is alternate then fish and rice are harvested alternately. One method is to harvest rice and fish once per year each. Another is a triple harvest in 1 year, two for rice and one for fish. Other methods, more complicated, can give five harvests of rice or fish over a period of 2 years.

Finally, methods differ according to the size of the fish produced and this will depend on the supply of water available and the size of fish used for restocking. It is possible to produce fingerlings starting with fry or to produce fish for eating starting with fingerlings. Sometimes rice fields are used as spawning ponds. But whichever kind of production is chosen the cultivation period is generally short and limited to a few weeks. It must correspond to the intervals between placing under water and obviously the time required for draining the rice field, for weeding for example, and the drying time required for the rice to flower and ripen. If facilities exist for storing fish in holding ponds the same fish can then be liberated during different and successive periods in the same rice swamps and their rearing can extend over several years. This is normal practice in Japan. Another system used in Taiwan enables a continuous method to be practised. Deep ditches are dug in the rice fields in which the fish take refuge during the drying of the field.

But whatever system is used rice is considered the principal product and fish cultivation is complementary and secondary.

III Rice Fields as an Environment for Fish Cultivation

Rice fields used for the production of fish must be considered and arranged in the same way as ponds. One fact must be noted at once, the water must be shallow because of the presence of the rice.

Considered as a pond the rice swamp must be so arranged as to permit the water supply and evacuation to be controlled and to help when the pond is being put under water and being drained. These operations should not be difficult. To keep the water at a sufficient level the field must be surrounded by a dike; this need not be high and should be built out of earth from the ditch dug at its base. In order to control water at both inflow and outflow, so as to stop unwanted fish getting in and farmed fish from escaping, different devices are used. They are generally simple and in the Far East are normally made of bamboo. Hollow bamboo sticks provided with bamboo lattice screens at one end, stop the fish escaping and keep out unwanted fish (Fig. 53). A by-pass is also necessary to avoid excess of water when floods occur.

The draining of rice swamps and the harvesting are helped by ditches dug round the swamp as well as by others crossing it and notably linking the points of inflow and outflow. These ditches should be at least 50 cm (20 in) wide and 30 cm (1 ft) deep (Fig. 312). At different points, notably at the water inflow and outflow as well as at the points where the principal ditches cross, slightly deeper holes should be dug (1 m at least) where the fish can take refuge during the draining and when the temperatures are too low or too high which could thereby imperil their lives.

The depth of water in rice fields is generally slight. If both fish and rice are produced then the depth will be determined by what the rice can best support: 5 to 25 cm (2 to 10 in). If fish are not reared at the same time as rice then the water can be deeper, from 20 to 60 cm (8 to 24 in). In the United States it is even 1·50 m (5 ft). It is always best and even essential to increase the depth of the swamp and the draining ditches in those regions where sudden cold snaps are likely. The width of the dikes depends on the depth of the water. If the water is shallow then the dikes should be 25 cm (10 in) high and 50 cm (20 in) wide at the bottom and 25 cm (10 in) at the top.

The water flow must be adequate and in many cases this limits the production of the fish. In Indonesia the estimated need is 1 or 2 litres per second (2 to 4 pt) per hectare. Generally fresh water for irrigation is used but in exceptional cases, brackish water can also be employed. The best system is to maintain a continuous flow of water to the rice field. Often the inflow is non-continuous with intervals more or less regular to compensate for loss through evaporation and leakage and so maintain the level of the water.

In tropical regions, at low altitudes, and in marshy countries, temperatures are high and the dissolved oxygen content can fall quite steeply and acidity can increase. Under these circumstances the choice of species for rearing is limited to those able to support such difficult conditions.

If fish and rice are produced at the same time, a variety of rice must be chosen which can support being in fairly deep water and at the same time resist the fish digging on the bottom of the swamp in search of food.

IV Principal Fish Cultivated in Rice Fields

Fish adapted to rice field rearing must meet the following conditions: they must be able to withstand shallow water, tolerate high temperatures and low dissolved oxygen content, have rapid growth to a saleable size, resist cloudy water and not show a strong tendency to escape (Coche, 1967).

On those farms which capture wild fish many different species are caught. In Malaya the principal one is *Trichogaster pectoralis*. Voracious fish such as *Ophiocephalus striatus*, *Clarias batrachus* and *Anabas testudineus* can also be harvested.

The production of farmed fish under conditions of controlled production in rice fields includes, at least in the Far East, such species as carp and *Tilapia mossambica* among the principal fishes used. In Madagascar the former species (carp) is the principal one, as it was in Europe at one time. In the United States two species of catfish are cultivated. In brackish water *Tilapia mossambica* is the most common along with mullets and other less abundant species.

Several species of cyprinids such as crucian carp, tench, Indian carps (*Catla, Cirrhina, Labeo*) as well as *Puntius javanicus* are occasionally reared.

V Principal Methods of Rearing Fish in Rice Fields

To succeed in the production of fish in rice fields several conditions must be met. These must be topographically favourable; there must be a regular and sufficient water supply; the fish must be adaptable to local conditions; the fish farmer must be well trained in the necessary techniques and there must also be a market capable of absorbing the saleable product (Coche, 1967).

A Production methods of fish in rice fields

The production of fish in rice fields can be carried out with either captured or farmed fish; production can be simultaneous or alternate.

1 Simultaneous production of rice and fish or rice-fish cultivation

Simultaneous production of rice and fish (Figs. 311 and 312) has evident advantages. It uses perfectly the available ground; it is good for the rice because it helps weed control, and it creates an hygienic medium through the control of molluscs and harmful insects; it produces animal proteins cheaply because the fish are, practically, a complementary harvest which costs nothing. This is very important in regions suffering from a weak rural economy.

It has also been noticed frequently that the presence of fish increases rice production by between 5 and 15 per cent. This can be explained by examining, among others, certain factors. These include the indirect fertilization of the rice thanks to the fish excrement; and also by the unutilized artificial food distributed. Also a better tillering of the rice seedlings is due to the activity of the fish digging in the mud and this also helps mineralization. Finally algae and weeds which compete with the rice are better controlled. However, some writers disagree and believe this action is unfavour-

able to rice production and that it actually reduces growth.

Simultaneous production, however, does present certain inconveniences. In the first place the water flow must be greater than would be necessary just for rice and this limits the spread of rice-fish cultivation. It also means having deeper water which all varieties of rice cannot tolerate. Dikes, draining ditches and capturing sumps take up space (estimated at around 5 to 7 per cent in Taiwan) but this loss is compensated for by the value of the fish. Also certain soils cannot be kept under water for prolonged periods.

Apart from the need for a higher water level the principal inconvenience is that rice-fish culture limits the use of modern agricultural techniques, notably mechanization, chemical fertilizers, herbicides and insecticides. In progressing agricultural areas rice-fish cultivation is being slowly abandoned in favour of the rotation of rice and fish crops.

Rice-fish production varies according to the regions. In Indonesia the following method is used.

After the planting out of the rice, the depth of the water should be from 2 to 5 cm ($\frac{3}{4}$ to 2 in). The first weeding comes 3 weeks later and 2 weeks after that the second weeding. During that interval the water level is increased to from 6 to 10 cm ($2\frac{1}{2}$ to 4 in) and after the second weeding the depth is gradually increased to 20 cm (8 in) and kept at that level until the flowering of the paddy. The water is then drained off to help the ripening of the rice.

Rearing of fish must be organized in connection with this type of rice cultivation. From the planting of the rice until the first or second weeding, about 5 weeks in all, the rice fields can be used for producing fingerlings. They are liberated 5 days after the planting out of the rice. During the first weeding the fields are drained and the fish take refuge in the ditches. The fry are harvested after the second weeding at about 3 to 5 cm ($1\frac{1}{5}$ to 2 in) in length.

After the second weeding and until the flowering, that is over a period of $1\frac{1}{2}$ to 2 months, the rice field can be used to produce fish for eating up to a maximum weight of 100 gm.

At the time of harvesting the fingerlings and fish for eating, care must be taken not to damage the rice. Drying out must be done slowly and carefully.

2 Alternate production of rice and fish

Most of the advantages and also inconveniences of

rice-fish cultivation disappear but the water requirements are still high.

Alternate production (Fig. 313) permits better care of both as well as the use of mechanical means, herbicides and insecticides for the rice. It also permits an increase in depth of the water during the production of the fish and this is favourable to both growth and reproduction.

Other advantages arising from this method have been given in the United States as helping when uncultivated land is brought under cultivation and the conservation of soil already under cultivation. This has also been stated in western Bengal when salt marshes were exploited for agriculture.

After the rice has been harvested the field is transformed into a temporary pond. At the same time, outside ditches and dikes are constructed. The dikes are built with soil dug up from the ditches. New dikes must be erected with each new crop. After harvesting the rice, the bottom of the pond should be cleaned of stalks which are cut up and piled in heaps. They decompose slowly and have a manuring action (Fig. 314).

The length of time the field is under water is more or less long according to the region and the methods used.

B Stocking rice field – Loss

1 *Stocking rate.* When production methods comprise only catching the fish it is not properly called stocking as wild fish will colonize the rice fields when they are being put under water.

But when rearing methods are used, stocking is carried out in the same way as with a pond and will depend on the type of production and particularly the time production will take. Customarily this is short, never being more than a few weeks. Stocking depends not only on the duration of the production, however, but also on the natural productivity of the water and the size of the fish to be produced.

The stocking rate differs considerably according to the rearing regions. Indications of what they are in Indonesia are given below. Other indications are given for other regions under heading six of this section.

In Indonesia, when production is alternate and the fish chosen are carp, then, if small fish are to be produced in a few weeks 100,000 fry measuring 1 cm should be liberated per hectare. If, on the other hand, larger fry is liberated with a view to producing more developed fingerlings, then only

30,000 to 50,000 should be liberated. But to produce fish of saleable size, that is, between 14 and 16 cm ($5\frac{1}{2}$ to $6\frac{1}{4}$ in) weighing from 40 to 100 gm ($1\frac{1}{2}$ to 4 oz), then 1,000 to 2,000 fingerlings from 8 to 12 cm (3 to 5 in) should be liberated per hectare. Rearing lasts from 2 to 3 months.

When the simultaneous method is used 60,000 1 cm fry are liberated per hectare for the production of fingerlings. But if the rice field is used for the production of fish for eating, which will take from $1\frac{1}{2}$ to 2 months, 1,000 to 2,000 fingerlings are liberated as for the other method.

If the objective is to produce *T. mossambica* by means of rice-fish cultivation then 1,000 to 10,000 fry some 1 to 3 cm (up to $1\frac{1}{5}$ in) are liberated along with a few hundred adult fish 10 to 15 cm in length (4 to 6 in), 1 week after the planting of the rice. Six weeks later there is a first harvest and the largest fish are sold while the rest are returned to the field for a further 6 weeks.

2 *Loss.* Cultivation in rice fields has a higher loss ratio than when ponds are used. It is generally between 40 and 60 per cent for fingerlings and 20 to 30 per cent for fish for the table. The principal cause of this is the presence of predatory animals such as cormorants, herons and otters. But in certain cases it is due to temperatures which are too high for the shallow water and leads to a deficit in dissolved oxygen.

The loss can be reduced by increasing the depth of the water and of the draining canals in which the fish take refuge, and also the depth of the capturing ditches. It is also recommended to use a sufficiently big area, 5 hectares at least, for this kind of cultivation.

C Manuring and artificial food for fish in rice fields

When the alternate method is used the pond can be fertilized with organic manure such as rice straw, compost, vegetable waste, various other manures including dung and liquid manure all of which increase, indirectly, the production of the field.

The artificial feeding of the fish also helps to increase production quite considerably. This is practised in Japan particularly, and also in Madagascar. Carp are fed when the temperature is over 12°C but when it goes above 27°C care must be taken. *T. melanopleura* are fed only if the temperature is over 20°C. Various foods are used such as cassava leaves, rice bran, cotton seed oil cake, silkworm cocoons and domestic waste.

Fig. 311 A general view of terraced rice fields used simultaneously for the growing of rice and the raising of fish, principally carp, either as fingerlings or for eating. Road from Tjiandjur to Soeka-bumi, Indonesia.

Fig. 312 Growing rice and raising fish simultaneously in a rice swamp. Cropping carp fingerlings. On the road from Tjiandjur to Soekabumi, Indonesia.

Fig. 313   View of rice fields used alternatively for growing rice and rearing fish. Depok, Bogor, Indonesia.

Fig. 314   Rice field prepared for rearing carp following the rice harvest. Piles of rice straw are laid out as organic fertilizer. Tjiandjur, Indonesia.

D Production level of fish in rice fields

Production of fish in rice fields varies considerably and depends on the methods of exploitation, the species cultivated, the depth of the water, the fertility of the soil and the water, and the amount of care given by the farmer to the cultivation of both the rice and the fish.

Methods which are based on captured fish generally give very small production. This is practised above all in Malaya and Indonesia. In this way, in Malaya, it is possible to harvest up to 40 kg (88 lb) per hectare per year. In Indonesia, where the method is far more widespread and covers 4 million hectares, harvests are very small, no more than 1·5 to 3 kg (3 to 6 lb) per hectare per year.

But production when rearing methods are used is much higher. According to Coche (1967) for Asia, Europe, Africa and America natural production is between 100 to 200 kg (220 to 440 lb) per hectare per year with an average of 120 to 150 kg (250 to 330 lb) per hectare per year. But this average is very relative, for when monthly production is calculated, it is found to be around 100 kg (220 lb) per hectare per month and, since production is limited to a few weeks in a year, annual production is much reduced (Huet, 1956).

However, with feeding, production can exceed 200 kg (440 lb) per hectare per year which is notably the case in Japan where production has reached 1,000 to 1,800 kg (1 ton to 1 ton 15 cwt) per hectare per year.

VI Development of Fish Cultivation in Rice Fields

Below is given, mainly according to Coche (1967), a brief idea of fish cultivation in rice fields in different parts of the world where it is practised.

A Indo-Pacific region

This region has the most developed fish culture in rice fields. It is believed to have been introduced from India into south-east Asia more than 1,500 years ago. Although it is practised over 136,000 hectares this only represents 0·64 per cent of the total area of rice cultivation.

1 *Malaya.* The cultivation of fish in rice fields is current usage on the west side of the Malayan peninsula. It is based on the capturing system. Rice fields are not stocked but are populated with wild fish which colonize the areas naturally when the field is put under water. The fish grow and reproduce. After the rice is harvested and the field is dried they remain in the ditches which cover 40 to 50 m² (50 to 60 yd²) and are 2 m (7 ft) deep. These ditches are already specially dug in the lower parts of the rice fields.

The principal fish harvested is the sepat siam (*Trichogaster pectoralis*) (Fig. 344) introduced from Thailand before 1921. Smaller numbers of *Anabas*, *Clarias* and *Ophiocephalus* are harvested. Harvest can reach 135 kg per hectare (300 lb) in from 6 to 10 months and as this is absolutely without cost the production from an economic point of view is very advantageous.

2 *Indo-China.* Rice-fish cultivation is largely followed in various forms according to the different regions.

In the mountains of Thailand only carp are cultivated. The rice fields are stocked in April (20,000 fry per hectare). After 3 months a first harvest gives from 80 to 100 kg per hectare (175 to 220 lb) of 40 to 50 gm ($1\frac{1}{3}$ to $1\frac{2}{3}$ oz) fish. But the harvest can wait for 6 or 7 months and then it will give 100 to 150 kg per hectare (220 to 330 lb) of fish weighing from 60 to 150 gm (2 to 5 oz).

In lower Vietnam the practice is to harvest after 5 or 10 months. In South Vietnam fish cultivation is one of capture, similar to that practised in Malaya though simpler. *Clarias* and *Ophiocephalus* are the principal fish harvested.

3 *India.* In the delta regions of western Bengal and in the Madras province fish culture in rice fields, similar to that of Malaya, is practised. This is a capturing method in brackish water; 100 to 200 kg (220 to 440 lb) are captured per hectare, of which most belong to the genera *Lates*, *Mugil* and *Mystus*.

Another practice is to develop this cultivation with Indian carps of the genera *Catla*, *Cirrhina*, and *Labeo*.

4 *Continental China and Taiwan (Formosa).* Despite the fact that serious efforts have been made in continental China to produce fish in rice fields the practice has never attained any extensive proportions.

On the other hand, on the island of Taiwan, after the Second World War, rice-fish cultivation developed well, especially in the south part of the island which is the hottest. The principal fish reared is *Tilapia mossambica* introduced from Java. Precautions must be taken against cold. Liberation should take place in warm temperatures above 15°C, the level of the water should be high, the draining and refuge ditches should be wide (1·20 to 1·50 m (4 to 5 ft) and deep (0·60 to 0·90 m) (2 to 3 ft). Stocking is with fry and fingerlings. Production lasts about 7 months and reaches 220 kg (480 lb) per hectare. Natural manure is used and the fish are lightly fed with artificial food.

5 *Indonesia*. The production of fish in rice fields is important in Indonesia where it has been practised for over a century.

The system of capture is used over an area of 4 million hectares but production is poor, around 1·5 kg per hectare (3 lb).

True rearing in rice fields covers around 60,000 hectares. It is particularly important in the western province of Java which contains about three-quarters of the fish-stocked rice fields. A rather supple system is employed and its essential characteristics are as follows: limited time production over a period of 1 to 3 months; specialized farms producing specific categories of fish such as fry, fingerlings and fish for the table, the latter measuring no more than 14 to 16 cm (about 6 in) and weighing 50 to 100 gm ($1\frac{3}{4}$ to $3\frac{1}{2}$ oz).

The principal fish reared are carp (*C. carpio* L. var. *flavipinnis* C. and V.) introduced from China several centuries ago. These represent around 85 per cent of total production in rice fields. The second most important species is *Tilapia mossambica*. Then come *Helostoma temmincki*, *Osteochilus hasselti* and *Puntius javanicus* used principally for the production of fingerlings in the rice fields.

Rearing is carried out either by the simultaneous system or by the alternate system. For the latter it is possible to obtain one production of fish and rice per year ("palawidja" method) or three harvests in 2 years ("panjelang" method). The latter is the most important along with the simultaneous production. According to the palawidja method, the period under water is 5 or 6 months and permits, during the interval separating two rice crop harvests, several successive fish harvests either of fingerlings or fish for eating, each one lasting 1 to 3 months. The panjelang method provides for a 2-month period under water allowing two successive harvests of fry or one of fish for the table.

6 *Japan*. The production of fish in rice fields prospered in Japan during a century. Now it is declining rapidly and fewer than 0·5 per cent of the rice fields carry fish. This is due to the intensi-fication of agriculture. The principal species reared is carp and it is produced by rice-fish cultivation.

The principal characteristics of rice-fish cultivation in Japan are as follows: intensive cultivation, the use of manure and the practice of artificial feeding both of which permit production exceeding 1,000 kg per hectare (1 ton per hectare); the production of fairly large fish around 100 to 250 gm (3 to 8 oz) in 2 years and 350 to 750 gm (12 to 26 oz) in 3 years which is a saleable size. Able and experienced farmers sometimes modify their cultivation methods in order to increase fish production.

B Europe

Only rice-fish cultivation is practised and it is unimportant.

1 *Italy*. Rice-fish cultivation was developed at the end of the nineteenth century, principally on the plains of Lombardy. It is now in steep decline because of certain technical problems but also because it is of no great economic interest. The principal species cultivated are carp and sometimes tench and goldfish. But whatever the species, production is practically confined to fingerlings sold either to owners of ponds or to fishing clubs.

2 *Hungary and Bulgaria*. Trials are developing. The species used is carp raised to fingerling size.

C America

1 *United States*. The cultivation of fish in rice fields is a recent development. It started in 1950 and there are still plenty of problems to resolve. It is only practised in two cases – first as a means of bringing into production unexploited wooded land. For, if this land is flooded for a sufficiently long period it helps deforestation. Secondly, the cultivation of fish in rice fields is one way of conserving land under cultivation. Used for rice production over too long a period the land becomes overrun with weeds which compete with the rice and the crops are reduced. The rotation method of cultivation for fish as well allows normal production of rice to be resumed.

Fish cultivation in rice fields is developing in the central, southern states of the U.S. in the lower valley of the Mississippi (Arkansas, Louisiana, Mississippi, and Texas).

This is alternate rice-fish production practised over wide areas and, currently, over several tens of hectares. The water in the submerged areas is deep, from 30 cm (1 ft) to 90 to 150 cm (3 to 5 ft).

Another feature of these fish-producing rice fields in the U.S. is that they stay under water for from 1 to 3 years. They are quite large and extend up to 30 hectares (75 acres) – a size which permits the use of machinery.

The two principal species raised are catfish: bigmouth buffalo (*Ictiobus cyprinellus*) and channel catfish (*Ictalurus punctatus*). Accessory production includes carp, bass and centrarchids. Old rice fields are stocked with fry and harvested 1 to 3 years later. They vary in size, are sometimes sold for consumption but also for transformation into fish meal or fish oil.

2 *Central and South America*. Fish production in rice fields is only slightly developed. Trials with carp and various tilapias have been carried out in Haiti and Brazil. In Argentina good rice-fish cultivation results have been obtained with pejerrey.

D Africa

1 *Madagascar*. This is the only African country in which fish cultivation with rice is really developed. It has in fact been practised for at least 80 years. It is characterized by mixing different fish such as carp, crucian carp and, since 1950, various types of tilapias such as *T. mossambica*. Production is based on the raising of fry and 1 year fish. Stocking is light and often reaches no more than 150 to 250 fish per hectare. There is also a holding pond situated at the side of the field.

Either simultaneous production (rice-fish cultivation) or alternate production methods are adopted. Manuring and artificial feeding are only rarely practised, but when they are production reaches 200 to 250 kg (440 to 550 lb) per hectare.

2 *Continental Africa*. Trials have not passed the experimental stage and took place after 1950. They have been carried out with various species of tilapias and sometimes with other cichlids. The principal trials took place in the central-south area of Africa – Zaire, Zambia and Rhodesia.

In Zambia the aim was to keep weeds down with the use of *T. melanopleura*. In the Katanga province of the Zaire interesting results were obtained. Weeds and molluscs were kept under control by mixing *T. macrochir* (planktonophagous), *T. melanopleura* (phytophagous) and *Haplochromis mellandi* (malacophagous).

SECTION III
FISH CULTIVATION IN BRACKISH WATER

Rearing fish in brackish water is fairly widespread in Asia. It is far less current in Europe and at the time of writing is scarcely practised in other parts of the world. It concerns only a few species which are well adapted to this special environment. The most important species used is *Chanos chanos* raised, above all, in the Far East. Then comes the rearing of several types of mullets. The rearing of eels in brackish water has already been described. The rearing of salmonids, principally rainbow trout in brackish water is a fairly recent practice.

I Breeding and Cultivation of Milkfish
(*Chanos chanos* Forskäl)

1 Brackish water ponds or "tambaks"

In the Indo-Pacific region brackish water ponds arranged along a flat shore of salt marshes was just another step towards the reclamation of land from the sea in order to turn it into arable soil after filling in the ponds and eliminating the salt. But in many places ponds have been constructed for the sole purpose of controlled rearing of fish and this has developed over many centuries. The ponds are known as "tambaks" in the Indonesian language and are important above all in Indonesia (128,000 hectares), in the Philippines (88,000 hectares) and Taiwan (13,000 hectares) (Schuster, 1960). In appropriate places tambaks extend along the flat coasts and then develop inland for several hundred yards and sometimes over several kilometres, forming an immense checked pattern (Fig. 315). This flat country with its ponds separated by low dikes and irregularly planted mangrove trees, is often densely populated with herons and other water birds. It is very typical and characteristic of these parts.

1. Choice of the site. Choice of site depends on several factors: topographical conditions, water

supply, the quality of both the soil and the water.

As far as the first factor is concerned the chosen site on the flat coast should have sufficient tide but not excessively so. It should be flat in order to permit an important area to be placed under water without the necessity of excavation nor the building of an important dike. On the other hand high tides should ensure the water supply. Also the drying of the pond should be practicable without difficulty at low tide. Consequently it is on these flat and low coastal lines determined by alluvial deposits left by high waters, which is frequent in tropical waters, that tambaks should be installed. Before the laying out of the ponds the sites would, normally, be submerged at high tide.

An adequate supply of water for the ponds, by means of the tides, is essential, This is easy enough near the sea but becomes more and more difficult inland where the height of the tides is less marked. The tides must be sufficient but never excessive in order to avoid the necessity of constructing high dikes. A high tide about 1·50 m (5 ft) is just right. Pumping water for supply is rarely planned. Naturally the site must be sheltered from flooding caused either by the sea or by rivers.

The soil should be sufficiently watertight and should favour the growth of algae. For this a pH between 6·5 and 7·5 is best.

2. The importance of the farm. In the Far East a fish farm in brackish water is generally self-supporting. Complete farming would include the production of fingerlings and fish for consumption. From time to time, however, such farms specialize in the production of fingerlings only.

An area covering 10 hectares (25 acres) is considered sufficient for the establishment of an autonomous farm. In the Philippines such an area would be divided as follows: 1 per cent nursery ponds, 9 per cent transitional or second rearing ponds, and finally 90 per cent growing ponds. The latter generally cover an area from 0·50 to 2·50 hectares (1 to 5 acres) and sometimes even more than that. Nursing ponds are smaller. The depths of growing ponds vary from 1 to 4 ft. Each group of ponds forms a tambak made up of small nursery ponds and of larger ponds for fattening.

3. Arrangement of brackish water ponds. On the whole the arrangements of tambaks is similar to that of fresh water ponds. As has been mentioned they are shallow. Generally the ponds are either square or rectangular. They are surrounded by a

dike which should stand at least 30 cm (12 in) above the maximum level of the water and be built out of soil excavated when digging the ditches required for draining the ponds. Besides these, secondary ditches are dug on the site of the ponds to ensure complete draining. The dikes built round the group of tambaks should be the strongest. They should rise at least 50 cm (20 in) above the highest tides. Secondary dikes separate the growing ponds and the weakest of them separate the nursing ponds. All dikes are planted, irregularly, with trees (Fig. 316).

Water is chiefly supplied by means of a channel linked with the sea but the latter is not the same as for fresh water ponds. In effect the supply and the evacuation of water is ensured by a single device installed at the deepest part of the pond and replacing the monk. It works like a sluice and is built like one (Fig. 320) in durable materials (brick or concrete) in the Philippines but in wood in Indonesia. The closing of the sluice device is so arranged that it admits sea water at high tide and evacuates it at low tide. The different operations which allow the sea water to enter the pond or permit draining must synchronize with the tides, the amplitude of which depends on the day, the season and the location. On the other side of the sluice, that is towards the interior, a screen, generally in bamboo lattice, is installed to prevent fish escaping or undesirable fish from entering. The catching device is situated even further inside the sluice. All the different ponds are tied into the main channel which, situated at the lowest level of the farm, controls the inflow and outflow of the water for the whole farm. This channel can also serve the ponds for the purpose of capturing or fishing out the harvest. The importance of the sluices depends on where they are built and how they work. The principal sluice controls all the ponds included in the tambak. Other sluices serve the growing ponds and are not so large while those serving the fingerlings ponds are smaller still. The most important are no more than 1 to 1·25 m (3 to 4 ft) wide. If necessary two sluices or more can be placed side by side.

2 Milkfish (*Chanos chanos* Forskäl)
Milkfish (*Chanos chanos*) (Fig. 317) known as bandeng, bandang and bangos in Indonesia, Malaya and in the Philippines, is a marine clupeid found over a large area in the Indo-Pacific region and inhabiting the tropical and sub-tropical zones.

It extends from the Red Sea and the east coast of Africa to the west coast of the United States (south of San Francisco) and Mexico and thus crossing the entire Oceania. Latitudinally they are distributed from the south of Japan to the south of Australia. They are warm water fish which live in temperatures superior to 15°C and will die in temperatures around 12°C. They can withstand 40°C. Even if they are very sensitive to low temperatures, a low dissolved oxygen content does not worry them.

These marine fish live near the coastline and are never caught in the high seas. They also frequent estuaries, grow very well in brackish water ponds but will not reproduce in captivity. Fry must be caught at sea and transferred to the ponds where they develop well and quickly, growing to 50 to 70 mm (2 to 3 in) in 1 month, 120 to 150 mm (5 to 6 in) at 2 months and in 8 to 10 months reaching 250 to 500 gm ($\frac{1}{2}$ to 1 lb) which is a commercial weight. They can easily reach 40 cm (16 in). It is possible to keep them in ponds over longer periods until they reach from 5 to 10 lb. At sea they can weigh over 15 kg (33 lb).

Milkfish have been raised in brackish water in Indonesia for 100 years as well as in the Philippines and Taiwan. They have been tried out in India, Thailand, the Fiji Islands and Samoa.

Milkfish can adapt themselves to fresh water and can be reared in ponds. This is practised sometimes in Indonesia and in India, in the State of Madras. The ponds are stocked with fry from 15 to 25 mm (up to 1 in).

In Indonesia, in the centre of Java, milkfish are reared in ponds fed with diluted sewage; one part sewage to two parts stream water. The ponds are small and never exceed 25 ares. Rearing lasts from 3 to 4 months. Stocking and production are high.

In a word, the chanos is sensitive to low temperatures, resistant to feeble dissolved oxygen content, capable both as fry and as adult fish of adapting itself to marked differences in salinity if the differences are progressive. They can adapt themselves quickly to a very varied diet.

3   Ecology of brackish water ponds

Soil is very important to tambak production. Its quality depends, for a large part, on the nature of the terrain drained off by the stream and the alluvion deposited. The best are those of volcanic origin. Needless to say the bottom of the pond must be watertight.

The two water supply sources are the sea and rain but sometimes it is possible to supply tambaks with fresh water brought from a water course. The relative possibilities of these water sources and the rhythm of the seasons determine variations in the salinity of the water and this can vary considerably. Evaporation determines the concentration of salt while rain and fresh water brought from a water course help dilution. Salinity will differ according to whether the ponds are by the sea or whether they are inland and also upon the ease with which fresh water can be supplied. But the supply of rain or stream water is interrupted during the dry season, and at this time of the year the level of the water falls and there is partial drying of the ponds.

Temperatures vary according to the seasons. In Indonesia it is always high and bearing in mind that ponds are shallow it can exceed 35°C. In Taiwan the fish can suffer from cold. To protect them from both heat waves and cold spells, certain sections must be dug deeper so that the fish can take refuge. In Taiwan the winter ponds used are quite deep. Brackish water is somewhat alkaline. Generally it is well oxygenated.

Milkfish prefer soft to hard, resistant food. During the first month they feed off blue-green algae (Myxophyceae) or already partially decomposed green algae. Myxophyceae develop abundantly on the surface of the mud on the bottom of shallow ponds provided they are well lighted and the mud has a good structure. Under these conditions Myxophyceae do better than Chlorophyceae. On the other hand aquatic phanerogams are not very abundant in brackish water ponds.

Under the conditions given above, that is, in shallow ponds, a living biological layer formed by micro-organisms develops on the surface of the bottom in quantities depending on the natural productivity of the ponds. This biological complex known as "lab-lab" in the Philippines is formed principally of algae and includes bacteria, blue-green algae (Myxophyceae) uni- or multi cellular, Oscillatoriaceae in particular, fragments of filamentous green algae, and diatoms. It also includes certain animals such as protozoans, entomostracans (cladocerans, copepods), worms, and also detritus and mineral particles. Fragments of lab-lab rise constantly from the bottom due to oxygen bubbles formed by photosynthesis. They become more or less planktonic or floating. They are eaten by the fish either in benthic, floating or planktonic form.

Milkfish also eat biological cover. Maximum production of benthic algae is found in shallow and well lighted water (5 to 15 cm) (2 to 6 in) but milkfish need water which is at least 20 cm (8 in) deep.

In order to help the growth of this biological complex certain conditions must be met. Good soil, preferably clay, is necessary; sufficient intense sunlight and heat. Water depth, slight at first, must increase and salinity should be between 10 and 40 per thousand. Reduced turbidity of the water is ensured thanks to the trees growing on the dikes since they cut out the harmful effects of wind and waves. Regularly renewed water bringing with it oxygen and nutritive salts is another condition.

When they grow in size, milkfish can feed off green algae and also more or less decomposed Characeae as well as, if necessary, filamentous fresh algae and higher plants. As the fish grow so this kind of food plays a more important part in feeding.

Before being stocked milkfish ponds demand very careful preparation which can take several weeks (2 weeks in tropical regions and up to 8 weeks in the sub-tropics). The bottom of the pond should be worked over and all unwanted vegetation removed. The pond is left dry until the earth cracks. This takes from 1 to 3 weeks (Fig. 375). Then it is placed alternately under water and dried, remaining dry for several days. This is necessary in order to destroy eels and other predators which dig themselves into the mud and are very dangerous for the fry. Finally the pond is put under water but at first with no more than a few centimetres (1 to 4 in). The depth is increased gradually in order to help develop the food supply. This can be accelerated by replanting from other ponds. To this end some Philippine farmers have special algae raising ponds. Shallow water, around 10 to 15 cm (4 to 6 in) will help the growth of blue-green algae and deeper water from 30 to 60 cm (1 to 2 ft) helps green algae. Algal growth is helped if the soil is slightly alkaline (pH from 6·8 to 7·5) and manured by organic or mineral fertilizers. When the fish are liberated the depth of the water should be at least from 20 to 40 cm (8 to 16 in).

The borders of the tambaks are lined with halophilic ligneous flora. Generally the dikes are planted with mangrove trees: *Avicennia* principally and also *Rhizophora* (Fig. 316). *Avicennia* leaves provide good material for green fertilizer.

4 Rearing milkfish in brackish water ponds

a The rearing phases

1. Catching and transporting of fry. Milkfish spawn at sea, near the coast and probably at a distance under 20 miles. The females are large and measure at least 90 cm (3 ft). They give from 2 to 7 million pelagic eggs about 1·2 mm in diameter. The larvae hatch within 24 hours and measure 6 mm. The fry caught along the coast measure from 11 to 13 mm and are supposed to be 2 or 3 weeks old. During this period they drift slowly from the place of spawning and hatching towards the coast.

The areas selected for fishing are sandy and flat and with clear water. Locations near the mouths of small rivers flowing into the sea are best. As favourable places for catching milkfish fry are few the fry have to be transported over long distances, sometimes several hundred miles, to rearing ponds. In the Philippines the principal locations of capture are situated on the north-west coast of Luzon and these sites are of such economic importance that they are leased out. In Indonesia, capture is localized to the east of Java and around the Island of Madura.

Capture is at specified periods and times. In Indonesia the principal fishing period is October and November. There is a second period in April and May which suggests that there must be two reproduction periods in that region. In other regions there is probably only one. The best fishing time is during full moon and new moon, and fishing takes place during the high tides providing the sea is not rough.

Different systems and tackle are used for catching. In Indonesia, small scoop nets are used. They are stretched with very fine linen and are employed on the beaches by fishermen or women who wade waist deep into the water. Frequently they use, at the same time, a kind of floating garland made of palm fibres under which the fry take refuge and are then captured with scoop nets. The garland is strung out perpendicularly from the beach or fixed at both ends or only at one end. In the latter case it is placed in such a way that the diameter should gradually be reduced (Figs. 318 and 319). Daily capture varies considerably from a few dozen to several thousand.

The fisherman hands the fry over to a trader who places them in circular flattish pottery containers or bamboo made watertight with tar (Fig. 324). Because they are flat the containers ensure good aeration of the water and can be placed one on top of the other and so are handled easily. They must also be circular because milkfish fry swim in one direction only and this movement must not be interrupted. Fry can remain for several days in the containers. They are then transported to their place of destination. The beach where they were caught and the point of sale can be separated by hundreds of miles so the time between catching and liberation in ponds can be as long as 10 days and even more. In certain cases transit takes from 2 to 4 weeks. Often the fry pass through several hands before they reach their final destination. From 500 to 1,000 milkfish fry per litre (pt) of water (Schuster, 1960) can be transported without artificial aeration at a temperature of 25°C.

During storage and transport they should receive great care. The quality of the fry at the time of release depends on this. The water in which they are transported is nine parts fresh and one part sea and it must be changed each day.

The fry are fed with very fine rice bran. Their resistance is quite remarkable when it is borne in mind that frail as they are, they can withstand transportation and storage over a long period and can change from plankton feeding to rice bran without ill effect. This is one of the great characteristics of milkfish culture. Nevertheless the fry are very sensitive to low temperatures.

At each operation, harvesting, transporting and sale, the fry are carefully counted. They are always handled in water and counting is carried out with a large mollusc shell (Fig. 323). During counting, fry of unwanted fish are eliminated. This operation demands a lot of practice and skill.

During the 5 to 20 days which is the normal period between catching and release, losses can amount from 2 to 20 per cent according to the care given during transport and to the length of the operation.

2. Nursing ponds. Before being liberated in the growing ponds the fry must first spend some time in the nursing ponds. This is the most delicate phase of rearing. Some farmers specialize in it and only produce fingerlings which they sell to producers of fish for eating.

In Indonesia, before the fry are dispersed in the nursing ponds, they are sometimes placed in a "baby-box" which is a very small pond. They stay there only a few days (Fig. 322).

Nursing ponds are small: from 1 to 10 ares. Before being stocked they are carefully prepared.

This takes from $1\frac{1}{2}$ to 2 months and the principal reason for this relatively long period is to develop the nutritive biological layer and eliminate predators and any possible competitors. It is very important that there should be abundant food in the ponds when the fry are liberated, for if there is not they will perish in large numbers. To protect the fry from too high temperatures palm leaves are placed on the banks of the ponds (Fig. 321).

Fry are liberated either early in the morning or at dusk, after the temperature of the water in the container and in the pond has been equalized. Normally from 30 to 50 fry are released per square metre ($1\cdot2$ yd^2). The fish remain in the nursing pond from 6 to 8 weeks and at the end of the period measure from 5 to 10 cm (2 to 4 in) and weigh from $1\cdot5$ to $5\cdot0$ gm. Losses are high and often reach from 30 to 50 per cent and even 80 per cent except on those farms which specialize, where they may not pass 20 per cent. Most of the loss is due to predators.

After rearing in the nursing pond the young fish are transferred to the growing ponds either directly, or after having spent a few weeks in a transition or second rearing pond (1 to 2 months) where they grow from 10 to 15 cm (4 to 6 in) in length. These transition ponds are larger than the nursing ponds and cover around 50 ares to 1 hectare. The fry are released at a ratio of from 15,000 to 20,000 fish per hectare ($1\cdot5$ to $2\cdot0$ per square metre).

3. Growing ponds. A tambak includes, generally, several growing ponds, each one covering from $\frac{1}{2}$ to 2 hectares. Some growing ponds are as large or larger than 10 hectares. When there are several ponds the harvest can be staggered.

The growing ponds must be carefully prepared in the same way as the other ponds. They are stocked at a ratio of 500 fingerlings per hectare in Indonesia, 1,000 per hectare in the Philippines and 3,000 per hectare in Taiwan, according to Schuster. According to Hora, Pillay, and Pillai the ratios are at least twice as large. The growing period lasts from 6 to 8 months. At the end of that time the fish weigh from 350 to 500 gm (12 to 18 oz) and are of saleable size. According to Schuster losses are around 20 to 30 per cent.

Sometimes a small percentage of older fish, those of 2 years, are mixed with the fingerlings.

b   Control and improvement of production

1. Maintenance measures. Brackish water ponds are subject to the same maintenance methods as

freshwater ponds. It goes without saying that the dikes must be looked after carefully, as well as the draining ditches and the water inlets and outlets.

One of the most important measures is drying the pond regularly. This allows good mineralization of the mud and helps the development of the algae on the bottom after the pond has been replaced under water. It is also necessary to keep an eye on the regularity of the water supply in order to avoid too high salinities.

2. Measures against predators and competitors for food. Milkfish can be wiped out easily by predators, especially when they are young. Among these, eels and *Lates calcarifer*, voracious fish, must be mentioned. They enter the ponds as fry through the water inlet. In the Philippines, tilapias, which multiply too easily, are considered undesirable competitors of milkfish. Measures must also be taken against aquatic reptiles.

Predators are dealt with by the installation of very fine mesh screens, by successive drying out and eventually by the use of poisons.

3. Manuring. This is not intensive and only organic manure is distributed. Often it consists only of *Avicennia* leaves from trees planted on the dikes and thrown into the pond in large quantities. If possible, leaves of other mangrove trees should be mixed with them as well as different grasses. This green fertilizer helps the development of blue-green algae. If possible, pig or other animal manure can be distributed.

4. Artificial food. This is not used in Indonesia. In the Philippines, food of poor value, such as algae harvested in neighbouring water or cultivated in separate ponds, is distributed. If other plants used in this way are not eaten directly, they decompose and play the role of organic fertilizer.

In Taiwan feeding is abundant, particularly rice bran but also soya oil cake and sometimes ground nuts.

5. Mixing the species. Sometimes other fish such as *Tilapia mossambica* are associated with milkfish. This practice developed considerably in Java during the Second World War because of transport difficulties which did not permit restocking brackish water ponds situated a long way from places where the fry were caught. Tilapias have the advantage of clearing up algae floating on the surface which serve as nests for mosquitoes.

In low salinity water, under 8 per thousand, tawes (*Puntius javanicus*) are also associated with milkfish. The advantage they present is that they eat algae and other submerged plants which they do not digest completely. This half-digested detritus can be eaten by the milkfish which they assimilate easily.

6. Secondary production. Apart from milkfish, brackish water ponds give a fairly good production of shrimps and crabs. Young shrimps and prawns are brought in by the sea water entering the ponds. In some ponds it is possible to produce 200 kg (440 lb) of shrimps and prawns per hectare. Shrimps have the habit of stringing out along the dikes of the ponds. This allows them to be caught with fykes fixed at the ends of bamboo screens installed perpendicularly and attached to the dikes (Fig. 325).

In the Philippines, siganids (*Siganus vermiculatus* Cuv. and Val.) are mixed with milkfish in very salty water but with a moderate temperature.

c Production level in brackish water ponds

Production in these ponds is variable and depends among other things on the natural productivity of the pond, the intensity of manuring and of artificial feeding and the care given to maintenance. In many places production is not high because the water is too shallow and at times partly dried out because of the low water.

According to Schuster annual production per hectare is 300 kg (660 lb) in Indonesia (it varies from 50 to 80 kg (110 to 180 lb) when the soil is poor and 250 to 450 kg (550 to 1,000 lb) when it is rich), from 350 to 500 kg (770 to 1,100 lb) in the Philippines and 2,000 kg (2 tons) in Taiwan thanks to intense manuring and feeding.

d Capture and marketing of the fish

The fish can be caught by draining the pond. Nevertheless, during growth, there can be several intermediary fishings, taking into account the gregarious characteristic of milkfish and their tendency to leave the ponds for the sea. This instinct is strongest during high tides and it has a strong influence on milkfish. Sea water is allowed to enter the sluice which is to be used at that time as the fishing device. The water flows into the pond and the fish swim against the current, massing together in the sluice after the screen separating it from the pond has been removed. In this way the fish are captured easily. Those which are too small are returned to the pond.

If the method so described cannot be used, then the level of the water in the pond must be lowered.

Fig. 315   General view of tambaks by the sea near Soerabaya, Indonesia.

Fig. 316   Tambaks at Watangrodjo (Soerabaya), Indonesia. Large, flat, shallow ponds, drained in order to mineralize the mud. *Chanos chanos* and *Puntius javanicus* are reared here together.

Fig. 317 Milkfish or "Bandeng" (*Chanos chanos* Forsk.) is the principal fish raised in tambaks.

Fig. 318

Fig. 319

Catching young *Chanos* larvae on the very flat and sandy beaches of the north coast of Eastern Java. A stretched, very fine-meshed triangular scoop net is used. Several methods are used. One, for example, is to push the net along in the water (Fig. 318) or another to use, at the same time, a loop made of palm tree fibres fixed at one end and progressively rolled spiralwise (Fig. 319). Situbondo, Eastern Java, Indonesia.

Fig. 320   Central device for controlling a tambak (complex of several brackish waterponds). In the foreground an example of sluice construction with a bamboo lath screen and a small clay dike. Behind the sluice is situated the central catching section. This ensemble permits, by manoeuvring, the flow of sea water into the ponds or the evacuation of water from the ponds, as well as capture of the fish in the fishing section. Kalanganjar (Soerabaya), Indonesia.

Fig. 321   Fry pond for rearing *Chanos chanos* alevins to the stage of 1 month-old fingerlings. Brandjangan, Soerabaya, Indonesia.

Fig. 322   Nursery pond with baby-box in foreground for *Chanos chanos* alevins. Krapjak fish farm, Djogjakarta, Indonesia.

Fig. 323   Counting young milkfish fry.

Fig. 324   Transporting milkfish fry in watertight bamboo cans.

Fig. 325   Bamboo laths and fykes for catching shrimps and prawns in the tambaks. Semeni, Soerabaya, Indonesia.

Personnel wade into the water and catch the fish by means of a cast net or a bamboo lattice. This is pushed into a deep part of the pond where the concentrated fish are easily caught (Fig. 464).

The captured fish are conserved on ice or sometimes smoked or conserved in brine.

II   Rearing Mullets

Mullets (family Mugilidae) are sea fish which enter estuaries and the lower reaches of rivers. They are found in many parts of the Indian Ocean, the Indo-Pacific region as far as Japan, the Philippines and Australia. They are also present in the Mediterranean. They can adapt themselves very well to brackish water and even to freshwater.

Mullets can be raised in ponds and are found, quite often, associated with other species in more or less brackish water in Israel, India, China, Hong Kong and Hawaii. In Israel they are associated with carp and tilapias in slightly brackish water.

According to the different regions, one or two species of mullets are used and reared together. The most frequent are *Mugil cephalus* L., *M. capito* Val. and *M. tade* Forsk. Mullet fry enter the tambaks as wild fish brought in by the sea water supply.

They spawn at sea and do not reproduce in fresh water or in streams. Artificial spawning has succeeded in Israel, but produced fry died after having been hatched. The eggs are 1 mm in diameter and float. After having been captured in estuaries, mullet fry measuring 15 to 18 mm are gradually acclimatized to slightly salt water.

The rearing of mullet is carried out in one or two phases. If there is only one phase the fry are released directly into the ponds in which they pass 1 or 2 years. Rearing in two phases calls first for nursery ponds where the fry remain 6 or 7 weeks until they reach the fingerling stage. In Israel, where fry are captured in winter at sea, they are first of all placed in storage ponds until spring at a ratio of 25 to 30 per square metre (1·20 square yard). They grow slowly and finally reach 2 to 4 gm. Loss is 20 to 30 per cent.

Mullet fry feed off animal and vegetable plankton and biological cover. Later they feed on the bottom off algae, different detritus and decomposing vegetation. They also accept artificial food such as is distributed to carp.

Growth is rapid on the whole but this depends on the species (*M. capito* grows more slowly than *M. cephalus*), the density of stocking, the amount of natural food available and the intensity of artificial feeding. Yashouv (1966) mentions 200 to 300 gm (7 to 12 oz) growth during the first year and 550 gm (20 oz) the second year. Under intensive conditions of cultivation they can reach 700 gm (1½ lb) the first year and even over 1 kilo (2 lb) in tropical regions. At sea, according to the species, mullet can reach from 50 to 70 cm (20 to 30 in) in length.

Mullet are mixed with other species such as carp and tilapias. In Israel, where fish cultivation is intensive, 700 to 1,000 fingerlings are spread over one-tenth of an hectare (1 dunam = ¼ acre). In Hong Kong, similar numbers or more, up to 15,000 per hectare, are placed.

III   Rearing other Species

Apart from milkfish and mullet, other species are reared in brackish water such as *Tilapia mossambica*, *Puntius javanicus*, and *Etroplus suratensis*.

Fig. 326   Grey mullet (*Mugil capito* Val.).

Details of the rearing techniques employed for these species are given in Chapters VI and IX.

Tilapia mossambica is cultivated in brackish water particularly in the Indo-Pacific region from India, Pakistan and Ceylon to the Philippines, Taiwan and Indonesia, passing by Thailand, Vietnam and Malaya. Rearing in brackish water started in 1939 in Indonesia. Sometimes they are reared alone, but more often are mixed with milkfish and other species.

Etroplus suratensis is mixed with other species in brackish water ponds in India.

Rearing eels in brackish water is practised in many regions throughout the world, notably in Japan and on the shores of the Mediterranean (Ch. VIII, Section I).

Appendix: Cultivation of Salmonids in Sea Water

The rearing in sea water of salmonids which normally live in fresh water or which spend a part of their life there, has been considerably developed in recent years in certain countries such as Norway. This rearing concerns the rainbow trout and the Atlantic salmon, and aims to produce large fish weighing one to several kilos, with salmon-coloured flesh to be commercialized as sea trout or salmon. Firstly, both species spend one or two years in fresh water where techniques are identical to those used in the cultivation of salmonids.

The idea of rearing salmonids in sea water is old, but was first successfully carried out in Norway around 1955. In 1977, there were about 200 farms producing over one ton of fish per year and the production of rainbow trout and Atlantic salmon amounted to 1,800 and 2,000 tons respectively (Edwards, D. J., 1978).

The broken-up nature of the Norwegian coastline provides numerous possibilities for salmonid cultivation in sites sheltered from wave and wind action. The water along the west coast is warmed up by the Gulf Stream. The growing season is longer in sea water than in fresh water, where it lasts for only five months, from mid-May to mid-October. The sea-water temperature is higher throughout the year.

Different rearing systems are used, where the salinity should not fall below 30 ppt. At first, ponds were often built in concrete into which sea water was pumped. This is possible in Norway where electricity is cheap. Up to 30 kg (66 lb) ot fish per square metre (square yard) can be reared in a pond 3 m (10 ft) deep, with a water flow of 0·5 litres (1 pt) per second and per kilogram of fish. A cheaper way is to fence off, with a metal screen, shallow coastal sea zones or marine bays. These zones can then be rather large, from one to several hectares, the water renewal being ensured by the tide. Today, floating cages are mainly used. Generally, their capacity varies from 100 to 500 m³ and the density at harvest time often reaches 10 to 15 kg (22 to 33 lb) per m³. This can reach up to 25 to 30 kg (55 to 66 lb) per cubic metre.

The fattening feed consists of fresh sea fish such as herrings, gadids and various wastes from fish factories, ingredients which are all easily obtainable in large quantities. However, as this feed is not available all the year round, some must be stored in freezers. Care should also be taken not to distribute the same kind of food all the time, nor to use food which is too fatty, such as herring, which can lead to nutritional diseases. Towards the end of the fattening period, the usual feed should be supplemented with about 10 per cent of shrimp wastes and other crustaceans to improve the meat taste and give the fish a salmon-pink colour. Fresh feeds are often replaced by pellets. The meat coloration is then obtained by adding either crustacean wastes or synthetic pigments.

The marine rearing of rainbow trout only concerns two-year old fish. These fish are reared first in fresh water and transferred to sea water when they measure 15–20 cm (6–8 in) or more. Sometimes the fish are acclimatized progressively but this is not indispensable. Growth in sea water is fast and fish weighing 200 g (7 oz) in May can reach up to 1 kg (2 lb) in October. Normally rainbow trout are harvested after 12 to 18 months when they often reach a weight of 2 kg (4 lb).

The smoltification of Atlantic salmon (*Salmo salar*) occurs in freshwater ponds in the spring of its second (about 30 g) or third year. It is at this stage that smolts are gradually acclimatized to sea water through intermediate salinities. This may take several months (Edwards, D. J., 1978). As with rainbow trout, salmon should be harvested before sexual maturity. Usually harvesting takes place after a two-year rearing period in sea water when fish average 4 kg (9 lb). Salmon sell tor a much higher price than rainbow trout, but their cultivation is more difficult and the fish are more sensitive to the quality of the fattening feed.

SECTION IV

FISH CULTIVATION IN FLOATING CAGES

The cultivation of freshwater fish in bamboo or wooden cages has been practised for many years in the Far East (Figs. 224 and 225). In recent years, the technology for the intensive rearing of fish in cages has been introduced and improved in several countries throughout the world, for coastal marine waters as well as for inland waters. The aim is to produce large quantities of fish on limited surface areas, in facilities less expensive and more manageable than ponds or tanks. The use of rearing cages makes the fish harvesting very easy. Another great advantage is the possibility of using water bodies in which classical fish culture cannot be carried out: large, natural or artificial non-drainable water bodies, slowly-moving water bodies with enough depth, thermal effluents. The feeding of fish is entirely artificial and cages should be well sited in order to avoid unfavourable sanitary conditions. In theory, intensive rearing in cages is possible for all species which accept complete artificial feeding such as trouts, carps, tilapias and American catfishes. Many details about cage construction and cage culture in freshwaters have been given by Dahm (1975) and Coche (1978).

Fixed cages made from nets or screen can be used, similar to those sometimes used for fish storage (Figs. 479 and 480), but cages floating at the water's surface are most commonly used. The principle of these floating cages is identical to the one applied for floating holding nets (Fig. 481).

Polyamide nets are often used, sometimes metal nets are also used, or a rigid construction made of metal or plastic screen. The mesh size must be large enough to allow good water renewal. This may vary from 6 to 25 mm according to the species and size of the fish. For trout, the mesh size should not be smaller than one tenth of the length of the fish. Floating is made possible by various means, such as barrels, usually in plastic, polystyrene floats, or closed PVC pipes. Floats can also be combined with working platforms which surround either a single cage or a series of them. Cage management is easier with such working platforms. Fish feeds are distributed either by hand or with automatic feeders.

Cages are often rectangular and dimensions may vary considerably. Dahm (1975) considers $3 \times 3 \times$ 1·5 m ($10 \times 10 \times 5$ ft) as rather typical dimensions for cage culture in European inland waters. Too large cages are difficult to handle, while too small cages may cause large food losses because of the wild swimming of the feeding fish. The cage walls should be at least 50 cm (20 in) above the water's surface. In order to prevent bird predation, the cages are covered with either a net or a screen. To protect the nets against rodents, a second net can be used, stronger and with larger mesh, or a light screen, or even a metal net. A cheaper and generally sufficient solution to this problem consists of protecting the upper part of the net down to a 50 cm (20 in) depth, with plastified wire netting. Floating cages ready to be used can be bought from several manufacturers.

Floating cages are set up either in rather large-standing waters or in slowly-running waters. Often thermal effluents of power stations are used, if the water is clean enough. The water depth below the cage bottom should be at least two or three metres, in order to avoid a lack of dissolved oxygen in case of an algal bloom. This will also reduce the risk of disease propagation, much higher in waters too rich in organic matter originating from fish faeces and waste feed. In fact, floating cages should not be set up in standing waters less than five metres (5 yd) deep.

To be successful, enough dissolved oxygen should be present in the cages. In standing waters, the water renewal is often due to wind-induced currents. To take full advantage of such currents, it is advisable to set up the cages perpendicular to the main currents. Net fouling organisms may also reduce water renewal. Regular checks and maintenance are therefore necessary. Anti-fouling preparations (copper salts, tar) or special paints can be applied to slow down the development of fouling organisms. If necessary, pumps or any other aerating systems can improve the water oxygen content (Chap. XII, Sect. III).

The size of the rearing facilities depends on many factors: fish species, water quality, water renewal, depth and surface of the water body, acceptable degree of eutrophication. It can be considered that the wastes load produced by the intensive rearing of 100 kg (220 lb) of fish is equivalent to that of five inhabitants. Mann (1974) reports that in an old 17·5 ha gravel-pit, 5–9 m (5–10 yd) deep, the distribution of about 100 tons of dry concentrated feeds for the annual production of 50 t of trout has not resulted in a real eutrophication. Such production has been obtained in 40 cages with a total volume of nearly 2,500 m³ which means an annual yield of 20 kg (42 lb) per m³ of cage.

Exceptionally high productions can be obtained in heated effluents of good quality and sufficient flow. Coche (1978) reports average monthly productions of 35 kg (78 lb)/m³ for common carp, 20 kg (42 lb)/m³ for *Ictalurus punctatus*, an American catfish, and 15 kg (33 lb)/m³ for rain-bow trout. In such cases, maximum densities may reach 220 kg (488 lb)/m³ for catfish and 140 kg (308 lb)/m³ for rainbow trout.

For rainbow trout rearing in non-heated waters, the normal stocking density is 300 fingerlings per cubic metre and the maximum final density is 40 kg (89 lb)/m³. For carp cultivation in cages in temperate climates and in non-heated waters, an annual production of 25 kg (55 lb)/m³ is normal. This production is attained with a stocking density of 10 kg (22 lb)/m³ and a maximum final density of 35 kg (78 lb)/m³. The conversion rate of dry pellets is about three.

SECTION V

FISH CULTIVATION IN HEATED WATER
RECIRCULATION OF WATER

I Fish Cultivation in Heated Water

In temperate or cold climates, the utilization of heated water allows fish culture to be partly or completely independent from the external temperature thus extending the growing period of indigenous fish species. One tries to reach the optimal temperature for growth. Water heating also allows the rearing of non-indigenous fish originating from warm climates. It also provides greater possibilities for the propagation of both indigenous and exotic species.

Water can be heated especially for the rearing needs, or else industrial cooling water can be used, mainly from electric power stations, either classic or nuclear. As energy is quite expensive, water heating for rearing needs can only be envisaged for small quantities of water and productions of high economic value: egg incubation, hatching and rearing of fry, maturing spawners. In addition, one endeavours to recirculate the heated water.

Industrial cooling waters, if free of charge and of good quality, offer considerable possibilities for intensive fish culture. Generally, water flows directly into the ponds or tanks, after eventually being mixed with cold water. Indirect heating of the rearing water is uncommon owing to the high cost of the facilities and its low efficiency. Floating cages are also set up in the canal discharging the warm water effluent or downstream, according to the temperature desired and the water flow.

The best species for cultivation in heated water are eels, catfishes, common carp, Asian carps, and tilapias. As such rearing is always intensive, using high fish densities and artificial feeds, water renewal should provide an adequate dissolved oxygen content and good removal of waste food and faeces. In addition to the problems specific to any intensive rearing system (hygiene, diseases), other problems may appear related to the utilization of industrial cooling waters: irregular warm water supply, unstable water temperature, and quality of the water which often originates from highly industrialized rivers. In the specific case of power stations, the water of the condensers may be supersaturated with gas (nitrogen) and the condensers are regularly cleaned. In nuclear power plants, the risk of radioactive contamination also exists.

Very high productions can be obtained under ideal conditions: good quality water, sufficient and regular flow, optimal temperature for the reared species. Coche (1978) reports monthly yields in floating cages of 35 kg/m³ for carp, 20 kg/m³ for the American catfish *Ictalurus punctatus*, and 15 kg/m³ for rainbow trout. In Italy, annual productions have reached 150 kg of common carp per square metre of cage area and in the USSR, productions of 80 kg/m³ have been obtained in four months. These high productions correspond to maximum densities up to 190 kg/m³ for common carp and 220 kg/m³ for catfish. In Japan, productions of eels in small ponds or tanks have reached 25 kg per m², for an 18-month cycle with a water temperature of about 21°C. An eel density of 40 kg/m³ should not be exceeded.

II Recirculation of Rearing Water

The short supply of good quality water in many instances has resulted in a growing interest for the total or partial recirculation of rearing water. Although water recirculation is particularly envisaged for intensive rearing systems, a reutilization of the water can also be considered in less intensive systems. This is sometimes the case in salmonid farms where water requirements are large. In this case, water can be succinctly treated (sedimentation, filtration, aeration) before being used again. The water can also be reutilized as it is, which reduces recirculation by pumping, always necessary, and by aeration of the water at the pond inlets. Sometimes cyprinid ponds are supplied with stream water with excessive nutrients and the water is partially purified in the rearing ponds themselves. Recirculation can then improve the water quality.

The reutilization of water is particularly interesting in intensive rearing systems. It results not only in a considerable saving of water but also in a saving of power, if the water had been heated for the cultivation needs.

The pollution resulting from the intensive rearing of one ton of fish is generally considered to be equivalent to that of 50 inhabitants. This pollution comes from the waste food and the fish faeces. This requires a treatment of the rearing water before its reutilization. In principle, this treatment is similar to the one applied in sewage-treatment plants which treat the organic sewages. It can be carried out in several ways and more or less intensively: sedimentation, biological treatment, filtration, aeration, eventually coupled with thermo-regulation. In addition, it always requires a pumping system and an additional fresh water supply whose amount depends on the rearing intensity, the degree of purification reached, and the water losses unavoidable during the cultivation period.

A more complete treatment includes a biological treatment. The latter consists in the degradation of organic matters and especially of nitrogeneous compounds through oxidation followed by reduction, resulting from micro-organisms action. Biological treatment is usually carried out using biological filters or activated sludge. Ammonia compounds, nitrites, nitrates and finally free nitrogen are formed during the biodegradation process. The complexity of this process and of the responsible living organisms is such that the biological treatment is a delicate operation which requires continuous control. Non-ionized ammonia is very toxic for fish but the toxicity of ammonia compounds depends on several factors: pH especially, temperature, alkalinity, dissolved oxygen. According to Westin (1974), concentrations from 25 to 35 ppm of nitrates and of 0·012 ppm of nitrites may be considered as safe for salmonids.

Although water recirculation can result in great water savings, and eventually in power saving, the necessary facilities generally require large investments. Its successful functioning often involves delicate techniques, especially when the fresh water supply is limited. Furthermore, the high fish density and the chemical quality of the recirculated water, often unfavourable, result in an environment more prone to disease. Although the more or less complete recirculation of rearing water is already used in commercial fish culture entreprises, it is always indispensable to make a techno-economic feasibility study prior to any investment.

Chapter IX
REGIONAL FRESH WATER FISH CULTURE

IN the chapter devoted to regional fish cultivation, the essential characteristics of the principal forms of cultivation in the main regions of the world not previously described are dealt with succinctly.

Fish cultivation in rice fields and in brackish water was dealt with in the preceding chapter. Consequently the present chapter is concerned with fish cultivation in fresh water ponds only.

SECTION I
DEVELOPMENT OF FISH CULTURE IN EUROPE

Fish cultivation in fresh water exists in most countries of Europe although, in fact, there is no particular regional fish cultivation on this continent.

European fish cultivation is divided into farming for consumption and farming for restocking.

Cultivation for the table is concerned with either salmonids or cyprinids. Salmonid cultivation for consumption is confined in Europe mainly to rainbow trout, or, secondly, to brown trout and brook trout. In almost all countries it is developed in cool water. Among the most important countries engaged in trout production for human consumption are Denmark, France and northern Italy. Cyprinid cultivation for consumption is principally the production of carp for the table, and, secondly, such other cyprinids as tench and Asiatic cyprinids. This is developed, most of all, in central and oriental Europe including Germany, Poland, Czechoslovakia, Hungary, Yugoslavia and

Rumania. The raising of secondary fish for eating includes eels, coregonids (Grosse Maräne) in Czechoslovakia and Danubian wels in Hungary, which is far more limited.

European fish cultivation for restocking and release in open waters (water courses, lakes) is intended to increase the number of catchable fish for sport or professional fishing. This kind of cultivation includes practically all of the European species. Salmonids come first, including salmon in Scandinavia, brown trout in fast running water courses all over Europe, Arctic char and coregonids in the sub-Alpine and Scandinavian lakes, and grayling in certain areas. It also concerns different cyprinids, notably for the restocking of canals or canalized water courses. Pike are released in calm running or stagnant waters; pike-perch especially, in certain canals, largemouth bass in waters which are sufficiently warm, and eels in most inland waters.

SECTION II
REGIONAL FISH CULTURE IN NORTH AMERICA

In the United States and Canada, fresh water fish cultivation is very varied. In cold water it includes first of all salmon, particularly the different species of Pacific salmon. These fish are cultivated for restocking. It is an economic form of cultivation. Salmonid cultivation also includes several other

salmonids living only in fresh water such as rainbow trout, brown trout, brook trout, lake trout and Arctic char and, secondly, grayling. The cultivation of salmonids is principally for the restocking of waters used by anglers. Nevertheless, salmonid cultivation for consumption is also widely practised, particularly with reference to the rainbow trout.

In slightly warmer waters the principal species of fish cultivated for restocking are walleye and several species of pike.

In warm waters in the United States, black bass and other centrarchids principally are farmed. This cultivation is primarily for restocking for angling. Natural water courses and lakes are repopulated as well as are fishing ponds, notably farm ponds used for many purposes including fishing.

In the warm waters of the United States, ictalurids and channel catfish in particular, are raised both for sport fishing and for eating.

Cultivation for repopulating and sport fishing is, in fact, often more important than rearing for eating. All forms of cultivation given above have already been described.

The extent of cultivation for sport has led to the development of a special kind of farming in the United States; the cultivation of bait fishes.

Breeding and Cultivation of Bait Fishes

1 Origin and aims

The cultivation of bait fishes in the United States springs from the considerable extension of sport fishing there. At one time it was not difficult to capture live bait for fishing in open waters. In fact, demand was so great that in certain States restrictions had to be imposed because the practice threatened the balance and the normal development of fish in certain natural waters. This gave birth to the cultivation of bait fishes.

According to Dobie *et al.* (1956) this kind of cultivation is beneficial; first, for the dealer who can always be sure of a regular supply, secondly, for the game angler who can be sure that the fish are of good quality and not weakened by too long a stay in ponds and storage tanks, and finally because the authorities need not fear depletion by over-fishing of these small fish in open waters.

The cultivation of bait fishes includes a certain number of cyprinids, generally small in size, among which the main ones are fathead minnow, golden shiner, bluntnose minnow, blacknose dace and goldfish.

In the United States the name "minnows" is used to cover all small cyprinids used for bait. For this reason the name should not be used for small fish of other families even if they are related. In Europe the name "minnow" has a more limited meaning and refers only to the species (*Phoxinus phoxinus* L.), used primarily as live bait for fishing salmonids.

Apart from cyprinids, a different type of fish called suckers, belonging to the catostomids, is also cultivated and used. In fact, suckers are as close in appearance as they are biologically, with European running water cyprinids, of which the principal representatives are the barbel, the chub and the hotu (*Chondrostoma nasus* L.) The principal sucker cultivated is the white sucker (*Catostomus commersonii*). Further bait fishes are cultivated from families other than the two mentioned above, such as members of the *Umbridae*.

It should be noted that although cultivated bait fishes are small they do not necessarily belong to small species; for example, adult suckers are large and vigorous fish.

The cultivation of bait fishes is carried out in one or two phases. In one phase this is done in one pond which serves, successively, for multiplication and growing. In two phases the first is carried out in a pond or in a hatching installation, and the second in a growing pond.

2 Breeding bait fishes

The multiplication of bait fishes is practised with the same techniques as for other branches of fish cultivation, that is, semi-controlled cultivation in ponds or multiplication with artificial fertilization and incubation. For the first the parent fish are often reared in ponds but for the second they are caught in open waters.

a Semi-controlled natural breeding of bait fishes in ponds

This method suits different species of minnows such as golden shiner, bluntnose minnow, fathead minnow, silvery minnow as well as goldfish. These fish normally reproduce in static water.

The species breed without difficulty in ponds. Certain species lay their eggs on submerged water plants while others lay them on the underside of plants, stones or other submerged objects (Figs. 327 and 328).

All these phases of rearing, from spawning to harvest, take place in the same water and the ponds can rightly be called breeding ponds. In size they vary generally from between 10 and 40 ares ($\frac{1}{4}$ to 1 acre), but sometimes they are larger and can cover 20 to 30 acres. According to the productivity, 1,000 to 2,000 adult minnows are released per acre. In small ponds this figure is as high as 4,000 minnows to the acre (100 per are), which must be considered as a great many.

Minnows are also raised in natural ponds and are fished with nets. In this case Dobie recommends extensive ponds of from 3 to 5 acres and from 5 to 8 ft deep.

b Breeding by artificial fertilization and incubation
This system suits creek chub, stoneroller, blacknose dace as well as suckers. Normally these fish reproduce in running water.

The method is more difficult than natural reproduction and uses the same techniques as for the artificial propagation of pike for example.

Spawning and fertilization is carried out in the normal way as for artificial reproduction. The eggs of some species tend to stick and they are then treated with clay or starch, but they could also probably be treated with tannin as is done for carp. Before starting incubation, sucker eggs must be allowed to harden for about 2 hours after fertilization and washing. This hardening is carried out in a pan and the water must be renewed from time to time.

As far as incubation is concerned, opinions tend to differ. Incubation is generally carried out in jars with 2 or 3 litres of eggs, and sometimes twice that quantity. The water current in the jars must be regulated so as to keep the eggs moving slowly. After hatching, the fry remain in the jars until the absorption of their yolk sacs. When they start to swim they will reach the collector tank and can then be liberated in the rearing ponds.

They can also be incubated on trays in the same way as salmonid eggs.

The best way is perhaps to hatch them successively in jars and on trays, as is practised with pike eggs. The first incubation phase takes place in jars. When the eggs are eyed they are transferred to trays where the hatching takes place and the yolk sacs are absorbed. This method has the advantage of allowing the removal of bad eggs from the jars which is easier than with trays. Also after hatching the fry might suffer if they are kept piled up at the bottom of the jars.

3 Growing ponds for bait fishes
The growing phase is carried out in one operation. If natural reproduction is used then the number of fry in the growing ponds, which was first used for reproduction, will be uncertain. If artificial reproduction is the system used, a variable number of fry is released according to the species and the intensity of rearing. This intensity depends to a large extent on the natural food in the pond, whether fertilization is more or less intensive, and also the frequency of feeding. For the latter, cotton seed and soya flour have given good results.

There is a considerable difference between extensive and intensive rearing. Extensive rearing is carried out in relatively large ponds, as much as 25 acres in area or more. Intensive rearing takes place in small ponds covering from $\frac{1}{4}$ to 1 acre (10 to 40 ares). Intensive rearing demands more work, but in the end the total production looks more economical.

In extensive rearing some 25,000 to 50,000 fry are released per acre, that is 6 to 12 per square metre (5 to 10 per square yard). For a long time 10 sucker fry per square metre was considered satisfactory. This corresponds to the release ratio for other species in good nursing ponds (trout and pike fry for example). The survival rate is between 20 and 40 per cent.

If intensive farming is the practice then there will have to be frequent and regular fertilization and feeding and it is possible to resort to far denser stocking, even up to 100,000 to 400,000 per acre (25 to 100 per square metre) for small species such as fathead minnow (Dobie, 1956). Production can reach 10 to 20 minnows per square metre or 40,000 to 80,000 per acre and even more than that; as much as 200,000 small fish per acre (50 per square metre) representing a weight in excess of 300 lb per acre.

If density is very important then the water flow crossing the ponds must be high in order to maintain sufficient oxygen and to eliminate poisons. If artificial feeding is used, then the food must be distributed in small quantities at a time so that it is eaten at once, otherwise it leads to serious sanitary problems.

According to the species, fry should reach from 2 to 8 in in 100 days of rearing. As soon as part of the fry reach sufficient size they must be partially fished out either with a seine net, a trap net or a drop net (Fig. 463). In this way the population is thinned out, allowing the remaining fry to grow. Intermediate fishing starts as the demand for minnows grows. The greatest number are sold between July and September. This intermediate fishing helps to increase production considerably. At the end of rearing the pond is dried out. If some minnows are to be held during the winter they should be kept in fairly low temperature water during that season and moderately fed.

4 Principal cultivated bait fishes – Choice of species

A cultivated bait fish must meet certain requirements (Dobie *et al.*, 1956). It must be prolific and, have an extended reproduction season; be of rapid growth and be resistant to handling, transport and disease. It must not be voracious. Finally, adults must be suitable for use as bait fish for the important carnivorous species of the region.

The choice of species often depends on the demands of the consumer. The producer, on the other hand, who has to sell his product, should choose several different species in order to be sure that he has sufficient appropriate bait for all game fish: small bait (6 to 8 cm (2 to 3 in) fathead and bluntnose minnows) for perch and small centrarchids; medium (10 to 15 cm (4 to 6 in) creek chub, golden shiner) for walleye, bass and pike; large size (20 to 30 cm (8 to 12 in) white sucker) for large pike and muskellunge.

Bait-fish farming should include many species, some of the most important being given below.

a Natural, semi-controlled breeding species

Fathead Minnow (*Pimephales promelas* Rafinesque) (Fig. 329). This is a small cyprinid minnow. The male is larger than the female but it still no longer than 10 cm (4 in). It is found in south-east Canada and in the east (north and centre) of the United States. It is a popular bait fish and is used after the golden shiner with which the season generally starts.

Fathead minnows are cultivated in a great many States excepting those in the neighbourhood of the Great Lakes where enough are found in open water to satisfy demand. It reproduces naturally in ponds at 1 year. Its life is short, often no more than 15 months. Fathead minnows spawn from the end of April until August in temperatures of at least 18°C but never higher than 29°C. They lay sticky eggs on the underside of submerged objects placed in the water and which are then guarded by the male. The fry hatch after 4 to 6 days.

Fathead minnows can be raised extensively in large ponds which can serve for spawning and also for growing. Extensive cultivation calls for 1,000 to 4,000 fish per acre (0·25 to 1 per square metre). After several weeks, intermediate fishing is neces-

sary in order to transfer part of the fry to the growing ponds.

Intensive cultivation in ponds of less than 1 acre is often preferred. Then it is best to have several ponds, some of which are used for reproduction and others for growing. In the first, 20,000 parents are released to the acre (6 per square yard). The stocking of growing ponds exceeds 400,000 fry per acre (120 per square yard) and production can reach 200,000 per acre. The young fish are sold in two sizes, 1½ to 2 in (4 to 5 cm) and 2 to 3½ in (5 to 9 cm). During the growing phase there are intermediate fishings.

Bluntnose Minnow (*Pimephales notatus* Rafinesque) (Fig. 330). These are small cyprinids no longer than 10 cm (4 in). The males are larger than the females. They resemble fathead minnows and are found in the same places. They eat the same food, spawn in the same way and are found in large numbers in different regions, as for example in Michigan. These fish spawn over a long period as soon as the water reaches 21°C. Their eggs are adhesive and are laid in water from 15 to 90 cm (6 in to 3 ft) deep. Females can spawn twice in one season. The young reach saleable size in the autumn. They are produced in great numbers by the same method as for the fathead minnow.

Golden Shiner (*Notemigonus crysoleucas* Mitchill) (Fig. 331) is a widely used bait. It is indigenous to the east of the Rocky Mountains from Canada to Florida where it lives in calm water, including lakes and slow currents. It reaches 25 cm in length (10 in) and, normally, is mature during its second year. Females grow more rapidly than males and lay their eggs on algae, other plants or anything submerged. The spawning period is long and takes place at temperatures between 20 and 27°C particularly throughout the month of May. The fry hatch out between 4 and 6 days after fertilization.

This bait fish is used throughout the cold season, and above all for catching pike and walleye. In high temperatures it is difficult to handle.

In the north the golden shiner reproduces itself in natural ponds. Elsewhere, artificial ponds are often used. When raising is in one phase, the same ponds are used for spawning and growing. When rearing is in two phases the two types of ponds are separated. In one-phase rearing, 2,000 to 3,000 parents from 9 to 20 cm (4 to 8 in) are released per acre. This is twice the number as for two-phase rearing. With one-phase rearing 60,000 fish of sufficient size can be harvested per acre (about 15 per square yard) but this quantity can be doubled when two phases are used by releasing 200,000 to 300,000 fry in the growing ponds (about 50 to 75 per square yard).

Goldfish (*Carassius auratus* L.) (Fig. 191), were introduced from Eurasia a long time ago. They are used a lot in the southern States but their use is prohibited in other States. They reproduce easily in ponds, provided the water is sufficiently warm, 18°C at least. Spawning is helped by placing artificial nests in the ponds just as for carp. Two hundred to 300 parents are released per acre. The fry hatch out within 6 or 7 days.

Rearing can be in one or two phases. In one phase the parents and the fry are kept together and 25 fry can be harvested per square yard. When the two phases are separated, that is reproduction and growing, then intensive production rearing can reach 50 to 75 per square yard. This requires both fertilization and artificial feeding. There are two variations of this method. In one (egg transfer method) the eggs, which stick to the artificial nests, are transferred to hatching and growing ponds. In the other (fry transfer method) three ponds are used successively – the first for spawning, the second for hatching and the third for growing. This is similar to carp rearing methods in the Far East.

b Species for artificial reproduction

White Sucker (*Catostomus commersonii* Lacepede) (Fig. 332). This is a rather large catostomid which can reach 35 to 50 cm (14 to 20 in) and sometimes more. It is found in the east of Canada and the United States and lives in clean lakes and water courses. It spawns at the beginning of spring on gravel beds.

The white sucker makes very good bait, particularly for walleye. It is easy to raise in ponds and its growth is rapid. Artificial fertilization is used for reproduction. The parent fish are caught in running water. At a temperature ranging from 12 to 15°C the incubation period is about 10 days. After hatching, the fry remain in jars until being released into growing ponds.

Young suckers are reared extensively in ponds of average fertility whether manured moderately or not at all. Ten fry are released per square yard

and the survival rate is around 25 per cent (10,000 to the acre). The fry eat natural food which includes an important percentage of chironomids. Growth is rapid, around 7 to 9 cm (2½ to 3½ in) in 2 months. Intermediate fishing, which starts in July, contributes towards an important increase of production.

Creek Chub (*Semotilus atromaculatus* Mitchill) (Fig. 333). These bait fish can reach an average length of 6 in for females and 11 in for males, and are found in relatively calm waters in the United States from the Rocky Mountains to the Atlantic. They lend themselves well to raising, and are considered excellent for catching pike. They are resistant to handling, transport and storage.

They reproduce between April and June on the gravel beds of running waters. Growth reaches 4 cm (1½ in) at 6 weeks and 9 cm (3½ in) by September.

Different systems of reproduction have been tried, one for example is more or less natural reproduction in a channel. This is a by-pass with a gravel bottom and alternating rapid current and depth such as exists in a natural spawning medium. But this method is inconvenient, for the parents concentrate in certain places and the nests are destroyed as a result.

In consequence, artificial fertilization is often used. Spawning is helped by the use of pituitary hormones which give good results. Second year males of 15 to 20 cm (6 to 8 in) and females of 12·5 to 15 cm (5 to 6 in) are used. The best temperatures for maturity are between 21 and 24°C. The growing ponds are more or less densely packed at a ratio of 50,000 to 600,000 to the acre (12·5 to 150 per square yard) according to the intensity normally practised by the farm.

Blacknose Dace (*Rhinichthys atratulus* Hermann) are small cyprinids never longer than 8 cm (3½ in) living in cold and clean water courses in the north of the United States. They reproduce in running waters in the months of April and May, make good bait for pike, bass and catfish, are reproduced artificially and their eggs measure 3 mm in diameter after swelling.

Stoneroller (*Campostona anomalum* Rafinesque) never measure more than 15 cm (6 in). They are common in the east of the United States, are very resistant and are good bait for catching bass.

Fig. 327 Ponds for breeding bait fishes. St. Marys Fish Farm, St. Marys, Ohio, U.S.A. (Photo: Cl. F. Clark.)

Fig. 328 A device formed of rough plates hooked onto a stretched wire across a pond to encourage natural reproduction of bait fishes. Lowell Gilbert Minnow hatchery, Celina, Ohio, U.S.A. (Photo: Cl. F. Clark.)

Fig. 329 The fathead minnow (*Pimephales promelas* Raf.). Above: male; below: female. (Photo: J. Dobie.)

Fig. 330 The bluntnose minnow (*Pimephales notatus* Raf.). (Photo: J. Dobie.)

Fig. 331   The golden shiner (*Notemigonus crysoleucas* Mitch.). (Photo: J. Dobie.)

Fig. 332   The white sucker (*Catostomus commersonii* Lac.). (Photo: J. Dobie.)

Fig. 333   The creek chub (*Semotilus atromaculatus* Mitch.). (Photo: J. Dobie.)

SECTION III

DEVELOPMENT OF FISH CULTURE IN THE NEAR AND MIDDLE EAST

The principal river systems of the whole region are the Nile, the Tigris, the Euphrates and the Indus. The region includes, among the African countries, Libya, the United Arab Republic, the Sudan and several Asiatic countries limited on the west by the Mediterranean and on the east by Afghanistan and Pakistan.

The region is predominantly arid and fish cultivation is limited. At this time it comprises experimental and demonstration stations set up in Pakistan, the Sudan, Syria, Iraq and the United Arab Republic. Because of wide variations in climate, from semi-arid zones in the northern steppes to the sub-tropical savannah in the south, fish cultivation is able to develop, thanks to a very large variety of species.

The cultivation of trout is possible and has been tried in cool water in certain countries including Pakistan, Afghanistan, Iran and Lebanon.

Warm water fish are the principal species for cultivation introduced in many countries in the region. In the first place there is the common carp associated with tilapias. Experiments have been carried out with, among others, *T. nilotica*, *T. zillii*, *T. rendalli* (*T. melanopleura*), *T. galilaea*, *T. macrochir*, *T. mossambica*. The cultivation of several Indian carps (*Labeo rohita*, *Cirrhina mrigala*, *Catla catla*) is practised in Pakistan and is being examined in the United Arab Republic. The latter has also introduced a Chinese carp (*Hypophthalmichthys molitrix*).

Other types of fish cultivation are more localized, and therefore have only limited development. For example, the cultivation of sturgeon in Iran, of eels in Egypt (for restocking lakes); trials with local catfish (*Clarias lazera*) in Syria and certain species of *Barbus* in Iraq; finally, in the Sudan, the cultivation of *Heterotis* is being studied.

Israel is considered separately. Geographically it is included in the region, but fish cultivation in that country is, in its general characteristics, European although its hot summer climate permits

Fig. 334 Important complex of rearing ponds, principally for carp, in the Jordan valley in the region of Hula, Upper Galilee, Israel.

the cultivation of sub-tropical fish. Cultivation is highly developed in Israel (Fig. 334), and has attained a high degree of perfection in an intensive form. The principal fish cultivated are carp (*Cyprinus carpio*); then come certain species of tilapias among which *Tilapia aurea* is the principal, then certain mullets associated with them in more or less brackish water, which are frequent near the coast.

SECTION IV

REGIONAL FISH CULTURE IN THE INDO-PACIFIC REGION

This region includes territories situated between Pakistan in the west, Japan in the north, and Australia in the south.

The Indo-Pacific region covers vast territories in which more than half the population of the world lives. Fish cultivation is very ancient, and started probably in China several thousand years ago. It includes, at the same time, the cultivation of carp and cultivation also, in association, of different cyprinids captured in rivers. With the characteristic patience and attention paid to details for which they are renowned, the Chinese have progressively perfected methods of cultivation which they took with them wherever they settled in the Far East, into which they introduced fish cultivation.

Fish cultivation in fresh water in Indonesia, India and Pakistan is also very ancient.

In the Far East one of the first forms of fish cultivation was probably in brackish water ponds or tambaks which started in Indonesia, where it was already a prosperous industry in the fifteenth century. The same type of cultivation then developed in the Philippines and today is extensively practised.

In Japan, with its restricted areas, fish cultivation has been directed to intensive production, particularly trout and eels.

The fish cultivation techniques now employed in the Indo-Pacific region have been improved recently thanks to the introduction of techniques for artificial spawning and incubation being applied to several species of cyprinids which originated in China and India.

Farming methods used in the Indo-Pacific region are essentially practised in fresh water ponds, in brackish water ponds and in rice fields. The last two were described in the preceding chapter.

I Principal Characteristics of Fish Cultivation in Fresh Water

1 Principal types of fish cultivation

In the Far East, cultivation in fresh water must be considered as an important factor in rural economy since it makes a precious contribution of animal proteins to the diet of rural populations. However, fish cultivation is rarely the only occupation of the farmers. In general, all fish culturists are farmers and all farmers cultivate fish.

The cultivation of fish in fresh water is divided into two categories: family ponds and commercial farms.

a Family ponds (Fig. 335) either aim at producing fish for eating, or are a hobby. A large number of small ponds of from 10 to 400 m² (approximately 12 to 500 yd²) which are spread out near and around the living quarters, must be included in this category. The peasants seed any kind and number of fry in these ponds. Quantities are predetermined by their financial resources rather than by the productivity of the ponds themselves. The fish are fed. The density of the fish diminishes as the result of fishing-out, so those remaining grow larger. The fry are bought from ambulant traders who buy the fish from commercial fish farmers.

b Commercial farms (Fig. 336) produce fry for restocking or fish for eating or, simultaneously, fry and fish for consumption. If the farm is in the second category (fish for eating) then the necessary fish are bought from fish farmers producing fry, or as fingerlings raised in rice fields.

Specialized cultivation of fry is very widespread and constitutes one of the characteristics of fish cultivation in the Far East. The transport and trading of these fish is carried out by ambulant dealers. They carry fish for eating 15 cm (6 in) long or fish for restocking which they sell in the markets or to the owners of family ponds and rice fields. The fish are carried in receptacles made of watertight bamboo lattice. According to the custom in the Far East, these receptacles are suspended from each end of a bamboo carried across one shoulder (Fig. 324). Fry for stocking are always transported in water. Fish for eating are carried sometimes in water or dry. During transport the swinging movement of the receptacles is sufficient to oxygenize the water and so losses are negligible.

2 Essential characteristics

(*a*) Ponds are often small, a few square yards, a few ares or a little more, but rarely a few hectares. They are generally found in the neighbourhood of the homes of the farmers which sometimes overhang the edge of the rearing pond.

(*b*) Artificial feeding is not used very much and when it is, it is only for brood fish or fry. However, for Chinese carps, food such as tender plants and grass cut up on the banks of the pond is distributed regularly. Common carp are fed soya meal, rice bran and other similar agricultural products. But in any case the food is of little economic value.

(*c*) Manuring is employed frequently, though mineral fertilizers are not used because they are too expensive and, to be satisfactory, must be employed frequently. Instead, manuring is organic and is distributed in the form of compost or in the form of animal or human faeces. Often the latrines and the stables overhang the ponds or are very near to them (Figs. 49 and 398). The most diverse organic matter of all kinds (rice, soya, different leaves) are thrown into the ponds and provide artificial food. If, however, this matter is not eaten by the fish it then decomposes on the bottom of the pond and manures it.

(*d*) The pond population is made up of a mixture of several species with different but complementary feeding habits and this gives higher production than a single species. The species chosen depend on the type of ponds, that is, ponds with clear and renewed water or ponds with stagnant water. In the latter, in Indonesia for example, a mixture consisting of 50 per cent *Helostoma*, 20 per cent *Osteochilus*, 20 per cent *Cyprinus*, and 10 per cent *Puntius* is chosen.

(*e*) Rearing is practised to produce small fish between 50 to 100 gm (about 2 to 4 oz) which corresponds to the economic potential of the consumers.

(*f*) The ponds are often densely stocked.

(*g*) The ponds are shallow, often less than 3 ft. Dikes are narrow and covered densely with herbaceous vegetation which gives stability and stops cracking. Often they are planted with banana and other fruit trees.

(*h*) Water inlets and evacuation are very simple and much use is made of bamboo.

(*i*) Intermediate fishing is general.

II Principal Fish Cultivated

In the Indo-Pacific region fish cultivation includes a great many different species: Hora and Pillay (1962) give 69, most of them raised in fresh water.

Cyprinids are the principal fresh water fish cultivated in the region. The common carp, *Cyprinus carpio*, which originated in China and Russia, comes first, as well as so-called Chinese carps: *Ctenopharyngodon idella*, *Hypophthalmichthys molitrix*, *Aristichthys nobilis* and *Mylopharyngodon piceus* introduced originally from China a long time ago and now on the way to being introduced into warm waters in many regions, notably in Russia and Central Europe. The cultivation of these species has already been described (Ch. IV).

Salmonid cultivation in Japan is particularly prosperous and is intensive in the case of rainbow trout (*Salmo gairdneri*). In other countries brown trout (*Salmo fario*) are raised for restocking.

Japanese eels (*Anguilla japonica*) are raised intensively in Japan and they are also cultivated in China, Vietnam and Taiwan.

Some indications are given below concerning the most important species which were not included in previous chapters.

1 Cultivated fish from the Cyprinid family

The principal among them are Indian carps: *Catla catla*, *Cirrhina mrigala*, *Labeo rohita* and *Labeo calbasu*, several other representatives of the genera *Labeo* and *Cirrhina*, as well as *Osteochilus hasseltii* and *Puntius javanicus*.

Indian carps (Major Indian carps) are the catla (*Catla catla* Ham. Buch.), the mrigal (*Cirrhina mrigala* Ham. Buch.), the rohu (*Labeo rohita* Ham. Buch.) and the calbasu (*Labeo calbasu* Ham. Buch.).

These four cyprinids possess common characteristics though their morphology and eating habits are different. In many respects the biology and cultivation of Indian carps resemble those of Chinese carps.

The distribution of Indian carps extends from Pakistan to Burma passing by northern India and Bangla Desh. Nearly all these fish have been introduced into southern India and some into Ceylon. All are large and can measure at least 90 cm (3 ft) or even more. Catla, which have large regular scales, can measure 1·80 m (6 ft) (Fig. 337: *Catlocarpio siamensis* Blgr.).

Indian carps are river fish which can be raised in limited areas (ponds); catla and rohu can withstand slightly brackish water.

In northern India, Indian carps spawn from June to September during the monsoon in the shallow flood water zones of rivers of the Ganges basin and of the Mahanadi and their tributaries. They do not reproduce in ponds but sometimes spawn naturally in large, artificial reservoirs in which artificial flood zones can be created. These reservoirs in which Indian carps spawn are called "bundhs". Since 1957 it has been possible to reproduce Indian carps artificially thanks to hormones injections (Chaudhuri and Alikunhi, 1957). The eggs are transparent, spherical in shape and large. They swell considerably after fertilization. Catla eggs swell from 2 to 4·5 to 5·2 mm; mrigal from 1·5 to 3 to 4 mm; calbasu reach 6 mm in diameter. They hatch quickly, from 16 to 19 hours after fertilization

Fry are resistant. For a long time the fry were only captured in rivers just as were Chinese carps, and transported over long distances for restocking ponds and reservoirs.

The eating habits of Indian carps differ from species to species which allows them to be mixed in ponds so that the most use can be made of the food resources of the medium.

Catla eat unicellular algae first, then zooplankton and then different algae (Myxophyceae, Chlorophyceae), rotifers, crustaceans and vegetable debris. Their growth is rapid. At 1 year they can reach 40 to 45 cm (15 to 18 in) and weigh 900 gm (2 lb) and at 2 years they can weigh 2 kg (4 lb).

When young, mrigal eat zooplankton. Later they become deep swimming fish feeding off blue and green filamentous algae, higher vegetation, decomposing plants, mud and various types of detritus. They can weigh over 300 gm (11 oz) at 6 months, 1 to 2 kg (2 to 4 lb) at 1 year and measure 45 to 60 cm (18 to 24 in).

Rohu eat unicellular algae and zooplankton when they are young, after which higher vegetation and decomposing vegetation takes an ever more important place in their feeding. At 1 year they can weigh 675 gm (1 lb 7 oz) and measure 35 to 45 cm (14 to 18 in). They reach maximum length at 3 years.

Calbasu eat unicellular algae first followed by phytoplankton, zooplankton (rotifers and crustaceans), organic waste, detritus and benthic animals.

They reach 30 to 35 cm (12 to 14 in) and weigh 450 gm (1 lb) at 1 year.

In fertile ponds with a mixed population it is possible to obtain natural production of 1,000 to 2,000 kg (1 to 2 tons) per hectare per year.

Several other representatives of the genera *Labeo* and *Cirrhina* are also raised in ponds. Different species of *Labeo*, raised principally in India and also in Vietnam, are one. Other *Cirrhina* spp. are cultivated in India, and *Cirrhina molitorella* Valenciennes (Mud carp) is cultivated in southern and central China, Taiwan, Thailand and Malaya. These are of relatively slow growth and weigh 25 to 75 gm (1 to 2½ oz) at 1 year, 300 gm (11 oz) at 2 years, 600 gm (about 22 oz) at 3 years. They are typical of fish living on the bottom and eat vegetable and animal debris, as well as microflora and microfauna living in the detritus. They clean the bottom of this detritus. They do not compete with the fish living in the body of the water or near the surface of the water, with which they are associated.

Osteochilus hasseltii Valenciennes (Fig. 338). These are small cyprinids never more than 35 cm (1 ft 2 in) long. They have two pairs of barbels and the caudal fin is strongly forked. They are found in Indonesia, Malaya, Thailand, Vietnam and Cambodia.

They live in running or static waters but can be raised in ponds and reach sexual maturity at the end of the first year when they measure 15 to 20 cm (6 to 8 in). Their eggs are transparent and 2 mm in diameter. They are laid at the start of the rainy season in rivers with gravel beds and a rapid current.

The fry live off phyto- and zooplankton. Later they eat the biological cover or periphyton on the submerged plants, and the tender parts of the latter. They can withstand high temperatures, even above 30°C, which do not suit carp. In such water they can weigh from 80 to 100 gm (about 3 to 4 oz) in 100 days and in 1 year can weigh 350 gm (12 oz).

Reproduction in ponds can be successful always provided the water is clean and has plenty of oxygen. This is essential. Spawning takes place in spawning ponds a few square metres in size with sandy or gravel bottoms traversed by a rather strong current. Fry hatch the second day after spawning and are transferred to nursing ponds or rice fields to become transformed into fingerlings. Fish for eating are produced either in freshwater ponds or in rice fields.

Puntius javanicus Bleeker (Fig. 339). These fish are important to cultivation in the Far East. They are raised in Indonesia, Thailand and Vietnam and have been introduced into Ceylon.

The body is never longer than 50 cm (20 in) and is covered with large regular silver scales.

Puntius are cyprinids which live in running water. They spawn there at the start of the rainy season when the rivers are in flood, and the sandy and gravel shores are submerged.

They develop well in both fresh water and brackish water ponds in which they are associated with milkfish. They are also raised in rice fields. They require well oxygenated water particularly when young. They are often associated with carp in ponds, and since they are surface fish more than are carp, they do not compete for food. Their natural food is, for the most part, leaves of submerged plants, and filamentous algae. Their excrement, only half digested, is eaten by the carp. They are of rapid growth and can reach 250 to 500 gm (8 oz to 1 lb) by the end of the first year. They need warm water, at least 15°C and their optimum growth is in water between 25 to 33°C.

Puntius spawn in ponds. The parent fish are at least 1 year old, and should measure 25 cm (10 in). Spawning ponds are from 2 to 5 ares and their bottoms should be hard. To ensure this the mud is left dry for several days until it has cracked (Fig. 340). Sometimes part is covered with sand or gravel. The pond should be fed by well oxygenated water at a ratio of 1 litre (2 pt) per second per are of surface. Six to 10 pairs of parent fish are liberated in the spawning pond. Reproduction follows rapidly. The eggs are transparent and measure 1 mm in diameter at spawning and 2 mm after swelling. They fall to the bottom of the pond and the fry hatch 2 or 3 days later. When they are 3 weeks old they are harvested.

2 Cultivated fish from the Anabantid family

These fish belong to the *Anabantidae* (labyrinth fish) and possess accessory breathing organs which allow them to take in oxygen in the atmosphere, and this protects them against low dissolved oxygen content of the water. Three species are important for cultivation.

Osphronemus goramy Lacépède (Fig. 341). Indigenous gouramies are raised in south-east Asian countries, particularly in Java where cultivation has become well developed. They have also been introduced and reared in India, Ceylon, the Philippines and even beyond the region of the Indo-Pacific. They have fine and much appreciated flesh.

They are fresh water fish and live in running and stagnant water. They are cultivated in fresh water ponds and can spawn if they find materials with which to make their nests.

They are herbivores and eat plankton. They accept cassava, pawpaw and colocase leaves, etc. They are slow growing and at the end of the first year weigh no more than 100 gm (4 oz) but they grow more quickly in the following years. Maximum length is 60 cm (2 ft). They are very sensitive to low temperatures and the water must be 15°C at least. Optimum temperatures lie between 24 and 28°C.

Gouramies are not good reproducers. A 5-year-old female gives only 3,000 to 5,000 fry per year with two or three partial spawnings in two monthly intervals. If the temperature is suitable the fish can spawn throughout the year.

The spawning ponds are small and the ratio is from 25 to 30 m² (30 to 36 yd²) per couple. Parents are from 3 to 7 years. Palm fibres and branches are thrown on to the surface of the water or bamboo sticks formed into cones are placed in the pond so that the female can build a nest generally 30 cm (12 in) below the surface (Fig. 342). The nests are spherical in shape and measure from 30 to 35 cm (12 to 14 in) in diameter. The opening is situated in the bottom section. Spawning takes 2 or 3 days. The eggs are deposited in the nest in layers and are separated by fibre beds. After spawning the female remains nearby and fans the nest with the caudal fin. Spawning is indicated by an oily substance which rises to the surface over the nest.

After spawning the fish farmer frequently removes the nest from the water and transfers the eggs to jars filled with clean water (Fig. 343). They float because they are lighter than water. After hatching, which takes place 5 or 7 days later, the fry remain in the jars for about 3 weeks. From the tenth day onwards they can be fed with very fine rice bran. They are then transferred to nursing ponds. At 3 or 4 weeks they start feeding off vegetation.

Trichogaster pectoralis Regan (Fig. 344). This species (Sepat siam) is indigenous to Thailand, Cambodia and Vietnam and has been introduced into several countries in the Indo-Pacific region, notably Malaya and Indonesia.

It is a small fish no more than 25 cm (10 in) in length and lives in marshy, static waters. It can be raised in ponds and in rice fields. It lives off phyto- and zooplankton, decomposing plants, detritus and when cultivated can be fed duckweed and ipomoea. Its growth is somewhat slow: at 12 months it measures from 16 to 18 cm (6 to 7 in); its weight never exceeds 130 to 140 gm (about 5 oz) and it is generally eaten after it has reached 50 gm (2 oz). It is best suited to temperatures between 25 and 35°C.

Trichogaster spawns in ponds. The eggs are laid in a floating bubble nest prepared by the male (Fig. 345). The spawning pond should be fairly deep: 70 to 100 cm (approximately 2 to 3 ft) but the water is not renewed though it should be partially covered with floating aquatic vegetation or straw. The parents are 7 months old at least and weigh 100 gm (4 oz). Over a period of 1 or 2 days the males prepare the nests which float and are made up of a mass of bubbles. The females then lay their eggs under the nests where they float. Hatching takes place 2 or 3 days later. There are rarely more than 4,000 fry per nest. The eggs and fry are protected by the male. Artificial spawning is encouraged during the dry season only, because the rains destroy the eggs and the nests.

Helostoma temmincki Cuvier (Fig. 346). Kissing gouramies are indigenous to Thailand, Malaya and certain large Indonesian islands. Like the species mentioned previously, they are small and do not exceed 30 cm (1 ft) in length but nevertheless meet the requirements of local markets. They have small, thick, protractile lips.

They live on both the surface and in the body of the water, are typically plankton eaters, and for this reason form complete mixed stocking with carp and *Puntius javanicus*. They grow rapidly if the temperature is satisfactory (from 25 to 30°C) but the plankton must be abundant. They can be given artificial food such as rice bran, and they are tolerant to water with a low dissolved oxygen content.

The fish can reproduce in ponds. The breeding pond should not be deep: 50 cm (20 in) and the water should be calm. The surface of water per couple, if the pond is manured, is around 3 to 5 m² (3½ to 6 yd²) but it should be six times greater if manure is not used. The brood fish, the sexes of which are difficult to recognize, are 1 year old at least and should measure 20 cm (8 in). If conditions are favourable they can spawn every 3 months.

Helostoma eggs are from 1·0 to 1·5 mm in diameter and float on the surface of the water. In order to prevent their getting caught up in some corner of the pond, rice straw or banana leaves are thrown on the surface of the water (Fig. 347). Spawning is of short duration and hatching takes place 24 hours later. The fry shelter under floating vegetation. Five days after hatching the pond is manured in order to help the plankton develop. This operation is repeated several times until the fry are harvested 50 days after hatching. A female can give between 1,000 and 4,000 eggs per spawning. Kissing gouramies can also reproduce in rice fields after the rice has been harvested, but only on condition that the field has been properly arranged for it.

3   Cultivated fish from the Schilbeid family

These fish are very similar to catfish (Ch. VII). Several representatives of the genus *Pangasius* (Fig. 348) are cultivated in Thailand, Cambodia and Vietnam. The principal species are *P. larnaudi* Bocourt, *P. suchi* Fowler and *P. micronemus* Bleeker.

Pangasius spp. are large fish and some species reach 50 to 150 cm (approximately 20 in to 5 ft). Their bodies are long and thin. The mouth has four barbels, the skin is naked, generally coloured blue-grey at the end, but the sides and belly are lighter in colour. The dorsal fin is short and an adipose fin is situated behind it.

They can live in running waters and lakes and can be raised in ponds, bamboo enclosures or cages. But as they do not reproduce in ponds the fry must be captured in streams. Until they reach 10 cm (4 in) they live in groups on the bottom but come to the surface regularly to take in air. They are caught in slow moving water shaded by bushes and trees.

Carnivorous in open water they are more or less omnivorous in captivity and can feed off kitchen waste, bananas, cooked maize, fish offal, bits of fish, chicken droppings, rice bran and tender vegetation (Fig. 349). They can grow to 1 lb. in 1 year and 2 lb in 2 years. If their basic food is fish they will reach 2 lb in less than 1 year.

4   Cultivated fish from the Cichlid family

Average size cichlids are raised in India. They are *Etroplus suratensis* Bloch (Pearl spot) and can be cultivated in either brackish or fresh water ponds.

They are brackish water fish living in estuaries at the limits of fresh water to which they can

Fig. 335 Family fish culture. Small ponds of no more than a few square yards situated near habitations. Tjibodas Road, Soekabumi, Indonesia.

Fig. 336 Commercial fish culture. Intensive production ponds with water enriched by sewage; production is as high as from 2 to 5 tons per hectare ($2\frac{1}{2}$ acres) per year. Bodjongloa, Bandung, Indonesia.

Fig. 337 *Catlocarpio siamensis* Boulenger; length: 1·30 m (4 ft 4 in); weight: 45 kilos (100 lb). Fish market, Bangkok, Thailand.

Fig. 338 *Osteochilus hasseltii* Valenciennes, small cyprinid feeding on biological cover (periphyton). Raised in fresh water and rice paddies with other species.

Fig. 339 *Puntius javanicus* Bleeker, cyprinid feeding on aquatic plants (submerged weeds, filamentous algae), reared in fresh water pond, in brackish water and in rice paddies.

Fig. 340 Spawning pond for *Puntius javanicus*. About 5 ares (five-fortieths of an acre) with the bottom hard and cracked, pond well supplied with running water. Tjangringham, Jogjakarta, Indonesia.

Fig. 341   *Osphronemus goramy* Lac., anabantid, vegetarian, raised in fresh water ponds. Fish with excellent quality flesh but of slow growth and sold only after weight reaches at least 1 lb.

Fig. 342   Artificial spawning nest of gourami. The nest is formed by a cone of bamboo laths to which the fish brings vegetable fibres and on which the eggs are laid. Tjangringham, Jogjakarta, Indonesia.

Fig. 343   Hatching jar for gourami eggs. The eggs float and are light in colour. Tjangringham, Jogjakarta, Indonesia.

Fig. 344 *Trichogaster pectoralis* Regan, plankton eating anabantid; slow growth; lives in stagnant swampy waters.

Fig. 345 *Trichogaster pectoralis* nests made of agglutinated floating bubbles under which the eggs are placed. The spawning pond is partially covered with floating leaves or with straw. Muntilan, Jogjakarta, Indonesia.

Fig. 346 *Helostoma temmincki* Cuvier, kissing gourami, plankton eating anabantid, slow growth in ponds and rice fields mixed with *Cyprinus carpio* and *Puntius javanicus*.

Fig. 347 Breeding pond for kissing gourami. Small pond with bits of vegetation on the surface in order to avoid the floating eggs accumulating in a corner by the wind. Muntilan, Jogjakarta, Indonesia.

Fig 348 *Pangasius pangasius* Ham. Buch. These fish breed in open water where the fry must be caught and transferred to ponds.

Fig. 349 In the rearing ponds, *Pangasius* are omnivorous. They can be fed with kitchen waste, bits of fish, tender weeds and poultry droppings. Bangkhem experimental station, Bangkok, Thailand.

become acclimatized. They spawn in ponds and even fresh water ponds. Eggs, laid by at least second year fish measuring 15 to 18 cm (6 to 7 in), are 1 to 2 mm in diameter. They stick to the underside of leaves and other submerged objects. At first the fry are watched over by the female.

Pearl spot are herbivorous and live off algae, vegetable waste and tender water plants. They are of slow growth: from 10 to 12 cm (4 to 5 in) in 1 year; 15 to 18 cm (6 to 7 in) in 2 years. Maximum length is 30 cm (12 in). They can be associated with other species.

SECTION V

REGIONAL FISH CULTIVATION IN AFRICA

Fish cultivation was imported into Africa. The raising of trout in cold water is limited and has been practised only since the end of the First World War in North Africa (Morocco) and Kenya, but for a longer period previously in South Africa. Cultivation is principally for angling.

Because the whole population of intertropical Africa lacks animal proteins part-time subsistence cultivation is practised by African farmers and has progressively developed since 1946 following the first successful raising of *Tilapia* in Katanga (Zaire). However, having started very well and developed markedly between 1953 and 1960, African fish cultivation has made no further progress and can even be said to have declined. The reason for this springs from the discouragement of farmers who found they were harvesting too many small fish in over-populated ponds (Meschkat, 1967). This over-population was caused by the general raising of mixed age groups both for reproduction and growing. It is necessary, therefore, to modify cultivation methods in which nursing and spawning are two separate operations (Lemasson and Bard, 1968).

African fish cultivation for food is principally a family occupation. Only rarely is it artisanal and it is essentially based on tilapia cultivation. Several methods have been tried out, including multiple ages mixed permanently or temporarily (this is met with most often); a single age method with separate age groups; a controlled reproduction method (monosex cultivation or association with a predator). Association is generally between a microphagous or omnivorous species such as *T. macrochir* and *T. nilotica* and with a macrophytophagous species such as *T. melanopleura*. The raising of tilapias is described in Chapter VI, Section V.

Common carp were introduced a long time ago into Madagascar and into certain other parts of Africa (South Africa and Southern Rhodesia), but with the exception of Madagascar have not, so far, been raised seriously in ponds. Their introduction into several countries of the African intertropical zone led to differences of opinion between 1949 and 1960. Carp have been more recently introduced into Nigeria, Uganda and Egypt.

Largemouth black bass (*Micropterus salmoïdes*) have been introduced into South Africa, Madagascar and Morocco and occasionally into other countries, but their development is limited as their production is considered essentially directed towards recreation.

Specific African cultivation, with the exception of tilapias, has scarcely developed.

Cultivation of *Heterotis niloticus* Ehrenbaum (*Osteoglossidae*) (Fig. 351). The possibilities of using *Heterotis* for cultivation was suggested in 1966 by Daget and d'Aubenton.

The area over which this fish is dispersed is relatively limited. It is found only in the middle and lower courses of the great north-equatorial rivers or in lakes included in the area.

It is of interest because of its eating habits, which are exclusively microphagous, and because of this it can produce fish of 1 kilo ($2\frac{1}{4}$ lb) in 1 year with an annual production of 2 tons per hectare if given good artificial food. This includes rice bran, ground nut seeds, ground nut oil cake, cotton seed oil cake, brewery by-products; other food such as maize should be used.

Another advantage of this species is its late spawning, that is when it reaches from 19 to 20 months. The danger of over-population does not therefore exist. It is impossible to recognize the sexes from their appearance. Spawning takes place during the rainy season. Nests are built in shallow water among dense vegetation. They are

Fig. 350 A group of individual ponds of from 1 to 2 ares (an are is one-fortieth of an acre) installed alongside the Tshama swamp, Kabare territory, Zaire. The ponds are built in terraces following the contour lines. Between the rows of ponds there are vegetable gardens, the waste of which is used for the ponds. Tilapias are reared here.

Fig. 351 *Heterotis niloticus* Ehrenb. Microphagous fish of the Osteoglossidae family. Growth in ponds is rapid but breeding is difficult. Pelbea Haffir, Province of the Upper Nile, Sudan.

round in shape from 1 to 1·50 m (3 to 5 ft) in diameter, are carefully cleaned and the bottom is slightly hollow and about 20 to 30 cm (8 to 12 in) deep. Eggs are spherical in shape, yellow-orange in colour and 2·5 to 2·8 mm in diameter. They are laid on the bottom of the nest and stick together. Hatching takes place 48 hours after spawning. The absorption of the yolk sac takes 5 or 6 days after which the fry start to leave the nest but return to it regularly. Until this stage is reached they are watched over by the brood fish, but this ceases the moment the fry abandon the nest definitely and disperse. Growth is rapid, reaching 30 cm (12 in) and 300 to 400 gm (11 to 14 oz) in 4 months, and 50 cm (20 in) and over 1·5 kg (3 lb) in 2 years.

However, the reproduction of *Heterotis* poses problems. It is easiest in large ponds of 1 or 2 hectares at least with special grassy and shallow spawning areas, rather than in small ponds. Breeding in small ponds has been practised but the method produced few fry.

Heterotis is raised in the Cameroons and in the Central African Republic. Because of its taste and the number of bones it is less appreciated than tilapias.

Cultivation of Nile Perch (*Lates niloticus* L. *Centropomidae*). This species, which is very voracious, has been used at experimental level, notably in Uganda, to control excessive reproduction of tilapias. It reduced profileration effectively. Unfortunately the reproduction of Nile perch in ponds is difficult. Reproduction has only been obtained in large ponds.

Diverse Cultivations. Representatives of the *Citharinidae* family, especially the genus *Citharinus*, which is typically planktonophagous, are found sometimes in rearing ponds. They grow quickly but unfortunately they do not reproduce in captivity. The same is the case with certain *Labeo* (*Cyprinidae*). Trials with the very voracious fish *Hydrocyon* (*Characidae*) did not give positive results.

SECTION VI

DEVELOPMENT OF FISH CULTIVATION IN LATIN AMERICA

Although there have been sporadic efforts to cultivate fish in several countries of this vast region there is no traditional fish cultivation comparable with that found in most of the other parts of the world. At first sight this is surprising, taking into account the considerable food deficiencies present in Latin America. Several reasons may explain this, the principal being that at all levels there is a shortage of technically specialized personnel.

Cultivation of salmonids has been slightly developed both for brown trout (*Salmo trutta*), introduced from Europe, and rainbow trout (*Salmo gairdneri*). The first brown trout were reared in the cold waters of Mexico, Central America and the Great Cordillera of the Andes. The rearing of rainbow trout for the table is rare.

Cultivation in warm water is only slightly developed and includes carp and tilapias, and, though secondary, black bass (*Micropterus salmoïdes*). All these fish were imported. Indigenous species which have been subject to cultivation trials are the *Cichlidae* and the *Characidae*, but they did not give encouraging results.

Mexico. Throughout Latin America it seems

that fish cultivation is widest spread in Mexico. It concerns, principally, the raising of common carp (*Cyprinus carpio*). In 1964 it covered 12,500 hectares (about 30,000 acres) with an average production of 700 kg (14 cwt) per hectare. Three species of *Tilapia* were also introduced as well as grass carp (*Ctenopharyngodon idella*) from China. Also an atherinid, *Chirostoma estor*, is cultivated for restocking purposes.

Central America and the Antilles. Since 1926 experiments have been carried out in Central America and the Antilles.

In Guatemala, in particular, it was first decided not to introduce exotic species but to try raising local cichlids and above all several species of *Cichlasoma*. These included both herbivorous, omnivorous and voracious fish as well as the characid *Brycon guatemalensis*. None of these trials, however, gave positive results.

Consequently the cultivation of imported fish was adopted and from 1952 carp and *Tilapia mossambica* were introduced into Haiti and the Dominican Republic. In Puerto Rico *T. mossambica* are reared, while black bass and bluegill (*Lepomis macrochirus*) were imported from the United States.

The principal fish farms of Central America are found in Guatemala. Carp have given the best results. In intensive rearing they have given 3,000 kg (3 tons) and even more per hectare. Tilapias, black bass and bluegill have also been introduced. Similar trials, though on a less important scale, have been tried in other central American countries including El Salvador, Honduras, Nicaragua, Costa Rica and Panama.

South America. On this continent fish cultivation is developed principally in Brazil. It is warm water farming and is particularly important in the states of São Paulo and Parana. Carp are cultivated in both. Since 1953, in the State of São Paulo, the cultivation of tilapias (*T. melanopleura*) has developed well but it started to decline because proper methods were not used. Improved methods are now being tried out. Black bass, associated with tilapias, have also been introduced and 9 tons per hectare were produced in warm water.

Warm water cultivation has also been tried in Colombia.

In Argentina the cultivation of the silverside (atheriniculture) (Chapter V, Section VI) which are cold and temperate water fish, is developed for restocking.

APPENDIX
RESTRICTED FISH CULTIVATION

In many cases, when there is one pond only, or a few small ponds, fish cultivation is restricted. A farm lacking space or with an insufficient number of ponds cannot include all the age classes from fry to fish for eating and also brood fish.

When this is the case, fish farming is either a secondary occupation which gives a supplementary income, produces extra food or is a hobby.

Among these restricted farms, of which the variety is considerable, there are notably small-scale salmonid cultivation farms, family pond farms, pond cultivation for angling and natural ponds which cannot be emptied.

I Small-scale Salmonid Cultivation Farms

These can be called, following an expression created by Léger, small fish cultivation farms restricted by space or water supply. They are confined to rearing fish of one or two age groups. This type of farming is intended to exploit a limited surface of water suitable for rearing trout.

Any farm with fresh and clean running water can raise trout. Naturally the production will depend on space, quantity and quality of the water as well as the food distributed. A current of 4 or 5 litres (about 1 gal) of water per minute should be sufficient for small amateur cultivation.

It goes without saying that all the rules relative to stocking, maintenance and water supply apply as much to a small farm as to a complete exploitation.

In certain cases the most interesting are those producing fish for the table, starting with fish which are sufficiently large to become marketable after 1 summer. In other cases it is economically viable to raise young fish which will be ready for restocking at 1 year. Thus, for example, it is possible to produce trout fingerlings of restocking age of 1 summer starting with yolk sac fry. This production is advantageous when it is certain the fish can be sold. Also, this type of rearing can be practised by individuals and angling clubs.

In Europe the rearing of trout fingerlings for restocking includes brown trout principally. Until now small-scale salmonid farming and amateur cultivation have been extensive; that is to say it relies on the use of natural food. But it is quite possible that semi-extensive or intensive farming will develop progressively in the future thanks to the advantages afforded by artificial feeding. In extensive farming the best production of trout fingerlings for restocking is obtained in spring ponds and in linear ponds. Details of this kind of pond are developed in Chapter III, which is devoted to salmonid cultivation.

In ordinary shaped ponds, though always extensive, 2-year-old trout can be raised for restocking. These should measure from 16 to 20 cm (6 to 8 in) and weigh around 50 gm (2 oz). For average ponds, 30 to 40 fish per are can be released and average loss should be around 25 per cent.

If the aim is to raise fish for eating by intensive or semi-intensive farming in small ponds, then rainbow trout should be raised rather than brown trout. Exceptionally, brook trout can also be produced provided the water supply stems from a nearby spring and the temperature never exceeds 18°C in summer.

Trout for eating can be produced in one or two stages. In one stage fingerling trout are bought, but in two stages, fry are bought to be grown into fingerlings the first year and into trout for eating the second year. In general, however, on small-scale salmonid farms cultivation starts with fingerlings. This permits harvesting in one season. Rearing can be extensive, semi-extensive or intensive.

In order that it should be worthwhile, *extensive rearing* needs a surface area of at least 10 ares. In such a surface it is possible to produce 10 to 15 kg (22 to 33 lb) (100 to 150 fish) in medium to good water. Stocking depends on the surface and the biogenic capacity which is calculated by the adequate formula (Ch. XIV). According to the productivity, 5 to 15 fingerlings are stocked per are in water of from average to good quality.

Semi-intensive or *improved extensive raising* aims at producing more fish per unit of surface than in extensive.

In general the aim is not to raise more than five or six times the natural productivity of the pond. Improved extensive production depends on natural feeding, with artificial feed added. This raising is carried out in ponds which are rather extended but not so large as for extensive farming – from 1 to 10 ares for example.

Intensive raising can be carried out in rather small ponds, no more than a few square metres in size, but there must be a sufficient current of water. It also demands large quantities of food. To determine the stocking ratio for intensive farming the litre-kg formula described earlier (Ch. III) can be used. One litre of fresh water per minute will permit the raising of 1 kg ($2\frac{1}{4}$ lb) of trout and with a sufficient water current it is possible to produce from 10 to 15 kg (22 to 33 lb) per square metre or cubic metre if the pond is 1 metre (3 ft) deep, which is more or less the possible minimum. If rearing is intensive then stocking trout for the table should be from 10 to 12 fingerlings per kg ($2\frac{1}{4}$ lb) of trout expected to be raised.

II Family Ponds

For a long time past in the tropical regions of Asia and the Far East, and more recently in the tropical regions of Africa, fish have been raised in limited spaces. These family ponds are often no more than 1 are in size, sometimes they are larger but they can be even smaller. Indeed it can be that the space is very restricted such as farming with the use of bamboo cages. Details of this kind of farming are given in chapters and sections devoted to cyprinid cultivation (Ch. IV), to the rearing of tilapias (Ch. VI, Sec. V) and to special cultivation in the Indo-Pacific region and Africa (Ch. IX). According to each case, Asiatic or Indian cyprinids, or tilapias, are raised. The ponds are well manured often by organic fertilizers. Generally the fish are fed kitchen waste and other food of small economic value.

III Angling Ponds

The use of angling ponds, in certain cases, is to produce something quite different from fish for stocking or eating. In fishing ponds the aim is to produce or to stock fish which will satisfy the demands of anglers for their pleasure rather than catching fish for eating.

In Europe, according to each case, angling ponds are for salmonids (trout) if the water is cold and well oxygenated, or cyprinids (carp, roach and associated predators) if the water is warm and not so rich in oxygen in summer. If the ponds are for salmonids then only brown or rainbow trout are fished. If the ponds are for cyprinids then the population will be mixed. They will contain different species of cyprinids: roach, carp, tench and several associated predators such as perch, pike, eels and, sometimes, rainbow trout which will be added at the end of the season if the water is deep enough and oxygenated.

These ponds (Fig. 352) are exploited by an angler, a group of anglers or a club. If run as a business, tickets for a day's fishing are sold.

In ponds of this kind one condition is indispensable to success; the fish must be relatively easy to catch and to ensure this, stocking must be clearly higher than in growing ponds. It seems that stocking should vary between four and six times the normal number of fish found in production ponds.

Generally these ponds are stocked with fish bought commercially. They are liberated when they are catchable or when they are almost catchable unless there is enough time for them to grow before the opening of the fishing season. A single release will not be enough before the opening of the season. There should be several at more or less regular intervals during the season.

In the United States fishing ponds are either for salmonids as in Europe or warm water ponds. In the latter the principal fish species is the voracious largemouth black bass with which are associated forage fish, generally bluegill. The aim is to keep the two species in equilibrium over several years. A description of this kind of rearing has been given in Chapter VI, section IV.

Fishing ponds should receive the same maintenance care as ordinary ponds. Upkeep often leaves much to be desired and sometimes the ponds are left under water for many consecutive years. In this case care and maintenance must be carried out while the pond is under water; notably the elimination of excessive vegetation. However from time to time the pond must be emptied in order to remove excessive mud, re-establish normal stocking and return to a balance between the species.

Fig. 352   Angling pond owned by one angler or a group of anglers. The pond is shallow and suitable for raising cyprinids and voracious fish.

Fig. 353   Farm pond in the United States. The water covers about 2 hectares (5 acres) and has a variety of agricultural uses. For angling it is stocked with a mixed population of largemouth black bass and bluegill (forage fish). Westfield, Wisconsin, U.S.A.

IV Non-drainable Ponds

The exploitation of non-drainable ponds, either natural or artificial (old quarries or mines) is concerned with fishing and not fish cultivation. Details of this, therefore, have no place in this work.

Aims differ according to regions. The goal may be the production of fish for eating or the ponds may be kept up simply for the sport.

In Europe, according to whether the summer temperature is low or high, the population of ponds which cannot be drained is mostly composed of salmonids or cyprinids. In the latter case, if fishing with a net is easily practical, then carp can be introduced. If it is not practical, as is often the case, several other species are favoured such as roach, tench, pike, eels and crayfish which can be caught with trap nets and hooks and lines. Old quarries are often very suitable for crayfish. Sometimes it is possible to pump the ponds dry if the volume of water is not too important.

Part Three

CONTROL AND INCREASE OF PRODUCTION IN FISH CULTIVATION

Introduction: The Concept of Production in Fish Cultivation

1 Different concepts of production in fish cultivation
The principal goal of fish cultivation in ponds is the production of fish of commercial value for eating or restocking, generally in as large quantities as possible and in the shortest possible time.

However, concepts differ, sometimes very much so according to the objective. Principally there are three: quantity, quality and economic production.

Quantity production aims at producing as great a quantity of fish as possible for eating or restocking. The principal aim is not to produce fish of high quality such as the production of graded fish. Quantitative production means production which will give the most weight. This is followed notably in Africa with tilapias raised by the system of mixed age groups.

Quality production seeks to produce as high a quantity as possible of graded fish either for eating or restocking. Production in terms of weight is never at maximum level although the fish of any given species produced are generally of uniform size and weight and of great commercial value. This kind of farming corresponds in Africa to the rearing of tilapias by separate age groups. According to Bard (1962) this method can give a production equal to two-thirds in weight of quantitative production obtained by mixing age groups.

Economic production aims to produce as great a quantity of fish of high commercial value as possible. The fish produced are either of high consumption value such as gourami in the Far East, or fish for restocking, not to be eaten but with a high market value. The production unit is not necessarily found in the weight but in individual fish. This is the case when fingerling pike or pike-perch of 1 summer are produced in carp growing ponds. Generally, in this case, the quantity produced in weight is not very high.

Methods of raising can be extensive, semi-intensive or intensive, depending on whether rearing is based on natural food only or whether it is more or less entirely artificial. Extensive farming produces a quantity of fish without artificial feeding, from rearing ponds which correspond to natural productivity. Intensive farming seeks to produce a maximum quantity of subjects or weight in a minimum of water by means of intensive or exclusive feeding. Semi-intensive farming is intermediary.

2 Means of increasing production
Whether production is quantitative, qualitative or economic, there are many systems of controlling and increasing output. These are either biological or non-biological.

a Non-biological methods for increasing production
These are numerous and include:

1 *General technical and sanitary methods.* Quite apart from the methods given below and before applying them, general sanitary and technical measures are indispensable if the farm is to be run normally. The following points, therefore, must be carefully observed; (1) different age groups must be represented normally and judiciously; (2) a sufficient oxygen content must be ensured; (3) action must be taken against disease and epizootics.

It is important to remember that all production factors play their part and that each one must be given even more attention as in these circumstances they are far from their optimum.

2 *Maintenance and improvement of ponds.* Quite apart from the upkeep of dikes, banks and other installations the maintenance and improvement of ponds include principally:

(*a*) Action against excessive water plants. This is carried out principally by mechanical means (cutting), or by chemical means, as well as by biological means (grass eating fish) in certain cases. These are discussed in Chapter XII.

(*b*) The improvement and restoration of the bottoms of ponds. The principal means used to this end are to empty the ponds and then when dry: to (1) leave the soil uncultivated or combine treatment with a crop; (2) the disinfection of the bottom; (3) work over the bottom either by breaking up the soil superficially or by ploughing more or less deeply; (4) dredge out the silted matter from the pond or ponds which are subject to invasion by swamp vegetation (Chapter XII).

Maintenance and improvement of ponds are necessary means of ensuring natural productivity and help to improve it.

3 *Liming and manuring of ponds.* The use of lime and fertilizers (Ch. XII) provides a simple and economic means of increasing the production of ponds. The use of both must be generalized. Liming has a variable action but is favourable to the hygienic conditions of both pond and fish. It also favours the biological factors of production. Manuring improves and accelerates the production of natural food in the ponds. From this fact, stocking and production are greatly improved often up to a proportion above 50 per cent when compared with the natural production of non-manured ponds.

4 *Feeding in fish cultivation.* Feeding or the distribution of so-called artificial food is one of the principal means of increasing production in fish cultivation. It is discussed in Chapter XIII.

The importance and intensity of feeding vary according to the types of cultivation. In certain kinds of cultivation, notably of salmonids, artificial feeding can form the exclusive basis of the food of cultivated fish which means, from this point of view, that they are independent of the natural pond medium and can give a production of over 100 tons per hectare. In traditional carp cultivation

natural feeding, which can be increased by manuring, represents an important part of the feeding of the fish. By using fertilization and feeding at the same time it is possible in carp cultivation to get a total production six times as high as the natural production. On the other hand it is still practically impossible to artificially feed certain species of fish such as pike.

b Biological means for increasing production
The principal biological methods for increasing production are:
1. A careful choice of species.
2. Control and adaptation of pond stocking.
3. Control of temperature and the oxygen content in ponds.
4. Improvement in reproduction and selection.
5. Mixing of species and age groups.
6. Successive productions throughout the same year.
7. Simultaneous productions, animal and/or vegetable in the ponds.
8. Intermediate fishing.
9. Action against enemies, parasites and diseases of fish.

The biological methods for increasing production are discussed in Chapter XI. Control, however, and the adaptation of stocking ponds are discussed in Chapter XIV and the struggle against natural enemies, parasites and diseases in Chapter XV. The practice of selection, control and improvement in reproduction and the use of special methods in cultivation are discussed throughout the book, notably in Chapter IV.

Chapter X
NATURAL PRODUCTIVITY OF PONDS

THE productivity of a fish cultivation pond is understood to mean its capacity to produce cultivated fish. It includes maximum normal production, which is the highest possible production obtained and maintained under normal farming conditions. In general, a fish farmer seeks to keep maximum production constant. It is always possible for him to operate below maximum production but this is not normally justified.

The evaluation of the productivity of ponds is fundamental for drawing up a working plan for the ponds. A rationally run farm relies, notably, on a predetermined and adequate stocking rate, that is, a fixed optimum quantity of fish to be released in the ponds. The calculation of this stocking rate rests on the predetermination of productivity.

In the first place, in general, the natural production of the pond must be predetermined. It is indispensable to do this for all farms where natural production is the most important, for in this case it will determine total productivity. This is the case for both extensive and semi-intensive farms but is less important for intensive production.

Consequently, in a great many cases, it is in the interests of the farm to predetermine natural productivity of the ponds as exactly as possible. However, this is both complicated and difficult, and generally remains an approximation. Frequently somewhat empirical methods must be used.

I Scientific Methods of Evaluation

Scientific methods for predetermining natural production in fish cultivation in ponds or in any other aquatic environment rely on the study of the physico-chemical elements of production, on the study of the aquatic flora and fauna, on the rhythm and intensity of production, and, in addition, in open waters, on a study of the fish population: the number and weight of different age groups and the annual growth rate of each of the age groups present.

The study and the predetermination of all the elements or production factors, their mechanism of action and interaction are far from being perfectly understood.

It is true that the physico-chemical characteristics of water in ponds, as well as the flora and nutritive fauna, are well known but what is less well known is their relationship to fish production. There have been many studies devoted to the establishment of primary productivity (aquatic flora) of different environments, but work on secondary productivity (aquatic fauna) and fish cultivation productivity are less numerous and are imprecise.

As a result of this it is not possible at present to meet the requests of anglers and fish farmers asking for sure and simple methods which will allow them to establish, fairly easily, a sufficient approximation of the productivity of fish in artificial ponds. For this reason one can only turn to more or less empirical methods.

II Empirical Methods of Evaluation

A The Léger-Huet formulae

The productivity of running water and of artificial ponds is given in the following formulae:

$$K = B \times L \times k \qquad K = Na/10 \times B \times k$$
for running water   for artificial ponds

The annual productivity per kilometer of a water course (K), expressed in kilos, is equal to the product of the biogenic capacity (B), of the average width of the water course (L) and of the coefficient of productivity (k).

The annual productivity of a pond (K) expressed in kilos, is equal to the product of the size of pond as expressed in ares and divided by 10 (Na/10), of the biogenic capacity (B) and of the coefficient of productivity (k).

It is necessary to consider the productivity of running water at the same time as that of artificial ponds, for when the latter are laid down the water supply must have been previously examined for productivity and in this way the probable productivity of the ponds which will be built is estimated.

The components of natural productivity formulae are the biogenic capacity, the size element and the coefficient of productivity. These components are analysed below.

It is very difficult, if not impossible, to appreciate the productivity of water from a distance; a visit is almost always indispensable. Some believe that a bottle of the water to be analysed is sufficient, but the conclusions which can be drawn from this are not enough.

1 The biogenic capacity (B)

a The notion of biogenic capacity

"*Biogenic capacity*" is the expression denoting the nutritive value of water examined for its feeding qualities for fish. This notion was created by Léger.

It is designated by the abbreviated sign "B" and is expressed in a form called "the scale of biogenic capacity" of which the degrees correspond individually to a given nutritive value from I (the weakest biogenic capacity) to X (the greatest biogenic capacity).

This enables fresh water to be placed in three categories: (1) *poor* water with a biogenic capacity between I and III; (2) *average* water with a biogenic capacity between IV and VI; (3) *rich* water with a biogenic capacity between VII and X. Sterile waters are given O and the richest X.

b The relationship between the biogenic capacity and the quality of the food present

The value and the richness of food present in a given medium depend not only on their quantity but also on their quality in relationship to the food required by the fish living (or which can be introduced) in the environment to be estimated. According to the quality and quantity of the dominant nutritive organisms this or that species of fish will grow more or less quickly and will be of very different quality.

There exist, therefore, as many bases for the appreciation of biogenic capacity as there are types of water for cultivation, and methods of using the water by fish of different ages and eating habits.

The bases of appreciation of biogenic capacity will not be the same for salmonid water as for cyprinid water, or in running as in still water. In consequence the biogenic capacity of running salmonid water (trout zone and grayling zone), of moderate running cyprinid water (barbel zone), of slow running cyprinid water (bream zone), of salmonid ponds, of cyprinid ponds, etc., will be appreciated differently.

c The quantity of food is the basic element of the biogenic capacity

The estimation of biogenic capacity rests on an evaluation of the quantity of nutritive organisms effectively present in the water examined, as opposed to the quantity present in ideally rich water of the same type which will be given a X rating for biogenic capacity.

The biogenic capacity of a given aquatic environment depends, in consequence, directly on the quantity of nutritive organisms (aquatic plants and animals) which the fish seek. Now, the animal organisms depend in their turn on the quantity of plant life present either living or dead (detritus). Living plants include macroscopic and microscopic organisms, the latter being present as phytoplankton and also as biological cover (Fig. 60).

Macroscopic vegetation serves as a support for the biological cover and shelters the nutritive fauna. The nutritive fauna living on the macrophytes is very important and it is necessary to understand fully the relationship between this vegetation and the nutritive fauna. It is known that the most interesting are the submerged and semi-submerged plants on condition that their growth is not excessive (they should not be more than 50 per cent). Floating and emergent plants are only relatively important (they should not be more than 20 per cent).

It is also known that the development of the water flora and nutritive fauna is that much greater in proportion as the positive physiographical factors, which are natural, are favourable. These factors are: the nature of the bottom; the physico-chemical characteristics; the rate of the current, etc. The negative factors are artificial for the most part (cutting, drying, harmful discharge, etc.). This means that an estimation of all these factors must not be neglected if the biogenic capacity is to be calculated.

d The physiographical characteristics of the environment which influence the biogenic capacity

It has been pointed out that certain physiographical characteristics influence the biogenic capacity. They can be grouped as physical, chemical, and mechanical, and are themselves dependent on geographical and climatic characteristics.

1 *Physical characteristics.* The principal are temperature and light.

1. *Temperature.* Temperature influences, considerably, the development of aquatic micro- and macro-organisms as well as the fish. It has strong action on multiplication, respiration and nutrition. This point was developed in Chapter II.

Temperature notably influences the growth of fish. Each species has a growth optimum which corresponds to a more or less wide and high temperature regime (Fig. 382). If that optimum is more or less established throughout the year then productivity will be near the maximum. This question is tied to the productivity coefficient k which will be discussed below.

2. *Light.* Light is essential for photosynthesis and in consequence to the production of micro- and macrophytic aquatic vegetation. In a shaded environment the biogenic capacity will be more or less reduced. Water courses and ponds in woods are often greatly influenced by this. It is notably the case when a small patch of water is situated in the middle of a forest dominated and densely covered by conifers or other trees with dense foliage (water courses in wooded tropical regions). Further, water arising from this environment is often acid.

The relative cleanliness of water or its turbidity also influences the biogenic capacity. It is the same for its colour. Greenish or bluish waters are generally good. Yellow or brown waters (dystrophic water) are not so good because they are acid and arise from moorland or marshy ground.

2 *Chemical characteristics.* The development of both the flora and nutritive fauna are greatly influenced by the chemical characteristics of the medium. The latter must be sufficiently rich in nutrient salts and free from toxins.

It is easily realized that mineral elements essential to plant and animal production must be present in sufficient quantities, and it is necessary that there should be plenty of calcium, nitrogen, phosphorus and potassium salts. If there is not, then artificial fertilizer can be used. Other salts of less abundant elements (sometimes called micronutrients) are equally indispensable.

The water supply and the ponds should be free from toxins. Many of these come from the discharge of industrial waste water and urban sewage but some exist in natural waters, such as water from peaty bogs or extensive coniferous forests. The biogenic capacity of such water is very low or even non-existent.

The chemical composition of spring water is closely related to the composition of the rocks through which it oozes. It influences the flora and aquatic fauna and consequently the biogenic capacity. In regions with water poor in calcium (less than 8 mg $CaO/1$), the fresh water shrimps and most molluscs, as well as such plants as watercress (*Nasturtium officinale* R. Br.) and other crucifers such as fool's watercress (*Apium (Helosciadium) nodiflorum* L.) are rare or absent which diminishes the nutritive value of the water in those regions.

Hard terrain (sandstone, granite) has clearer water than flaky soil (schists, clay) where the water is that much more cloudy as it is fast flowing. Colloidal waters are not toxic for aquatic animals but they diminish the biogenic capacity of the water.

3 *Mechanical characteristics.* These characteristics must be taken into account for running water. Generally, all things being equal, the more the bottom is stable the more the fauna is rich. Stony bottoms with fast flowing water and muddy-clay bottoms with calm water are the richest. For running water sandy bottoms are the poorest; often nearly sterile as is the case with moving pebbles. Gravelly terrain has less abundant fauna than when the bottoms are stony, for there are then fewer crannies in which animals hide.

For mechanical reasons, the biogenic capacity of running water with moving sandy bottoms or rolling stones, is reduced. The proverb "a rolling stone gathers no moss" fits them. Aquatic moss therefore is very rich in nutritive fauna.

On stabilized bottoms plant life establishes itself, grows easily and the nutritive fauna find both shelter and food and can reproduce more or less abundantly. On moving bottoms plants cannot do this and the animals, lacking shelter and food, are rare. In any case the environment is always somewhat poor.

In running rheophile waters the ideal bottom is stabilized and made up of mixed stones, pebbles and gravel. A sandy bottom is very poor.

In running limnophile waters and in ponds the ideal bottom is formed of organic mud of excellent quality while the ideal development of plant life forms a circular fringe of semi-submerged and emergent vegetation, and also submerged vegetation covering about 50 per cent of the total surface.

e   The determination of biogenic capacity

Few factors necessary for the determination of the biogenic capacity can be appreciated in absolute value, and it is even more difficult to appreciate their relative value.

Can the appreciation of biogenic capacity, therefore, which depends on so many different physiographical and biological factors, offer any precise value at all? The answer is yes, but it is more a matter of judgement than measure.

Certainly experience is necessary and a great deal of judgement in order to appreciate and co-ordinate all the positive and negative factors which in the end will give the biogenic capacity. However, the difficulties must not be exaggerated. Conclusions can be drawn from a collective number of impressions gathered as to the nutritive value, these being drawn from a specific examination of the water. In sum the naturalist must give marks for biogenic capacity in the same way as an examiner called on to judge a candidate. With a little experience it is not difficult to judge in either case.

To help in the estimation of the biogenic capacity a drawing showing the biological cycle of fresh water has been given (Fig. 60). This will enable an estimation to be made. According to each case, in relationship to the whole, the part due to the fauna on the bottom, to the phytophile fauna (living among the higher plants), to the plankton must be considered and also the eating habits of the fish to be raised in the environment under examination.

In rheophile running waters (rapid or rather rapid) there is no plankton and detritus is rare, but the higher vegetation as well as the biological cover living on these plants and on the bottom are plentiful. In ponds and limnophile running water courses plankton is important and the biological cover on the bottom is replaced by detritus.

In rheophile water, the nutritive fauna on the bottom and living among the higher vegetation must, principally, be taken into account. It can, indeed, be considered equally important. Plankton is non-existent or of very secondary importance

whilst the exogenous fauna, by no means negligible, depends a great deal on the nature of the banks.

When appreciating the biogenic capacity of limnophile water (slow running water and ponds) account must be taken of the nutritive fauna living on the bottom, of the phytofauna and of the plankton. These would be given equal or less equal importance according to the fish to be stocked in the environment.

In certain cases the importance of submerged, tender vegetation or that of plankton might preponderate if the pond is used for the raising of herbivorous fish or fish which feed principally off plankton (young fish and planktonophagous fish).

When appraising the biogenic capacity it is important to correctly evaluate the dominant nutritive fauna which will predominate in the food of the fish in the water being examined for use. From the same point of view organisms which are rare are of no practical importance.

The general rules for salmonid and cyprinid waters as well as running and still water are as follows:

1. Poor water is practically deprived of phanerogamic vegetation (Fig. 354); or it is slightly developed.

2. Average water is richer and includes phanerogamic vegetation particularly well developed along the banks and composed of semi-submerged and emergent vegetation. Completely submerged plants are rare (Fig. 355).

3. Rich water has plentiful phanerogamic vegetation, though not excessively so, composed of favourable species, submerged above all, and developed not only along the banks but also in the middle of the water courses and ponds (Fig. 356).

From a first approximation it is possible to determine the nutritive group (poor, average and rich) to which any given water for rearing fish belongs by using as a base the characteristics of the macroflora (moss and phanerogams), because the nutritive fauna living on the bottom and among the vegetation – as well as plankton development in calm waters – often reproduce at a similar rate to that which determines the development of macrophytes.

An appreciation of the biogenic capacity based on higher plant life is not applicable to still water in which plankton is the preponderant food, given the species of fish which will live in the environment.

Once the degree of the first approximation has been established, in order to give an exact definition of the biogenic capacity it is necessary to appraise the actual richness in relation to the ideal richness. Adequate knowledge of the various fish raising waters likely to be met, is indispensable. Beginners will run into serious difficulties, though these will be far fewer for the experienced hydrobiologist. The appreciation of the biogenic capacity is above all else a question of judgement and experience.

The production of ponds in direct relationship to their degree of biogenic capacity will also depend on their location. There are differences between village ponds, ponds located in the middle of fields and crops and ponds surrounded by woods or marshes. Their productivity will decrease.

Village ponds (Ex.: pond at Dhlavi, Moravia, Czechoslovakia) (Fig. 397) are enriched by water trickling across farms and watering places. Ponds located in meadows and crops are enriched by fertilizers spread over the terrain (Ex.: pond of the Ferme des Quatre-Bras at Sart-Dames-Avelines, Brabant, Belgium) (Fig. 358). Ponds situated in woods or in the middle of marshes are made unfavourable by the waters which flow from them (Ex.: Voxemeer pond, Rijkevorsel, the Antwerp Campine) (Fig. 357).

f The real and potential biogenic capacity

In order to raise fish in ponds it is necessary to secure water of a reasonably satisfactory quality. It is necessary therefore to calculate the probable or the potential biogenic capacity.

If an estimation of the probable capacity of a pond yet to be constructed at a given place is to be established, then the biogenic capacity of the water course which will feed the pond must be carefully predetermined. It may be presumed that the biogenic capacity of future ponds will belong at least to the same group (rich, average, poor) as the water course. In effect, as far as the physiographical conditions are equally favourable to both the water course and the pond, then the productivity basis – that is, the nutrients in solution in the water – will be the same. Consequently it is logical to expect the biogenic capacity to be the same.

But it is possible that the biogenic capacity of the ponds to be built might be higher than that of the water which will feed them, for it is possible that the water course could be influenced unfavourably by mechanical factors, e.g. if the bed is not stable due to a defective substratum, or if the water flows too fast. It is the same if a physical factor has an unfavourable influence on the water course. It might not receive sufficient light because of being too densely covered. The poverty of water courses and ponds could also be the result of unfavourable chemical action such as too low a content of indispensable mineral elements.

Nevertheless for ponds it is possible to remedy most of these inconveniences and so improve the biogenic capacity and the productivity. From the fact that the bottom of a pond is stable the mechanical inconveniences can be greatly reduced or eliminated altogether. By adopting a sufficient surface area and an adequate depth it is possible to correct all the inconveniences due to lack of light. Finally, liming and fertilization can increase the mineral element content. However, if the water contains toxins then it will be almost impossible to eliminate them and it is better that such water should not be used at all.

In the predetermination of the biogenic capacity of the water supply it is necessary to establish a distinction between the *actual biogenic capacity* which can be calculated from the water course at a given place where it was examined (for the calculation of its direct value), and the *potential biogenic capacity* which is that which the water course would possess if better physical and mechanical conditions obtained. Generally, when prospecting a water course carefully, spots are found, here and there, which are well lighted and stabilized and where it is possible to evaluate with sufficient exactitude the potential biogenic capacity of the water. For example, if the substratum of a water course moves or varies it should be possible to find a submerged root of a tree or a river plant which will have the stability characteristic required. These improved stability conditions can presumably be carried out in a pond.

2 The surface area

It goes without saying that the productivity of an aquatic medium, whether running water or still water, is proportionate to the extent and size of the surface of the water.

The surface area plays its part in the productivity formula of running water as well as for artificial ponds. In the latter case the surface area is expressed in ares (1 are = 100 m^2, 1 ha = 100 ares) and is divided by 10 (Na/10). If the surface

Fig. 354 Poor water: The Wamme at Mochamps (Belgian Ardennes). Brook situated in the Trout Zone. Very irregular water flow. For the most part the watercourse runs through conifer forests. Acid water with only low alkalinity (SBV usually around 0·15 but often less). Stabilized stony-gravel bed. Biological cover only slightly developed. No higher aquatic flora. Nutritive fauna rare. B = I.

Fig. 355 Medium productive water: The Sure at Radelange (Belgian Ardennes). Small stream belonging to the upper type of the Barbel Zone. Water course bordering meadows. pH slightly acid or neutral; medium alkalinity: SBV = 0·45. Stony, partially moving bed. Biological cover well developed. Higher flora rather abundant, in the form of a continual border of semi-submerged and emergent (paludal) vegetation (*Sparganium ramosum*, *Potamogeton natans*, *Menyanthes trifoliata*); submerged vegetation rare. Nutritive fauna relatively abundant: insect larvae (Trichoptera, Ephemeroptera, Diptera), molluscs. B = V.

Fig. 356 Rich water: The Semois at Jamoigne (Belgian Lorraine). Large river belonging to the lower type of the Barbel Zone. Water course bordered by meadows and a few trees. Water alkaline; pH superior to 7·0; high alkalinity: SBV = 2·0. Pebble-stony stabilized bed. Biological cover very abundant. Higher flora well developed, formed by a border of semi-submerged and emergent vegetation (*Glycerieto-Sparganietum*) and by a lot of submerged vegetation (*Ranunculetum fluitantis*). Nutritive water fauna very abundant: insect larvae, molluscs, crustacea. B = IX.

Fig. 357 Voxemeer pond at Rijkevorsel (the Antwerp Campine). Very acid water: pH = 4·6; very low alkalinity: SBV = 0·2. Water of low productivity. Ponds located in wooded and swampy region. Water flora fairly abundant particularly *Sphagnum;* nutritive fauna poor (particularly insect larvae). Natural production not exceeding 50 kg of cyprinids per hectare per year (50 lb per acre).

Fig. 358 Pond of the Ferme des Quatre-Bras at the source of the Ry d'Hez at Sart-Dames-Avelines (Belgian Brabant). Alkaline water: pH = 7·7; high alkalinity: SBV = 4·4. Very productive water. Meadow banks. Invading aquatic flora; nutritive fauna rich and abundant. Natural production in cyprinids up to 300 kg per hectare per year (300 lb per acre per year).

area is expressed in acres (1 acre = 40 ares; 1 ha = 2·5 acres), the productivity is found by replacing the factor Na/10 by 4 Nac (four times the number of acres).

For running water the surface area factor also intervenes for as the width of the water course is expressed in metres (L) and as the formula gives the productivity per kilometre, it then implicitly refers to 1,000 m in length.

3 The productivity coefficient k

a The need for introduction of the coefficient k

The productivity formula given above for running water and for certain still water (artificial ponds) is derived from the first formula of Léger (1910) which only concerned running mountain waters of the Dauphiné (France) expressed by: $K = B \times L$.

Replacing the factor L by Na/10 permits application of the formula to salmonid ponds in the same or similar regions. Nevertheless the formulae were not applicable to very rich alkaline waters, to cyprinid waters or to waters in warm temperate regions, or to tropical and equatorial regions. The results they gave were feeble or very feeble. Léger himself estimated that the value of his formulae as applied to cyprinid water ought to be doubled.

The introduction of the k factor in the formulae (Huet, 1949 and 1964) permits the generalization of the formulae and their application to artificial ponds in all regions of the world regardless of the nature of the water, the species of fish raised or the climatic conditions.

b The components of coefficient k

The coefficient k is composed of four secondary coefficients designated as k_1, k_2, k_3, k_4, which correspond respectively to temperature (k_1), to the chemical characteristics of the water (k_2), to the species of fish (k_3) and to the age of the fish (k_4). The product of the four secondary coefficients gives the value of k ($k = k_1 \times k_2 \times k_3 \times k_4$).

Coefficient k_1: temperature. Temperature is preponderant in fish cultivation and production. This is discussed throughout the book, notably in Chapter XII, Section III (Fig. 382). Concerning the effect of temperature on the growth of fish and the productivity of fish cultivation it is easy to discover with fish that can live in both temperate and tropical regions. Carp provide an excellent example; their growth, though feeble in western Europe, is better in central Europe where the summers are warmer, better still in Israel which has hot summers with prolonged periods of high temperature, and finally best of all in Indonesia where the temperature is constantly in the neighbourhood of the optimum of 25°C which carp needs. Similar conclusions have been observed for grass carp (*Ctenopharyngodon idella*).

The value 1 has been given to the coefficient k_1 for temperate western Europe where the average annual temperature is around 10°C. It seems that the coefficient should be doubled and trebled respectively for values of 16 and 22°C. It is also probable that the value 3·5 can be given to the average annual temperature of 25°C which can only be supported by warm water fish. Below 10°C the coefficient k_1 must be reduced to 0·5 for a temperature of 7°C and practically to 0 for an average temperature of 4°C.

Coefficient k_2: the alkalinity or acidity of water. When the biogenic capacity conditions are equal it is found that production is higher in alkaline than in acid water. The values given below have been suggested for k_2: 1·0 for acid water (pH inferior to 7·0); 1·5 for alkaline water (pH superior to 7·0). The difference in production is particularly striking in running water where, with equal surface areas, running alkaline waters are deeper and allow greater living space which permits faster growth for the fish.

Coefficient k_3: fish species. The values 1·0

and 2·0 have been suggested respectively for salmonids or cold water fish and for cyprinids or warm water fish. These values have been implicitly known in fish cultivation for a long time and can be experimentally verified in ponds. These different values result from the fact that so long as the conditions of growth are normal, warm water fish do better than cold water fish on a given quantity of food.

Coefficient k_4: the age of fish. This coefficient is only used for productivity calculations relative to artificial ponds. It is not used for running water. It is known that in artificial ponds quantitative production is greater when young fish are raised rather than older fish. This results from the fact that the maintenance ratio of young fish is not so high as that of adults. The action of this factor is reinforced by stocking rate which is numerically higher per unit of surface area in nursing ponds. The value 1·0 has been given to fish of 6 months and over; the value 1·5 to fish below 6 months.

c Principal values of coefficient k

The principal values of the secondary coefficients are given below:

k_1 = Average annual temperature	k_2 = Acidity or alkalinity of the water
10°C − k_1 = 1·0 16°C − k_1 = 2·0 22°C − k_1 = 3·0 25°C − k_1 = 3·5	Acid waters: k_2 = 1·0 Alkaline waters: k_2 = 1·5
k_3 = Fish species	k_4 = Age of fish
Cold water fish: k_3 = 1·0 Warm water fish: k_3 = 2·0	Over 6 months: k_4 = 1·0 Under 6 months: k_4 = 1·5

The extreme values of the productivity coefficient k ($k = k_1 \times k_2 \times k_3 \times k_4$), under conditions given above are between 1·0 (1·0 × 1·0 × 1·0 × 1·0) and 15·75 (3·5 × 1·5 × 2·0 × 1·5), rounded off to 16·0.

4 Some examples of productivity calculations

If for reasons of simplified calculation the natural productivity K in artificial ponds calculated by the above formulae is given per hectare, it is seen that this productivity can vary between the extreme 10 and 1,600 kilos (22 lb and 1½ tons).

When considering the productivity per hectare by means of the formula applicable to artificial ponds the factor Na/10 = 100/10 = 10. The productivity K can be equal to 10 kg (22 lb) only if B = 1 and k = 1. This would be the case for ponds raising second year trout in very poor acid water where B = 1 and k = 1; K = 10 × 1 × 1 = 10. On the other hand if the example is a pond with young warm water fish in an equatorial region with very rich alkaline water where B = X and k = 16 the annual production per hectare will be 1,600 kg (1½ tons); K = 10 × 10 × 16 = 1,600. These extreme values have been confirmed by practice.

The table below gives the principal values of the *formula for natural productivity of growing ponds in temperate western Europe.*

Types of ponds	Acid water	Alkaline water
Salmonid ponds	$K = \dfrac{Na}{10} \times B \times 1{\cdot}0$	$K = \dfrac{Na}{10} \times B \times 1{\cdot}5$
Cyprinid ponds	$K = \dfrac{Na}{10} \times B \times 2{\cdot}0$	$K = \dfrac{Na}{10} \times B \times 3{\cdot}0$

These values must be multiplied by 1·5 (co-efficient k_4) for fingerling ponds (fish under 6 months) in temperate western Europe. They must also be multiplied respectively by 2·0, 3·0 or 3·5 (coefficient k_1) in temperate warm or tropical climates or where the annual average temperatures are respectively 16·0, 22·0 and 25·0°C.

Here is a first example for calculating the natural productivity of a 40 ares (1 acre) pond in western Europe (average temperature: 10°C; k_1 = 1·0), supplied by acid water (pH = 6·7; k_2 = 1·0), at which the biogenic capacity is valued at VI and in which the aim is to raise second year trout (k_3 = 1·0). Productivity will be 40/10 × 6 × 1·0 (k = 1·0 × 1·0 × 1·0 × 1·0 = 1·0) = 24 kg.

A second example predetermines the natural productivity of a pond 60 ares in size situated in the Far East (average temperature 25°C; k_1 = 3·5), fed by alkaline water (pH = 7·8; k_2 = 1·5), of which the biogenic capacity is valued at VIII and in which the object is to raise carp for eating and other warm water fish (k_3 = 2·0 and k_4 = 1·0). Productivity will be: 60/10 × 8 × 10·5 (k = 3·5 × 1·5 × 2·0 × 1·0 = 10·5) = 504 kilos rounded off to 500 kg (½ ton).

B German tables for productivity of artificial ponds

The German productivity tables below given by Schäperclaus are based on the classification of ponds in four categories or classes which give productivity according to species and ages.

Natural productivity of ponds given in kg per hectare (According to Schäperclaus, 1933)

	Class I	Class II	Class III	Class IV
Cultivation of saleable carp (K_{2-3})	400-200 (8-4 cwt)	200-100 (4-2 cwt)	100-50 (2-1 cwt)	50-25 (1-½ cwt)
Cultivation of young carp (K_{1-2})	560-280 (11-5½ cwt)	280-140 (5½-3 cwt)	140-70 (3-1½ cwt)	70-35 (1½-¾ cwt)
Cultivation of trout for eating (R_{1-2} or B_{1-2})	240-120 (5-2½ cwt)	120-60 (2½-1¼ cwt)	60-30 (1-½ cwt)	30-15 (½ to ¼ cwt)

The productivity values given for third year carp (K_{2-3}) are average and can be obtained with subjects of normal weight (350 gm) (12 oz) at stocking, and for which average growth is the aim (1,250 to 1,000 gm (2 lb 12 oz to 2 lb 3 oz) for class I, 1,000 gm (2 lb 3 oz) for class II, 1,000 to 750 gm (2 lb 3 oz to 1 lb 10 oz) for class III, and 750 gm (1 lb 10 oz) for class IV).

Schäperclaus estimates that productivity is 40 per cent higher for ponds in which second year carp (K_{1-2}; individual growth 300 gm (10 oz)) are raised. In ponds in which second year brown and rainbow trout are raised (B_{1-2} and R_{1-2}; individual growth 120 gm (4 oz)) production is 40 per cent lower than that calculated for raising saleable carp.

Production is higher for ponds stocked with fish of 1 year than with 2 year fish, because the 1 year fish are more numerous than those of 2 years in an identical space, which means that better use is made of the natural food because it is consumed by a larger number of fish. And besides, maintenance rations for small fish are less important than for large fish.

On the other hand, for an equal space with the same productivity, the production of trout scarcely passes 50 per cent of the weight of carp of the same age produced, as the growth of the former is slower than that of the latter.

Without giving identical results the Léger-Huet formulae and the Schäperclaus tables give, nevertheless, very similar results which may be considered satisfactory in a field as difficult as this one, that is to say the estimation of natural productivity of ponds.

According to formulae based on Léger the natural productivity of cyprinid growing ponds in the temperate climate of western Europe can reach a maximum of 300 kg (6 cwt) per hectare whereas Schäperclaus gives a higher maximum of around 400 kg (8 cwt) which without doubt is a great deal. Léger gives three classes of production: rich water, average and poor. Schäperclaus gives four: I, II, III, IV.

Formulae for natural productivity given for growing cyprinids permit the following conclusions: (a) productivity of salmonid ponds are reduced by 50 per cent according to Léger-Huet and 40 per cent according to Schäperclaus, and (b) young cyprinid ponds show an increase of 50 per cent according to Léger-Huet and 40 per cent according to Schäperclaus.

Chapter XI
BIOLOGICAL MEANS OF INCREASING PRODUCTION

BIOLOGICAL methods likely to help increase production were referred to in the introduction to Part Three, and are developed further in the present chapter.

I Choice of Species

Choice of species is the first biological element which must be taken into account in order to increase production. The choice is predetermined by the aims of the enterprise – whether quantitative, qualitative or economic. These classifications of production were defined in the introduction to Part Three under: concepts of production in fish cultivation.

The type of production depends to a great extent on the species to be cultivated. Requisite conditions to ensure the suitability of a fish for cultivation were given in the Preliminary Notes. Sometimes it is possible to increase production notably by the substitution of other species.

1 Different levels of production

Under established ecological conditions, quantitative, qualitative and economic production can reach different levels according to the species cultivated and the methods of farming.

Thus a pond in Europe with summer temperature around 20°C will be suitable for rearing salmonids and also cyprinids. Under the same conditions of production it should be possible to obtain, without feeding, 100 kg/Ha (110 lb per acre) of salmonids and 200 kg/Ha (220 lb per acre) of cyprinids.

Swingle (1968) gives similar indications relative to pond production in Alabama (U.S.A.), where the ponds receive uniform fertilization. The production of omnivorous fish (e.g. *T. mossambica*) reaches 900 to 1,400 lb per acre, insectivorous fish (catfish, bluegills) 200 to 500 lb per acre and voracious fish (largemouth bass) 70 to 150 lb per acre. Hickling (1962), on the other hand, mentions that the quantitative production of plant-eating

fish is superior to those species with other eating habits, at least if there is no artificial feeding.

In the Campine (Belgium) it was found that in second year carp ponds, three young associated carnivorous fish reached in October, at the end of their first summer, an average weight of 14·3, 32·6 and 158·8 gm ($\frac{1}{2}$, 1, and $5\frac{1}{2}$ oz) respectively for perch, pike-perch and pike. Individual weights were in the following proportions: 1·0, 2·3, and 11·1.

The choice of species is therefore very important in achieving the highest rate of production. In general it would be in the interests of the farmer to choose a rapidly growing species likely to give a high quantitative production.

It is also important to bear in mind that a species which is satisfactory for one type of cultivation is not necessarily so for another. Further, species should be subjected to experiments under different conditions such as in a fertilized and unfertilized environment as well as with or without artificial feeding.

2 The introduction and substitution of species

Quantitative production is highest for fish with a short food chain, i.e. herbivorous, omnivorous and plankton eaters as well as those which eat detritus. It is also important that they should accept abundant, cheap artificial food. An additional interest, regarding non-voracious species, is that they are tolerant of other species and can therefore be associated with them.

In the light of what has been written above it would be of interest to replace a low productive species by a species with better growth. This explains the efforts made by certain European countries to introduce Asiatic herbivorous species.

Those looking for economic production should find small voracious fish of interest as a secondary production because of their individual value.

Foreign species should be introduced only with the greatest care and under the control of an expert since they can present real dangers such as the introduction of parasites or new forms of disease, while their presence in open water could become a nuisance and even a pest. Such thoughtless introduction is unfortunately too frequent as, for example, catfish (*Ictalurus melas*) and pumpkinseed (*Eupomotis gibbosus*) brought into Europe from North America. On the other hand, excessive precautions are also inadvisable since the introduction of fish for cultivation has been found to be necessary and has shown itself to be beneficial. In effect, if there are thousands of fresh water fish only a few species are, in reality, adaptable to the requirements of cultivation in ponds and it is a pity that there are so few plant-eating species.

3 Choice of species according to regions

1. The few remarks below concern fish which support *high temperatures*, and give a glimpse of a situation of which greater details have been given in Part Two, and particularly in Chapter IX, which is devoted to regional cultivation.

In Europe the carp is the principal fish cultivated in such ponds in association with tench as well as young carnivorous species such as pike and pike-perch for increasing economic production. The cultivation of common carp in association with Asiatic carps, either herbivorous or plankton-eating species, is being developed in central and eastern Europe.

In North America the aim in certain cases is the production of fish for eating and in other cases for angling. In the first case omnivorous species are chosen (buffalo fish), plankton eaters (*Tilapia*) or insectivorous species (bluegill, channel catfish). In the second case they are associated with predators which become the principal species and these are generally largemouth bass.

Carp are the principal fish cultivated in Israel. They are associated with tilapias (*T. aurea, T. galilaea*) and mullet (*Mugil cephalus*).

In Asia, in the Indo-Pacific region, fish cultivation is very ancient, widespread and also very diversified because of climatic differences. It essentially comprises the cultivation of omnivorous, herbivorous and plankton or benthos eating species to the exclusion of all carnivores. Milkfish are found in suitable brackish waters (Chapter VIII, Section III). In China and neighbouring countries, different groups of Chinese carps are raised mixed. In Japan, carp and other cyprinids are raised, in India and Pakistan important groups of Indian carps and in Indonesia, carp, other cyprinids and anabantids.

In Africa, where the raising of fish is relatively recent, cultivation is concerned principally with a mixture of two tilapias, one herbivorous and the other microphagous-omnivorous. In certain African countries carp and *Heterotis niloticus* are also raised with success.

In South America fish cultivation is only moderately developed and concerns principally the pejerrey or silverside. *T. rendalli* (= *melanopleura*) has given good results in certain States of Brazil.

2. In waters with *low temperatures*, that is, in cold water for most continents, different species of salmonids are raised either for eating or re-stocking. The principal salmonid for eating, cultivated in cold water throughout the world, is the rainbow trout. Salmonids for restocking are more numerous but their cultivation is rather localized. These include Atlantic and Pacific salmons, brown trout, brook trout, Arctic char and grayling raised for stocking in running water. For restocking the lakes of the northern hemisphere, especially in Europe and America, coregonids are cultivated.

Further comment on the choice of species can be found in the different chapters devoted to the cultivation of the principal types of fish.

II Controlling the Stocking of Ponds

The stocking of ponds seeks to determine the optimum quantity of fish to liberate so that the result will be maximum quality and quantity under the most economic conditions. Stocking depends on the productivity and the size of the ponds.

The total productivity of ponds is equal to the sum of natural productivity and productivity due to fertilization and artificial feeding, all of which have been dealt with in Chapters X, XII, and XIII. Calculations for stocking were described in Chapter XIV.

III Temperature Control in Ponds

The influence on fish production of temperature and the requirements of fish regarding temperature and the dissolved oxygen content upon which it largely depends, as well as the proper measures of control of these two factors, were considered in Chapter XII, Section III.

Some *connected elements* which influence the productivity of ponds are:

1. It has been found, in Europe, that when

summers are wet, even if the average temperature is low production is relatively good. This is probably due to abundant rain which brings to the ponds, along with the trickling water, nutrients drawn off from the terrain over which it has flowed. Further, the level of water in the ponds remains high and the shallow sections do not dry. The complete surface therefore remains under production.

2. An increase in the *depth of ponds* is advisable provided that permissible norms are present. Hickling (1950) reports in this respect that Chinese fish farmers used to say that one extra inch in depth means one inch more in the length of the fish. An increase in depth means an increase in the living space necessary to fish. In any case, insufficient depth must be avoided because partial drying can be brought about by dry weather leading to the lowering of the water level. But the moment sufficient depth is established it should not be increased, for if a pond is too deep it will not warm and this can have an adverse effect on cultivation.

IV   Improvement in Reproduction and Selection and Special Cultivation Techniques

1   Control and improvement of reproduction

The development of fish cultivation is closely tied to the control of reproduction of cultivated fish. Some species reproduce without difficulty in ponds, but this is rather rare. Others reproduce badly or not at all so it is necessary to provoke reproduction. Other species reproduce too easily and this phenomenon has to be slowed down.

a   Induction of reproduction, artificial fecundation and incubation

Many species reproduce badly or not at all in ponds. Considerable progress was made in the nineteenth century thanks to the discovery and general use of the artificial fertilization and incubation of brown trout and other salmonids (Chapter III). The application of these techniques was responsible for the strides made in salmonid cultivation throughout the world.

Artificial fertilization and incubation were applied successfully to coregonids and pike and to several other species.

There are still difficulties to be overcome for common carp and Asiatic cyprinids. The use of pituitary extracts which determine the spawning of these fish, and the use of techniques permitting artificial incubation of the eggs, have led to an important extension of the cultivation of these species. The techniques employed were described elsewhere (Chapter IV).

b   Slowing down reproduction

Other species, especially perciformes and notably tilapias, reproduce too easily and too frequently in ponds and this leads to over-population and dwarfing. Remedies exist such as lowering the temperature which is not always possible, or using special methods such as monosex cultivation and the cultivation of hybrids, though these are not always easy.

2   Selection

Selection offers real and important possibilities for increasing production, but such possibilities are not sufficiently well known and are insufficiently used.

It is well known, however, that considerable differences in growth exist within the progeny, even when the parents are the same. However, methods which allow choice, at the time of stocking which, later on, will show better growth, are not well established as yet. Presumably, as in other fields, the differences which show themselves in the young will emerge and will be maintained.

Selection has been practised, albeit empirically, with carp and rainbow trout. For carp it has produced races of different colours: brown, yellow, orange and white or with different squamation: scaled, mirror or naked carp, or with different shape indices, long or short. With rainbow trout selection has given early or late spawning strains as well as particularly rapid growth.

3   Special cultivation techniques

Different techniques are also used to improve the qualitative production of such species as tilapias and carp which reproduce too abundantly and prematurely in growing ponds. In such cases the economic production is appreciably reduced though the fish, for the most part, are small and not of great commercial value.

Undesirable reproduction can be controlled notably by the association of voracious with a non-voracious species (see Section V below). It is also possible to limit, at production level, fish which are too young to reproduce but in that case the fish produced are small.

Monosex and hybrid cultivation is also possible.

1. *Monosex cultivation* has given results in Africa with tilapias, and notably with *Tilapia nilotica* in Kenya. It has been found that the males of this species have superior growth to the females. This is also true of other tilapias. For *T. nigra* when the young fish are 10 cm (4 in) long, it is possible to distinguish two size categories and the largest, at least in theory should be males. In practice this separation is very delicate and not very sure.

For other species such as pike, tench and eels, females grow faster than males.

2. The production of *hybrids* of one sex is more certain. It has been found that some matings give birth to issue of one sex. The best results have been achieved in Malaya where *Tilapia hornorum* Trewawas males were crossed with *Tilapia mossambica* Peters females. All of the progeny were male. Good results were also reported from Uganda when *T. mossambica* males were crossed with *T. nilotica* females. All of the hybrids were also male.

V Mixing Species and Age Groups

The mixing of species and age groups is widely practised in fish cultivation. It was first practised, in all probability, in China. The general aim is to improve quantitative, qualitative and economic production. It has been found that a well planned association is beneficial to each of the groups which make it up.

The particular aims of mixing species and age groups can be very different, just as the methods used to achieve them can differ. Research is necessary to find the species from different regions of the world best suited for association.

The principal examples worthy of consideration are given below:

1. An association of different age groups of the same species or the mixing of non-voracious species with a view to increasing the quantitative production of fish for eating.

2. Mixing non-voracious species with small-sized voracious species with a view to improving economic production owing to simultaneous and complementary production.

3. Mixing non-voracious and voracious species in order to increase qualitative production of the non-voracious fish. This is achieved thanks to the control by the carnivorous species of excessive reproduction of the non-voracious species.

4. Mixing non-voracious species to act as forage fish for larger-sized voracious species which are the principal object of the operation.

1 Mixing different age groups of the same species or mixing non-voracious species

These associations are undertaken in order to increase quantitative production principally of fish for eating and by using, as far as possible, the nutritive resources of the pond.

It goes without saying that the first precaution, in this kind of cultivation, is to prevent unwanted species from getting into the pond, for they will compete for food and could also be somewhat dangerous. Among these, in Europe at least, are common perch, pumpkinseed and catfish; in Africa, *Haplochromis* and other cichlids. The best way of keeping these unwanted species out is to place a horizontal, fine mesh screen at the water intake. Sometimes a double screen is used, the lower section having a very fine mesh.

a Mixing different age groups of the same species

This kind of association is one of the simplest ways to increase the quantitative production of a pond. Fish of different age groups, but of the same species, have different feeding habits, so if they are mixed, better use is made of the food present in the pond and production is increased. This method, which is known as the mixed age groups method or the mixed method, is widespread in Africa for the production of tilapias. The different age groups do not appear to compete with each other and therefore do not hinder each other's growth. Nevertheless production is generally made up of small fish. The mixing of age groups for carp is also practised in several countries.

b Mixing different non-voracious species

This association is often recommended and practised. The general opinion is that even in intensive cultivation a single species cannot use, completely, all the nutritive resources of a pond since one species is bound to show food preferences. This is the case, notably, for carp which do not eat sufficient plankton and do not touch water vegetation. Therefore, by mixing species, most of the food resources of the pond will be used and production increased.

Yashouv (1968) laid down rules to follow in the choice of species for association. They should have complementary eating habits and live in different ecological niches. The result of this will be mutual growth of the associated species. But it is also necessary that different species should reach marketable size at the same time. There are other considerations which may be taken into account including provisioning possibilities for fry, local climatic conditions, the consumer market requirements and artificial food resources.

c   Principal examples of application

The association of non-voracious species is widespread over different regions of the world. The mixing of age groups of the same species is frequent in the cultivation of carp, milkfish and tilapias.

At the start of fish cultivation in Africa, herbivorous and microphagous-omnivorous tilapias were mixed. The tilapia species, which can be recommended, vary in Africa according to regions.

In North America the association of non-voracious species is not widespread.

In Asia and the Far East, on the other hand, the mixing of species is frequent and has been traditional for many centuries. It is not rare to find four and even six non-voracious species in association. Suggested associations must differ according to the different regions. In China and in neighbouring countries various Chinese carps are associated. In Japan the latter are associated with common and crucian carp. In India and Pakistan different Indian carps are raised at the same time. In Indonesia common carp are associated with various native anabantids and with *Tilapia mossambica*.

In Europe the mixing of cyprinids for eating has been practised for a long time. However, there have been certain developments in the suggested associations. Formerly carp and tench were mixed. Tench are losing favour because, despite its appreciated flesh, its growth is slow and its feeding habits are too close to those of carp. In central and eastern Europe mixing in cyprinid ponds includes more and more two Asiatic cyprinids, one herbivorous and the other plankton eating, which benefit from the nutritive resources not used by the carp. In cyprinid ponds a secondary cultivation, that of the golden ide, an excellent ornamental fish, will improve the economic production of the pond.

In Israel carp is the principal cultivation, mixed with tilapia and mullet.

2   Mixing non-voracious species with small voracious fish as a complement

This kind of association is often found in Europe and aims at producing a principal fish for eating, generally carp, with a secondary species such as young pike, pike-perch or 1-year-old black bass.

The young predators cannot be considered as serious competitors for food. On the other hand they eliminate unwanted fry which have hatched prematurely as well as small wild fish which enter the ponds and compete with the carp.

Such a mixture can increase economic production seriously. This interest is underlined by Steffens (1960) who found that in one pond which simultaneously produced per hectare 520 kg (1,150 lb) of carp and 50 kg (110 lb) of 1-year-old pike-perch, the commercial value of the latter was slightly higher than that of carp whose weight was 10 times greater.

3   Mixing non-voracious species and controlling voracious species

Such a mixture can increase the production of the non-voracious and the voracious fish.

1   *Mixing in favour of non-voracious species.* In this type of association, found above all in Africa, the object is to improve the qualitative production of tilapias by controlling excessive reproduction which is responsible for the large number of small fish of no commercial value. In order to avoid this proliferation they are mixed with a voracious fish. Encouraging results have been obtained with *Hemichromis fasciatus*. Experiments have also been carried out with *Lates niloticus*. In both cases there was room for improvement.

2.   *Mixing in favour of voracious species.* This type of association is practised in the United States in ponds in which the object is to produce a voracious fish, generally largemouth bass. When the voracious fish are liberated the pond is insufficiently stocked, so in order to remedy this a small non-voracious species is liberated. It reproduces rapidly and feeds off the disposable food. When the voracious fish is large enough to feed off the small species it is then eaten in turn and practically eliminated. The small species are generally mosquito fish (*Gambusia*) or, preferably, fathead minnows (*Pimephales promelas*).

4 Balanced mixing of voracious and non-voracious species

This kind of mixing is intended principally to obtain voracious fish of good size either for eating or for angling. It differs from the foregoing for the aim here is to equlibrate, over a prolonged period, both the voracious and non-voracious population. To feed the carnivores, one or several species of non-voracious fish (forage fish) are mixed in. Their value as fish for the table is slight or even non-existent.

Widespread in the United States, this association is generally made up of largemouth bass (*Micropterus salmoides*), for the voracious species, and bluegill (*Lepomis macrochirus*) or different similar centrarchids and other kinds of forage fish. Similar associations are not practised very much in other parts of the world. However, some experiments have been carried out in Thailand and they have also been considered in Africa.

To maintain this association it is necessary to ensure an equilibrium between the voracious and non-voracious species. The theory of this equilibrium, which is somewhat difficult to obtain and to maintain, has been outlined by Swingle (1950). One of the principal values which must be predetermined is the A_T (total availability value) which gives the percentage of the global production represented by fish of sufficiently marketable size. In a balanced population A_T is found between 33 and 90, and a value higher than 85 indicates that the pond is overstocked with predators.

VI Successive Productions throughout One Year

In order to ensure maximum production it is necessary to leave the pond under water and in a condition to produce for as long as possible. Further, to avoid non-employment of the pond which might be prolonged, it is worth while producing several fish crops in one year. It is also of interest to make the most of the dry period separating the two periods when the pond is placed under water. Periodic drying, over more or less prolonged periods, is indispensable for cultivation for it favours the mineralization of the mud and improves the sanitary condition of the pond.

1 Successive fish productions throughout a year

As the growth of fish is tied closely to temperature the organization of successive productions largely depends on climate.

1. *Successive productions in intertropical regions.* In these regions production does not stop nor even. in some cases, does reproduction cease. If there are one or two dry seasons, which are colder, reproduction slows down and reproduction may be momentarily stopped, and this is not always a bad thing. By limiting the period of raising to 6 months or even less, such as is done in Africa and the Far East, it is possible to obtain several successive productions in one year with the same species and the same age group.

2. *Successive productions in warm temperate regions.* In these regions production slows appreciably and reproduction stops during the cold season. The growing season extends over from 6 to 9 months. During this time it is possible to obtain successive productions in the same pond, with the same species of the same age groups just as in tropical regions, provided the growing period is limited to 3 or 4 months. This is practised in Israel.

3. *Successive productions in cold temperate regions.* The problem is very different in these regions because of the climate, because the species only spawn once a year, and in the case of warm water fish, late in the year. Under these climatic conditions, ponds devoted to the production of fish of 1 year are not employed for prolonged periods, and there is therefore a considerable loss in production.

In order to avoid such unproductive periods it is possible to obtain several successive productions by rearing a species of somewhat earlier spawning before the production of the warm water fish. It is, for example, possible to produce 6- or 8-week pike fingerlings in these ponds before the production of carp fingerlings, and so in an appreciable way increase the economic production of the pond throughout the year. It is also possible to use the same pond for the successive production of 4- to 6-week carp followed by 1-year-old carp; and where the climate is favourable it is possible to achieve three productions: pike of 6 weeks (end of April to beginning of June) followed by carp of 6 weeks (beginning of June to mid-July) and finally 1-year-old carp (mid-July to the following spring).

2   Alternation of animal and vegetable productions
The advantages and even the necessity of the periodic drying of ponds is recognized. This is developed in Chapter XII, Section II. If drying out cannot be managed each year but only at fairly long intervals then it is possible to prolong the dry period for a whole year. This can be done by simply leaving the soil as it is or putting it to agricultural use. This latter practice of crop rotation is less frequent now than previously since, because of the wider use of fertilizers, it is no longer justified.

However, rotation crop has not been abandoned altogether. In Hungary it is recommended once every 4 years. Lucerne or other green food is grown that year (Fig. 379). One advantage of this rotation of fish and agricultural productions is that it helps in the fight against fish diseases.

In certain tropical regions the rotation of agriculture and fish cultivation is current usage. A rice crop is produced alternatively with one or several productions of carp fingerling or with the production of small carp for eating (Chapter VIII, Section II). The duration of the fish cultivation should, however, not exceed 6 months.

VII   Simultaneous Production of Animals and (or) Vegetation in Ponds

Such production in ponds, either on the surface or nearby, can considerably increase both quantitative and economic production.

1. *Fish and ducks.* The advantages of simultaneously rearing fish in a pond with ducks are very real (Fig. 400).

1. The regular spread of duck droppings on the surface of the pond provides excellent organic manure for the water and the bottom of the pond. The consequence of this is an abundance of plankton and other food by which the carp can profit. Ducks therefore can replace manure or at least supplement it.

2. By digging the bottom in shallow sections, ducks contribute to the reduction of water vegetation and particularly to the development of duckweed (*Lemna* sp.).

3. Those parts where the ducks dig the mud are stirred up and the nutrients which they contain are released in solution and increase natural productivity.

4. Food distributed to the ducks but which is not eaten and falls into the water is consumed by the carp, or if not it decomposes and the ponds profit by it as an indirect manure.

5. In tropical regions ducks eliminate molluscs in depths of up to 40 or 50 cm (16 to 20 in) and so help to fight bilharzia.

6. To the advantages given above must be added the fattening of the ducks.

These varying advantages result in an increase in fish production. The percentage of the increases under conditions of natural production vary according to different writers. According to Eppel (1961), the intensive feeding of 300 ducks per hectare increases fish production by 100 kg (220 lb) on the average. The number of ducks recommended by European authorities varies between 200 and 400 per hectare (in general 250).

The rearing of ducks in ponds is practised on a large scale in central and eastern Europe, especially in Hungary, East Germany, Poland and the Soviet Union. Trials considered to be encouraging have been carried out in Africa and especially in Zambia and Rhodesia.

The combination of fish and poultry is a good way of integrating fish cultivation and agriculture and is largely practised in the Far East where all farmers are fish farmers and vice versa.

The raising of pigs on the banks of ponds follows the same line as for ducks.

2. Rice-Fish cultivation. In rice fields the cultivation of rice can be in association with the cultivation of fish. Rice is the principal harvest and fish production is complementary. This question was studied in Chapter VIII, Section II.

3. Molluscs, crustaceans and batrachians. Besides fish it is possible to produce and harvest crustaceans, molluscs and frogs in cultivation ponds. They can give very important economic production.

The harvesting of crustaceans is practised in brackish water ponds in the Far East and can be very important (Fig. 325). In central Europe, notably in Poland, the harvesting of crayfish and frogs is responsible for an appreciable supplementary income.

4. Raising nutrias. The raising of nutrias (*Myocastor coypus*), sometimes practised in central and eastern Europe, can increase the economic production of ponds. The advantages of raising nutrias which do not gnaw into dikes in the same way as musk rats, are as follows: (1) they reduce the aquatic vegetation, notably reeds and cat's tail;

(2) they accelerate mineralization of the mud thereby increasing the productivity of the pond; (3) they increase food for the carp which live off nutria's droppings; (4) their meat is of economic interest and their fur particularly is valuable. The raising of nutrias also ensures an increase in the quantitative production of the ponds.

5. *Accessory and diverse productions.* Among production which helps the economic production of a pond, reeds (*Phragmites* notably) growing in shallow parts of the pond must be mentioned.

VIII   Intermediate Fishing

1. *Opportunity and advantages.* Intermediate fishing is necessary when the density of the population is too great and almost at maximum stocking for a pond. This condition will be noticed when production shows a steep decline. The fish must be thinned out. Those remaining will then have more space and more food. They will start growing again until the pond, once more, reaches maximum population level. At this point another intermediate fishing can take place or the pond can be dried. By means of successive intermediate fishing it is possible to keep the ponds at maximum capacity while maintaining a high rhythm of production which, without it would cease each time maximum stocking is reached.

Intermediate fishing cannot be repeated indefinitely however. If it were, then the cultivation of fish would in fact be replaced by open water exploitation. Drying out periodically is necessary for the control of population, to avoid overstocking, to remove unwanted species which will have entered the pond and to restock the pond rationally.

2. *Conditions of utilization.* Intermediate fishing is only justified when the population is too dense. This can result from two very different causes right from the start which, therefore, means different methods must also be used.

1. In the first case *the population will be too dense because the number of fish stocked was voluntarily too great.* The population will then be composed of several species, often of unequal growth (as exists in the Far East) or it will include two age groups of the same species, carp for example. The fish will not spawn in the ponds.

The object of intermediate fishing is to remove fish of saleable size and to leave alone species or age groups which are not yet marketable.

2. In the second case the *density will be too great as a result of uncontrolled and exaggerated spawning.* This is what happens with the production of tilapias of mixed age groups. It is necessary to thin the fish out as is done in Africa and Madagascar where the largest are removed. But this practice is not perfect for, by the removal of large fish only, the possibility of obtaining others is diminished not only because some have been fished but also because the small fish, which remain and are already too numerous, will start spawning and thereby accentuate the density of the population. This will finally lead to dwarfing. Intermediate fishing must therefore also include small fish.

3. *Effects on production.* From the fact that the density was high at the start, maximum stocking or close to it will be reached rapidly. Consequently the food will be used completely within only a short time after stocking, and production will rapidly reach a rhythm very close to the maximum. Besides, intermediate fishing permits greater intervals between draining and reduces loss of food and production which result from it. Full and precise information on the exact evaluation of production increases, thanks to intermediary fishing, is not available. Yashouv (1959) mentions that in Israel, in one year, it led to almost double the production of the maximum stocking of the pond.

4. *Utilization.* Intermediate fishing is current in Israel, Africa, Madagascar and the Far East. It can be carried out by several methods including lines, trap nets, seine nets, cast nets, or bamboo lattice which is common in Indonesia (Fig. 464) and is quite possible provided the ponds are not too deep. The methods are described in Chapter XVI, Section I.

IX   Control of Fish Diseases

Before trying to intensify production by biological or non-biological means it is necessary to ensure that the fish produced are healthy. Chapter XV is devoted to the study of parasites and fish diseases in ponds.

Chapter XII

MAINTENANCE AND IMPROVEMENT OF PONDS

THE maintenance of ponds, coupled with a close study of the stocking density and the use of artificial feeding, constitutes one of the principal means of increasing their productivity.

It comprises three principal objectives which are closely linked together:

1. The conservation and improvement of installations by maintaining dikes and other constructions as well as eliminating or reducing excessive vegetation and soil deposits.

2. By assuring the best possible hygienic conditions, increasing the dissolved oxygen content, and destroying parasites and germs of diseases.

3. Increasing production by various methods to improve the biological cycle of the pond.

Among the general maintenance measures which are current is the conservation of dikes and banks. If a break or a crack is noticed it must be repaired at once. Such a break can be caused by trees growing on dikes, so it is best to remove them. Trees can also stop water from warming. They should be retained only for keeping trout ponds cool or for protecting cyprinid spawning and fingerling ponds from the wind.

When the ponds are dried out all parts must be well cleaned, including the principal and secondary drainage ditches as well as the inlet, evacuation and by-pass ditches. Holes in the bottom of the pond must be filled in, and the dikes, and the monk in particular, must be kept in good condition. Any cracks in a monk made of durable materials (stone, bricks or concrete) should be repaired immediately in order to assure that they will not increase in size when it freezes.

As far as possible steps should be taken to improve the dissolved oxygen content. This is particularly important in trout ponds and especially so if intensive production is the chosen method.

The principal measures which must be taken to assure the maintenance and improvement of the ponds are: control of excessive aquatic vegetation, improvement and restoration of the bottom, the control of various physico-chemical factors of production, liming and fertilization.

SECTION I

CONTROL OF EXCESSIVE AQUATIC VEGETATION

In fish cultivation, aquatic vegetation when moderate and sufficient is useful and even necessary. The importance of aquatic vegetation was underlined in the study of the biological cycle of ponds (Chapter II, Section II).

From the point of view of fish cultivation the following classification of aquatic plants is logical.

A Partially aquatic weeds

1. *Shore* weeds: they do not live in the water but on the fringe. They need a high degree of humidity, particularly for their roots, and they are not affected when partially or temporarily covered with water.

They are herbaceous shore-plants (Ex. *Filipendula ulmaria*) and woody shore-plants (Ex. *Alnus glutinosa*).

2. *Paludal* weeds germinate and the roots grow in submerged soil or in damp soil but most of the stems, leaves, flowers and fruits grow out of the water (Ex.: *Phragmites communis*).

3. *Semi-emergent* weeds have roots in submerged soil. A considerable part of their stems and leaves also grows in water. They can often live either completely out of water or submerged. They are generally small plants (Ex.: *Apium (Helosciadium) nodiflorum*).

B Completely aquatic weeds

4. *Floating* weeds: the greater part or all of the leaves develop on the surface of the water, while remaining in contact with the atmosphere through their upper sides and with the water on their lower sides. They are rooted floating plants (Ex.: *Nymphaea alba*) and free floating plants (*Ex.: Lemna minor*).

5. *Submerged* weeds: stems, leaves and sometimes the flowers live in the water (Ex.: *Callitriche stagnalis, Potamogeton crispus, Ranunculus aquatilis*).

Scientific and English names of the principal aquatic weeds in temperate Western Europe which can grow excessively in ponds.

A Woody shore plants (Ex. : *Alnus glutinosa*

Gaertn., Glutinous Alder) and *herbaceous* shore plants (*Ex.: Filipendula ulmaria* Maxim., meadow sweet; *Epilobium hirsutum* L., great hairy willow-herb). Shore plants never invade when the ponds are sufficiently deep.

B Paludal or emergent weeds

Phragmites communis Trin. Reed
Typha latifolia L. Great reedmace
Baldingera arundinacea Dum. Reed-grass
Glyceria maxima Holmb. Float-grass
Carex sp. Sedge
Juncus sp. Rush
Sparganium ramosum Huds. Bur-reed
Equisetum limosum L. Water horsetail
Equisetum palustre L. Marsh horsetail
Scirpus lacustris L. Bulrush
Sagittaria sagittifolia L. Arrow-head
Alisma plantago-aquatica L. Water plantain
Iris pseudacorus L. Yellow flag
Acorus calamus L. Sweet flag

C Semi-emergent weeds

Nasturtium officinale R. Br. Watercress
Apium (Helosciadium) nodiflorum Lag. Fool's water cress
Glyceria fluitans R. Br, Flote-grass

D Floating weeds

Rooted
Nuphar luteum Sm. Yellow water-lily
Nymphaea alba L. White water-lily

Potamogeton natans L. Broad-leaved pondweed
Polygonum amphibium L. Amphibious bistort
Free floating
Lemna sp. Duckweed
Hydrocharis morsus-ranae L. Frog-bit
Nymphoides peltata O. Kuntze. Fringed water-lily

E Submerged weeds

Elodea canadensis Mich. Canadian pondweed
Ceratophyllum sp. Horn-wort
Myriophyllum sp. Water milfoil
Callitriche sp. Starwort
Potamogeton sp. Pondweed
Ex.: *P. lucens* L. Shining pondweed
　　P. crispus L. Curled pondweed
　　P. pectinatus L. Fennel-leaved pondweed
Ranunculus sp., *Batrachium* group Water crow-foot
Ex.: *R. circinatus* Sibth. Circinate water crowfoot
　　R. peltatus Schrank Peltate water crowfoot
Trapa natans L. Water caltrops
Hottonia palustris L. Water violet

Many aquatic plants are harmful when they become excessive (Figs. 359 to 361). This applies especially to emergent weeds and to floating weeds and, in certain cases, submerged weeds.

Excessive vegetation accelerates silting, makes difficult the movement of the fish and stops the water from warming. It also provides refuge for competitors seeking nourishment and to enemies of the fish and creates a bottom rich in cellulose and, in general, diminishes the productivity of the pond.

Just as the weeding out of harmful plants is necessary for agriculture, so the control of aquatic vegetation is part of the maintenance necessary for the cultivation of fish. It is cheaper to weed out excessive vegetation regularly than it is to restore a pond in which vegetation and silting up are excessive. Maintenance is far easier if it is carried out regularly.

The elimination of excessive vegetation will improve production due to an increase in natural food and an improvement in the sanitary condition of the pond.

It should nevertheless be borne in mind that aquatic vegetation only becomes a nuisance when its development is really excessive. A fringe of emergent plants and a string of gramineous plants as well are indispensable for the protection of banks and dikes.

Fig. 359 A cyprinid pond, fairly deep in the centre with a border of emergent, floating and submerged vegetation, well developed but not excessive. Ponds at Hofstade, Belgium.

Fig. 360 Large cyprinid pond, shallow in the centre and overgrown, for the most part, by emergent vegetation. Pond becomes shallow by accumulating organic and inorganic sediments. Urgent need to control invading weeds every year. Large pond at Rhode-Sainte-Agathe, Valley of the Dyle, Belgium.

Fig. 361 Old pond completely invaded by emergent aquatic vegetation and now without any value for fish culture. Gemioncourt, Belgium.

The methods employed to control excessive vegetation are mechanical and chemical although in certain cases biological methods can be used.

I Biological Methods

It is sometimes possible to counteract the development of harmful plants by the introduction or propagation of other less harmful or more useful plants. For example *Elodea canadensis*, which is often very excessive, can be supplanted by introducing *Callitriche* which has only a loose hold, is less dense and more easily controlled. However, the principal biological methods used in fish cultivation are herbivorous animals or fertilization.

1 Herbivorous animals

In the first place *herbivorous fish* can be used. Fish which principally feed off plants are rare. The principal species are grass carp (*Ctenopharyngodon idella*) (Fig. 226), of Chinese origin and herbivorous tilapias (*Tilapia rendalli* (= *melanopleura*) (Fig. 279), *T. zillii*), of African origin. But these are warm water fish and are not really capable of reducing the aquatic vegetation unless the temperature is above 20°C. Grass carp have the advantage of being able to resist severe winters whilst herbivorous tilápias die quickly when the temperature of the water falls below 12 to 13°C. The use of these fish is limited, therefore, to intertropical regions or hot temperate climates.

In climates which suit them, tilapias reproduce easily, while until 1960 grass carp only spawned in China. Thanks to the use of hypophysation it is now possible to propagate them outside their natural areas.

Grass carp have been introduced during the last few years into the Soviet Union and into several European countries and other parts of the world. It is too early to judge just how useful they are in the fight against excessive vegetation in the countries into which they have been introduced. Positive results can only be expected in those places where the summers are hot and sufficiently long. The food which grass carp prefer is given in Chapter XIII, Section IV.

Herbivorous tilapias live off submerged and semi-submerged plants, floating plants and even the submerged part of emergent plants provided they are not too tough.

The development of submerged and floating plants can be stopped indirectly by releasing a large number of 2- or 3-year-old carp. Four hundred carp weighing 250 gm (9 oz) or more per hectare are necessary, but higher figures have been given such as 2,500 to 3,000. As they search for food the carp dig into the bottom, stir up and cloud the water and stop the penetration of light, and in consequence the development of the vegetation. By the use of a large number of fish it is possible to eliminate lightly-rooted plants. It has been noticed that sometimes in carp-nursing ponds there is a significant growth of filamentous algae, but these disappear if sufficient carp of 1 or 2 summers are released in addition.

Swans and *ducks* can also help control submerged vegetation when they search for food in those parts of the ponds which are not too deep. Ducks feed happily off duckweed (*Lemna* sp.).

Nutria (*Myocastor coypus*) feed mainly off emergent vegetation. Their cultivation (about 15 per hectare) is not widespread and the interest in them depends on the price obtainable for their fur.

In a shallow pond with a sufficiently hard bottom, *cattle* can live off young emergent plants.

2 Fertilization

Fertilizers help the development of phytoplankton to form a screen against the light on the water and hinder the development of higher plants. When fertilizers are to be used, the submerged, floating and emergent vegetation must be removed first, otherwise it will stimulate them. Compared with ponds where the weeds have not been cut and which have not been manured, those which have showed a production increase of from 40 to 80 per cent. However, climatic conditions can be a determining factor if this method is to succeed. American authors recommend the use of a mixture of ammonium sulphate, superphosphate and potassium chloride at a ratio of ten parts nitrogen, five parts phosphorus and five parts potassium. About 100 kg (220 lb) of this mixture is used over 1 hectare.

More delicate but more effective is the use of organic manure: dung, compost and liquid manure, etc. Sometimes it is possible to supply cyprinid ponds with water containing organic matter (sewage water for example). By this method the sewage water is purified, but its use necessitates a careful mixture of fresh and polluted water as well as regular control. In any case there will be a danger of disease and of lack of oxygen.

II Mechanical Methods

In fish cultivation mechanical methods, principally cutting, are used to control excessive vegetation.

German authors differentiate between "Entkrautung" for cutting submerged weeds and "Entschilfung" if cutting concerns emergent plants commonly known as reeds.

In fairly large ponds one section is often particularly favourable to the development of this emergent vegetation. It is in the upper and generally shallow part where silting is easy. Here reeds become excessive. The size of the invaded part should then be reduced.

A Weed cutting

Weed cutting means the mowing of aquatic vegetation in the water.

1 Rules of cutting

Cutting is effective against emergent vegetation as well as against rooted plants with floating leaves. It is less effective against submerged vegetation which, after cutting, can rapidly become excessive again. If the weeds are cut regularly then control is much easier. If cutting is carried out early in the season it is possible to get rid of *Trapa natans*, which is a floating plant.

It is best to cut the emergent weeds at least twice yearly; first in mid-June when the vegetation is still tender, and a second time before flowering. The best time to cut the different species of vegetation is not well known. Cutting at the start of the season gets rid of sedge and much later, reeds.

If the vegetation is cut only once at the beginning then the plants will grow again. If cutting is late the plants will still be harmful as they will shade the water, and during a variable period of time, reduce the growth of the fish.

The manner in which cutting is carried out is also of importance. Generally the plants are cut above the soil, but it has been found that cutting close to the soil is more effective against certain plants such as *Sagittaria sagittifolia*. This kind of cutting is only possible, however, when instruments are handled directly on the ground when working.

To deal satisfactorily with emergent plants, the water in the pond must be kept at as high a level as possible both before and after cutting. This means that the inconvenience of the water cooling is compensated by better maintenance and a greater usable area.

After cutting, the vegetation will float on the surface. It is inadvisable to leave too great a quantity of cut vegetation in the pond for it can cause a lack of oxygen or favour conditions leading to gill mycosis (*Branchiomyces*). The hard vegetable tissues rich in cellulose will not rot easily. It is advisable, therefore, to remove the cut vegetation when it is blown towards the banks by the wind. It is then piled up in heaps in the pond and left to rot slowly, or removed from the pond, piled up in heaps and turned into compost.

2 Methods of cutting

The different techniques and the apparatus used are numerous and varied. These can be either simple instruments or mechanical weed cutters.

When machines are used the ponds should be sufficiently large and the total surface area should justify their use. Small farms can share a machine or can call on cutting specialists. It is also necessary that the ponds should be easy to reach and the bottoms should be free of obstacles such as sunken tree stumps.

For small ponds a *scythe* is used (Fig. 362) from the bank or in the water. The workmen then wear long rubber waders. The scythe should not be too long nor too light. The space between the shoulder and the blade should not be too great in order to avoid the latter catching and accumulating plants. It is also possible to set up a *double scythe* attached to a swivel fixed to the back of a boat, or tie together two boats, equipped with double scythes.

Jointed articulated scythes (Fig. 363) have several blades, rounded at the ends and held together one to the other by rivets or bolts. The blades are from 50 to 80 cm (20 to 32 in) long and from 9 to 12 cm (4 to 5 in) wide. The scythe cuts on the bottom and is operated by two men either standing in the water or on the banks or working from two boats firmly bound to each other. The scythe is pulled from side to side and forward.

Instead of jointed scythes *special saws* which consist of a steel band saw with teeth on both sides are sometimes used. They are from 10 to 15 m long and generally turn round and round. If they are short they have to be weighted with spindle-shaped metal weights.

Mechanical aquatic weed cutters are boats provided in front with cutting beams or other cutting instruments (Fig. 364). They are operated by a motor installed at the back and which also drives the boat.

There are many different models of mechanical weed cutters, some of which have multiple usage such as tilling, cutting and spreading fertilizer.

A good weed cutter must be light, robust and have only small water displacement (20 cm (8 in) or less) and it must also be possible to regulate the depth of the cutting. It should be easily handled by one man and removed from the water without difficulty. It can be transported on a trailer.

Mostly, weed cutters comprise two cutting beams (the principle of a farm mower) one horizontal and the other vertical. The latter cuts vegetation which is so thick that it can hinder the progress of the boat. The width of the cutting varies between 2 and 2·50 m (6 and 8 ft) and the depth from 0·20 to 1·50 m (8 in to 5 ft). According to conditions of work a good weed cutter can cut from 2 to 5 hectares (5 to 12 acres) per day.

Schuberth (1954-1956) has described other cutting systems which are situated in front of the boat which is driven by a motor. One of them comprises four circular saws, three of which are horizontal and one vertical. This permits the cutting of very thick and hard vegetation to a depth of 1 m (3 ft) with a capacity of from 3 to 8 hectares per day (7 to 20 acres). However, this machine is very large and heavy and draws too much water.

Floating weed cutters are generally driven by paddle wheels (Fig. 364) and rarely by means of a propeller. In France, certain boats are equipped with what are called antiplant propellers. Schuberth (1960) has described an amphibious weed cutter mower (Fig. 365) the cutting system of which is normal but the paddles are replaced by large air-filled tyres. If the depth is less than 30 cm (1 ft) the wheels rest on the bottom but if it is greater they float and work like paddles. This mower can be used in water less than 10 cm (4 in) deep.

The removal of cut vegetation is difficult and tedious. In the Soviet Union mechanical cutters are equipped with straw choppers, which chop the reeds up very small. The quantity of reeds left in the pond should not exceed 1,500 kg (1½ tons) per hectare. Schuberth (1958) has described an apparatus for the collection, on the banks, of cut vegetation floating on the water. This is a large wooden rake, 4 m (13 ft) wide provided with 15 teeth 30 cm (1 ft) long. It is mounted on a light float and its working depth can be regulated by means of a lever. The whole is attached to a long cable and it is pulled by a motor-driven winch on the bank. The 4 m (13 ft) wide bands are cleaned one after the other. It takes two men to work, one for the winch and the other on the float.

The principle of the machine described for gathering cut vegetation can also be used for *pulling up plants*. The wooden rake is replaced by a smaller instrument which rests on the bottom. It may consist of a heavy chain, a cutting blade or an iron rake. When working on a pond bottom on which there are obstacles it must be so arranged that the winch can be stopped easily when necessary while two men in the boat lift the instrument over the obstacle. It is also possible to use a heavy chain, resting on the bottom, and drawn by two tractors moving along the banks. These machines are more suited for pulling up submerged and floating vegetation than for controlling emergent plants.

In certain ponds *islands of floating* emergent plants (*Glyceria*, *Typha*) sometimes form. These are drawn to the banks by means of a winch possibly after they have been cut into pieces.

B Other methods of weeding

Certain weeding methods are very simple. They include the removal of filamentous algae (blanket weed) and water duckweed (*Lemna*) by means of perforated receptacles fixed on to the ends of poles.

Vegetation can also be drawn to the outlet by the current of the water or by a water jet. Slightly rooted vegetation (*Callitriche*) can be removed by hand rakes or hooks attached to ropes. Movable screens (planks, black plastic) can also be placed on the surface of the water but all these methods are only suitable for small ponds.

In winter, or at the beginning of spring, when the plants are quite dry, emergent plants can be burned. This, however, will have no influence on their future growth but it will diminish the amount of organic matter of little interest to the pond. After draining the pond a hard winter, with plenty of frost will also diminish the growth of some plants such as *Elodea canadensis*, *Nuphar luteum* and *Nymphaea alba*. Another good method of control is to turn the soil of a dried pond over during a dry summer.

III Chemical Methods

In certain parts of Europe and North America herbicides are used to control aquatic vegetation. However, this has not developed to the same extent as for agriculture.

The use of chemicals in the fight against emergent

Fig. 362 Cutting aquatic weeds in a small pond with a scythe.

Fig. 363 Articulated scythe (after Maier-Hofmann).

Fig. 364 Mechanical weedcutter with horizontal and vertical beams. It is motor driven and equipped with paddle wheels at the back. Valkenswaard fish farm, the Netherlands.

Fig. 365 Amphibious weedcutter. Milicz fish farm, Poland.

vegetation now presents few problems and is often far more economical than cutting.

This is not so, as yet, for the control of floating and submerged plants although significant progress has been made over the past years. However, some plants have proved very resistant. One of the causes which complicates the method and gives divergent results is the varied chemical composition of the water to be treated. In the case of certain excessive or unwanted water plants it is indispensable to repeat the treatment, and this can increase costs seriously.

The use of herbicides is more delicate than cutting, and in fish cultivation they should only be used by competent and skilled personnel.

The success of good chemical treatment depends on the choice of the herbicide in relation to the type of vegetation to be controlled and the security for the fish. Account must also be taken of the time of the year and the technique of the treatment, as well as the care with which it is used.

1 The herbicides used

In agriculture a number of herbicides are used after each one has been systematically studied and placed on the market. Several of these herbicides have also been employed, more or less successfully, to control aquatic vegetation. Only those which offer satisfactory results and sufficient safety for the fish when used in the normal way are mentioned here.

According to their action they can be subdivided into herbicides which act on leaves or those which act on roots. The distinction is not absolutely rigid, however, as some herbicides can be effective against both leaves and roots.

A total herbicide will destroy most plants while a selective herbicide can kill certain plants only.

1 *Herbicides which are absorbed by leaves.* These are divided into: (a) *Contact herbicides* with local action: calcium cyanamide, copper sulphate, sodium arsenite, sodium chlorate; (b) *Systemic herbicides* of which the action is translocated including synthetic growth regulators: MCPA, 2,4-D, 2,4,5-T; dipyridilium herbicides: paraquat and diquat; dalapon; dichlone; amitrol; amitrol-T. These substances are absorbed by the leaves and reach the growth centre; they injure or kill the susceptible plants.

2 *Herbicides absorbed by the roots:* monuron, diuron, amitrol, amitrol-T, simazine, TCA, dalapon, sodium chlorate and chlorthiamide.

Synthetic growth regulators are used, particularly against broad leaved marginal emergent and floating weeds (dicotyledons). Dalapon is effective against grasses and other monocotyledons. When applied, sprayed on to the leaves, amitrol and amitrol-T may be considered total herbicides. Paraquat and diquat act quickly, but are less lasting as contact herbicides on the aerial parts of the plants. They have a systemic but slower action against submerged vegetation if sprayed in the water to act against those submerged weeds.

2 The toxicity of herbicides

The quantities of herbicides and their concentration given below are expressed in terms of active ingredients; as such they are not directly toxic to the fish under the given conditions. However, it sometimes happens that certain commercial products based on the same active ingredient have different formulations and this can influence their toxicity.

When using herbicides all the usual precautions must be taken, as much for the sake of the personnel as for neighbouring crops. The surface or the volume of water must be correctly measured and the herbicide distributed as evenly as possible. When the aerial parts of the plants are to be treated, care must be taken against too much of the herbicide falling into the water. The treatment of spawning and nursing ponds can only be undertaken when it is certain that there is no danger to either eggs or young fish. The water treated must not be used for watering plants or irrigation, particularly if the herbicide has a high persistency such as diuron, amitrol or simazine.

Generally there are fairly good data on the direct toxicity of the herbicides to fish, but this does not apply to all aquatic nutritive fauna. The long term effects on fish of the repeated use of herbicides are not well known, nor the possible accumulation of herbicides in the fish themselves.

An indirect danger from the use of herbicides can arise from the mass of dead vegetable matter, particularly if it is submerged, for it will rot rapidly and this could reduce considerably the dissolved oxygen content which could prove fatal for the fish. The oxygenation of the water must, therefore, be carefully watched and action taken if necessary. It can be remedied by an extra supply of oxygen being given by abundant renewal of water during the period of decomposing of the killed vegetation.

3 Treatment techniques

It is possible to use herbicides on the bottom of dried ponds or in filled and even stocked ponds, possibly after having lowered the level of the water. The quantities given below should be considered as doses of active ingredients of which the percentage can differ from one commercial product to another.

For small surfaces or over long shaped ponds, and also for canals, the herbicides can be sprayed from the bank. For large surfaces it is necessary to use a boat. The spraying materials chosen will differ according to each case. It is possible to use a knapsack sprayer, an atomizer or a motor sprayer pump installed either in the boat or on the bank, provided it has an extending arm (Figs. 366 to 371).

The quantities indicated are for 1 hectare (2·5 acres) and the amounts given are diluted at a ratio of 1,000 litres (220 gal) for ordinary spraying (high volume application) or 250 litres (55 gal) for low volume application.

a   Treatment of dry ponds

Ponds which have been dried are treated with the following products: diuron (10 to 15 kg per hectare), monuron (15 to 20 kg per hectare), simazine (10 to 15 kg per hectare), chlorthiamide (10 to 15 kg per hectare), TCA (100 to 150 kg per hectare) and sometimes sodium chlorate (300 kg per hectare). The same quantities of products expressed in lbs are used per acre. Larger quantities must be used if the bottom of the pond has deep mud and the plants are deeply rooted.

Application is by spraying or by spreading, if possible after the old vegetation has been cut and removed. When it is spread it may be necessary to mix the herbicide with sand in order to increase the volume and assure a more even distribution. To this end some herbicides are sold as pellets containing only a small percentage of the active ingredient.

The bottom should be damp when treated just before or at the start of the growing season. It is preferable that the pond should, previously, be left dry for 4 or 5 weeks. After treatment it is generally best to place the pond under water slowly in order to obtain, temporarily, a high concentration of the herbicide. During the growing season the water should be renewed as little as possible.

The herbicides just mentioned can also be used on the bottoms of newly constructed ponds, in which case it is best to use products which remain active for as long as possible, e.g. diuron, monuron, chlorthiamide and bromacil.

The spreading of 750 kilos (15 cwt) of calcium cyanamide will control *Elodea canadensis* and duckweeds. After spreading, the level of the water should be raised from 20 to 25 cm (8 to 10 in) but the pond cannot be stocked with fish until this water has been removed and replaced. Bank (1958) mentions the use of ammonium nitrogen (5 cwt of nitrogen per hectare) to destroy *Elodea canadensis* and ammonium sulphate (5 cwt of nitrogen per hectare) to control *Fontinalis*.

b   Treatment as ponds are filled

The indications given below are used against vegetation in ponds as well as being usable for canals, ditches and other natural or artificial waters with an excess of similar vegetation. The treatment of ponds under water and with fish present is often the only way to deal with submerged and floating vegetation.

1   *Control of emergent and semi-emergent weeds.* The control of this vegetation means destroying that part of it which is out of the water. This is less dangerous for the fish than the control of submerged vegetation.

Treatment is best carried out from the end of June to the middle of September. A spray is used and it is important that the herbicide should be well spread over the aerial parts of the vegetation being treated. The leaves should be well covered by it. The use of a mirror-jet or better still a "vibrajet" is advisable. The volume of the solution should be around 750 to 1,000 litres (165 to 220 gal) per hectare. A low volume application (very fine droplets) is only possible in the absence of wind and when the atmospheric pressure is high.

For the control of *Phragmites communis* (Fig. 373) the following products are used. These generally assure an effective control which lasts over several years: amitrol 10 to 15 kg (22 to 33 lb) per hectare, amitrol-T: 8 to 12 kg (18 to 27 lb) per hectare, dalapon: 20 to 25 kg (44 to 55 lb) per hectare, sodium dichlorobutyrate: 20 to 25 kg (44 to 55 lb) per hectare, dalapon + amitrol: 15 + 5 kg (33 + 11 lb) per hectare, dalapon + amitrol-T: 15 + 3 kg (33 + 7 lb) per hectare.

Those herbicides considered effective against *Phragmites communis* are also satisfactory against *Typha latifolia, Typha angustifolia, Glyceria maxima, Glyceria fluitans* (after the lowering of the water level), *Baldingera arundinacea.*

Carex sp., *Juncus* sp., *Acorus calamus, Iris pseudacorus* and *Scirpus lacustris,* are more resistant plants and for that reason larger doses are often necessary.

A mixture of dalapon and amitrol as well as dalapon and amitrol-T is often considered to be the most effective. Amitrol and amitrol-T also permits, at the same time, the control of broad-leafed aquatic vegetation.

Juncus sp. can also be controlled by 2,4-D and MCPA (3 to 4 kg or 7 to 9 lb) per hectare as it can for *Stratiotes aloides*.

Sparganium ramosum, Sagittaria sagittifolia and *Alisma plantago* are controlled with 2,4-D or 2,4,5-T (3 to 5 kg or 7 to 11 lb per hectare).

Equisetum limosum and *Equisetum palustre* are difficult to destroy by means of an herbicide, and it is often necessary to repeat the applications. 2,4-D, 2,4,5-T, MCPA or MCPB (4 to 5 kg or 9 to 11 lb per hectare), amitrol (10 to 15 kg or 22 to 33 lb per hectare), amitrol-T (8 to 12 kg or 18 to 27 lb per hectare) are utilized.

Certain emergent plants: *Juncus* sp, *Sparganium ramosum, Sagittaria sagittifolia* are very sensitive to dipyridylium herbicides such as paraquat and diquat (3 to 4 kg or 7 to 9 lb per hectare). These can be used against most of the emergent plants but the results are not so good and do not last so long as when the products noted above are used. The use of these herbicides may be considered as a partial chemical eradication of aquatic weeds.

It is possible to buy a large number of different herbicide mixtures capable of controlling a wide range of plants: amitrol + 2,4-D + simazine, dalapon + 2,4-D + MCPA, diuron + amitrol, TCA + amitrol + 2,4-D, etc. These mixtures can be used to control emergent and semi-submerged vegetation.

2 *Control of floating weeds.* Treating floating plants with herbicides is difficult as it is necessary to assure that, at the time of treatment, the product is not washed off the leaves by the movement of the water. This means that some form of adhesive product should be added which will prolong the action of the herbicide since it will stick and thereby stop the leaves being washed either by rain or by waves in the pond. If possible high pressure sprayers permitting treatment from a greater distance should be used. Also if possible the level of the pond should be lowered particularly if growth regulators are used. In that case there should be an interval of 5 or 6 hours at least before raising the water level again.

Nuphar luteum, Nymphaea alba, Nymphoides peltata and *Hydrocharis morsus-ranae* can be effectively controlled by products with an ester formulation of 2,4-D and of 2,4,5-T or by mixing two of these products (2 to 4 kg or 4 to 9 lb per hectare). Because they are very volatile, esters are herbicides which must only be used with the greatest caution. It is advisable not to use them when the wind is strong nor to employ a high pressure sprayer.

Besides, these products are generally more toxic than other growth regulators. But in practice, in any case, there is no direct danger to the fish if the treatment is carried out normally, that is to say if the chemical is not washed off the leaves completely and if the depth of the water is over 50 cm (20 in).

It is also possible to use other formulations of 2,4-D and 2,4,5-T as well as MCPA or MCPB (3 to 5 kg or 7 to 11 lb per hectare) against the floating plants mentioned above. These products are also suitable for controlling duckweeds but their application is difficult.

Hydrocharis morsus-ranae is very sensitive to diquat and to paraquat or to a mixture of both (1 to 2 kg or $2\frac{1}{4}$ to $4\frac{1}{2}$ lb per hectare), whereas *Potamogeton natans* and *Ranunculus aquatilis* can often be controlled throughout a growing season with these same products. As far as the use of diquat and paraquat is concerned, further indications are given at the same time as those on control of submerged vegetation. These herbicides also destroy duckweeds (*Lemna minor* and *L. trisulca*) in ponds under water, but spraying must be carried out with the greatest care. Control of duckweeds (*Lemna*) is complicated because it is often composed of several layers. Treatment destroys the top layer which disappears rapidly, but it reveals a lower layer which turns green quickly and so it is necessary to start the treatment all over again.

Trapa natans is controlled by the sodium salts of 2,4-D ($3\frac{1}{4}$ to $5\frac{1}{2}$ lb per hectare). *Polygonum amphibium* is a very difficult plant to control and against which growth regulators and dipyridylium herbicides give only partial results. The best results are obtained by spraying the leaves in June and July with dalapon + amitrol (25 + 5 kg or 55 + 11 lb per hectare) or amitrol-T (10 kg or 22 lb per hectare) after the ponds have been dried.

3 *Control of submerged weeds.* The control of submerged vegetation is complicated and depends on the species to be controlled. These are varied and numerous.

In Europe sodium arsenite is not used as a herbicide in the aquatic environment but it is used in the United States. Copper sulphate is only employed for the control of algae. Pellets containing synthetic growth regulators or the spreading of simazine in the pond under water have rarely given entirely satisfactory results. The same is the case when growth regulators are used after the level of the water has been lowered in such a way that part of the vegetation normally under water is dried.

Good results against submerged vegetation are often obtained with the use of dipyridylium herbicides: paraquat and diquat. They are normally used in ponds under water containing fish. They often permit the control of a large number of different plants during the growing season and their excessive growth during the following season is diminished. In practice a mixture of the two products is advised: $1 + 1$ kg ($2\frac{1}{4} + 2\frac{1}{4}$ lb) per hectare or $5 + 5$ litres ($1 + 1$ gal) of commercial products. If diquat or paraquat is used alone, then the ratio would be 10 litres (2 gal) of the commercial product per hectare.

Elodea canadensis, *Potamogeton crispus*, *P. pectinatus*, *P. oppositifolius*, *Ranunculus circinatus*, *Ceratophyllum* sp. and *Myriophyllum* sp. are sensitive to diquat and also to paraquat. The results obtained against *Potamogeton lucens* are not so good, and the effect on *Hottonia palustris*, *Callitriche* sp., *Chara* sp. and the greater part of the filamentous algae seems to be rather weak, if not completely ineffective.

It is strongly recommended that dipyridylium herbicides be used before the pond is completely invaded and it is also important that its distribution should be carried out with the greatest care to ensure that it is sprayed on all the plants coming under control. If the mass of vegetation is too compact the herbicide will not penetrate deeply. It is not necessary, therefore, to lower the level of the water before treatment but it is necessary to ensure that there is no organic matter in suspension, for if there is it will rapidly hinder the action of the herbicides. It is advisable to carry out treatment when the sky is overcast. The slight toxicity of these products, and the fact that their destruction in the water is rapid, are advantageous.

In ponds containing fish it is also possible to use chlorthiamide pellets. This is effective against a large number of submerged plants at concentrations of 2 to 3 parts per million (2 or 3 gm per cubic yard of water to be treated). The pellets must be spread out early in the season (end of April, May) and the water supply for the pond must be cut off for several days. Nevertheless, filamentous algae often develop later.

Diuron pellets also control several species of submerged vegetation at a concentration of 0·4 ppm. Diuron can be used later than chlorthiamide, even when the plants are well developed but it is really better to start the treatment early in the season. Diuron has the advantage of being effective against filamentous algae and stoneworts. It is necessary to curtail the water supply to the pond for several days after treatment, which is not always possible especially in spring-fed ponds. Instead of using pellets it is also possible to spread wettable powder mixed with 10 parts sand or to spray a solution of wettable powder.

In a word, it is often possible to eliminate submerged vegetation over a fairly long period after treatment has been given with the herbicides mentioned. However, it often happens that certain submerged plants start growing again during the season and even become excessive. This happens more readily as the water remains clear after treatment. In cyprinid ponds in which fertilizers, or manure, and artificial food are distributed, as soon as the water warms sufficiently it often happens that there is an appreciable growth of unicellular algae which hinders the penetration of light and, thereby, the development of submerged vegetation which had been already controlled with herbicides at the start of the season.

American authors recommend composite fertilizers and notably a 10–5–5 mixture of ammonium sulphate $(NH_4)_2SO_4$, superphosphate $Ca(H_2PO_4)_2$ and potassium chloride (KCl), in such a way that in weight it comprises 10 parts nitrogen, five parts phosphorus and five parts potassium. This mixture is used at a ratio of 100 kg (220 lb) per hectare.

4   *Control of algae.* Copper sulphate is an algicide which is effective and cheap for the control of *Spirogyra*, *Cladophora*, *Chara* and other algae. It has been used to this end for a long time. The concentrations necessary are 0·5 to 1·0 gm of copper sulphate per cubic metre (1·3 yd³) in acid water and 1·5 gm per cubic metre in alkaline water. As these concentrations are near to being toxic concentrations for the fish, the whole pond is not treated but only a part, e.g. a half or a third. It is important to measure the volume of water carefully and to use the copper sulphate regularly either in the form of a solution or as crystals placed in a wide mesh sack suspended from an air-filled tube attached to a cord (Fig. 372) by which it is drawn from the bank. It can also be pulled from the back of a boat.

Fig. 366 Spraying with a knapsack sprayer while wading in the pond. Linkebeek fish culture station, Belgium.

Fig. 367 Spraying floating weeds with a jet spray. The pump and the container cannot be seen in the photo but are installed on a vehicle on the bank.

Fig. 368 Treatment from the bank with a motor pump and a jet spray. Pond at Hofstade, Belgium.

Fig. 369 Spraying from the bank with equipment mounted on a vehicle. Linkebeek fish culture station, Belgium.

Fig. 370 Using a pneumatic boat for spray with a portable atomizer. Antitankkanaal, 's Gravenwezel, Belgium.

Fig. 371 Destruction of emergent vegetation (*Equisetum limosum*) with herbicides. Plots treated alternatively to permit study of treated and untreated plots. Ponds at Groenendaal, Belgium.

Fig. 372 Treating a pond, overgrown by algae, with copper sulphate. The crystals are in small sacks with a wide mesh and suspended from an inflated rubber tube.

Fig. 373 The banks of a canal showing a border of reeds 2 years after treatment with a mixture of dalapon and synthetic growth regulators. Along the two banks, treated stubs alternate with non-treated stubs. In this way fishing is made easier and sufficient shelter for the fish is assured. In order to avoid erosion a border of sedges is left untreated. Canal at Damme, Hoeke, Belgium.

Smaller quantities of copper sulphate will permit control of unicellular algae and algal bloom without endangering the fish. The ratio is 2 kg (4½ lb) per hectare. If the copper sulphate is in powdered form it can be distributed over a part of the surface of the water. If it is in crystalline form it can be put in two or three small wide-mesh sacks fixed to pegs a few centimetres below the level of the water. The copper sulphate is spread slowly by the wind or a slow current. It may be necessary to repeat the treatment after a few weeks.

Bank (1962) advises the use of copper oxychloride for the control of filamentous algae in ponds containing fish (1 kg (2¼ lb) of copper per hectare) as well as malachite green (0·3 gm per cubic yard), or a fungicide derived from a dithiocarbamic acid (2·5 kg or 5½ lb per hectare) or the spreading of powdered superphosphate (16 per cent phosphoric acid) at the rate of 600 kg (12 cwt) per hectare. All these products, very probably, are not so effective as copper sulphate and cannot control characeae. Nevertheless they are less toxic for the fish.

The following products are also used for the control of filamentous and unicellular algae in ponds containing fish: monuron (0·5 to 1·0 ppm), diuron (0·2 to 0·4 ppm), dichlone (0·05 to 0·15 ppm). Diuron is effective against characeae.

In small ponds, powdered quicklime can also be used for the destruction of masses of algae floating on the surface of the water. The aim is to produce a pH of 10·0 to 10·2 within a very short time, but this could have certain dangers for the fish. The amount used depends on the chemical characteristics of the water to be treated.

SECTION II

IMPROVEMENT AND RESTORATION OF POND BOTTOMS

The principal measures which should be taken for the restoration and improvement of the bottoms of ponds are as follows: draining and drying out, working the soil, dredging.

Dredging is intended to restore those ponds in which drying out or working the soil is not sufficient.

The drying out of ponds and working the soil is intended to improve fish production by looking for one or more of the three objectives given below. These are intended to improve the physical, chemical and biological condition of the soil.

1. The creation of organic, fertile and finely colloidal mud necessary for biological production. The progressive reduction and mineralization of excessive mud particularly when it is rich in cellulose.

2. The reduction and destruction of excessive roots of emergent vegetation which transforms into substances that are lost or are of no interest to the biological cycle of the pond a part of the nutrients.

3. To obtain good hygienic conditions by consecutive destruction, while drying, of the encysted stage or of intermediate hosts of organisms which cause fish disease; and, furthermore, the destruction of larvae and adult aquatic insects which are harmful and other enemies of fry and fish (leeches, fish-lice); good hygienic conditions are particularly important for farms practising intensive cultivation of trout and other species.

Drying the pond, which provokes mineralization of accumulated mud and a general improvement in the bottom is the best way to achieve objectives 1 and 3. If the upper layers of the soil are worked over slightly this is the best way of bringing about this action.

The same is not the case when the soil is worked in depth. This is necessary to achieve the second objective but this action buries the top productive layers which are often very thin, and brings to the surface the deep sterile layers. There is therefore some contradiction between goals 1 and 2.

I Drying the Pond Bottoms

In temperate regions the drying of pond bottoms generally only takes place in winter. It starts in the autumn after the pond has been drained and is continued until it is again placed under water. In general the latter is carried out in the spring in March and April, or in June and July if it concerns cyprinid nursing ponds. In intertropical regions the ponds can be dried out at any time of the year. The period must be short, for if it is prolonged it will lead to a loss of production. It should last only long enough in order to obtain the mineralization of the mud, which will be evident when the surface cracks (Fig. 375).

Drying out must be effective (Fig. 374) and this will only be possible if the pond has a good network of drainage ditches (both principal and secondary) which are cleaned out after the water has been drained off.

In certain special cases if the drying out is not undertaken each year but only at longer intervals then the pond can be left dry during the summer so that it will remain dry for over 1 year. This can consist of simple drying or it can be combined with agricultural cultivation. This method is less practised than it formerly was because of the growing use of fertilizers. It has therefore lost most of its purpose.

A good way to practise alternate fish and agricultural cultivation is as follows. During 3 or 4 years, the pond is placed under water in the summer and simply dried out in the winter. At the end of this period, the winter drying is not followed by filling the pond but by ploughing and then cultivating cereals: rye, oats and barley mixed with clover. The following year it is put under pasture for livestock until June–July (Fig. 378) and then placed under water slowly for the cultivation of young carp.

Another way of proceeding is to sow leguminous plants (lupine, pease, clover) (Fig. 379) the first year. The pond is then put under water in June for the cultivation of small carp. The following year cereals are cultivated. This last crop will benefit from green fertilization by the leguminous plants and the absence of competing weeds. This rotation of crops favours each of the harvests. To be exploited in this way the pond must be so conditioned as to permit perfect draining thus avoiding any excess of moisture for the cereals. In the case of green manure it is just as advantageous to sink it into the pond at the time of flowering as it is to plough it into the soil.

Maier estimates that in temperate regions one third of the total production is lost as a result of keeping the pond under water during the winter. Ponds which are never dried out gradually lose their value, and their exploitation finally ceases to be profitable. Nevertheless it must be recognized that the regular practice of drying ponds in winter is prejudicial to the development of a number of aquatic organisms which are destroyed. However, the advantages of drying out are greater than the

inconveniences provided that the first objective, which will benefit healthy production, is to obtain mud in good condition.

Necessary repairs which cannot be carried out when the pond is under water are carried out during the drying period. These include verification of the watertightness of the dike, control of the monk, rectification of the banks, deepening of the upper part of the pond, destruction by pick and fire of tangled reed rhizomes, and the liming of mud zones. Screens and fences are repaired and painted.

Excessive mud and endogenous or exogenous detritus (dead leaves) are removed. This is done in dry winter weather. The mud can be used in neighbouring agricultural or horticultural land. Some of the mud can be eliminated by means of a very strong current of water traversing the pond.

II   Soil Cultivation

If soil cultivation has to be done then it is, in principle, superficial. The deep working of the soil should only be carried out in a few well determined cases, that is to say when the pond is largely or totally covered with excessive reeds or when the productive humus layer is very thick and there is no problem mixing it, which can only come about when ponds are very productive. In average or feebly productive ponds, this is not the case and working the soil would be harmful. The simple drying of a pond, followed by frost, is sometimes prejudicial while it destroys the humus layer. However, for this last category, working the soil can be favourable if it is followed by a rather long dry period or if it is cultivated.

The soil can be worked from the moment it is sufficiently dry and firm which is often only in the spring.

A first method is to use a plough, especially if aquatic plants are to be destroyed (Fig. 376). For the ploughing to be well done there should be a 180° turn-over. If it is only 90° the vegetation will rapidly grow again and the ploughing will make future cutting of the vegetation, which will be indispensable, more difficult. When excess of vegetation is total there should be no fear of deep ploughing, up to 25 cm (10 in), for bringing sterile layers to the surface is better than leaving the pond totally covered with emergent weeds and render it practically unproductive.

In all cases where the bottom of the pond is not completely covered by emergent vegetation, working must avoid turning over which will bring to the surface unproductive layers. Working the soil but not turning it over calls for a variety of tools, the best of which is the rotary cultivator (Fig. 377). It is available in several models. This tool breaks the soil up well, opening it without burying the fertile top layers and bringing to the surface the deep sterile layers. At the same time it levels the soil and destroys tangled roots of undesirable plants.

The machines used for this work are drawn either by animals or tractors or caterpillars.

It is also possible to work the soil under water. This ensures that the bottom of the pond is kept clean and liberates the nutritive elements of the top layers. The effect can be compared with that which would be obtained by a dense carp population (at least 1,000 K_2 per hectare) digging into the mud for seeking food. The same work on the bottom can be carried out with instruments similar in shape to rakes or harrows. In Hungary an instrument has been built which blows compressed air on the mud in order to turn the top layers over.

III Dredging the Ponds

The dredging of ponds is necessary when mud accumulated on the bottom is so considerable that ordinary measures such as drying out and working the soil are not effective.

After a more or less long period this measure becomes necessary for ponds which have been neglected over a number of years and have never been left dry. The mud accumulates little by little and raises the level of the bottom, thus encouraging the progressive growth of emergent vegetation. From the moment when this happens, silting up begins and the useful area of the ponds and its productivity diminish rapidly. If nothing is done the pond will become a swamp and will be of no further use for fish cultivation (Fig. 361).

The silting up of the pond occurs that much faster if the pond is shallow and the water rich in nutrients which favour the growth of aquatic plants and give abundant detritus. After several years the thickness of the mud will exceed 50 cm (20 in).

The dredging of ponds is a burdensome task caused through neglect and failure, over a long period, to carry out proper maintenance. Often the cost is so high that owners are unable to meet it and this leads to the final abandonment of the ponds. An additional complication is what to do with the mud which has been extracted. If there are only small quantities the problem is not serious, but if there is a lot then transportation will increase the cost considerably. One good way of disposing of mud is to use it to fill in horticultural and agricultural land.

Dredging ponds is the removal of mud either after simple drying or after liquefaction, followed by sedimentation and drying.

1. *Dredging after the drying of the mud.* After the pond has been dry over a prolonged period, that is, at least during the winter months, the mud should be sufficiently dry and firm. For small areas and for ponds which are not extended, it is possible to use a shovel for removal and a barrow or an endless belt or even a small gauge rail and trucks for the transportation of the mud. A dragline can be used for the more important work and a bulldozer if the bottom is sufficiently hard.

2. *Dredging by liquefaction of the mud.* In ponds which have been previously dried the mud is sprayed with a strong jet of water under pressure. It liquefies and dilutes the mud which is driven downstream (Fig. 380). The pressurized water is drawn from a water course, or a near-by reservoir which could be a neighbouring pond.

The liquefied mud which runs in a dip in the pond is pumped out and then directed to a hollow outside the pond, or to a dump laid out in the pond and surrounded by small dikes. It pours off and stays there or is removed later after it has dried.

A similar system can be used in ponds which cannot be drained off. The apparatus used for the removal of the mud is a suction dredger (Fig. 381) which comprises a small boat, a motor pump and a rotating cutter-head attached to the end of a suction tube as well as a discharge pipe. The mud, detritus and aquatic vegetation are pulled up and cut up by the rotating cutter and then sucked up in the form of liquid mud and discharged outside the pond. This discharge is by means of flexible pipes which float on the water (they may be held up by barrels for example), and also by static pipes on the land.

These methods are more rapid than dredging after the drying of the mud but they call for expensive apparatus. They can only be used by specialists and are only economical for large ponds. The installation and deplacing of the apparatus also takes a long time.

Fig. 374   After the autumn draining, practise of the winter drying in European cyprinid ponds. Note the very good network of draining ditches for emptying and, as a result, the perfect drying. Bokrijk fish farm, Campine, Belgium.

Fig. 375   Mineralization of the bottom of a pond in a tropical region, by drying up for several days until the mud cracks. Tjimindi, Bandung, Indonesia.

Fig. 376   Winter drying up of a pond and ploughing the soil. Pond at Gemioncourt, Brabant, Belgium.

Fig. 377   Rotary cultivator for working pond bottoms. Bellefroid fish farm, Heverlee, Belgium.

Fig. 378 Drying up of ponds followed by soil cultivation; young cereals as pasture for cattle; note the principal draining ditch on the periphery. Vaesen fish farm, the Netherlands.

Fig. 379 Cyprinid pond used alternately for fish culture and growing leguminous plants (lucerne). On the dry bottom of the pond growing squares of lucerne covering part of the pond bottom. Szarvas . fish culture station, Hungary.

Fig. 380 Cleaning up mud with three strong water jets used simultaneously to liquefy it. Ponds at Groenendaal, Belgium.

Fig. 381 Dredging the pond with a suction dredge. The rotating cutter-head in the foreground detaches the mud and debris which are sucked in and then pumped back. The apparatus is mounted on a boat. Angling pond, Viane, Belgium.

SECTION III

CONTROL OF THE PRINCIPAL PHYSICO-CHEMICAL FACTORS OF FISH PRODUCTION

The water used for the cultivation of fish will not give maximum production if the conditions are not optimal for the fish and for the other aquatic organisms. This supposes that the physical and chemical factors are present at their optimum level to determine a good initial production. The latter means the production of vegetation which must be present in an optimum quantity (Fig. 60).

In this respect the quantity of nutrients in solution in the water varies considerably and determines the very different natural productivities (Chapter X). That productivity depends on the physico-chemical characteristics and the principal ones must be known in order that they can be controlled.

On all fish farms and new installations it is necessary from the start to make chemical and physical analyses in as detailed a manner as possible in order to determine the quality of the water which will be used. The determination of the water temperature, pH and alkalinity is essential, and must be carried out regularly. This can be done quite easily by the fish farmer. To determine other chemical characteristics, it is possible to use specialized laboratories or to buy very practical kits with which it is easy to measure dissolved oxygen, ammonia and even several other parameters. The accuracy obtained with these kits is generally sufficient for the needs of a fish farm. Although they are more expensive and more fragile, electro-chemical instruments are very easily and quickly used for the accurate determination of pH and dissolved oxygen content. They are particularly convenient when numerous measurements have to be made.

I   Temperature Control of Fish Ponds

1   Temperature needs of the fish

Temperature has a considerable influence on the principal and vital activities of the fish notably their breathing, growth and reproduction (Fig. 382). This great influence of temperature is due to the fact that the body temperature of fish varies with and is almost the same as that of their environment.

Breathing. Fish breathe dissolved oxygen in the water. They can only breathe normally in an environment with sufficient oxygen. In pure water the dissolved oxygen content is first of all tied to the temperature according to a precise relationship shown in the graph figure 382. It is seen that the dissolved oxygen content is 14·62 mg/litre at 0°C, 11·33 mg/litre at 10°C, 9·17 mg/litre at 20°C, 7·63 mg/litre at 30°C.

Certain species of fish, salmonids in particular, are very demanding with regard to the dissolved oxygen; they need normally 9 mg/litre, which more or less corresponds to a temperature of 20°C in pure water.

Most salmonids can withstand temperatures above 22°C for only a short time, and with the exception of the special case of water which is particularly rich in oxygen, they die if the temperature is over 22 to 25°C according to the species.

Other species, cyprinids for example, are satisfied with a lower oxygen content, and in consequence live in water with higher temperatures. For this reason salmonids are kept in cool water. Cyprinids and species which need only small quantities of dissolved oxygen can endure or need higher temperature waters.

Growth. The action of temperature on growth, which is effective between variable limits from one species to another, is considerable. It is only necessary to mention, for example, the influence on carp and tilapias. As far as salmonids are concerned the influence of temperature on their growth was discussed in Chapter III.

Carp can withstand wide variations of temperature. In Europe the most favourable temperatures for growth are between 30 and 20°C, the optimum being around 25 to 28°C. This optimum is higher in tropical regions where there probably exist geographical forms better adapted to high temperatures. Carp grow moderately well between 20 and 13°C, feebly between 13 and 5°C and not at all below that, although young carp can still eat at 4°C.

The optimum development of tilapias is also above 20°C but these fish do not withstand temperatures as low as carp. Many species suffer and die below 12°C while certain species (*T.*

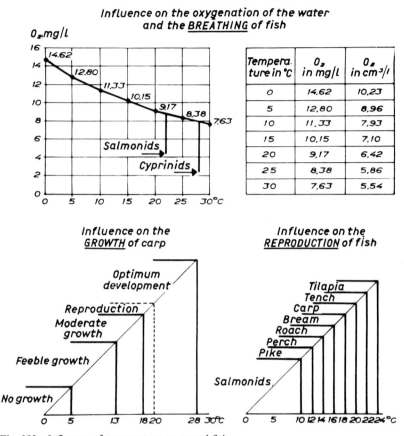

Fig. 382   Influence of temperature on reared fish.

sparrmani) withstand water the temperature of which falls as low as 8°C.

Reproduction. Fish only spawn when the temperature of the water suits them. It is therefore essential that the required temperature should be reached at the right time which will be at the properly-called reproduction period as well as during the preparatory period; if not, the fish will not spawn. Cyprinids and other fish withstanding or requiring high water temperature do not spawn if the water does not warm up sufficiently. Pike spawn at the first warming around 10°C; perch a little later around 12°C, and for most cyprinids 15 to 20°C are necessary and even more than that according to the species. From 14 to 16°C are required for roach, 18°C at least for carp in Europe, higher than that in tropical regions, and 20°C for tench. For tilapias the minimum reproduction temperature is 21 to 23°C.

Salmonids spawn in low temperatures below 10°C. They do not reproduce if the water is not cold enough and this factor limits their reproduction in tropical regions even at high altitudes where cool water exists from 15 to 20°C. This suits their growth perfectly but not their reproductive capacity because the water is not sufficiently cold.

2   The influence of temperature on fish production

Temperature is one of the most important among the external factors which influence fish production. Each species of fish is adapted to a scale of temperatures including minimum and maximum temperatures outside which they cannot live. There is also a scale of intermediate temperatures which corresponds to the maximum growth of the fish and the maximum production of the pond.

Within supportable limits high temperatures are

best for the production of cultivated fish. It has been noticed that when summer temperatures are high the years are good. However, higher production in waters in tropical regions is not due to the fact that the water is richer in food than the water in temperate regions. The differentiation between lakes and ponds in the two regions stems from the fact that the rhythm of production is higher in warm regions because the temperature is higher and because the rhythm of production is more frequent in tropical regions and is maintained throughout the year. It is only present during a part of the year in temperate regions, however.

Based on findings of real production obtained with the same species, notably carp in regions in which annual average temperatures differ considerably, Huet (1964) proposed the following values for the coefficient k_1 (temperature factor) of his productivity coefficient k (Chapter XI). $k_1 = 1 \cdot 0$ if the average annual temperature is equal to $10°C$, $k_1 = 2 \cdot 0$ if this temperature is $16°C$, $k_1 = 3 \cdot 0$ if it is $22°C$ and $k_1 = 4 \cdot 0$ if it equals $28°C$. If this proposition, which needs to be verified, is exact it signifies that the production of two similar ponds, stocked with the same species and quantity of fish will be three times greater in a pond with an annual average temperature of $22°C$ (tropical region) than in a pond with an annual average temperature of $10°C$ (cold temperate region).

3   Best ways of controlling the temperature of ponds
It is evident that there is a major interest in stocking fish for rearing under the conditions which best correspond to their needs for breathing, reproduction and growth. At the same time it is far more difficult to influence the temperature of a pond than it is to improve other means of bettering production such as fertilization or artificial feeding for example.

If the possibilities are limited they are nevertheless real. It is possible to influence the temperature of a pond by increasing or reducing the quantity of water admitted. If reduced during the warm season the temperature will rise, but if increased the temperature will remain relatively low which is a good thing for salmonids.

For a long time warming ponds have been in use in Europe as well as windbreaks intended to increase the temperature of spawning ponds for carp. During the growing period it is a good thing to increase the temperature, but there may also be reasons for reducing it. This is the case if the temperature is too high and exceeds 30 to 33°C

which would bring about a reduction in the growth of the carp in the pond. The same applies to tilapias when the intention is to lower the temperatures to below 20°C in order to stop reproduction. In similar cases the reduction in quantitative production will be more or less compensated for by an improvement in quality.

Under certain circumstances, for example where it is possible to use the cooling water of an electric generating station, of which the temperature is already high, production can be increased because of the increase in temperature. In this way it is possible to benefit from the high temperature water in the ponds to raise tropical fish in ponds of the temperate region.

The evacuation of water from the bottom of the pond by means of a Herrguth type of monk (Fig. 25) assures the general rewarming of the water of the pond.

In warm temperate regions the raising of tilapias is possible, for example in the southern states of the United States. However, as winter temperatures are too low, the tilapias have to pass that season in artificially warmed ponds.

4   Measuring the temperature of ponds
The temperature of ponds should be taken, preferably, with a precise mercury thermometer. Generally it is taken at the surface, near the monk for example.

In a pond there can be relatively wide differences of temperature between the surface and the depths. It is difficult to take the temperature of the deep water with an ordinary thermometer since the reading will change as it is drawn up from the bottom to the surface. However, the approximate temperature can be taken if the thermometer is drawn up very quickly. It is also possible, of course, to take the exact temperature deep down with a reversing thermometer, but this is an expensive apparatus.

II   Control of the Dissolved Oxygen Content in Ponds

1   Importance of the dissolved oxygen content
The considerable importance of this was mentioned above when the temperature demands of fish were considered.

In the first place the dissolved oxygen content depends on the pond water temperature linked with it. It also depends to a considerable degree on the quantity of organic matter present and the submerged aquatic vegetation. Neither should be too great.

If decomposing organic matter is in too great a proportion, it will absorb too much of the dissolved oxygen in the water. The content of the latter could then fall below the minimum level acceptable to the fish. This danger is particularly great in warm and thundery weather, for example when the first rewarming of the water starts in the spring. It is also present in winter under ice if the water supply is not sufficiently abundant, and above all if the ice is covered with snow, for then the penetration of light will be reduced as well as the photosynthesis action of the submerged vegetation.

This vegetation can also have a considerable influence on the dissolved oxygen content in the ponds if it is too dense. During the day it can give off excess of oxygen to a point of saturation beyond 150 per cent of the normal level, and this over-saturation can endanger the lives of the fish, and the fry in particular. At night this same vegetation absorbs oxygen, and can bring the oxygen content down below the minimum level acceptable to the fish. The danger is greater in warm and thundery weather, and when there is a superabundance of aquatic algae.

2 The dissolved oxygen needs of different species of fish

The dissolved oxygen needs of the fish vary according to the species. This was explained above when temperature was discussed. For salmonids in general 9 mg/litre are necessary, which corresponds to a temperature of around 20°C in pure water. Exceptionally and temporarily the amount can fall to below 5·5 mg/litre. For cyprinids in general the dissolved oxygen content should be around 6 to 7 mg/litre; temporarily the amount can fall to 3 mg/litre. Other species, cichlids and silurids even more so, can withstand a much lower content and in consequence, high temperatures and water rich in organic matter, both of which reduce the dissolved oxygen content in the aquatic environment.

3 Troubles caused by an insufficient dissolved oxygen content

The behaviour of fish lacking oxygen is typical. They come to the surface in an effort to breathe air. Other fish group together near the fresh water inlet where they behave normally. Fish killed by asphyxia have raised gill covers and their gills are wide apart (Fig. 450).

4 Possible measures to be taken to increase the dissolved oxygen content

To ensure that there is a sufficient oxygen content from the start, cascades, or waterfalls can be built into the supply channel, or water wheels can be installed.

When the water reaches the pond it can be divided and spread out by means of passing over a perforated screen (Fig. 383) which will permit good contact with the air and in this way enrich the water with oxygen. Good results can also be obtained by introducing submerged plants in the pond or semi-submerged plants (watercress, fool's watercress) in the water supply channel. Here also are other steps which might be taken to improve indirectly the dissolved oxygen content. The soil can be worked and limed and this will reduce the organic matter content; cutting of emergent weed will promote penetration of light and fertilizing will favour the development of submerged plants.

If there is a lack of oxygen in a pond then as much good quality water as possible must be supplied. It is possible to try to aerate water by projecting it into the air with a pump (Fig. 385), or using an outboard motor on a boat circulating on the pond, or a compressor for injecting air into the water (Fig. 386). The water can also be oxygenated with an aerator-mixer (Fig. 384).

When ponds freeze over it is advisable to get rid of the snow by sweeping it to the sides. This will permit light to penetrate, and in this way the submerged vegetation will give out oxygen. In wintering or storage ponds it is possible to install a wattle fence (Fig. 190) (Chapter IV: Cyprinid cultivation).

5 The determination of the dissolved oxygen content

The methods used for the determination of dissolved oxygen are based on the ability of certain chemical compounds to absorb all the dissolved oxygen in the water sample. These compounds are capable of oxidizing iodine ions introduced into the solution into atomic iodine, the quantity of which can be determined by titration. Knowing the equivalence of the iodine to the oxygen it is possible to determine the initial quantity of dissolved oxygen in the water sample.

Fig. 383 Enrichment in dissolved oxygen content by a water supply through a perforated conduit. Limal fish farm, Belgium.

Fig. 384 Electric aeration-mixer installed over a pond for salmonids. Newin Hatchery, Madison, Wisconsin, U.S.A.

Fig. 385 Automatic aerator with "whirligig" on the surface of a growing pond. Ma'agan Michael fish farm, Israel.

Fig. 386 Storage pond with automatic aeration under water, functioning discontinuously. San Samuel Kibbuts, Israel.

The chemical compound is manganous hydroxide $Mn(OH)_2$. A modified Winkler method is used. After the addition of potassium hydroxide (KOH) to obtain an alkaline solution, the dissolved oxygen reacts quantitatively with the manganous hydroxide.

In a medium acidified by the addition of sulphuric acid (H_2SO_4), the hydrated manganese dioxide formed (MnO_2H_2O), liberates the iodine of an iodide (KI: potassium iodide). The liberated iodine can then be determined by the normal method with the aid of a titrated solution of sodium thiosul-phate ($Na_2S_2O_3$) using starch as an indicator. It is possible to buy very practical kits for the chemical determination of dissolved oxygen.

It is also possible to determine the dissolved oxygen approximately. Before adding the H_2SO_4 a precipitate is formed which grows darker as the dissolved oxygen content is high. The precipitate is then compared with a colourscale and, in this way, the dissolved oxygen content is estimated.

Electro-chemical instruments, of which numerous models are commercially available, are rather small and provide rapid and accurate measurements of temperature and dissolved oxygen. However, they require regular tuning-up and maintenance.

The determination of the oxygen consumption is important for establishing the degree of pollution of the water. As far as pollution is concerned the B.O.D. (the biochemical oxygen demand) should be pre-determined. This test generally takes several days.

III   Control of the pH of the Water in Ponds

1   Notion of pH

Like any other liquid, water can be acid, neutral or alkaline. It is not only possible to determine the nature of its reaction but also to measure intensity by determining the pH.

One must know that water, even when it is chemically pure, is weakly dissociated into hydro-gen ions H^+ (positively charged) and hydroxyl ions OH^- (negatively charged).

The water used in fish cultivation is not chemically pure and contains, in solution, different substances which give it an acid, neutral or alkaline reaction. The intensity of these characteristics is measured by determining the exact quantity of H^+ ions. Quantities are always very small and expressed in very small numbers. It is agreed for convenience to use simply the logarithm of the number of hydrogen ions, without the negative sign. This hydrogen exponent is called the pH, according to the notation of Sørensen.

The pH values vary between 0 and 14. If the pH is equal to 7·0 the water is called neutral. If it is higher than 7·0 the reaction is alkaline; it is acid if the pH is lower than 7·0.

2   Determination of pH

Many methods are used to determine the pH. The most accurate and the fastest ones are electro-metric methods which call for relatively expensive apparatus. The manipulation of these is relatively simple. However, the current needs of fish cultivation can be met by colorimetric methods, although only approximately. A simple inexpensive apparatus, which can be used on the farms is needed.

Among the colorimetric methods are the Hellige or the Czensny comparator (Fig. 387). The latter is more fragile but easier to handle in the field.

Two or three drops of Czensny reagent (Merck indicator with phenolphthalein) are taken in a glass

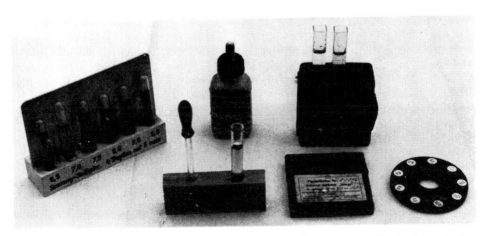

Fig. 387   On the left, Czensny comparator and on the right, Hellige comparator for the determination of pH.

tube containing 5 cm³ of the water to be analysed. After mixing it is compared with coloured tubes giving per 0·5 unit colour types from 4·5 to 9·0. The intermediate values are then estimated by approximation. This apparatus permits determination of the pH to within an approximation of ± 0·1 of a unit, which is quite sufficient for current fish cultivation. Using two measuring discs the Hellige apparatus allows the pH evaluation from 4·0 to 8·0 and from 7·0 to 11·0.

3   The importance of pH to fish cultivation

The pH is the result of the interaction of numerous substances in solution in the water and also of numerous biological phenomena. What counts most is the stability or the variability of the pH. This is of great importance, for most of the aquatic organisms are adapted to an average pH value and do not easily withstand sudden and strong variations of the latter. Biological conditions are better when the pH of the water is quite constant rather than in water undergoing considerable variations.

The best water for fish cultivation is that which is neutral or slightly alkaline, with a pH between 7·0 and 8·0. The more the alkalinity is high the less the pH varies. This point is now further examined.

Marshy and peaty water with weak alkalinity have an acid pH. The same is true for water springing from a soil poor in calcium. If the water is very poor in this mineral then the fish risk death from a sudden drop in the pH. This can happen, notably, in winter following abundant rain or through the sudden melting of snow. For most fish the decisive pH value is around 4·5. When there are great quantities of iron it cannot fall below 5·5.

As opposed to this an excessive increase in the pH can occur, not only through the flow of effluents with high alkalinity but also through a natural process following assimilation by the plants of all the carbon dioxide reserves in a water with low alkalinity. According to Bandt (1936) the top dangerous limit of pH varies according to the species, between 9·2 for trout and 10·8 for pike, carp and tench.

4   Possible ways of controlling pH in pond water

pH varies according to a number of chemical and biological factors. It is closely related to alkalinity as explained above.

In water with a low alkalinity, liming (Section IV) will permit the latter to be increased and ensure sufficient pH stability.

IV   Control of Methyl Orange Alkalinity (SBV) in Pond Water

1   Notion of SBV

Different alkaline elements indispensable for fish cultivation are found in solution in the water. The most common are calcium and magnesium, the first of these normally being preponderant. It is found principally in solution as calcium bicarbonate in the water.

The determination of the alkalinity in the water covers all the carbonates and bicarbonates of alkaline and alkaline earth metals in solution. In practice it almost equals, in normal waters, the quantity of calcium bicarbonate.

The definition given by the Germans to methyl orange alkalinity is "Saürebindungsvermögen" which is abbreviated as SBV. This is also used in works on fish cultivation written in French. One unit of SBV corresponds to 1·0 milliequivalent (1·0 me) per litre of water or to 50 mg $CaCO_3$ per litre or also to 28 mg of CaO per litre. By multiplying the SBV value by 2·8 or by 5·0 it is possible to obtain the calcium hardness of the water expressed respectively by the German or French degrees of hardness. The determination of the SBV is equivalent to that of the calcium hardness of the water, although expressed in another form.

The higher the alkalinity the more stable is the pH. These regulation phenomena are complicated.

2   Determination of SBV

Determination of SBV (methyl orange alkalinity) is performed by a titrimetric method by adding to a given quantity (100 cm³) of water a strong mineral acid, such as hydrochloric acid (HCl), chosen by convention. The acid is added right up to termination of the reaction which takes place the moment all the bicarbonate is neutralized. This happens when the pH falls to 4·4. There is then a change in the colour of the methyl orange indicator added previously. It changes from a yellow to a rose colour.

The following appliances are necessary for determining the SBV: (1) a graduated glass holding 100 cm³, (2) a burette with 0·1 N(1:10 normal) HCl, (3) a dropper with a 0.1 per cent solution of methyl orange, (4) a flask containing about 200 cm³ serving for titration.

100 cm³ of the water to be analysed is poured into the flask and three drops of methyl orange solution are added. By using the graduated burette at levels

Fig. 388 Czensny apparatus for the determination of the alkalinity (SBV).

of 0·1 cm³ and shaking the titration flask constantly, the HCl will fall drop by drop until the yellow methyl orange turns a rose colour.

The Czensny appliance for titration is easy to handle (Fig. 388) since it combines in very practical form: a graded titration flask at 100 and 50 cm₃ and a burette with a tap containing 5 or 10 cm³ 0·1N HCl. A graduated cylinder is not necessary.

The methyl orange alkalinity is expressed by the number of cubic centimetres of HCl normally used per litre of water analysed (x cm₃ HCl/l). The use of 0·1N HCl for 100 cm³ corresponds to the use of normal HCl for 1 litre of water. A SBV of 1·0 corresponds to 2·8 degrees of hardness of the German scale or 5 of the French scale (1 degree of French hardness = 0·56 German).

3 The importance of SBV to fish cultivation

This is of great importance from the fact that if the SBV is high it assures the stability of the pH. Up to a certain point, and as far as it might be established below 3·5, productivity will improve at the same time as the SBV increases.

Waters with SBV below 0·1 should be considered very poor; those between 0·1 and 0·3 as poor; between 0·3 and 1·5 as average, and an SBV above 1·5 as rich. Water over 3·5 may be considered somewhat less good because of the calcareous fur which forms and counters the development of the biological cover.

The pH and the SBV should be controlled regularly in acid water arising from moorland or marshes. In alkaline regions sheltered from important fluctuations of pH a very regular control is not indispensable. The pH must, however, be controlled again when the water or the bottom of a pond has been treated with quicklime (disinfection, liming) for the pH can then exceed the value which fish can tolerate.

4 Possible way of modifying SBV in pond water

It is principally by liming ponds (Section IV) that the alkalinity (SBV) of the water can be increased and brought to a level considered the best.

SECTION IV

LIMING OF THE PONDS

1 The action of liming

Liming is part of the maintenance for ponds. It has a varied and favourable action on the health of fish on the one hand, and on the biological factors of production on the other. This action varies and has been excellently described by Schäperclaus, a résumé of which is given.

1. The use of quicklime or calcium cyanamide for liming the pond bottom has an anti-parasitic action. It destroys the parasites in the water or the infected fish. The parasites are destroyed at the bottom or in the water either directly or as encysted organisms. They can also be destroyed at intermediate stages in their temporary hosts. Liming also destroys for a short time algae and aquatic plants where their roots are not too deep. Finally it destroys water insects and their larvae which are enemies of the fish.

2. One of the consequences of the liming of acid water is an increase in the pH to more desirable levels – that is to say, slightly alkaline (pH = 7·0 to 8·0). It is at this level that the fish enjoy the best health and it is also a preliminary necessity if the biological cycle of the ponds is to remain under ideal

conditions and the intensification of fish production is to be effective.

3. The consequence of liming is to increase the alkalinity (SBV). If the latter is higher than 0·5 to 1·0, the pH will not be subject to strong variations either up or down. There will be a reserve of CO_2 sufficient to avoid biological decalcification and to permit assimilation by the plants. There is always sufficient calcium on which the vegetation can live and also for the shells of molluscs and crustaceans. The calcium present in sufficient quantities neutralizes the harmful action of magnesium, sodium and potassium salts.

4. The liming of the bottom produces quick and considerable improvement. This improvement comes from the liberation of bases, the neutral reaction, the increase in biological activity, the acceleration of decomposition of the mud and its cellulosic components, and the obtainment of crumb structure. At the same time, because of the mineralization of the organic matter, the risks of spreading certain parasitic and bacterial diseases is reduced. On the other hand, for the same reason, the danger of an oxygen deficit is limited.

5. Liming brings about the precipitation of excessive organic matter in suspension in the water. For the same reasons as given above, the danger of spreading certain diseases is reduced and there is less risk of an oxygen deficit as the content will, indirectly, be higher.

6. The nitrification of ammonium compounds into nitrites and nitrates demands the presence of sufficient quantities of lime.

2   When should liming take place?

Liming is indispensable:

(a) whe the pH of the water is too low;

(b) when the alkalinity (SBV) is too low;

(c) when the bottom is too muddy or neglected (through ponds not being regularly dried out each winter);

(d) when the organic matter content is too high and there is a danger of lack of oxygen;

(e) when there is a threat of, or contagious diseases are noticed. Preventive measures which should be taken include systematic liming after emptying of the growing ponds in which rearing has been intensive.

As far as the first two are concerned (acid water and low alkalinity) liming is indispensable as a preliminary measure before all other measures are taken in order to maintain and improve the ponds.

It is particularly useful before fertilizing the water.

As the purpose of liming is to improve the bottom it will only be effective if the soil is covered with a layer of mud.

The effects of liming on the production of a pond will be small if the bottom is already rich in lime and if the pond water is calcareous. Liming can even be harmful in water which is very rich in calcium, for under these conditions phosphorus forms insoluble calcium phosphate which deposits on the bottom.

3   Fertilizers containing lime

The principal calcareous fertilizers to be used are: powdered limestone, marl, quicklime, caustic lime, and calcium cyanamide.

1. *Powdered limestone and marl.* These substances contain calcium in the form of calcium carbonate ($CaCO_3$, agricultural lime), which is practically insoluble in water. The carbon dioxide (CO_2) transforms it little by little within 1 to 2 months into soluble calcium bicarbonate [$Ca(HCO_3)_2$].

The powdered limestone contains about 90 to 95 per cent of $CaCO_3$; the marl contains $CaCO_3$ in variable proportions, the maximum being between 80 and 90 per cent. As 100 parts of $CaCO_3$ are the equivalent of 56 parts of quicklime (CaO), it is easy to see that the products carrying the calcium in the form of $CaCO_3$ have an action equal to one half of the efficacity than when they are in the form of CaO. In consequence double the amount must be used.

Limestone and marl are that much more effective if they are finely ground. The grains should be less than 1 mm in diameter.

Calcium is used in the form of $CaCO_3$ when the purpose is to control the acidity of the water (pH and SBV too low) or when the use of quicklime is a danger to the fish.

2. *Quicklime.* When fresh, quicklime (CaO) has a more or less strong and toxic action according to the intensity of the liming and the low alkalinity found initially in the pond. If it is low, the pH can rapidly reach and even pass 10·0; but if it is high it will not possibly pass 9·0.

The quicklime combines with carbon dioxide of the water or the air according to whether the water or the bottom is limed, and transforms rapidly into carbonate which deposits on the bottom of the pond. In 1 or 2 months it is transformed into bicarbonate.

Quicklime is found in two forms, either in lumps or as a finely ground powder.

Quicklime in lumps can only be used for the preparation of lime milk which, if it is used immediately, is particularly effective for disinfecting and killing parasites in small ponds.

Finely ground quicklime is used just as much for liming as for the destruction of enemies of the fish or against diseases. It is also used for liming heavily mudded pond bottoms and to produce the precipitation of an excess of the organic matter in suspension in the water.

3. *Caustic lime*. This, also called slaked lime or powdered hydrated lime [$Ca(OH)_2$], is obtained by stirring water on quicklime or leaving it exposed to the air.

Caustic lime is used commercially. It contains at least 65 per cent CaO. It is often incorrectly considered as quicklime – probably because it is relatively toxic for fish.

It is possible to prepare caustic lime from lumps of quicklime which are first broken up into fragments about the size of a fist. The quicklime is then spread out in layers of about 15 cm (6 in) and sprayed successively at a ratio of 12 litres (25 pt) of water per 100 kg (220 lb) of quicklime. The heaps are covered with earth and the quicklime transforms into powdered lime. This is always a fine powder.

4. *Calcium cyanamide* ($CaCN_2$) comprises 60 per cent calcium cyanamide and 17 per cent CaO. Its action is toxic not only because of the quicklime it contains but above all the cyanamide. The former remains toxic over 2 or 3 weeks and the latter for several months. The use of this product is advisable for the control of encysted and very resistant parasites such as those of whirling disease in salmonids. Great care is therefore necessary, and it is important to ensure that none of the water which it has been used to treat should enter other ponds without its having been completely neutralized.

4   Methods of liming

Liming can be carried out in three different ways: (1) liming the bottom of a dried pond, (2) liming the pond water, and (3) liming the water flowing into the pond.

In general, one or another of these methods can be used indifferently. Certain cases, however, demand a particular method. If the problem at hand is the control of gill rot by provoking the precipitation of organic matter, it will be necessary to lime the water in the pond. If it is a matter of controlling parasites or improving the soil then the latter should be limed. Liming the soil should. preferably, be carried out in the autumn and the bottom should be damp.

At the time of liming, the lime should be spread as uniformly as possible over the complete surface either of the bottom or of the water.

The spreading of powdered quicklime or of caustic lime must be carried out with all the usual precautions. Protective clothing and goggles should be worn. Those parts of the body which are unprotected should be covered with grease. Distribution should not be carried out against the wind. Care must be taken not to leave lumps of quicklime about for they will be active for a long time particularly within the mass and could cause, ultimately and even as long as a year after use, disagreeable surprises such as the killing of fish when they are moved about or split open. The spreading of lime milk is carried out with the aid of a watering can when treating small areas. Careful disinfection demands two applications with an interval of 8 to 15 days between them. Disinfection is carried out in the autumn after the drying, or in the spring. It should not be carried out in rainy weather in order to avoid the liquid being washed about. After spreading quicklime, there should be an interval of from 10 to 15 days before restocking the pond.

1. *Liming pond water*. A boat is used. If the operation is with ground limestone no special precautions need be taken. If quicklime is used it can be distributed up to 200 kg per hectare (220 lb per acre) per day even over several days. Nevertheless, it is necessary to keep an eye on the pH variation which might result, especially when the water is poor in calcium. The pH should not exceed 9·5.

2. *Liming the pond bottom*. The quantities to use for liming the bottom of a pond are very variable and depend on the aim and the nature of the soil.

1. If it is for the control of parasites then 1,000 to 1,500 kg (1 to 1½ tons) of CaO or 1,000 kg of calcium cyanamide (Fig. 391) are used per hectare. The product is spread over soil which is still damp.

2. If it is to improve the soil (Fig. 390) before using other fertilizers, the ratio is 200 to 400 kg (440 to 880 lb) of CaO per hectare provided the pond is not acid.

3. If the aim is to increase the pH and the alkalinity of an acid pond the quantities used vary considerably according to the degree of acidity

Fig. 389 (left)   A lime shed on the banks of a pond. Draganici carp farm, Yugoslavia.

Fig. 390 (below, left)   Brood fish pond limed before being put under water. Zdencina carp farm, Yugoslavia.

Fig. 391 (below, right)   Salmonid pond disinfected with calcium cyanamide. Linkebeek fish culture station, Belgium.

Fig. 392   Lime mill. Mirwart fish culture station, Belgium.

and the nature of the soil. In principle, 200 kg (440 lb) of CaO per hectare should be enough to increase the SBV by one unit, but it is also necessary to increase the pH of the bottom so that the improvement is constant. This can be done more or less empirically by using 2,000 kg to 250 kg (2 tons to a ¼ ton) per hectare according to the pH value from 4·0 or even less to 6·0 and according to whether the soil is heavy or light. It is more logical to control the alkalinity after liming in order to obtain an SBV of 0·5 to 1·0 and even if possible 1·0 to 2·0. For poor ponds in which these values are to be obtained, Schäperclaus estimates 1,000 to 1,500 kg as sufficient.

In poor ponds which have a thick coat of mud (20 to 40 cm) (8 to 16 in) an appreciable quantity of lime can be distributed. Only small quantities are used when the bottoms are sandy and there is only a thin coating of mud.

3. *Liming the water flowing into the pond.* This avoids the necessity of spreading. A lime mill (Fig. 392) placed in the water channel is used. It comprises a funnel into which the limestone or the caustic lime is poured. The bottom of the funnel can be regulated. A water wheel turned by the current drives an endless chain connected with a hammer which strikes the outside of the funnel, causing a certain quantity of lime to fall through the funnel at each blow.

The mill can be operated by a water current of 1 litre per second. It uses from 50 to 1,000 kg (110 lb to 1 ton) per week and it should be easy in this way to double or even triple a SBV of 0·5. Consumption can be regulated by modifying the opening and the working speed which depends on the water flow admitted to the mill.

These mills have done great service by increasing the alkalinity of acid waters which are dangerous for the fish especially if they are used to feed wintering ponds for cyprinids.

SECTION V
FERTILIZATION OF THE PONDS

I General Notes

1 Aims and action of fertilization

Apart from artificial feeding, fertilization provides the best means of increasing fish production in ponds. It is also the simplest and most economic way of achieving this. It assures a more hygienic intensification of production than artificial feeding. Fertilization increases production without the risk of dietary diseases. On the contrary, it improves the hygienic condition of the pond. The use of fertilization is as important for fish cultivation (exploitation of ponds) as it is in agriculture (exploitation of farm land).

The aim of fertilization is to aid the production of fish by increasing the quantity of natural food present. The latter controls, as is generally known, the productivity and stocking of most ponds exploited extensively or semi-extensively. Now the quantity of natural food depends on the production of vegetation, which depends in its turn on the quantity of nutrient substances present. In the final analysis, provided there is enough heat and light, fish production depends on the relative quantity of inorganic nutritive substances available (Fig. 60).

Liebig's law of minimum intervenes here. The basis of total production in plant life depends on an indispensable nutrient when it is present in a minimal amount, taking into account its relative utility. The aim must be to achieve an optimum relationship between the various necessary nutrients. Experience has shown that phosphorus, potassium and nitrogen are generally found in ponds in minimum quantities, and from this fact springs the necessity of using fertilizers containing them. The practice is specially recommended for ponds with poor soil and water. By intensifying their productivity, they can turn unprofitable ponds into profitable ones.

It is, above all, due to their favourable action on the mud which, in truth, is the production laboratory of the pond, that the fertilizers act and not directly on the water mass or the organisms in suspension. A satisfactory action of the bottom determines the productivity of the pond. Mud in a good state of decomposition contains within it nutrients which it liberates according to the needs of the plants by dissolving them little by little. The fertilizing substances in the fertilizers are therefore absorbed by the mud which returns them

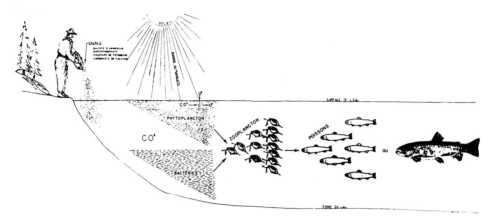

Fig. 393   Schematic diagram showing the action of fertilization (after Prevost, 1946).

little by little. This explains the prolonged action of fertilizers. Animals, therefore, have no direct connection with the mineral nutrients, that is to say with the mineral fertilizers. As for the aquatic vegetation (from a fish cultivation point of view, the best of them, have for the most part slightly developed roots), the mineral matter cannot be used except as a solution in the water. The same holds true for the phytoplankton.

Mineral fertilizers are recommended for all ponds in which natural fish raising is practised extensively and semi-intensively, and especially for ponds containing brood and young fish. They can only be usefully used in a pond which is not traversed by a strong current unless the latter can be regulated. For sanitary purposes Walter also recommends the use of fertilizer in intensive growing ponds, for they encourage the decomposition of excrement and alimentary residues.

A new pond placed under water generally gives a high production which diminishes rapidly until it reaches a lower level. It will from thence on vary scarcely at all, and then only according to atmospheric conditions and the kinds and quantities of fish species liberated, etc. Its high initial production will have been due to the nutrients in reserve in the soil used and exhausted more or less rapidly.

2   Determining the quantity of fertilizer to be used
The type of fertilizer and the quantities to be used differ considerably from one region to another and even between different farms. Fertilization must be carried out only to compensate for those substances found in too small quantities. Nevertheless there are limits to the growth of productivity due to

fertilization by a single mineral substance for this will be followed by another substance falling to a minimum. A chemical equilibrium will become established in a pond and if, for example, too much phosphate fertilizer is used it can form iron and aluminium phosphate precipitates. Physical factors such as temperature and light can also limit the increase in production. Finally, economic factors must be taken into account. Fertilizers should not be used when cost is equal to or is greater than the extra production obtained.

Types of fertilization suitable to Europe have been established thanks to tests made by two German fish farm stations – one at Wielenbach and the other at Sachsenhausen. The tests showed that many of the findings resulting from this research could be used for other soils and climates. Nevertheless there are still many points to be clarified and each farmer should determine for himself the quantities of fertilizer to be used on his own farm.

To this end, the results to aim for must be to establish and follow rules laid down by Schäperclaus (1933), though they may be difficult to achieve.

1. The test ponds must not be too small. They should all be the same size and should have been exploited for at least 3 years under identical conditions and should have a similar productivity.

2. Natural productivity should be known for several years.

3. Only some of the ponds should be fertilized; others should not.

4. Artificial feeding should not be practised.

5. The tests should include normal stocking and the normal growth of the fish should be assured. In consequence, the stocking should take into account

the fact that the total of the expected production is equal to the natural production plus a supplementary production due to fertilization (in general 50 to 100 per cent of natural production).

The tests should be carried out with fish of the same size belonging to a single species of the same race, and if possible coming from the same line of descent. Schäperclaus recommends as subjects a mixture of 1 year (K_1) and 2 years (K_2) carp. On this basis he suggests the following plan for the experiment:

Pond 1: No fertilization.
Pond 2: 40 kg/ha (88 lb per hectare) P_2O_5.
Pond 3: The same as 2 + 50 kg/ha (110 lb per hectare) K_2O.
Pond 4: The same as 3 + 500 to 1,000 kg/ha ($\frac{1}{2}$ to 1 ton per hectare) CaO or 50 kg/ha N.

3 General rules for using fertilizers

a Prerequisites

The following conditions must be met if the use of the fertilizer is to have a favourable effect.

1. The water and the soil must have a neutral reaction or be slightly alkaline. It is through the absorption facility that the soil plays its part. Soil showing an acid reaction has a reduced absorption capacity. If the soil and the water in the pond are acid they must be limed before fertilizer is used.

2. The bottom must be covered with good quality mud rich in colloids, not too thick and constituted, principally, by fine detritus of submerged plants and algae. Bad mud resulting from emergent plants is too rich in cellulose and decomposes badly. It gets too thick and is only slightly productive.

3. Vertical vegetation must be disposed of by repeated cutting or be treated with herbicides. If reeds are left they will compete indirectly with the fish for food, using for their own growth the fertilizers placed in the water and to the detriment of the nutritive organisms of the fish for which they were intended. When this vegetation is present the parts of the pond in which they are found should not be fertilized. The floating and submerged vegetation should be kept in such proportions that it will not interfere with the penetration of light and heat.

In many warm water fish ponds the essential aim of fertilization is the development of the plankton.

b Rules for the distribution of mineral fertilizers

1. Fertilizers are used at the time of restocking, on the bottom of the still dry pond, or immediately after it has been put under water. In the latter case a boat, preferably equipped with a motor, should be used in order to ensure uniform spreading (Figs. 394 and 395). The fertilizer must be very fine. The amounts used normally should not be harmful for the fish.

2. Mineral fertilizers can be used in one or several spreadings.

It has been found that in certain cases a single application is better than several applications of small doses regularly spaced. With a single application the soil, in the spring, contains nutritive reserves which the aquatic vegetation particularly needs. Nevertheless it is often advisable, under these circumstances, to apply a further dose of the fertilizer during the summer as soon as the phytoplankton starts disappearing.

In several parts of Central Europe, the United States and Israel, fertilizers are applied in small doses spaced out regularly. This method is to be preferred when the ponds have sandy bottoms and little mud.

Organic manure is always distributed several times in small quantities.

3. To keep costs down it is possible to mix, just before use, basic slag and potassium fertilizer. Lime or fertilizers rich in calcium (basic slag, Rhenian phosphate) should not be mixed with ammonium sulphate and organic manure rich in ammonium ions (dung, liquid manure). Eight to 15 days should be allowed between the spreading of superphosphate and the lime, as the latter makes it difficult for the former to dissolve.

Fertilizers which dissolve easily (superphosphates) can be spread as soon as the water starts to warm. For nursing ponds the fertilizer is spread 2 or 3 weeks before stocking in order to help the development of the natural food.

II Phosphate Fertilizers

1 Action of phosphate fertilizers

Phosphate fertilizer is by far the mineral fertilizer with the most favourable and effective action in fish cultivation. It seems that nearly all waters lack phosphorus, and consequently mineral phosphate fertilizer is nearly always worthwhile. This is the case above all for ponds which are rich in lime with mud in a good state of decomposition. The favourable action of phosphate fertilizers can often be seen with the naked eye as the water turns green as a result of the multiplication of certain algae. Often "water bloom" is seen to form due to the appearance of an irregular green coloured film (Fig. 399) made up of unicellular algae, the prolification of which in the water is caused by the fertilizer. The algae are floating on the surface. The wind and the waves destroy this formation which, in any case, is only temporary.

The increase in production due to phosphate fertilization varies frequently between 50 and 125 per cent. The annual variations depend on numerous circumstances, notably atmospheric ones. The better the summer, and particularly the spring, the better the action of the phosphate fertilizers. Walter, who carried out tests with phosphate fertilizers at Wielenbach (Bavaria) over a period of 20 years, reached the conclusion that 1 kg (2 lb) of P_2O_5 will give an average production supplement of 2·13 kg ($4\frac{3}{4}$ lb) of carp.

The action of the phosphate fertilizer is noticeable not only during the year in which it is applied, but also during the following year. This late action varies considerably and depends on the atmospheric conditions and the nature of the bottom. In practice it is best to fertilize each year.

The fertilizer being held in the soil and liberated gradually, according to requirements, makes it possible to fertilize salmonid ponds in which the water is constantly renewed. The production supplement is variable. To avoid the phosphorus being carried off it is a good idea to stop the renewal of the water during 5 days after spreading, especially if superphosphate is used.

2 Types and quantities of phosphate fertilizers used
Different phosphate fertilizers have, approximately, equal value: superphosphate, basic slag, Rhenian phosphate, di-calcium phosphate. The use of one or the other depends among other things on the state of the market. Phosphorus from basic slag, which contains at the same time 40 to 50 per cent CaO and other nutritive elements (magnesium, manganese, cobalt, etc.) dissolves less easily. This fertilizer is particularly suitable for acid or light soil, or for water which is poor in lime. Superphosphate, where the phosphorus is most soluble, suits heavy soil and water which is naturally rich in lime. Other fertilizers have intermediate qualities.

The optimum quantity seems to be 30 kg (66 lb)

P_2O_5 per hectare, which corresponds to 100 to 200 kg (220 to 440 lb) of the fertilizer given above (an average of 150 kg (330 lb)). But it is also the minimum, for smaller quantities are not very effective while very much larger quantities are not economic. However, certain authors advise quantities of up to 60 kg (132 lb) of P_2O_5.

III Potassium Fertilizers

1 The action of potassium fertilizers
The favourable action of potassium fertilizers has not been clearly established. It is in any case doubtful for average and high productivity ponds, for potassium is generally present in sufficient quantities in the soil and in the water supply. On the other hand it is satisfactory in the following cases: (1) for ponds poor in potassium, (2) for ponds with a low alkalinity, (3) for ponds in moorland and peaty areas, (4) for hard bottom ponds poor in aquatic plants. This is in consequence the general rule for all poor ponds. Even if it is not directly economic, potassium fertilizer is good for rearing ponds for it develops the natural food and improves the hygienic conditions.

Potassium fertilizer seems to cause a reduction in vertical plants which are harmful, in favour of submerged plants which are beneficial.

The extra production due to potassium fertilizer varies considerably.

2 Types and quantities of potassium fertilizers to use
The different types of potassium fertilizers appear to have the same action but preference is often shown for kainit which contains other elements such as magnesium salts. On the average, Walter suggests 30 to 40 kg (66 to 88 lb) of pure potassium (K_2O) (200 kg (440 lb) of kainit per hectare, for example) except on moorland and peaty ground which are very poor in potassium, where he recommends double the amount. Potassium fertilizers can be mixed with phosphate fertilizers.

IV Nitrogenous Fertilizers

After tests in Europe it is still not clearly established whether mineral nitrogenous fertilizers are economic. It has, however, given extra production which can reach 50 per cent used either alone or combined with phosphate or potassium fertilizers. According to Wolny (1967), it is above all the relationship between the phosphorus and the nitrogen which is important, and the best P/N relationship is 1/4. It is the phosphorus deficit which stops the full use of the nitrogen present in the water. In well mineralized water with an alkaline bottom the P/N relationship can reach 1/8.

The employment of a nitrogenous fertilizer is advised for new ponds which are either poor or without mud. If the pond has a good layer of colloidal mud it will produce nitrogen itself and the use of a nitrogenous fertilizer will be unnecessary.

However, it is not necessary that the use of a fertilizer should be directly economic to be advantageous. In effect, its indirect advantages could be sufficient to justify its use, and the principal of these is improvement in the health of the fish, a particularly important factor in nursing ponds.

The different nitrogenous fertilizers: sodium nitrate and ammonium fertilizers have, however, shown themselves to be almost identical in results. The quantities of fertilizer normally used in Europe correspond to 50 kg (110 lb) of pure nitrogen per hectare.

In Israel the use of nitrogen in the form of ammonium sulphate is current. In that region, with warm temperate water rich in calcium and relatively rich in potassium, the phosphate fertilizers give a very interesting production supplement. Nitrogen also has an effect which is by no means negligible. Generally 60 kg (132 lb) of superphosphate and 60 kg of ammonium sulphate are distributed per hectare every 2 weeks during the 7 or 8 months of growth (Figs. 394 and 395).

In Louisiana (U.S.A.) where, as in Israel, the summers are long and hot, nitrogenous fertilizers are currently used. According to Summers (1963) they are used at each distribution at a ratio of 100 kg (220 lb) per hectare and as a mixture containing 8 per cent N, 8 per cent P and 8 per cent K. This fertilization starts at the first signs of the water rewarming and is continued during the summer and autumn. It is distributed at intervals of 1 week to 10 days until the water becomes green or brown following the development of the plankton. The use of this compound fertilizer is stopped when that goal has been reached and is used again when the water becomes clear.

V Micronutrients

A sufficient number of satisfactory experiments has not been carried out to demonstrate the use of micronutrients in fish cultivation. Certain fertilizers which are currently used, e.g. Thomas slag and kainit, contain micronutrients in quantities which are by no means negligible.

VI Organic Manure

1 The action of organic manure

Organic manure gives a favourable action well established by the high production of ponds fed with diluted liquid manure from farms or agricultural communities (Fig. 397).

This favourable action is due to the fact that organic manure: (1) brings with it nearly all of the nutritive substances indispensable for the biological cycle; (2) at the same time it often has a favourable action on the structure of the soil; (3) it favours the multiplication of bacteria in suspension in the water which have a favourable action on the development of zooplankton if it is not excessive; (4) organic matter is indispensable to the action of phosphate and potassium fertilizers.

Organic manure, however, presents certain inconveniences. (1) There is a risk it might cause an oxygen deficit, above all in the early hours of the morning and in warm weather. The ponds in which it is used must be watched attentively. The risk is high, especially, in salmonid ponds which are only suited to this kind of manure on exceptional occasions. (2) It favours certain diseases (gill rot).

2 Practice and usage of organic manure

Most of the organic manure which might be used is often found only in small quantities. This manure must be reserved for ponds holding the most important age groups, that is, ponds with fry and young fish.

In principle, organic manure is distributed frequently in the water from once to three times weekly and in small quantities at any one time. The organic fertilizer is best distributed evenly and regularly in several specified spots along the banks. If possible it is best spread more or less uniformly over the surface of the pond.

3 Organic fertilizers used

The best are those which will be in a fairly advanced state of decomposition when suspended in the water.

Liquid manure is distributed in small doses once or twice every 8 days. It causes the plankton to reproduce and multiply but the ammonia it contains is a dangerous poison. Each distribution should comprise 1 m³ (1⅓ yd³) of liquid manure per hectare in the first rearing cyprinid ponds.

Faecal matter and abattoir waste are distributed in small quantities at a time.

Dung and compost can be deposited at a ratio of 20,000 to 30,000 kg (20 to 30 tons) per hectare on the bottom where it can be incorporated before the pond is placed under water (Fig. 396). It will help development of the cladocerans. Dung should not be spread evenly, as this can have a harmful effect on the biological activity of the soil, but in heaps or in rows. It is only spread uniformly if the soil is virgin and sterile, and in need of a fertile coat of colloidal mud.

In nursing ponds the production of plankton is helped by placing here and there heaps of dung and compost. This is done notably when fingerling carp (Fig. 209) and fingerling pike are reared.

4   Different types of organic manure

Fertilization with organic manure is capable of giving good results when the ponds are fed with sewage water. This manure, when used, demands special conditions. At Munich, over a total surface of 200 hectares (500 acres), sewage water is used in fish ponds in a ratio of 2,000 inhabitants per hectare. This sewage water which cannot contain toxics, is first cleaned mechanically and aerated at the inlet of the pond where it is mixed with fresh water at a ratio of at least 1 to 3. The production of these ponds, without artificial feeding, is on average 500 kg (½ ton) per hectare.

Manuring with the aid of cut water plants, particularly submerged and young emergent vegetation left where it was cut or heaped up to decompose slowly, also gives good results. In the U.S.S.R.

mechanical weed-cutters are provided with blades which cut the emergent vegetation up into small pieces.

Manuring with cut cultivated plants can also be practised. This is current in rice-fish cultivation. When the harvests are alternate the stalks of the rice are laid out in heaps before stocking with fish (Fig. 314).

From the time of drying, good organic manure can already be provided by allowing pigs access to the bottom of the pond.

The raising and feeding of ducks on ponds (Fig. 400) produces good organic manure. It is estimated that the extra fish production amounts to ½ kilo (1 lb) of fish per duck. Two hundred and fifty ducks per hectare of water are the maximum in order to avoid the inconveniences of too much organic manure. The regular drying of these ponds is necessary.

The alternate animal and vegetable cropping can also be considered as a kind of organic manuring because of the remains of precedent vegetable crops or the manure used for them which benefits, indirectly, fish cultivation. It is the same for plants (cereals, rice, leguminous) (Fig. 379) cultivated in the pond before filling it, for the young submerged vegetation provides good organic manure in the form of green fertilizer.

The raising of cattle and poultry on the banks of a pond as well as the setting up of latrines and stables near the pond all provide a regular dosage which is sometimes high, and even too high, of organic manure for the pond (Fig. 398).

Artificial feeding of fish also contributes organic matter and constitutes indirect manuring thanks to uneaten food which decomposes on the bottom, as well as undigested food rejected by the fish and mixed with the excrement.

Fig. 394   Distribution of liquid fertilizer from a motor boat. Ma'agan Michael fish farm, Israel.

Fig. 395   Moving a fertilizer distributor mounted on a drawn trailer. Ma'agan Michael fish farm, Israel.

Fig. 396   Working the soil and incorporation of dung. Bussche ponds, Lubumbashi, Katanga, Zaire.

Fig. 397 Village pond enriched with domestic and farm waste water and by raising geese and ducks. Dhlavi, Moravia, Czechoslovakia.

Fig. 398 Fertilization of ponds with organic manure from latrines as well as the droppings of domestic animals. This method of manuring is widespread in Indonesia. Brandjangan, Soerabaya, Indonesia.

Fig. 399   Water bloom (algae floating on the surface of a pond) consecutive to organic manuring. On the right, a feeding installation square. Bangkhen experimental station, Bangkok, Thailand.

Fig. 400   Simultaneous raising of fish and ducks, the latter fertilize the pond. The ducks are fed. Also the central shelter and feeding troughs spread out at regular intervals over the pond which covers 173 hectares (about 430 acres). Biharuga fish farm, Hungary.

Chapter XIII
ARTIFICIAL FEEDING OF FISH

SECTION I
ARTIFICIAL FEEDING IN GENERAL

I Importance of Artificial Feed The Conversion Rate

Artificial food or feeding is one of the principal methods of increasing production in fish cultivation. Its importance varies according to the intensity of cultivation, for the latter can be extensive, semi-intensive or intensive.

In certain types of fish cultivation, for example in intensive salmonid culture, artificial food can be the exclusive basic food of the fish which, from this point of view, become independent of the medium in which they are cultivated.

In traditional cyprinid cultivation it is generally considered best that natural food should represent about one half of the diet and that artificial food should only be used as an addition – although it is very important and economically indispensable. It is possible, however, to raise carp exclusively with dry concentrates containing all the necessary nutritional requirements (Meske, 1966). It is too soon to write that in ponds and on a large scale the artificial intensive feeding of cyprinids will be possible and economical, for trials are only just starting.

The artificial feeding of fish allows high stocking and, from this, the better use of natural food which itself profits from the unconsumed artificial food and the excrements of a denser population – which in turn acts as a fertilizer.

The practice of more or less intensive feeding is simply an economic question. This depends on the cost of the foods used and their conversion rate.

The rate of food conversion or "conversion rate" expresses in kilos the amount of food necessary to produce 1 kilo of fish. It is necessary to distinguish between the "absolute conversion rate" and the "relative conversion rate".

The absolute conversion rate known as the "conversion rate" is obtained by dividing the quantity of food distributed by the extra growth obtained and believed to have been obtained only by that food. No account is taken of error due to

loss through death if it is normal. Nevertheless this "absolute conversion rate" is not exact, not only from the fact that it does not take into account loss through death but also because if food is given the stocking density can be increased. This increase results in a better use of the natural food and in consequence an increase in natural production.

The "relative food conversion rate" is obtained by dividing the quantity of food distributed by total production (production due to natural food plus that due to manuring plus that due to artificial feeding).

The rate of food conversion depends not only on the food distributed but also on a number of other factors such as the density of the stocking, individual weight, age class of the fish, their state of health, the temperature of the water, the methods of feeding (quantity, the spreading and frequency of distribution).

II Principal Foods used in Fish Culture

A Vegetable foods

The principal vegetable foods distributed on a fish farm are: pulse foods (lupin, soya) and grains (maize, various cereals either whole or ground or in the form of meal or bran); leaves of water and land plants distributed to herbivorous fish, as well as several other foods such as yeast.

Lupin. Lupin is one of the best foods to give carp and other farm cyprinids. It is simple to prepare (it is easily ground) and possesses a narrow conversion rate (average, 4), giving fish with good quality flesh.

One inconvenience is that it is light and this makes distribution difficult. It is often too expensive to make its use economically viable. Besides, lupin seed does not keep very well when the winter is wet if it has been harvested before being absolutely ripe.

Soya. Soya competes with, or can replace lupin

as it is, or in the form of waste after the extraction of its oil. It possesses the same advantages as lupin. Soya waste is not as expensive. As it is already crushed no other preparation is needed for young carp. Being heavier than lupin, soya is distributed more easily. Its conversion rate is, on the average, 4.

Other pulse foods: peas and beans can be used for feeding cyprinids, but their price is usually too high.

Maize. Maize is used a lot for feeding cyprinids. It is cheaper than those mentioned above and it can be ground more easily. It has a conversion rate which is slightly less advantageous (average, 5), but without being too much so.

It gives white and rather fat flesh and it does not keep for long. If it is used for fattening its distribution must be stopped a few weeks before the fish is eaten.

Allowing for those factors maize is best suited for fattening third year carp. It is also used for fry more than 5 cm (2 in) long and for second year carp. For fry the mixture is 75 per cent crushed maize and 25 per cent fish meal.

Corn meal can also be used at a ratio of 20 per cent of total food weight for the fattening of trout. It is, however, difficult to bind.

Cereals (rye, barley, oats, wheat). These provide excellent food for carp and other cyprinids and fish of other families which eat grain. Their conversion rate is advantageous: 4 to 6. They give good quality flesh without a disagreeable taste but their use is not always economic.

Meals and brans of cereals, maize and other grain food are good for fattening cyprinids and other fish which feed off vegetable food. Their conversion rate is from 4 to 6. Nevertheless they are that much less satisfactory as they are fine, and a binder must be used in order to obtain a curd or cake after cooking. Meal and bran are used in salmonid cultivation as an extra food mixed up to 35 per cent with animal food (meat, fish).

Leaves of water and land plants. In intertropical regions herbivorous tilapias do well on sufficiently tender land vegetable leaves such as: cassava, sweet potatoes, eddoes, banana trees, pawpaw, maize, canna and different vegetables and grasses (Fig. 288). The conversion rate of the leaves of these plants is not well known but it could be around 15 to 20.

These same land plants as well as submerged aquatic plants, floating duckweeds or stalks and leaves of young reeds can be given to grass carp (*Ctenopharyngodon idella*) (Section IV).

Yeast. Dried yeast which is kept in the form of flour is a rich food with assimilable proteins and B group vitamins. It is a supplementary food of considerable value, as much for salmonid cultivation as for cyprinid cultivation.

Others. Formerly, among others, cooked potatoes mixed with animal meat were used to feed carp. The result is far too watery however, and its use is unsatisfactory. Certain sawdusts, used as fillers and for binding have also served for salmonid cultivation.

B Foods of animal origin

The principal foods of animal origin distributed to fish are: Fish – fresh sea fish, sea fish meal, fresh water fish; shrimps, fresh and dried. Meat – fresh meat; meat meal; fresh abattoir waste; dried abattoir waste; dried blood. Certain organisms of nutritive aquatic fauna. White cheese and other products.

Until a few years ago food of animal origin was the basic food in salmonid culture either in the form of sea fish for farms near the sea, or meat for inland farms. These foods have been replaced more and more by dried concentrates in the form of meal or as pellets. The concentrates are for the most part prepared from fresh animal food mixed with vegetable meals.

Fresh sea fish. Raw and very fresh, this is the best fresh food which can be given to trout for fattening.

Fish used for this purpose are those which have either been injured during handling or stale fish kept in stock. Provided fatty fish are eliminated (herring and smelts) most species can be used, as well as small fish and fish heads.

It is a very good food which, according to the conditions of use, has a conversion rate from 6 to 8 which can rise to 9 in the most unfavourable cases. This food must be eaten fresh and it should not cost too much. These two conditions limit its use to salmonid farms which are not far from the sea or which, at least, have rapid communications with the coast. To ensure a regular supply the best way is to sign a contract with a fishing port sales organization. Fish which are not sufficiently fresh have to be cooked.

Fresh sea fish are used particularly for fattening but they can also be used from the first year. For young fish 2 months old the fish can be mixed with spleen, at 3 months the sea fish can form 50 per cent of the food distributed but this should rarely be more than 60 per cent.

Fish meal. Fish meal is prepared with spoilt fish, fish of practically no value, and heads and waste left over from the production of fish oils. They are cut up, dried, ground and sieved. Drying should not be carried out in too high a temperature otherwise the proteins will be spoiled and will not be digestible.

According to Schäperclaus, good fish meal should not be too fat (less than 3 per cent fat), too salty (less than 3 per cent and even less than 1 per cent for fry) nor have too many bones (less than 30 per cent of calcium phosphate). Meal which does not meet those specifications cause dietary diseases, and especially inflammation of the intestines. Meal made from herring is too fat. Often also meals contain impurities such as sand, etc.

Good fish meal forms one of the best artificial food for fish. The conversion rate varies between 1·5 and 3.

In cyprinid cultivation fish meal can be incorporated at a ratio of 25 to 30 per cent with the food distributed to fry and second year carp. In salmonid culture it provides first class extra food and can become the principal food if other food is in short supply. It should be mixed with vegetable meal. It plays an important part in the preparation of dry concentrates.

In practice fish meal is always prepared from sea fish. Naturally it could be prepared from fresh water fish which have the same nutritive value but it is exceptional to find them in sufficient quantities to make the operation economically viable.

Fresh water fish. Fresh water fish is a good food for trout especially brood fish. Essentially coarse fish (forage fish) (roach, bleak, etc. in Europe) are used. It is, however, only a supplementary food as regular supplies can almost never be secured. After having been chopped, it is distributed, mixed with other complementary food.

Shrimps. Shrimps are one of the most important foods in the cultivation of salmonids. They have a conversion rate of 4 to 6.

Fresh shrimps are a first choice food for brood fish. It is a food closest to the natural food of fish. As shrimps remain fresh for only 2 to 3 days they can thus be used advantageously only on farms which are not too far from the coast.

Dried shrimp is an excellent supplementary food which can be mixed with other fresh foods (fish or meat) and also with dry concentrates. It is rich in minerals and proteins and suits young fish just as it does fish for fattening. It gives good quality flesh.

In order to avoid sharp practices (for example, such as mixing in crab shell) which could lead to a too salt meal and which could cause intestinal diseases, it is best to buy whole dried shrimps and to chop them up one's self with a meat chopper at the same time as the principal food.

Because cost is high their use has been limited.

Fresh meat. Before the generalized use of dry concentrated food and, previously before methods of transport permitted inland farms to obtain easily supplies of fresh sea fish, fresh meat was used to a large extent for inland salmonid farms. Horse meat and cattle were used. Supply came from healthy animals bought in markets, as well as abattoir waste and animals from knackers yards. In the latter case, if the animals slaughtered were diseased then it was necessary to cook the meat to avoid the spreading of contagious diseases. Cooking has the inconvenience of partially destroying the vitamins and of making the proteins difficult to digest.

Fresh meat has, for a long time, been the basic food of inland farms which cannot, economically, use fresh fish from the sea or food of superior quality. Meat has an average conversion rate of 5 to 8 which can increase to 10 for cooked meat. Meat is a food which can be easily minced and which is not quickly diluted when thrown into water. Trout take it easily.

The fattening of trout with fresh meat has two principal drawbacks: it easily causes inflammation of the intestines and it gives only poor quality flesh. Besides, fish fed in this way have abundant excrement, with a lot of sinew, which can cause putrefaction in farm ponds.

These inconveniences can be overcome by avoiding too uniforme a diet. To this end the meat is mixed with fillers such as bran. The diet should be varied and substances rich in vitamins, such as spleen and yeast should be added to the mass of food. To improve the taste of the trout fed in this way it is necessary to suppress the meat diet or reduce it considerably 3 or 4 weeks before the fish are harvested.

Meatmeal. Meatmeal is often prepared from good quality animals. This is notably the case for meals coming from South America. When this is so it can be used in fish culture in the same way as fish meal but its role is far less important. Certain meals brought to too high a temperature (up to 180°) in order to destroy possible disease germs cannot be well assimilated by the fish.

Spleen, liver and other quality slaughterhouse

waste. These were largely used in salmonid cultivation before the employment of concentrated meals.

Spleen is rich in proteins and vitamins. It is used in pulp form. Calf and cattle spleen are preferred to that of pigs, though when mixed with the former, the latter is quite good. It is also possible to use horse spleen. Spleen has a conversion rate of 8 but if account is taken of the loss of food through non-utilization and the loss of fry during raising, the physiological conversion rate is nearer 3.

Spleen can be used exclusively for trout fry up to 4 or 6 weeks but later it is mixed and then replaced gradually with fresh sea fish, fish meal, shrimps and dry concentrates.

Dried spleen mixed in equal quantity with fresh spleen is as good a food as the latter. It is also possible to deep freeze fresh spleen.

If salmonid fry are fed with fresh food, which was the general practice for a long time, spleen is an excellent food for them. It is not always possible to obtain regular supplies and the preparation of spleen pulp takes time. For this reason its cost price is high. It is possible to stock spleen or prepared pulp and to keep it in a deep freeze. It is thawed as required and then used immediately.

Liver is an excellent food for trout fry and can replace spleen advantageously, but it is often too expensive.

Blood. Blood is a good food rich in proteins but very concentrated and poor in minerals.

It can be used raw, or mixed with white cheese for trout fry. When mixed with yeast and spleen, blood provides a good food for fry.

In a mixture with vegetable meal, shrimps or bran it is used to fatten trout. For feeding carp it can be cooked and then passed through a meat mincer. It can also be dried and turned into meal to be used in the same way as other meals. Overcooking must be avoided.

White cheese. White cheese was used for feeding salmonid fry. It is poor in iron and vitamins and its conversion rate is 10 to 15.

It is not used during the first weeks for feeding fry but later mixed with blood, spleen and natural food. This food must be used fresh otherwise it will ferment rapidly and become dangerous.

White cheese is prepared with skim milk curdled with a few drops of rennet or vinegar while its temperature is slightly raised. It is then slightly salted.

Small water animals. A distribution of small natural fresh water fauna rarely pays and can only be considered for young fish of high value such as salmonids and pike. This fauna can be cultivated in a closed environment or harvested from open waters.

Fig. 401   Concrete tanks 4 × 2 × 0·75 m. (about 13 ft × 6 ft × 2 ft 6 in) for the culture of plankton by addition of organic manure. Obrh-Vrhnika fish farm, Yugoslavia.

Fig. 402   Salmonid ponds installed with plankton tanks at their upper part. Gamelyne fish farm, Ljubjana, Yugoslavia.

Only some organisms of the fauna can be cultivated: water fleas, chironomid larvae. They can be raised in small ponds or concrete tanks with more or less stagnant sunny water so that the temperature is relatively high (Figs. 401 and 402). These tanks contain no fish and are only filled with water during the food production period. Before being used they must be disinfected with quicklime.

Before being filled with water, or immediately afterwards, the tanks are well manured with pig or cow manure or with meat and fish meal. A wheelbarrowful of cow dung can be tipped in over every 10 m² (12 yd²). Fresh blood should be added regularly as well as liquid and other organic matter so that the water is coloured a deep green where there are algae, and red where there are daphnia. The daphniæ and chironomid larvæ are harvested with the aid of a very fine mesh net. They are washed in clean water and distributed to the fry in small quantities.

It is also possible to harvest natural food in open water. In lakes and ponds which are rich in zooplankton, the plankton is collected by means of a net drawn by a boat. This system is used notably for pike fry (Fig. 247). It is also possible to harvest in certain water courses if the current is rather slow and the water is polluted (though not excessively) with organic matter. *Tubifex* and chironomid larvæ of the *Thummi* group are found there.

Others. Other foods have been used in fish cultivation. Normally they cannot be obtained in sufficient quantities and they cost too much. For example, frogs passed though a meat mincer and land and water molluscs were given to salmonid fry.

During the good season it is possible to obtain extra natural food for the trout by placing an electric lamp from 10 to 20 cm (4 to 8 in) above the water level at the water inlet. A lot of winged insects gather, fall into the water and are eaten by the trout. In fact this system offers a small supplement to natural food.

C   Dry concentrated foods

The use of dry concentrates as foods is relatively recent in fish farming. It is gradually becoming generalized after having been developed, largely, for feeding cattle and poultry. The rules and methods of preparation for cattle and poultry are also used for the preparation of food for fish and, in fact, many manufacturers make food for different kinds of farming. Generally the makers prepare different foods according to the species of fish and the stages of raising.

First of all the aim was to develop and improve the feeding of salmonids with dry concentrates in order to make salmonid farming independent of other types of food. The purpose above all was to reduce labour costs, ensure regular supplies and easy storage and also to use automatic food dispensers. Tests were carried out in North America and Western Europe. In both, dry concentrates have been used and developed since 1960. There is also a tendency to use concentrated food for cyprinid culture and other types of farming, such as catfish, in which it is becoming normal.

Dry concentrates call for a large variety of products according to local supply possibilities and price, but the trend is towards complex mixtures which satisfy all the food requirements of fish at various stages of rearing. There are fresh and dry foods used for feeding such as meat and offal, fish, blood, fish meat and blood meal, hydrolyzates of fish meal, vegetable meal (soya and algæ for example), yeast, minerals, vitamins and even medicaments.

As for other artificial food the conversion rate of dry concentrates can vary considerably, notably according to the species and the age group of the fish, the density of stocking, the temperature of the water and the methods of distribution. On the whole a good dry concentrate used under normal conditions give a conversion rate of 1·0 to 2·5.

SECTION II
ARTIFICIAL FEEDING OF SALMONIDS

I   Principal Foods

The food distributed in salmonid culture differs according to age and whether the fish are being fattened or whether they are brood fish. Both fresh food and dry concentrates are used.

Formerly, salmonids were only fed fresh food often mixed with meals as a supplement. Feeding with fresh food, the greater part of animal origin, poses certain supply as well as conservation and preparation problems. Also, since about 10 years ago, feeding with dry concentrates has gradually replaced fresh natural food.

Dry concentrates are easier to stock and distribute than fresh food and this leads to important economies. Some installations, such as for example refrigeration and handling sheds are not necessary. Long unpleasant preparation of spleen and meat or fish pastes is avoided. Also, distribution is easier, and especially when mechanized permits an appreciable economy in labour.

The feeding of concentrates give the fish complete, good quality food which is often better than fresh or ordinary dry food, of which most are only partially complete foods. If the farmer wants to use fresh, complete food he must prepare his own mixture. This takes time and is not always easy.

The distribution of the food is better when dry concentrates are used than for fresh paste. In effect, as the distribution of fresh food is not uniform, the strongest fish tend to get too much at the expense of the weaker. It has been noticed that fish populations are more regular when dry concentrates are used.

Dry concentrates offer another advantage – that of avoiding the introduction of certain diseases which can follow when food is based on fresh fish. The risk of polluting the water is also reduced because the food is more easily and completely digested.

Nevertheless, for fattening purposes, the use of fresh food continues to be justified on farms located near the sea where it is possible to obtain cheap, high nutritive food of good value.

1. *Fresh food.* For fry the principal fresh food is spleen. In order to increase its richness in vitamins and amino acids it is a good plan to add from 2 to 5 per cent yeast and finely ground cooked shrimps. Instead of spleen it is also possible to use liver, brain, blood and white cheese with added yeast.

For fattening fish the best fresh food is sea fish and meat. As a supplement or alternative, dried fish and dried animal meal (fish meal, meat meal) can be used. For food of lesser value for filling or binding, dried shrimps are good as well as vegetable meal (rice and cereals) or bran.

For feeding brood fish the following are the principal foods in their order of importance: fresh fish, fresh shrimps.

2. *Dry concentrates.* This food is distributed either as meal or as pellets of different sizes according to the size of the fish being fed. The general composition of this food has been described in the first section. More details will be given later in this section.

II Preparation of Foods

If fresh food is distributed to salmonids then its preparation is the responsibility of the farmer. The essential rules of preparation are given below. On the other hand if the salmonids are fed with dry concentrates, these are bought already prepared and preparation is not the responsibility of the farmer so the method of preparation is not given in this work.

A Fresh Food

When feeding with fresh food it is essential that the food to be distributed should be very fresh and well prepared.

Storage, mincing and mixing are very important. These operations are generally carried out in a near-by room leading on to the hatching shed and called the "kitchen". It should be clean, light and well aired and should include (Fig. 405) a table, butcher's knives, a meat mincer and an installation for cooking or boiling the meat or fish. Next to the kitchen there should be a refrigerator or deep freeze installation (cold store).

1 Storage of fresh food
The fresh meat or fish should be kept in a refrigerator or in a deep freeze. If the food is only going to be kept a short time, it can be placed in the supply water in the hatching shed which is always cool. In this same water the frozen meat to be used can be defrosted slowly.

2 Cooking the fresh food
Cooking should only be carried out when it is absolutely necessary, and then only when the meat and fish are not considered sufficiently fresh or when the meat of sick animals is used. Also when preparation and mincing includes large fish or butchers meat with large bones, before the food passes through the mincing machines.

These are the only advantages gained from cooking, for it does not help digestion. On the contrary, it reduces the utilizable raw proteins, partially destroys the vitamins, leaches out the nutrients in the food, reduces the volume in which the nourishment is concentrated and necessitates the use of fillers if inflammation of the intestines is to be avoided.

If it is necessary to cook the food then it should be done in a pressure cooker at 120°C for 10 to 15 minutes which means a shorter cooking time. There is then no washing of the nutrients and the fat is separated and can be set aside. Finally the salt in the salty food is eliminated for the greater part.

3   Mincing and preparation of fresh food

For fry, spleen pulp is prepared raw. This is done by laying the spleen out flat (the cattle spleen is about 40 cm (16 in) long and 15 cm (6 in) wide) on a cutting slab. Three or four longitudinal incisions are made, the spleen is scraped in the same direction until it is emptied of all its pulp which emerges as a clammy blood-tinged mass. On average the spleen of an ox can produce about $\frac{1}{2}$ lb of pulp. For fry which have reached 4 cm ($1\frac{1}{2}$ in) the spleen can be fed unshredded and simply passed through a meat mincer. Supplementary food is added gradually.

The most important parts of large sea fish are removed from the bone, but this can only be done after cooking. Fish is treated in the same way as meat. For both meat and fish, small bones must not be removed, for trout digest them perfectly. In their natural diet they can swallow fish whole. The bones also enrich the food with minerals.

The meat of warm-blooded animals must be boned and cleaned of sinew and fat. The flesh is then minced and the binder and supplementary food (different meals, bran, blood) are passed through at the same time.

Mincing machines are provided with perforated plates in which the holes vary in size from 2 to 8 mm in diameter. The fineness of the mince should correspond with the size of the fish. For fry the holes are between 2 and 4 mm; for trout to be fattened they need only be from 5 to 8 mm. On small farms, machines are worked by hand but they can also be turned mechanically. This is necessary in large farms.

For young fish the food must be minced very finely (2 mm holes). As there may be obstructions the first mincing should be fairly coarse before being passed through the machine a second time with the binder and supplementary food. Shrimp meal has been found to be very good, and it can be replaced later by fish meal. For fry, all of the large bits must be removed.

4   Composition of mixed fresh food

In practice, mixed foods for cultivated trout are without limit. The basic and supplementary foods are both numerous and their relative proportions can, to a large extent, be varied according to their economic value under local supply conditions. The cost of food is, of course, a factor in the viability of the farm.

Pure spleen pulp is distributed to the young fry over a period of 4 to 5 weeks. It is a good thing to add from 2 to 5 per cent yeast which is rich in digestible protein and vitamins of the B complex. The addition of yeast gives better growth, reduces mortality and improves the health of the fish.

Wiesner recommends the following mixtures for small trout over 4 to 5 cm ($1\frac{1}{2}$ to 2 in):

> I – 40 per cent spleen pulp
> 50 per cent sea fish
> 10 per cent sawdust
> II – 33 per cent white cheese
> 33 per cent blood
> 33 per cent shrimp meal

Blood from slaughterhouses is easy to obtain and can be used to replace spleen if this is not available. It can be given raw or lightly cooked in a double boiler which allows it to be kept for several days if slightly salted (3 gm per litre) and if it is kept in a cool place.

It is a good idea to vary artificial feeding for fry by using planktonic organisms, principally *Daphnia*, which contribute to the natural food of fry.

For second year trout fed with fresh food, trout farms in the vicinity of the sea use principally marine fish while inland salmonid farms formerly used abattoir waste. To the principal basic foods: fish or meat, at least 15 to 20 per cent binders and fillers, such as vegetable meal, fish, meat and shrimp meal are added in variable proportions. These permit easier preparation of the food, stimulate food consumption and avoid loss during distribution. Besides, the vegetable meal reduces the excess concentration of nutrients in the basic food, thereby avoiding dietary diseases and providing sufficient indispensable and easily digestible food at the same time. The two following mixtures have been recommended:

> I – 67 per cent abattoir waste
> 33 per cent fish meal and rye meal in equal proportions;
> II – 50 per cent fresh sea fish
> 30 per cent fish meal
> 10 per cent dried shrimps
> 10 per cent ground rice

The last three foods of the second mixture are minced, mixed in a machine through which the fresh fish have been passed a first time. Blood can be added to the mixture.

Arens recommends a mixture of 25 per cent rye meal, 25 per cent dried shrimp, 25 per cent fish meal, 25 per cent meat meal. The last two can be replaced with fresh food.

It is always a good thing to add 5 per cent yeast to the feeding mixture prepared by the farmer. Einsele (1965) advises the addition, per kilo, of the following vitamins: vitamin A: 6,000 to 8,000 I.U. (international units); B_1: 20 to 30 mg; B_2: 12 to 15 mg; B_6: 5 mg; B_{12}: 0·03 mg; choline: 300 mg; niacine: 0·65 mg; pantothenic acid: 20 mg; biotin: 0·3 mg; folic acid: 4 mg.

If the composition of the mixtures is varied then this should be done gradually and not drastically or suddenly in order not to stop growth.

B Dry concentrated foods

1 Composition of dry concentrated foods

Dry concentrated foods (pellets and meals) cover the entire requirements of trout during their raising.

The composition of dry foods differs according to the species, the size, the age and the category of fish (fattening or breeding) for which the food is intended. The composition varies according to the producer and there are as many formulæ as there are manufacturers.

It is, therefore, difficult to give a typical mixture of this food of which the principal components, referring to the mixtures of different European manufacturers, are as follows:

Rough protein:	22·0 to 58·0 per cent
Fats:	1·2 to 8·0 per cent
Carbohydrates:	2·0 to 41·0 per cent
Cellulose:	1·0 to 6·0 per cent
Minerals:	10·4 to 22·0 per cent
Water:	6·5 to 11·0 per cent
Vitamins:	several

The basic component of these products comes from the protein which gives the fish the essential amino acids.

The percentage of proteins is often rather high and, in certain diets, reaches 58 per cent for young trout. Animal protein is better than vegetable protein and gives a better conversion rate. Nevertheless the danger of overfeeding increases with the quantity of proteins. In order to avoid this danger Deufel (1964) advises no more than 28 to 32 per cent protein, while admitting up to 40 per cent for fry.

In trout nutrition the carbohydrates would only

be important as ballast to aid digestion and to make the most of the protein. Fats, which seem to be necessary, should only be present in small quantities (4 to 5 per cent) and should not exceed 8 per cent. It is a good thing to mix in from 3 to 5 per cent cod liver oil which is rich in vitamins A and D.

As for the mineral and trace element requirements, not a great deal is known about them but it is certain that calcium, magnesium and phosphorus are indispensable. The food should not contain more than 2 per cent salt. Trace elements are often added in the same way as for concentrates fed to cattle. The trace elements are found in sufficient quantities in fresh food such as liver, meat and fish.

In concentrated dry foods the vitamins, of which the needs of fish are not sufficiently well known, play an important part. This refers to vitamins A, C, D, E, H (biotin) and those of the B complex.

In salmonid nutrition the food intended for rainbow trout must be less rich in proteins. Nevertheless, just as for the food used for salmon, and other salmonids for restocking (brown trout brook trout, Arctic char), as for food intended for rainbow trout, the food for the young fish is richer in proteins and with easily digestible ingredients.

In fact, compound foods can be prepared from a number of ingredients. Phillips *et al.* (1964) considers the following mixture as a complete food for trout.

Bone meal:	5·0 per cent
Fish meal:	24·0 per cent
Skim milk powder:	1·5 per cent
Cotton seed meal:	15·0 per cent
Wheat middlings:	19·0 per cent
Brewer's yeast:	10·0 per cent
Dried distiller's solubles:	21·0 per cent
Mixture of 14 synthetic vitamins:	1·5 per cent
A-D feed oil:	3·0 per cent

Besides the food intended for normal feeding, certain manufacturers include additives such as calomel, sulphonamides, oxytetracycline, nitrofurans or chloramphenicol to combat hexamitiasis, furunculosis and other bacterial diseases respectively. To combat furunculosis 10 gm of sulphamerazine and 3 gm of sulphaguanidine are added to the daily food for 1 week and for each 100 kg (220 lb) of fish. Against *Hexamita salmonis*, 0·2 gm of calomel per 1 kg ($2\frac{1}{4}$ lb) of food distributed three or four times per day over three consecutive days. The addition of antibiotics such as oxytetracycline di-*n*-butyl does not increase the growth of the fish. Anthelmintics such as tin oxide may also be added to the diet to combat intestinal parasites such as cestodes.

2 Conservation of dry concentrated foods

The conservation of dry concentrated foods is limited by the effective life of the vitamins, which is no more than a few months (6 months at the most). It is not therefore advisable to accumulate large stocks of dry concentrates. They must be kept in a cool, dry place and protected from the light.

3 Presentation of dry concentrated foods

Dry concentrated foods are produced in different sizes according to the fish to be fed (Fig. 404). These include meals for young fry, crumbs for larger fry and young trout and pellets for larger trout, second year trout and brood fish. The current diameters of these foods are: 0·8 mm for meal (one or two thicknesses: 0·25 to 0·4 and 0·4 to 0·8 mm); 0·8 to 1·4 (one or two thicknesses: 0·8 to 1·1 and 1·1 to 1·4 mm) and 1·4 to 2·4 mm for crumbs; 2·5 to 4·0 mm, 4·0 to 6·0 mm for pellets intended for the larger trout fingerling and second year fish; 6·0 to 10·0 mm for pellets intended for brood fish.

For young fry, floating food is advantageous. It also seems a good idea to colour the food, brown and red pigments being the most commonly used at present.

III Distribution of the Food

1 Principles

In the culture of salmonids, artificial food must be distributed frequently and with care. The rules given below should be followed for distribution.

1. Artificial food must be eaten immediately after it has been distributed, that is to say before it falls to the bottom. It should therefore be distributed frequently but not too much at a time.

2. It is possible to feed trout exclusively with artificial food. The total quantity for distribution can then be calculated exactly.

3. For each distribution the amount given should be as much as the trout can eat, that is to say they must be fed up to a normal level of saturation. In this way rapid growth is assured. The faster they grow so the maintenance ration is smaller and fattening becomes more economical. Nevertheless overfeeding must be avoided, especially when the quality of the food is not first class, or, with dry

concentrates rich in proteins, for this can lead to intestinal diseases, lipoid degeneration of the liver, hepatomas or hepatic lesions. It is necessary to regularly verify the state of the liver, intestines and kidneys of farmed trout. If any abnormalities are noticed, the artificial feeding must be suspended for 8 to 15 days, when it may be restarted gradually.

4. As a measure of preventive hygiene feeding is suspended from time to time, 1 day per week for example. Feeding should also be varied for fish being fattened by giving them some natural food, as far as possible and according to local conditions. Feeding is suspended, or at least reduced, when the weather is thundery and the temperature is high.

When dry pelleted food is distributed the amount fed should not be greater than that indicated by the manufacturer. It is also advisable to suspend the feed, or to reduce it considerably, when the temperature of the water falls below 6°C. The ration is also reduced when the temperature of the water rises to 20°C. Feeding is suspended 3 or 4 days before the fish are to be handled either for grading or transportation.

Overfeeding for a period of 3 to 4 weeks before the end of the fattening period is not harmful if it is moderate and gradual. It is not the same for fish which must still be kept for several months nor for those intended for restocking water courses or fishing ponds.

2 Quantity of food to be distributed

a The daily ration

The daily ration depends on the water temperature, the size of the fish and the type of food used. Normally the daily ration is expressed as a percentage of the weight of the fish being fed.

1. *Temperature.* The temperature limits to choose for commercial fish farming vary between 5 and 18°C to 20°C according to the species. Between these temperature limits, growth is that much more rapid as the temperature is high, and if rapid growth is required then a temperature near the upper limit is chosen. In practice, intensive feeding in salmonid culture is best carried out with temperatures of between 10 and 16°C. Until a maximum is reached at 19°C for rainbow trout, the daily ration to distribute for fattening 2-year-old trout increases with the temperature.

2. *The size of the fish.* Proportionally the daily ration is that much higher as the size of the fish is small. For normal fattening at temperatures above

10°C during the summer growing months, it is necessary to count on a fattening ration equivalent to 5 or 6 per cent of the weight of the fish being fattened for first year fish and from 2·5 to 3·0 per cent for 2-year-old fish. These amounts refer to fresh food. They are less for artificial concentrates.

3. *The food.* The nature of the food considerably influences the quantity to be fed. The conversion rate of fresh fish is 6·0 to 8·0 but it falls to between 1·0 and 2·0 for dry pelleted food. Under normal farming conditions the conversion rate of dry pelleted food is between 1·0 for young fish and 2·0 for fish for fattening.

The daily ration to be distributed depends on the nutritive value and on the bulk of the food or on the mixture used. It goes without saying that daily rations of pelleted dry foods are lower than those of fresh foods.

The manufacturers of dry concentrates give feeding schedules for the food they produce. These give the percentages of the daily food to be distributed according to the size and weight of the fish and the temperature of the water. It is not advisable to exceed the quantities indicated by the makers.

Here, according to Einsele, is a feeding schedule for salmonids fed with dry concentrates. If feeding is with fresh food (fish or meat) the percentage quantities indicated are increased by one or two units.

b Total amount
The total amount of food which will be distributed is obtained by multiplying the weight representing the desired growth by the conversion rate of the food used. The total quantity of food necessary under these conditions, on a large commercial trout farm, is astonishing. It can reach several tons daily (Fig. 416).

3 Feeding schedule
Except on farms with a very abundant water source and flow, and in consequence always having fresh water at relatively constant temperature, artificial feeding should be considerably reduced or even totally suspended in winter. Feeding generally starts in the months of March-April and continues until November. In winter if the temperature does not fall very low only feeding for maintenance is necessary. Schäperclaus has drawn up a feeding schedule for second year trout:

March	1 per cent	August	18 per cent
April	4 per cent	September	17 per cent
May	7 per cent	October	14 per cent
June	13 per cent	November	10 per cent
July	16 per cent		

4 Methods of distribution
These differ according to the age of the fish being fed. In any case distribution must be carried out

Daily feeding schedule for salmonids fed dry pelleted food
(According to Einsele, 1965)

Temperature of the water (°C)	Length of the fish in centimetres (inches)											
	3 (1⅕)	4 (1⅗)	5 (2	6 (2⅖)	7 (2⅘)	8 (3⅕)	10 (4	12 5)	15 (6	20 8)	25 (10	30 12)
	Daily ration as a percentage of weight of fish											
4°	5·0	4·5	3·5	2·5	2·0	2·0	1·5	1·0	0·8	0·5	0·5	0·4
6°	7·0	5·0	4·0	3·0	2·5	2·5	2·0	2·0	1·0	1·0	0·8	0·6
8°	8·5	7·0	5·0	4·5	3·5	3·0	2·5	2·0	1·5	1·0	1·0	0·9
10°	10·0	9·0	7·0	5·0	4·5	3·5	3·5	2·5	1·5	1·2	1·1	1·0
12°	12·0	11·0	9·0	7·0	5·0	4·0	3·5	2·5	1·5	1·4	1·2	1·2
15°	14·0	14·0	12·0	10·9	7·0	5·0	4·5	3·5	2·0	1·7	1·5	1·4
Weight of 1,000 trout in kg (2¼ lb) or 1 trout in gm	0·2	0·5	1·0	2·0	3·5	5·0	9·0	17·0	35·0	90·0	175·0	300·0

For fry, Raveret-Wattel estimates the quantity of spleen pulp necessary as 10 gm per day for 1,000 fry of 1 week and 75 gm (2 oz 10) for 1,000 fry of 6 weeks.

very carefully and in such a way that the fish derive maximum profit from all the food distributed. Loss of food which causes loss of money and possibly provokes disease must be avoided.

a   Feeding fry

The most difficult problem is to persuade the fry to take artificial food. Success is more likely in tanks and rearing troughs than in ponds. It is more difficult to persuade brown trout, which are very timid and stay at the bottom, than rainbow and brook trout which are less afraid and live more on the surface.

It is possible to distinguish two stages in the feeding of fry: pre-feeding aimed at persuading the fry to take artificial food, and feeding *sensu stricto*.

Fry are fed with fresh food such as spleen pulp or with dry food presented in the form of meal. In either case, by varying the proportion of water it is possible to obtain a pap which can be distributed as a kind of cloud or as a mash which can be spread out on a submerged food holder.

Pre-feeding. If spleen is used then the pulp is placed in a strainer and pressed at the water inlet. The finest nutritive particles pass through. If meal is used it is mixed into a pap with a little water and tipped into the rearing trough. In one way or another the fry find themselves plunged into the nutritive cloud and are practically forced to feed. The food clouds the water and falls slowly to the bottom and it is at this moment that the fry take it. In rearing troughs the remains of the uneaten food fall to the bottom and must be siphoned off daily. From the start of feeding the trough covers are removed.

Feeding sensu stricto. Whether it is practised in troughs, tanks or rearing ponds fry are fed with spleen or meal. The spleen is distributed in two different ways – either through a very fine strainer or spread on a mesh or on a flower pot. The meal can be distributed either by hand or by means of an automatic food dispenser.

The spleen pulp can be placed at the bottom of a fine mesh strainer through which it passes and falls into the water through stirring or pressing. If this method is used for small ponds the strainer is attached to a long handle.

It is also possible to spread the spleen pulp out on pieces of metal mesh or on flower pots suspended by hooks on wires stretched across the water. The pots and metal pieces are suspended about 10 cm (4 in) below the surface of the water (Fig. 407). The fry get used to this method very quickly for it gives them a permanent food supply. Two series of pots or pieces of mesh are necessary, one under the water spread with spleen pulp and the other out of the water ready to be cleaned and then spread with pulp. This method produces greater

differences in growth between individual fish than the former method. To remedy this a large number of pots or pieces of mesh are used.

If meal is distributed to fry it is done on the surface of the water. It can be distributed by hand.

Whether the food be spleen or meal, five or six distributions are necessary every day for the young up to 2 months. After that the number is reduced.

Meal can also be distributed with the aid of an automatic dispenser (Fig. 410). The apparatus used consists of reduced models of the same type of fixed dispensers that are successfully used for feeding large fish. In this case the frequency of distribution must be increased considerably and even raised to 50 times per day. The amount given at each distribution is, obviously, reduced in consequence.

As the fry grow the kind and frequency of the distributions are modified progressively to suit the stronger small trout and the second year fish.

b   Feeding fingerling and second year trout

This is easier to carry out than for fry. The fish can be given as much food as they can eat but it must be consumed before falling to the bottom. This provokes violent disturbance among the trout which pounce on the food where it falls on the water. To encourage the trout to copy each other when the food is distributed, the pond should be well stocked. It is important to ensure an equal distribution of food on the surface of the pond or tank to make certain that the food does not always fall at the same place, for in this case the strongest will occupy the spot to the detriment of the weaker fish.

The distribution of food to trout fingerlings, second year and older fish can be done manually or mechanically.

1   Manual distribution. Until a few years ago feeding was always by hand. If fresh food is distributed the mixture must be prepared in advance and distributed to the fish by throwing the food from the banks of the pond (Fig. 408). Manual distribution of pellets can also be carried out from the banks. Such a method of distribution takes time if the farm is large. Two or three distributions per day are necessary for 4- to 6-month-old trout and two distributions for older trout. Only one daily distribution need be made for 1 year trout, but during the intensive summer growth period at least two are necessary.

To facilitate transportation of the food from the

kitchen to the feeding ponds, large farms have narrow gauge rail trucks (Fig. 409) or some other form of transport.

2 *Mechanical distribution*. Problems arising from the use of labour, which is becoming more and more costly and leads to irregular and infrequent distribution, have encouraged the adoption of automatic methods. First employed in Scandinavia and the United States, they have now become more wide spread. Fixed and mobile distributors are used. Both work automatically. The food is held in a holder or a hopper under which is installed a mechanical distributor which ensures the distribution of a fixed quantity of food at regular and adjustable intervals. These automatic distributors can only handle dry pellets although in exceptional cases, they have been adapted to distribute food pastes.

(*a*) *Fixed dispensers*. Ingenious distribution mechanisms have been built, generally driven by electricity. Certain distributors can also function by hydraulic action, the water being brought by gravity to the device which regulates the distribution frequency (Fig. 410).

The dispensers are placed at one or several spots around the pond or reservoirs, but principally at both ends if they are rectangular (Fig. 412). This method concentrates the fish at the distribution points.

On important farms equipped with electricity an electric clock control can regulate all the details of distribution. It fixes the time the distribution starts and ends and the intervals between each distribution. Quite often the equipment is regulated in such a way as to ensure feeding each quarter of an hour over a working period of 12 hours, which corresponds to about 50 distributions per day. It is also possible to regulate the control so that distribution is interrupted when the temperature of the water falls below or rises above the fixed limits.

All the food dispensers can operate simultaneously thus serving all of the tanks or ponds at the same time. Each dispenser can also have its own clock. This system allows the regulation of different quantities and frequencies for each tank or pond. But this elaborate equipment costs more money.

Dispensers which work hydraulically are simpler and much less costly. They also function reliably. The principle of these dispensers (Figs. 410 and 411) rests on the filling of a dissymetric shaped bucket. The water is brought by gravity to the bucket. When it is filled, or almost filled, it tips over and this activates a piston installed beneath the hopper which holds the food. Each time the piston works a ration is distributed to the fish. After each movement a spring returns the bucket to its original position. A tap controls the water supply brought to the bucket and the length of time it takes to be filled and consequently the frequency of the distributions.

(*b*) *Mobile dispensers*. Certain dispensers move on rails laid on the ground or by a monorail suspension system. Small trucks travel over the rails carrying the distribution apparatus or transporting extended conduits with the distribution apparatus fixed at the end (Figs. 413 and 414). An electric clock regulates the advance and return of the mechanism, which can be installed between two series of ponds so that they can be served simultaneously.

Another system of distribution which is both simple and very mobile is a carried distributor (Fig. 415). This includes a hopper and a blower worked by a compressor installed on a trailer. The tractor moves slowly along the banks of the ponds while the driver controls by hand the projection and distribution of the food for each pond. He chooses the spots for distribution and the duration of distribution.

Mobile dispensers ensure an equable distribution of food, they help the spreading out of the fish in the water and in consequence favour a rational and intensive use of the food.

Fig. 403   Feeds for salmonids: spleen (maintained frozen) 65 per cent; pellets 35 per cent; mixed with water (25 per cent of food volume).

Fig. 404   Pellets of different sizes.

Fig. 405   Preparation of fresh fish for fish food. Fanure fish farm, Roscrea, Ireland.

Fig. 406   Preparation of dry concentrated food. Lieburg Hatchery, Oregon, U.S.A.

Fig. 407   First feeding, non-automatic, in rearing troughs for trout alevins. Pulp made from spleen or mash made from dry feed is stuck either on flower pots or on small pieces of wire mesh.

Fig. 408   Distribution with a trowel of fresh food for fattening trout.

Fig. 409   Distribution by hand of fresh food. Bringing up feed by small gauge rail truck.

Fig. 410 Fixed dispenser working hydraulically for distribution of food in circular fry rearing tanks. Linkebeek fish culture station, Belgium.

Fig. 411 Nursery pond with automatic dispenser of the same type as Fig. 410. Linkebeek fish culture station, Belgium.

Fig. 412 Electrically driven automatic dispensers ranged in series. The dispensers switch simultaneously by means of a programmed electric clock. (Photo Steenberg; document Dansk Orredfoder, Brande, Denmark.)

Fig. 413

Fig. 414

General view (Fig. 413) and detailed view (Fig. 414) of an automatic and mobile dispenser of dry concentrates. Nimbus Hatchery, Sacramento, California, U.S.A.

Fig. 415 (above)   Automatic distribution of feed by tracted pellet blower. Canizzano fish farm, Venice, Italy.

Fig. 416 (right)   Silos for stocking dry feed in a large industrial fish farm.

SECTION III

ARTIFICIAL FEEDING OF CARP

I   Natural Food and Artificial Food

In traditional cyprinid culture it is generally thought that natural food must constitute an important part in feeding carp. It is generally given at around 50 per cent.

It is possible, however, to raise carp intensively and this is sometimes done in the Far East (Figs. 224 and 225). Trials on a reduced scale have also been made in Europe. It is, however, premature to conclude, under existing conditions, that the exclusive use of artificial food for carp will be economic one day in classical type ponds. Such a technique would produce problems other than those being solved now, such as the quantity and the quality of the water. Nevertheless it is possible that the dry pelleted foods already being used in cyprinid culture will play, in the future, a more and more important part.

By limiting the present normal conditions of farming to the principle that each fish should consume 50 per cent natural food and 50 per cent artificial food, theoretically it should be possible to raise, in the same pond, double the number of fish assuming that the individual growth of the fish remains unchanged. Considering that natural food is never entirely used up even in an over-stocked pond, the increase in stocking permits a better use of the natural food and, in consequence, a more important distribution of artificial food. In practice, artificial feeding gives a total production which is more than double that of natural production for identical individual growth (Chapter XIV, Section I, I, A, 3).

The increase in production due to artificial feeding is proportionately greater in poor ponds than in rich ponds. In effect, because of the low stocking density in poor ponds, the percentage of unused natural food is greater than in rich ponds when production is based solely on natural feeding and the aim is to obtain identical growth in both cases.

The possibility of increasing, at the same time, the density of the fish and the use of natural food, thanks to the distribution of artificial food, then becomes relatively greater in poor ponds than in rich ones. The growth percentage caused ostensibly by artificial feeding is therefore higher in the first than

in the last instance, even though the natural food in each case remains about 50 per cent of the total quantity of food eaten.

Comparing the initial production of ponds classified in four categories, Schäperclaus assumes, according to research by Walter, the following proportions between natural growth (growth due to natural food) and artificial growth (growth due to artificial food).

Productivity class	Natural growth	Artificial growth
I	50%	50%
II	33·3%	66·6%
III	25%	75%
IV	20%	80%

On this basis, Walter gives the following productivity figures which have become standard practices in western Europe:

Natural, artificial and total growth in cyprinid ponds according to the productivity classes (in kilos per hectare or in ponds per acre) (According to Walter, 1934)

	I° class	II° class	III° class	IV° class
A – Natural growth	400-200	200-100	100- 50	50-25
B – Artificial growth	400-200	400-200	300-150	200-100
C – Total growth	800-400	600-300	400-200	250-125

II   Principal Food for Carp

The most important factor to be taken into account when feeding carp with artificial food is economic. The food distributed must be simple and cheap and the fish must rely to the greatest possible extent on natural food which contains all the constituents of a complete diet – amino acids, vitamins, minerals, etc.

During the first year of rearing the aim is, above all, to increase the natural food by drying out in winter, working the soil and good fertilization. This does not exclude the distribution of artificial food, which finely ground, can be the same as that given during the second and third years of growth. There is an advantage in feeding young fish with proteinaceous foods such as lupin and soya or to add to inexpensive vegetable food high value foods, such as meat and fish meal, blood, distillery wastes, yeast, and good quality pellets. The yeast contains valuable proteins and vitamins and helps digestion, which is important if food rich in starch is distributed, as is often the case in cyprinid culture.

A complete pelleted food can be given above all in the spring to fish of 1 summer to 1 year; it should also be given in the autumn, during wintering and the following spring, periods during which natural food can be scarce. Either pellets only can be distributed or used in proportions of a quarter or a third as an addition to the ordinary ground food. With good quality pellets it is possible to obtain a conversion rate of 2·2.

The feeding of second and third year carp is current practice, and is even indispensable from an economic point of view. Simple and cheap foods according to local supply conditions are generally distributed. The use of foods which are richer in proteins would be necessary when there is danger of the fish weakening, for example, after being handled, as well as during the winter.

The principal foods of vegetable origin given to carp for fattening are lupin, soya and cereals such as rye, barley, maize, wheat and oats (Fig. 417). Different animal and vegetable waste is also used. Food employed for carp is also suitable for tench and other accompanying cyprinids.

In order to reduce losses due to infectious abdominal dropsy (Chapter XV, Section II), certain manufacturers of pelleted foods incorporate antibiotics. According to Bank (1967) intensive feeding with food containing antibiotics (192 mg of oxytetracycline per kilo of food) leaves no trace of the antibiotic in the carp flesh after from 2 to 3 days. They can be eaten by humans without danger. The same applies to intra-peritoneal and other parenteral antibiotic injections.

III Preparation of Food

The preparation of fresh food of vegetable origin given to carp is quite simple. According to each individual case it is ground, soaked, dried or cooked. In many instances the food can be distributed without preparation (Fig. 418).

It is only ground when the grain is large (lupin, maize) and in so far as it will be distributed to fish of small size. Large grain is only ground for fish weighing less than 250 gm (9 oz). Otherwise there is no advantage in grinding the food. On the con-

trary it is a costly operation and the crushed grain does not keep so well as when it is whole. Only small quantities are prepared at a time.

The need for soaking is to stop the ground food floating on the surface when it is distributed, and thus risking loss. Germination of the grain is accompanied by an enrichment in vitamins.

Drying the food is not always beneficial. It aids conservation and lowers the conversion rate but this latter advantage only means compensation for a loss of water.

Cooking may be necessary to stop the decomposition of certain animal or vegetable materials. These cooked materials are then passed through a meat mincer. For animal meals only light cooking is necessary, for its object is to bind the food which is to be mixed with vegetable meal, transforming all of it into a sufficiently coherent paste so as not to disperse at the time of distribution, which would mean a loss of food.

IV Distribution of Food

1 Total amount of food to be distributed

The total amount is easily calculated by multiplying the artificial growth expected, expressed in kilos per hectare, by the conversion rate, according to the following formula:

Amount of food to distribute per hectare	=	Growth per hectare due to artificial feeding	×	Food conversion rate

Example: The plan is artificial feeding with lupin (conversion rate, 4) in a 1·60 hectares (4 acres) pond in class II Schäperclaus (natural production = 150 kg/Ha (150 lb per acre)). What is the total quantity of lupin to be distributed?

Natural growth = 33·3 per cent; artificial growth = 66·6 per cent or twice the above which makes: 150 × 2 × 1·60 = 480 kg (1,058 lb).

The quantity of lupin to be distributed equals 480 × 4 = 1,920 kg (1,058 lb × 4 = 1 ton 18 cwt).

2 Feeding schedule

With the total quantity of food to distribute known, a feeding schedule should be established in order to distribute the food wisely.

The first aim of the schedule must be to determine the quantities of food to be distributed monthly. This will depend on the size of the fish desired, their relative size during the months of growth, and the distribution periods.

If the object is very good growth, then this will take place particularly during the hot months, July and August, when the greatest amounts of food should be given. The fish, being larger at the end of the growing season than at the beginning, the monthly amounts should be greater at the end of summer than at the beginning. At the start of the growing period quantities are reduced taking into account the fact that the fish must get used to artificial feeding gradually.

On this basis, Walter gives, in the table printed below, the monthly percentages of food to distribute according to the individual growth desired.

therefore have certain fertilizing effects. Pollution of the water must be avoided, for it can lead to a lack of oxygen in the summer and to the occurrence of diseases such as the gill rot.

2. Distribution is preferable in the morning.

3. The food is not spread out over the whole surface of the pond but is placed at different marked spots indicated with poles which stick out of the water and can therefore be easily located. From 4 to 6, well spaced out, are chosen per hectare so that the weaker fish can find a place to eat. All the fish eat natural food in the meantime.

4. The spots chosen for distribution should be clean and from 0·60 to 1 m (2 to 3 ft) in depth. There should be no vegetation. The bottom should be hard and not muddy. From time to time the places are

Sharing out the total quantity of food to distribute in carp ponds during fattening months according to individual growth (Walter 1934)

		Relationship between the weight at the time of stocking and the weight at harvest, as 1 to:					
		For K_2				For K_1	
		2	2·5	3	4	10	20
Percentage of the annual	May	15	13	11	9		
amount to distribute during	June	20	18	16	14	10	
the months of	July	25	24	23	21	20	25
	August	30	32	33	36	45	50
	September	10	13	17	20	25	25

As the monthly amounts are known, so the quantities to be given at each distribution are calculated according to the number of distributions per month.

A certain flexibility is necessary with respect to the amount of food to be distributed at a time and the length of the distribution period. Much depends on the temperature of the water. If October is still sufficiently warm, feeding can continue.

Some are satisfied with one weekly distribution but it is better to feed twice or three times per week. Frequent distributions give smaller and better conversion rates but call for more labour and this reduces its advantages.

3   Rules for distribution
Distribution rules to be followed:

1. A quantity of food is given which will have been eaten entirely before the next distribution.

Carp do not take their food immediately in the same way as trout. The food which is distributed can

changed. If ducks are raised at the same time they are not allowed access to the feeding places, which are protected by twigs or in some other way, and a sufficiently deep spot is found for distribution.

5. If varied artificial feeding is the method used, then the variation should be introduced gradually and not drastically or suddenly.

6. The amount of food distributed at one time depends on the temperature. Fish are not fed, or hardly so, below 13°C nor when the temperature is very high, over 27°C. In Europe carp grow best between 18 and 25° C.

In wintering ponds, the young fish are fed for as long as the temperature is above 4°C. They are given good quality food, though not more than 1 per cent of their weight at each distribution. It is necessary to ensure that all the food is consumed before the next distribution.

7. In the case of epizootics feeding ceases, or the amount of artificial food is considerably reduced.

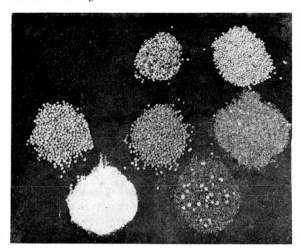

Fig. 417 Possible feed for carp: sorghum, lucerne and soya, lupin, peas (dried or green), rye, bran and diverse grains. Dinnyes fish culture station, Hungary.

Fig. 418 Mixed maize and rye for feeding carp; the food is distributed with a shovel from a boat. Poljana fish farm, Zagreb, Yugoslavia.

Fig. 419 (above, centre) Feeding carp; feed distributed from a boat at plots marked with indication poles. Poljana fish farm, Zagreb, Yugoslavia.

Fig. 420 (above) Automatic dispenser for feeding carp. 22 hectare pond (55 acres) at Dinnyes, Hungary.

Fig. 421 Using a cast net at feeding points to control growth of carp. Biharugra fish farm, Hungary.

8. If the food contains antibiotics in order to prevent or reduce infectious abdominal dropsy, then it is distributed as early as possible in the spring. It should be evenly shared between a sufficient number of points and distributed at least once every 3 days.

4 Different forms of distribution

In carp culture the food is generally distributed by hand. More recently, some distribution has been mechanical.

Manual distribution is sometimes carried out from the bank of the pond. But in most cases, as fattening ponds are very large, the distribution is from a boat (Fig. 419). The boat is filled with food and taken to the places of distribution. At each spot the food is distributed by means of a shovel. From time to time the growth of the carp is investigated or controlled, for example, by throwing a cast net into the water at the place of feeding. (Fig. 421).

Fixed automatic dispensers are now starting to be used (Fig. 420). They are electrically operated, and of the same type as those used in salmonid culture, but because of the significant amounts of food which are distributed in large carp ponds, their size is very much greater.

It is also possible to use mobile automatic distributors. These comprise a blower installed on a trailer of a type similar to that used in salmonid culture (Fig. 415).

SECTION IV

ARTIFICIAL FOOD FOR GRASS CARP

Grass carp (*Ctenopharyngodon idella*) can be fed with submerged vegetation, floating plants, certain emergent plants and even sufficiently tender land plants.

In the Far East different plants are used for feeding, notably duckweed. To avoid its being blown all over the pond by wind, the food is distributed on square trays formed by floating bamboo stalks (Fig. 422) and placed near the banks of the pond. This helps the distribution of the food. But it is also possible just to throw the plants onto the surface of the water, on which they will float.

As a result of experiments carried out in aquaria in Hungary, Penzes and Tolg (1966) classified 22

Fig. 422 A floating square for feeding grass carp. The square is formed by bamboo stalks. Duckweed (*Lemna* sp.) is distributed for food. Banghkem experimental station, Bangkok, Thailand.

Fig. 423 Failing to find more tender vegetation, grass carp feed off reed leaves (*Phragmites communis*). The fish first feeds off the tips of the leaves. Dinnyes fish culture station, Hungary.

species of aquatic plants and two land plants as food for grass carp:

1. Plants eaten in 8 hours with great appetite: *Ceratophyllum demersum*, *Chara* sp., *Cladophora* sp., *Elodea canadensis*, *Myriophyllum spicatum*, *Najas marina*, *Potamogeton lucens*, *Potamogeton natans*, *Potamogeton perfoliatus*, *Spyrogyra* sp., *Lactuca sativa*, *Medicago sativa*;

2. Plants eaten within 24 hours with average appetite: *Hydrocharis morsus-ranae*, *Phragmites communis* (hard parts partially eaten), *Potamogeton crispus*, *Potamogeton pectinatus*, *Schoenoplectus tabernamontani*, *Trapa natans*, *Sium latifolium* and *Typha latifolia* (the last two, hard parts partially eaten);

3. Plants eaten in 48 hours with slight appetite: *Iris pseudacorus* (hard parts uneaten), *Polygonum amphibium*, *Typha angustifolia* (hard parts partially eaten);

4. Plants uneaten after 72 hours: *Ranunculus trichophilus* (only 20 to 30 per cent eaten).

A narrow connection between temperature and appetite was established. No food was eaten below 14°C and from 15 to 16°C the fish only ate the soft parts of the plants. Above 20°C they ate with a good appetite.

The leaves and stalks of soft plants such as watermilfoil and starwort are completely and simultaneously devoured. The leaves of emergent plants with hard stalks, such as reed, are eaten by the fish first starting from the tip (Fig. 423). The stalks are only rarely eaten. The food passes across the pharynx and is masticated by the pharyngeal teeth.

SECTION V
ARTIFICIAL FEEDING OF OTHER FISH

The artificial feeding of fish other than salmonids, carp and grass carp was described above when methods and techniques for the cultivation of these species were considered (Chapter V to IX).

This also applies to tilapias (Chapter VI, Section V), which do very well on different wastes of vegetable origin (Figs. 288 and 290). Besides, herbivorous tilapias can be fed with land plants (Fig. 289).

In the Far East, eels were formerly fed with fresh food but this has been replaced by dry pellets (Chapter VIII, Section I; Fig. 310).

In the U.S.A. certain catfish, notably channel catfish, are also fed with meals and pellets (Chapter VII)

Chapter XIV

TOTAL PRODUCTIVITY AND STOCKING OF PONDS

SECTION I

TOTAL PRODUCTIVITY OF PONDS

THE total productivity of ponds is equal to the sum of natural productivity, of productivity due to fertilization and also to productivity due to artificial feeding.

Total productivity = Natural productivity + Productivity due to fertilizing + Productivity due to artificial feeding

Natural food ⏞   Artificial food (+ indirect natural food)

1 Components for Total Productivity of Ponds

A Productivity resulting from natural food

1 Initial natural productivity
The appreciation of natural productivity K (according to Schäperclaus (1933) "Naturzuwachs") of ponds was examined in Chapter X. This productivity is very important for methods of production in which natural food plays a part. In these cases the improvement of natural productivity is the first goal to reach in order to increase total productivity.

2 Natural productivity resulting from fertilizing
The increase of productivity resulting from fertilizers relies on an increase in the quantity of natural food initially present, as well as on an increase in the stocking resulting from it. A study of the fertilization of ponds and the consequences for fish production was considered in Section IV, Chapter XII. According to the circumstances of each case the extra production due to fertilization is between 0 and 100 per cent of natural production *sensu stricto*. If, for example, in a specified case the increase in production due to fertilization represents a two-thirds increase of the initial natural production, the natural production will be $\frac{3}{3}$ K + $\frac{2}{3}$ K = $\frac{5}{3}$ K.

Based on experiments with European carp culture Schäperclaus (1963) makes a distinction between the true "natural production" (Naturzuwachs) and "calculated natural production" (rechnerische Naturzuwachs) which corresponds to the sum of the properly called natural production and natural production due to fertilization.

3 Natural productivity through artificial feeding
The increase in production due to the use of artificial food brings with it a parallel increase in natural production. This latter increase is the result of a significant increase in stocking density (below, II, 2), and also to the indirect action of manure due to the use of artificial food. According to Schäperclaus (1963) the increase of calculated natural production is in the order of 60 per cent which almost triples the initial natural production as seen in the following calculations.

The natural global production is equal to the sum of: K (initial natural production) + $\frac{2}{3}$ K (fertilization) + 6/10° of $\frac{5}{3}$ K (indirect action of artificial feeding = 30/30°K or 1 K), or to the total K + $\frac{2}{3}$ K + K = 2 $\frac{2}{3}$ K. Schäperclaus indicates this as "natural global production" called "effektive Naturzuwachs". In the above example, which is relative to intensive farming of carp in central Europe, the following is obtained:

K = natural initial production,
1 $\frac{2}{3}$ K or $\frac{5}{3}$ K = natural calculated production,
2 $\frac{2}{3}$ K or $\frac{8}{3}$ K = natural global production.

By the simultaneous employment of fertilizing and feeding, the initial natural production can then be increased considerably on condition, nevertheless, that the stocking is increased proportionally.

B Productivity resulting from artificial feeding
On most farms productivity due to natural food

can be increased by productivity due to artificial feeding if the latter is used, which is advisable whenever it is technically and/or economically possible. The use of artificial foods for fish is considered in Chapter XIII.

In certain cases, productivity due to artificial food is of small importance but in other cases it is considerable.

It is relatively slight in traditional carp culture. It is also estimated that productivity due to artificial food can be equal to one, or to one and a half times, the productivity (initial, calculated or global) resulting from natural food.

In certain cases such as for tilapias, certain salmonids and eels, which can be farmed on an almost intensive scale, production can be totally due to artificial feeding. In this case, natural food plays a secondary part or even no part at all and fertilization is then superfluous. The production per unit of surface area can be considerable, exceeding 10 tons per hectare and even much more in the case of rainbow trout. Such production does not depend on natural productivity.

C Total productivity

To determine the total productivity of a pond it is necessary (see the formula on p. 359) to calculate the sum of natural productivity, of productivity due to fertilization and productivity due to artificial feeding. Each of these terms of production has its own value, as specified in Chapters X, XII (Section IV) and XIII. The relative value of each term of total productivity is very variable. Extreme cases are as follows:

(a) total productivity is equal to simple, natural productivity if there is no recourse to either fertilization or artificial feeding;

(b) total productivity is, in the case of intensive farming, equal to the productivity due to artificial feeding only, for in this case natural productivity and that due to fertilization is non-existent or negligible. Between these extremes there is a very varied scale of intermediate cases.

1 Extensive or semi-intensive rearing

On an extensive or semi-intensive farm, in which natural productivity is more or less preponderantly important (in carp culture for example), the production of a pond can be equal to the following values according to the case in question and to whether one wishes to considerably intensify production.

1 *Without artificial feeding* productivity is equal to:

(a) Natural productivity alone if the pond is not fertilized (the fish are not fed); productivity equals K.

(b) To the sum of initial natural productivity (K), plus productivity due to fertilization ($\frac{2}{3}$ K in the preceding example) if the pond is fertilized; in this case the calculated natural productivity equals $\frac{3}{3}$ K $+ \frac{2}{3}$ K $= \frac{5}{3}$ K.

2 *With artificial feeding:*

(c) Productivity can equal natural productivity (K), plus productivity due to feeding if not accompanied by fertilization. In carp culture, and in similar cases, it is possible to envisage production due to feeding as equal to one and a half times the natural productivity. In this case the total productivity is equal to K $+ 1\frac{1}{2}$ K $= 2\frac{1}{2}$ K.

(d) Productivity can be equal to the sum of natural productivity, productivity due to fertilization and productivity due to feeding if both are undertaken.

In carp culture, if the aim is to establish the relation 1/1·5 between natural global production and production due to artificial feeding, the total production can represent $1 + 1·5 = 2·5$ times the natural global production. Now, the latter can be equal (as was written above) to $\frac{8}{3}$ K, so in the hypothesis envisaged above, total production can equal $\frac{8}{3}$ K (natural global production) $+ \frac{12}{3}$ K (production due to artificial feeding $= 1·5$ times natural global production) $= \frac{20}{3}$ K, that is near seven times the natural initial production ($\frac{3}{3}$ K) and four times the natural calculated production ($\frac{5}{3}$ K).

Under conditions a, b, c, and d given above, a pond in Europe able to produce 120 kg (265 lb) of third year carp per hectare without fertilization or artificial feeding, will be able to give, in the different cases taken into account, respectively: 120 kg (265 lb), 200 kg (441 lb); 300 kg (661 lb) and 800 kg (1,764 lb) of fish.

The data given above relate to carp in Europe; the data furnished by Van der Lingen (1957) relate to the raising of tilapias in Africa, and can be considered together. He mentions that the natural productivity can be tripled by fertilization and increased up to seven times by the simultaneous use of fertilizers and feeding. Higher values have been obtained.

In the intensive farming of tilapias, the intensified

feeding of these fish accelerates growth so well that they reach marketable size at the moment they start to reproduce, and when they do not meet competition for the available food from their fry. Another result of this is also an improvement in the production quality.

2   Intensive rearing

When farming is intensive (trout and eels for consumption) production rests entirely on feeding and can be very high. The roles of natural food and fertilization, as well as that of the productivity which results, are non-existent or negligible.

II   Variation in the Productivity of Ponds

In all cases where productivity depends on the use of natural food, the total productivity of ponds can vary because of different reasons, all of which must be understood and used to the best advantage in order that production should be as high as possible. It is influenced notably by the biological methods of increased production, by the density of stocking, by the size of the fish which are raised.

1   Biological methods for increasing production

These methods were dealt with in Chapter XI. It is certain that the choice of species, the selection of races, the use of special farming techniques, the control of rearing pond temperatures, the mixing of species and age groups, successive and simultaneous productions, intermediate fishing and the control of the hygiene of the ponds and fish all constitute as a whole proper methods of improving the productivity of a pond.

2   Stocking density

High stocking with respect to the number of fish results in a better use of natural food than reduced stocking; the number of fish seeking food being higher in the first case. This leads to a higher quantitative production.

The advantages of stocking in large numbers were brought to light a long time ago in Germany by research at Wielenbach (Bavaria) by Walter (1931, 1934) and at Eberswalde (Prussia) by Contag (1931). They were developed further by Schäperclaus (1960b, 1963).

Stocking in large numbers brings with it, at the same time, an increase in total production and a reduction in the weight and size of the fish produced.

Conversely, reduced stocking results in an increase, often quite important, in individual growth of the fish. It is not accompanied by an increase in the total production of the pond if the number of fish is too small.

The advantages of high stocking densities are as follows (Schäperclaus, 1960b):

1. Thanks to the large numbers of fish seeking natural food, a larger proportion of the latter is used.

2. The accelerated consumption of the food available brings about a correlated development of the remaining food which is much faster, and from this an increase in the global production of the food.

3. In carp culture, the digging of the bottom of the pond by the fish is intensified which is very important.

(a) It results in better and faster use of available nutrients in the mud, thanks to digging the soil, which brings them into solution in the water. This results in an increased production of food.

(b) It also disturbs the water, and this helps to control the higher aquatic vegetation.

4. High stocking is of interest especially if intensive artificial feeding of the fish is carried out. The result is a more or less important indirect action of manuring. A portion of the food distributed can be lost and decompose in the pond, but above all the fish, by means of their excrement, release significant amounts of mineral nutrients and organic matter which have a fertilizing action.

3   Size of cultivated fish

Productivity increases inversely to the size of the fish stocked and cultivated. The release of small-size fish leads to higher production per hectare than can be obtained with large fish.

The maintenance rations of the fish being that much greater as the size of the fish is larger, the use of natural food will be that much better, so far as weight increase is concerned, if the average weight of fish stocked is small and if the average weight of fish harvested for sale will also be small.

From this fact Schäperclaus has recommended, with the aim of increasing the economic productivity of the ponds, that in Europe the weight of carp produced for eating should be clearly lower than that of fish produced previously, which was approximately 1,500 gm ($3\frac{1}{2}$ lb). It is also noticeable that in the Far East a lot of small edible fish are produced.

Bard (1962) also points out that the production of tilapias by the separate age groups method, in which fish of fairly large size dominate, only reaches two-thirds of the production obtained in ponds in which age groups are mixed and in which small fish dominate.

Besides, it has already been pointed out that the natural productivity of ponds stocked with first year carp is higher than that of ponds with older carp.

4 Ponds under water for the first time

Ponds filled for the first time have, during 1 or 2 years, a higher productivity than old ponds. But this is due to the temporary liberation of nutrients not as yet exploited.

SECTION II
STOCKING RATE OF PONDS

A The object of stocking

The stocking of ponds seeks to determine the optimum quantity of fish to be released in such a way as to obtain the highest quantitative and qualitative production under the most economic conditions. Stocking depends on the size and productivity of the ponds.

Stocking tries to adapt exactly the species and above all the number and weight of the fish released to the predetermined physico-chemical and biological conditions in the pond. This is one of the principal means of improving the quality and quantity of production.

On a number of farms, and notably in cyprinid culture, the quantity of fish released must be carefully adapted to the natural productivity even when the fish are fed artificially, for the rational and economic use of the food distributed depends generally on the quantity of natural food produced and consumed in the pond.

Stocking aims to release, in the farm pond, a number of fish of each species and age classes so that each fish reaches a weight as near as possible to that previously decided. It is, in consequence, very important that stocking be carried out as precisely as possible. This is particularly true of fish for eating which are sold at a specified weight.

If production is intensive and concentrated, and depends essentially on the degree of intense artificial feeding, stocking is independent of natural productivity and must above all be adapted to chemical, physical and hygienic conditions. It depends, evidently, on the quantity of artificial food distributed.

B Determining the stocking rate

1 General notes

The stocking of a pond must be calculated in such a way as to obtain maximum production, or the maximum population in a minimum of time, after which the fish are harvested either at once by emptying the pond or in several times by intermediate fishing. The calculation of the stocking density is established in different ways according to whether the fish do not reproduce or reproduce in the pond during growth. Stocking must be carefully adapted to the productivity of the pond and must aim to profit as much as possible from the period during which the speed of growth is at its maximum. This growth is slow at the start. It then increases considerably before slowing down and finally ceasing.

It has already been pointed out that the production of a given area of ponds grows with the stocking in numbers and inversely. On the other hand, individual growth increases when the numbers stocked diminish, and *vice versa*.

In practice it is of interest to go for a rather high stocking density and a slight average size rather than a limited density and a higher average size. In ponds which are only slightly productive the number of fish released is reduced simultaneously, as well as the growth target.

2 General formula for stocking

The general formula for stocking ponds in which the fish do not reproduce during the rearing period is as follows:

$$\text{Stocking rate (Number)} = \frac{\text{Growth target or total productivity (in kilos)}}{\text{Individual growth target (in kilos)}} + \text{Loss (Number)}$$

In the first place, the growth target or the total productivity (natural productivity + productivity due to fertilization + productivity due to artificial feeding) should be predetermined.

The individual growth, obtained by the difference between the average weight at harvesting and the average weight at stocking, is fixed by the farmer within the normal limits of growth of the species raised, and according to the method of farming which is used. It was mentioned above that it is best to produce fish of a somewhat small marketable size.

The average rate of loss of the different species of the principal fish which are farmed is well known (see table point E).

3   Examples of calculations for stocking rates of a pond

1. Take a pond of 3 hectares ($7\frac{1}{2}$ acres) in temperate Europe, suitable for raising carp and of which the biogenic capacity (B) is given as VI. The water is alkaline. The object is to raise carp of 3 years of which the individual growth target is 1 kilo ($2\frac{1}{4}$ lb) and the rate loss estimated at 10 per cent. Increase of natural production due to fertilization is expected to be around 66 per cent and the aim is to quadruple the calculated natural production by means of artificial feeding.

The natural productivity K of the pond is as follows (see formula, Chapter X, page 273), the coefficient k being equal, under the circumstances, to 3·0:

$$K = \frac{Na}{10} \times B \times 3\cdot0 = \frac{300}{10} \times 6 \times 3 =$$
$$30 \times 6 \times 3 = 540 \text{ kg} \,(1{,}191 \text{ lb}).$$

Productivity due to fertilization is given as $540 \times \frac{2}{3} = 360$ kg (793 lb) which brings natural calculated productivity to $540 + 360 = 900$ kg (1,984 lb).

Total productivity equal to the quadrupled natural calculated production will be $900 \times 4 = 3{,}600$ kg (3 tons 11 cwt). Productivity due to artificial feeding will be $3{,}600 - 900 = 2{,}700$ kg (2 tons 13 cwt).

Stocking ought to be $3{,}600/1 + 10$ per cent $= 3{,}600 + 360 = 3{,}960$ K_2.

The quantity of artificial food to be distributed will be, supposing the conversion rate is equal to 4, $2{,}700 \times 4 = 10{,}800$ kg (10 tons 13 cwt).

2. For example take a pond of 10 ares ($\frac{1}{4}$ of an acre) in Western Europe with a biogenic capacity given as VIII and utilized for raising trout for eating. Brown trout are stocked and the individual growth target is 120 gm (4 oz) with rate of loss estimated around 10 per cent. No fertilizer is used and the fish are not fed. The coefficient k equals 1·5.

Natural productivity (K) of the pond will be: $\frac{10}{10} \times 8 \times 1\cdot5 = 12$ kg ($26\frac{1}{2}$ lb).

Stocking will be 12: 0·120 + 10 per cent = 100 + 10 = 110.

If the brown trout are replaced by five times their number of rainbow trout fed with pellets (conversion rate = 1·5) the production target will be increased by 48 kg (106 lb) ($12 \times (5-1) = 48$) and a five times greater number of fish will be released, that is $110 \times 5 = 550$ trout. Throughout the course of the year $48 \times 1\cdot5 = 72$ kg (159 lb) of artificial food (pellets) will be distributed.

4   Stocking with fish reproducing during the rearing period

In ponds with fish which reproduce during rearing, stocking is calculated differently. It is not necessary to measure the fish by number, but by weight. This is notably the case for farms raising tilapias in mixed age groups. The fish reproduce very quickly, and soon there are too many, predominantly small fish without much commercial value. Under these conditions quantitative production is easy to achieve, but qualitative production of marketable fish is difficult.

There are few data available giving the quantity of fish, by weight, which should be stocked, although it is recommended that the average weight of fish stocked should be more or less equal to 10 per cent of the harvest target. Certain workers (De Bont, 1952) have suggested high stocking densities of up to 20 per cent in weight of the target aimed at. Van der Lingen (1957) mentions stocking of up to 25 per cent of the harvest target in unfertilized ponds, and stocking of over 25 per cent in fertilized ponds where the fish are fed. Lessent (1959) suggests that stocking should not be less than 5 kg (11 lb) per are for family ponds using the mixed ages method.

C   Limits to stocking

There is a limit to stocking. This limit varies according to the species, and for each species according to the age and size of the individual fish as well as the methods of more or less intensive farming.

When farming partially depends on natural production, then an exaggerated stocking density in terms of numbers will result in a higher production than with normal stocking, but individual growth will be far below normal. If stocking is well above normal, then dwarfing due to lack of growth will result, which will lead to a waste of food for growing. Schäperclaus (1933) found that in carp cultivation, growth was practically nil when stocking was 16 times above normal. In a more recent study he (Schäperclaus, 1960a) states that stocking six times above normal (compared with stocking corresponding to initial natural production), with fertilization and feeding at the same time, is still good farming. In consequence, he strongly recommends high numerical stocking. Nevertheless, as far as the upper limits are concerned, which should not be surpassed in carp culture, they must be determined in such a way that the individual weight growth is never less than 200 per cent of weight at stocking.

It has been found that stocking which is clearly too high first of all leads to a weakening of the fish which are insufficiently fed. Secondly, the rate of loss is definitely higher among the weakened fish. Losses in such ponds can reach double the wastage in ponds which are not overstocked. In consequence stocking a pond as high as possible is to be recommended, but overstocking must be avoided.

Stopping growth due to overstocking is due to several reasons. Among them the reduction of food available is the most important. Growth limitation also results from competition for minimum life space. In this respect the size of the pond can influence the maximum size of the fish, which can grow larger in large ponds.

In water which is not renewed, as is the case at certain times of the year in Israel, it has been noticed that carp stop growing. Yashouv (1959) estimates that, under these circumstances, the possibility should not be excluded that cessation of growth might be caused, at the same time as other factors, by an increase of concentration in the pond of carp excretions which would be more or less toxic for the fish. For other species the same excretions could be responsible for the fish not reproducing when a pond is overstocked.

For intensive farming, such as rainbow trout, the stocking limit is determined on the one hand by the amount of food available, which can generally be found in sufficiently large quantities as needed, and on the other hand by the dissolved oxygen content which is linked to the temperature and the supply of water. This question was considered in Chapter III: salmonid cultivation.

D Total productivity of a pond, maximum (carrying) capacity and maximum standing crop of a pond

In an effort to reach maximum productivity either quantitative, qualitative or economic, many European writers hold notably to the view that the total productivity and the stocking density on which it rests, are determined by the method described above.

Others (Charpy, 1957) tend rather to utilize the "carrying (or maximum) capacity" of a pond which is reached when the maximum density of the fish population in a pond is obtained in terms of weight. At this stage the total weight of the fish in the pond will not increase further. The capacity of the pond is linked directly to the species of the fish which are raised and the artificial food distributed.

Others again (Hickling, 1962) prefer to refer to "maximum standing crop" defined by Van der Lingen (1957) as the maximum weight of fish which a pond can support without gain or loss of weight, taking into account the artificial or natural food available. Hickling has analysed different factors, both biological and non-biological, which cause variations in the maximum population of a pond.

E Table of medium stocking rates for European ponds

The following table gives average stocking of some European ponds and is based on average productivity, normal weight when stocked and normal individual growth. The table shows average weights at stocking, individual growth, and average weights of fish at harvest, the number of fish to be stocked and normal loss.

The figures shown are only relative to those farms practising natural production. They can be increased by fertilization (Chapter XII, Section V) and above all by artificial feeding (Chapter XIII).

Species and age classes	Individual weights at stocking (gm)	Individual growth (gm)	Individual weight at harvest (gm)	Number and age of fish stocked per hectare (2½ acres)	Normal rate of loss %
Carp (K)					
K_v		2 to 5	2 to 5	50,000 to 100,000K_0 (5 to 10/m²)	75
K_1	2 to 5	50 to 55	50 to 60	2,000 to 4,000 K_v	20
K_2	50 to 60	250 to 400	300 to 450	400 to 800 K_1	30*
K_3	300 to 450	700 to 1,000	1,000 to 1,500	100 to 200 K_2	10 to 25*
Brown trout (B)					
B_1		5 to 10	5 to 10	30,000 to 60,000 B_0(3 to 6/m²)	50 to 75
B_2	10	80 to 120	90 to 130	750 to 1,500 B_1 (7·5 to 15/are)	10 to 30

*Account is taken of important losses due to infectious dropsy.

Table of Formulae used for Calculating the Productivity and the Stocking of Ponds

I   TOTAL PRODUCTIVITY

Total productivity = Natural productivity + Productivity due to fertilization + Productivity due to artificial feeding

II   NATURAL PRODUCTIVITY

$$K = \frac{Na}{10} \times B \times k$$

III   PRODUCTIVITY DUE TO FERTILIZATION

Between 0 and 100 per cent of natural productivity

IV   ARTIFICIAL FEEDING

Amount of food to distribute   =   Productivity due to artificial feeding   ×   Food conversion rate

V   NUMERICAL STOCKING RATE

Stocking rate (in numbers)   = $\dfrac{\text{Growth target or total productivity (in kilos)}}{\text{Individual growth target (in kilos)}}$ +   Loss (in numbers)

Chapter XV

ENEMIES AND DISEASES OF FISH

SECTION I

ENEMIES OF FISH

IN fish culture fish have many enemies. They are found among insects, fish, amphibians, reptiles, birds and mammals. They can be divided between permanent enemies (of eggs, fry and adults), occasional enemies and those competing for food.

I   Harmful Insects

The principal harmful aquatic insects are found among the water beetles, water bugs and dragonflies. Considering their respective sizes they can only attack eggs and fry. They also compete for food. Sometimes only the larvae are harmful but sometimes both larvae and adults can be equally destructive.

Water Beetles. These are abundant principally in water rich in aquatic plants. Among the most harmful are (Figs. 110 and 111):

Great diving water beetle. The adults (30 to 35 mm) eat fry and are very voracious. The larvae have no mouth but possess two powerful hollow mandibles with which to suck victims, leaving only the skin. They can do a lot of harm in nursing ponds.

Black water beetle. As these are vegetarian the adults (40 to 50 mm) are not harmful but the larvae (60 to 80 mm) are strong predators. They can destroy a lot of fry.

Prevention. 1. Do not place nursing ponds under water more than 15 days before stocking in order that harmful larvae have no time to develop. 2. Clean up swamps and grassy ditches in the neighbourhood of the fish farm.

Water Bugs. Water bugs, which are also very numerous, are all at different degrees enemies of fish and in any case are undesirable competitors for food. Only the small species serve as food for the fish. Among the most harmful, *Naucoris cimicoides* (15 mm), *Notonecta glauca* (15 mm) (Fig. 104), *Nepa cinerea.*

Dragonflies. Dragonfly nymphs (Figs. 94 to 97) are aquatic. The adults fly. Nymphs are recognized by an extending lower jaw folded like a hinge. Nymphs of *Agrionidae* have long and slender bodies but are harmless. Nymphs of *Aeschnidae* however, which have long robust bodies and those of *Libellulidae* which have short stout bodies, are more dangerous.

The lengthy, annual drying of a pond helps to stop the multiplication of Odonata nymphs which generally live 1 year or more.

II   Voracious Fish

Voracious fish (pike, pike-perch and perch) live off fish more or less exclusively. Thanks to their occurrence in open water they ensure a natural balance between the voracious species which generally have limited reproduction, and the non-voracious which are very often more prolific. It is said that the latter have a high resilience (the faculty of reproduction and of natural preservation of the species) and that of the voracious is feeble.

Voracious fish which penetrate nursing ponds can cause great damage particularly in nursing ponds. The danger is slight if the pond is properly installed (sunken horizontal screen at the water inlet and the possibility of complete drying out). Puddles of water which may remain after drying can be treated with quicklime.

Equally as undesirable as voracious fish, are those which are somewhat less voracious but in particular are competitors for food (black bullhead (*Ictalurus melas*) in Europe and *Hemichromis* in Africa). They get into unprotected ponds not installed with a good control screen at the water inlet, or (and) badly run (intervals between drying too long and irregular).

III   Harmful Amphibians

Either as larvae or adults, amphibians are all indirectly harmful to fish as competitors for food.

On the other hand they also serve as food for those fish which eat tadpoles, or even in some cases adults.

The most harmful amphibians are frogs. Certain species live off fry (Fig. 424). Tadpoles, which are sometimes harvested in very large numbers when carp nursing ponds are emptied, make the grading of fry difficult.

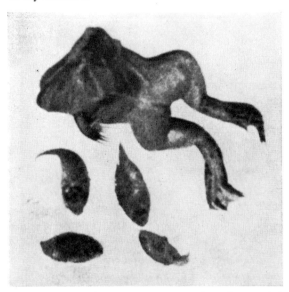

Fig. 424 Toad of the genus *Xenopus* disgorging partially digested small tilapia fingerlings. Province of the Kivu, Zaire. (Photo: Halain.)

One way of trying to get rid of amphibians is to destroy the eggs. This is done from the bank of the pond by means of a scoop net. It is also possible to destroy them with quicklime. The adults can be caught in wire trap nets (Fig. 425).

IV   Harmful Reptiles

Certain reptiles which swim easily and live for the most part in water can also destroy many fry and even good size fish. In the tropics it is true of crocodiles. The destruction they cause is higher in open waters than in ponds.

V   Harmful Birds

There are a number of ichthyophagous water fowl which can do serious damage in farms especially at the time of emptying when the fish are concentrated in a little shallow water. Among the most harmful are kingfishers and herons.

The **Kingfisher** (*Alcedo athis* L.) (Fig. 426). This is a very beautiful blue-green bird with irridescent glints. It can measure from 16 to 17 cm ($6\frac{1}{4}$ to $6\frac{3}{4}$ in) in length and has a strong, straight pointed beak. It is also a fast flyer and skims over the water. Hiding itself as well as it can, it perches on posts, branches or tree trunks and from there plunges like lightning

Fig. 425 Baited trap for *Xenopus*. Nyakabera breeding centre, Kivu, Zaire. (Photo: Halain.)

Fig. 426 Kingfisher (*Alcedo athis* L.). (Diessner-Arens.)

on its prey which are generally small fish from 4 to 7 cm (1¾ to 2¾ in) long which it swallows in one gulp. It can swallow from 10 to 12 a day and do a lot of damage to nursery ponds, above all in salmonid culture. More common than is generally thought, it nests in old burrows made by rats in banks.

Destruction. If the destruction of kingfishers is authorized by law, they can be shot or a spring trap can be placed on posts sticking out of the water by about 1 metre (3 ft) and on which the bird perches and observes. These posts should be fairly isolated otherwise the birds will prefer the near-by branches of trees.

Grey Heron (*Ardea cinerea* L.) (Fig. 427). This is a large bird with long legs, neck and beak. It can measure up to 1 m (3 ft tall). Its wing-spread is 1·70 m (about 5 ft 6 in). Heron feathers are ash grey with some parts almost white and others black. It nests in flocks on high trees.

Herons live round shallow waters in which they walk. They hide, remain perfectly still, standing on one leg with the water up to the knee and the neck folded between the shoulders. When a fish approaches the neck shoots out rapidly. These birds fish during the day and when the nights are clear. They can swallow whole fish between 15 and 20 cm (6 to 8 in) and they destroy great quantities of fish. Sometimes they go for very large fish which they cannot swallow. The fish then carry the mark of the attack across their backs.

Destruction. Herons are very wary and it is difficult to approach them. The best way is to shoot them (when this is legally permitted). If their destruction is permitted in other ways, they can be trapped either with or without the use of bait; they can also be poisoned (dead fish injected with strychnine or phosphorus) or their nests can be destroyed or the young shot there.

Many other **Water Birds** live occasionally off spawn, fry, young fish and are more or less harmful. They must be kept away from spawning beds and nursing ponds. Water birds are also indirectly harmful because they can serve as intermediary hosts for certain parasites which can be dangerous for the fish.

Ducks can be harmful because they disturb the fish with their unceasing coming and going and they also destroy the spawn and fry on the spawning beds and in fry ponds. On the other hand they can

Fig. 427 Grey heron (*Ardea cinerea* L.). (Diessner-Arens.)

be useful in growing ponds thanks to their excrement which acts as fertilizer (Fig. 400).

Swans live off water vegetation and are not harmful even in nursery ponds. Their excrement fertilizes the ponds but dirties the banks.

Water Hens are vegetarian and do not destroy the fry.

Fig. 428   Otter (*Lutra lutra* L.) with captured fish. (Diessner-Arens.)

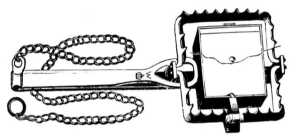

Fig. 429   Otter trap (Diessner-Arens.)

Fig. 430   Musk-rat (*Ondatra zibethica* L.). (Diessner-Arens.)

VI Harmful Mammals

Otters (*Lutra lutra* L.) (Fig. 428). Otters are large Mustelidae which can measure over 1 metre (3 ft) of which 40 cm (1 ft 4 in) is the length of the tail. They can weigh 10 kgm (22 lb). They have very short legs and are webb footed. Their heads are small and very flat, the mouth large and with strong teeth. The snout has a long stiff hair moustache. The body is covered with thick dark brown fur. They stay exclusively in the immediate vicinity of water and burrow into the banks under the roots of trees. The burrow has several openings one of which at least is under the water. In April-May the female gives birth to two to four young.

Otters principally live off fish but supplement their food with other aquatic animals. If food is abundant they destroy more than they can eat, and because of this they are harmful especially for wintering ponds. They eat the best parts of their prey and leave the rest. Otters always attack relatively large fish which means that nursing ponds are almost out of danger.

They live a nocturnal life and roam and hunt mainly at clear night. They are very wary and tricky, swim and plunge with ease and can remain a relatively long time under water. They move about over long distances between ponds or following water courses but return from time to time to the same spots. They can easily be traced thanks to their webbed foot-prints. Their excrement is also recognizable as it is mixed with scales and bones. There are more otters than is generally believed.

Destruction. The best means of combating them is with traps. This is because otters generally follow the same track, when they change their surroundings, for eating their prey and dropping their excrement. Solid toothed traps (Fig. 429) with a 18 cm (7 in) opening should be used. Otters can escape from smaller traps easily. The traps are placed under the water if it is shallow or if not they are laid out of the water in the passages generally used for entering and leaving the water. The traps should not be baited. They should be held by at least a 3 metre (10 ft) long chain.

Wintering ponds, which are particularly vulnerable, can be protected by suitable fencing.

Musk-Rat (*Ondatra zibethica* L.) (Fig. 430). Imported from America in 1905 to be raised for fur, musk-rats frequent stagnant and slow current water rich in aquatic vegetation. They are vegetarian, live off roots and aquatic twigs and rarely attack fish in wintering ponds. They are very harmful because of the large burrows they dig in banks and dikes. Capturing musk-rats is best done with traps and trap nets. They can also be shot.

Other aquatic rodents such as **Brown Rats** and the **Water Rats** can do similar damage. They are omnivorous and destroy eggs, fry and food reserves.

Water Shrew (*Neomys fodiens* Schreber). This little rodent measures from 10 to 13 cm (4 in to 5 in) in length of which from 5 to 7 cm (2 to $2\frac{3}{4}$ in) are the tail. It has a long mobile snout, swims perfectly and, when in the water, the air bubbles held by the hair gives the animal a silver appearance. It destroys eggs and fry and goes for trout eggs and so can ravage fry farms. It is destroyed by poison, traps and by fumigating its burrows with sulphur.

SECTION II
FISH DISEASES

Fish can die through disease or pollution of the water. In the case of pollution, death is more or less rapid for fish of all sizes belonging sometimes to very different species. In the case of disease a large number of different subjects often belonging to one species or even one age class, once attacked can die rapidly within a few days or over a period of several weeks.

The difference between disease and pollution is not absolute for bad quality water can cause more or less serious diseases. In some diseases certain exterior signs permit them to be identified.

In fish culture the fish can be subject to many diseases especially when farming is intensive. Certain diseases cause considerable loss such as infectious dropsy of carp and the viral haemorrhagic septicaemia in trout.

The frequency and importance of diseases which occur on fish farms can be the result of density of stocking, poor conditions of farming, and the small number of species farmed which favours the development of specific diseases. In a wild state fish are widely dispersed and diseases are often not noticed as the risks of contamination are fewer and the losses less.

Fish disease may be classified according to the organs affected skin, gills, liver, etc.; the species: carp, tench, trout; age: fry or fully grown; and also the seasons: spring, summer and autumn diseases. There are also diseases caused by parasites, diseases from stocking, nutritional diseases or those which only affect some fish or the whole farm (epizootic).

The diseases from which fish suffer have been described in specialized books. Here, only the most frequent found in fish culture will be discussed. Before describing them, certain general remarks on prophylaxis, hygiene and disinfection are given.

Fish diseases cannot be transmitted to man with the exception of certain parasites such as tapeworms (Cestodes of the *Diphyllobothrium* = *Botriocephalus* genus), which are found for example in the open waters of the Gulf of Bothnia.

Prophylaxis, Hygiene and Disinfection in Fish Culture

Means exist of fighting several diseases effectively. However, the best means at the disposal of the farmer are phophylaxis and hygiene. In certain cases ponds and growing basins must be disinfected.

1 Hygiene and prophylaxis

In the fight against fish disease the best thing is to try to prevent them for only a very small number can be cured.

First of all the pond water supply must be sufficiently abundant and of good quality in order to avoid dangers from lack of oxygen and pollution.

The ponds must be well maintained and silting and harmful plants kept down. The bottom and ditches must be arranged to ensure complete evacuation of the water when the pond is emptied. The ponds must be regularly dried out and then possibly disinfected with quicklime. The penetration of wild fish must be prevented by means of adequate screens.

Fish must be kept in the best condition possible by avoiding too high stocking densities, lengthy storage, and unnecessary handling and transport. In carp culture it is very important that the fish should be released in the spring in healthy ponds rich in natural food. In salmonid culture artificial feeding must be carried out with particular care and overfeeding should be avoided.

It is always a good thing for one farm not to depend on another for supplies of eggs or fish for restocking, but if this is necessary then only healthy farms should be approached for supplies. It is equally important that the water supply should not come from a water course which is already feeding one or several farms.

When there is an outbreak of disease the dead fish or those seriously affected must be removed from the ponds and buried with quicklime. Infected ponds and the material used must be disinfected. Nets and waders etc. may be routinely disinfected by benzalkonium chloride solution containing 600 to 1,000 ppm of the active ingredient.

2 Destruction of external skin and gill parasites

While it is difficult to fight internal parasites it is possible to free fish of parasites affecting the skin and the gills, by different kinds of baths.

The principal baths and their concentrations will be given when diseases caused by *Argulus*, *Piscicola*, *Costia*, *Dactylogyrus* and *Saprolegnia* are discussed. According to each case the products in general use such as quicklime, common salt, copper sulphate, potassium permanganate, malachite green, formalin and lysol are employed as well as other products which have more special uses such as quinine, lindane, trypoflavine, chloramine, benzalkonium chloride and others.

The proper use of well dosed baths generally produce healthy fish or at least fish that with further treatment can be cured. Those fish which are seriously affected will die but they are condemned anyway.

Einsele (1963) and Bank (1964) consider that a bath in *common salt* (sodium chloride) as a prophylactic measure is important before the ponds are stocked as much for salmonids as for cyprinids. Such a bath will kill several species of external parasites (*Costia*, *Trichodinc*, *Chilodonella*). Furthermore it will fortify the fish and stimulate the appetite. This practice is particularly good after wintering for it permits the fish to meet the beginning of spring, a critical period, in good condition.

Salt baths are given either in large receptacles or in holding tanks or in the vats in which the fish are transported. The addition of air or oxygen is necessary during treatment.

For a bath lasting 1 to $1\frac{1}{2}$ hours 1·5 kg ($3\frac{1}{4}$ lb) of salt are dissolved in 100 litres (about 22 gal) of water for cyprinids and 2 kg ($4\frac{1}{2}$ lb) for salmonids which are less sensitive. If the treatment is given in transport tanks the fish can be transported in the water used for the bath after fresh water has been added to the ratio of 1 kg ($2\frac{1}{4}$ lb) of salt for cyprinids and 1·5 ($3\frac{1}{4}$ lb) for salmonids per 100 litres of water.

3 Disinfection of contaminated tanks or ponds and material

Generally, disinfection of tanks or ponds is carried out with quicklime and sometimes with calcium cyanamide. Potassium permanganate can also be used. For disinfecting equipment, commercial sodium hypochlorite *"eau de Javel"*, quaternary ammonium compounds, iodophores or benzalkonium chloride can be used.

a Disinfection with quicklime or cyanamide. This is recommended for large tanks and ponds with natural bottoms. The water must be emptied, and while the bottom is still wet roughly crushed quicklime should be spread at a ratio of 100 gm ($3\frac{1}{2}$ oz) per square metre (1,000 kg per hectare) (1 ton per hectare). The water should than be let in slowly until it becomes milky, left for 15 days, evacuated again and replaced by fresh water. It is necessary to ensure, at least at the start, that water which might be strongly alkaline does not flow into other ponds containing fish for that would be fatal. Sometimes quicklime, which is difficult to spread uniformly, is replaced by limewash (one part of freshly hydrated quicklime to four parts water). The solution should only be prepared when it is about to be used.

Disinfection with calcium cyanamide (Fig. 391) is sometimes used against whirling disease in particular.

b Disinfection with potassium permanganate. This method, at a ratio of 1 gm per 100 litres (22 gal) of water is suitable for small tanks. If the fish cannot be removed then a weaker bath is given, 1 gm per 200 litres. The fish can withstand this for 1 hour.

c Disinfection of equipment. Nets and waders etc. may be routinely disinfected using a benzalkonium chloride solution containing 600 to 1,000 ppm of the active ingredient. Gerard (1974) recommends using commercial sodium hypochlorite, quaternary ammonium compounds or iodophores for the disinfection of the fish farming equipment. Three litres of water are added to one litre of commercial sodium hypochlorite. This solution is used at 5% for the disinfection of non-metallic objects. However, the chlorine of the sodium hypochlorite evaporates. A solution of quaternary ammonium compounds (1/5,000) or iodophores (250 g/m^3 of water) is used for disinfecting equipment which would be damaged by chlorine.

4 Sending fish for pathological examination. Samples of water

It is desirable that fish sent to laboratories for examination should be alive when they arrive in order to show the characteristic signs of any disease. If they are dead then it is essential that they be fresh when they arrive. Even when transported for only a few hours, dead fish should never be carried in water or packed in paper or plastic but should be wrapped in moss or fresh leaves.

For longer transport they can be packed in material impregnated with 4 per cent formalin. The best method, however, is to pack them in moss or fresh leaves and to place them in a case cooled with ice. All shipments must be as rapid as possible and fish should never be sent over the week-end, when delivery may be delayed.

The shipment of diseased or dead fish should be accompanied by a detailed and precise report including a description of the water (supply water, possible pollution, stocking), the artificial food used, the species and sizes attacked, the clinical signs and behaviour shown by the fish, the start and development of the disease and of the mortality.

In cases of pollution a chemical analysis of the water is necessary. In such cases the examination of the dead fish does not often help very much. Samples of the water should be taken as soon as possible, poured into clean numbered bottles holding 3 litres ($5\frac{1}{2}$ pt). If possible the water samples should be taken up-stream from the source of pollution, then immediately down-stream and then from water in which death took place. The temperature of the water must be noted. There should also be a very detailed report on possible causes, the behaviour of the fish, and the more important signs (mouth, gills, skin).

I Bacterial and Viral Diseases

1 Infectious abdominal dropsy of carp (spring viremia of carp—erythrodermatitis of carp)

This is currently the most feared disease in carp culture. It started ravaging Central Europe in about 1928 and caused great losses to carp farms, but also attacked other cyprinids such as roach.

The literature referring to this disease is very extensive. In fact, two forms of infectious dropsy can be distinguished which are probably different diseases: an acute form (spring viremia) and a chronic form (erythrodermatitis). Spring viremia would be caused by a Rhabdovirus and erythrodermatitis by bacteria, but it is often accepted that a primary viral infection is accompanied by a bacterial invasion. The relationship between these two forms of dropsy is not clear and they can be found together, with all of the intermediate stages. According to E. Otte (1972), the swimbladder inflammation in carp could be another form of infectious dropsy.

Signs. Generally in the spring, when the water starts to warm, the first signs are noticeable and the fish start to die. The intensity of the disease diminishes gradually during the summer. The spring viremia (Fig. 432) is characterized by a swelling of the belly caused by the accumulation of a yellow-pinkish liquid in the abdominal cavity. When the infection starts in springtime, the fish jump. The erythrodermatitis (Fig. 431) reveals itself by bloody spots on the body, which can deeply affect the muscles. At the end of the summer fish which have been cured show dark grey scars. The fins are sometimes partially destroyed. Loss among second year carp can average 30 to 50 per cent but this number can be higher if the infection is serious.

Treatment. It is possible to use preventive measures and to effect a cure.

a Preventive measures. As infectious abdominal dropsy in carp probably arises from the action of both viruses and bacteria, prophylactic measures help to diminish the mortality rate considerably. Everything which might weaken the fish must be avoided such as bad wintering, long storage, unnecessary handling and lack of natural food, especially in the spring.

In Poland the Zator method has been tried with success. Small carp of one summer are not fished in the autumn and remain throughout the winter in the same ponds in which they passed the summer. Consequently the density is not so high as is usual in wintering ponds. Fishing takes place late in the spring, and second year stocking is in ponds well prepared and rich in food. As carp for eating are produced in 2 years and not 3, stocking is fairly low. To avoid a sudden change of environmental conditions in the spring it is also possible to stock ponds in the autumn in which carp of two summers are to be reared next year.

When the disease presents itself the dead fish must be removed carefully and then destroyed. After emptying, the pond should be left dry throughout the winter and disinfected with quicklime. Stocking in the spring should only be in ponds which have been well prepared with fertilizers and in which natural food has had time to develop. Artificial food of good quality can be distributed, though not in excess, early in the season in order to allow the fish to get over this most critical period with less danger.

Fish of different origins should not be mixed and on each farm an effort should be made to obtain, by means of selection, carp which are able to resist dropsy.

b Curative measures. Antibiotics are effective against the bacteria but scarcely affect a virus, which explains why it is that with the medicinal products used a real improvement can be obtained but a lasting cure is more rare. The employment of antibiotics is useful in fish farming in which carp are reared for eating. If it is a question of carp for restocking (angling waters) there is always a danger of a relapse after a change of environment.

If the carp are diseased when they are stocked or if there is a danger of infection, antibiotics can be used: injected, in a bath or mixed with food. Most of the time chloramphenicol is used for it is effective against *Aeromonas punctata*, a bacterium generally found in great abundance among infected fish. If the infection is due to other similar bacteria or to resistant strains of *Aeromonas punctata*, then oxytetracycline and streptomycin are more effective. Both are administered in identical doses. A mixture of oxytetracycline and oleandomycin can be useful if incorporated with the artificial food.

For injection purposes (Fig. 433 to 436) syringes have been designed to allow the treatment of several

Fig. 431   A mirror carp showing characteristic signs of the ulcerative form of infectious abdominal dropsy. The muscular tissue shows open sores, deep red in colour and encircled by a whitish edge. The dorsal fin is attacked.

Fig. 432   Roach badly suffering from the intestinal form of infectious abdominal dropsy.

Fig. 433

Fig. 434

Figs. 433 and 434   Young carp being given antibiotic injections. Four teams of paired workers work simultaneously. Königs-wartha, German Democratic Republic.

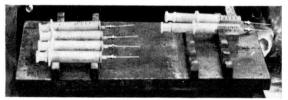

Fig. 435   Syringes prepared for antibiotic injection in series of young carp threatened by abdominal dropsy.

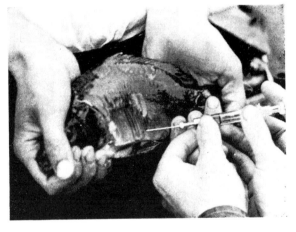

Fig. 436 Injecting an antibiotic solution into a mirror carp.

fish without the necessity of refilling. The injection is given in the body cavity at a ratio of 1 to 1·5 mg of chloramphenicol for each 100 gm (about 4 oz) of fish body weight. The necessary quantity of chloramphenicol is dissolved in from 1 to 2 ml of water according to the size of the fish.

For K_1, Schäperclaus advises a 10 hours chloramphenicol bath (60 mg of chloramphenicol per litre of water).

It is also possible to mix antibiotics with food in such a way that fish weighing 100 grm (about 4 oz)

will absorb 1 mg per day. This method is advised for the ulcerative form of infectious abdominal dropsy. If, however, it concerns an intestinal form the affected fish will eat very little, or stops eating. Food is distributed as early as possible in the spring and at least every 3 days during 2 or 3 weeks, preferably after the fish have become used to the same food without antibiotics. It is recommended to distribute food in several different plots for all the fish must be able to feed off food containing antibiotics.

Fig. 437   A rainbow trout showing an acute stage of viral haemorrhagic septicaemia. (Ghittino, 1966.)

According to German research (Schäperclaus, 1961; Bank, 1967; Mann, 1967), antibiotics are not found in the flesh of the carp a few days after the injection or after even prolonged feeding with antibiotics.

2   Viral haemorrhagic septicaemia of trout (VHS)
This is a viral disease found frequently on intensive trout farms especially in rainbow trout. It is known by several names: INuL-Krankheit, infectious anaemia, entero-hepato-renal syndrome, Egtved disease, Danish virus. It is now commonly known as viral haemorrhagic septicaemia of trout (VHS).

As early as 1941 Schäperclaus considered that the disease was caused by virus. This was confirmed later and Zwillenberg (1965) has isolated the virus. The environment, and above all the feeding regime, also seemed to play a part in the appearance of the disease and its development.

Signs. When latent the disease is difficult to discover. It can become chronic, suddenly, among apparently healthy trout and a high mortality rate will follow when the environment is altered or the fish are handled.

It can last for months in its chronic form and the growth of the fish will be small. Everything seems to indicate that any unfavourable element is sufficient to kill off the infected fish rapidly. If the temperature of the water remains sufficiently high and feeding is moderate, mortality is not very important. During the cold seasons (autumn and spring), particularly if the fish are fed intensively, losses can be severe. This is also the case in the summer if there is a sudden fall in temperature or in the quantity of dissolved oxygen in the water.

The signs of the disease, which causes grave metabolic changes, are numerous, varied and only slightly characteristic. As with most viral diseases, at one time one organ and at another time another is attacked, though most often the kidneys and the liver are affected (Fig. 437). Other diseases have similar symptoms and can be present at the same time as VHS.

Of the numerous signs mentioned above, the following are the most obvious. First is apathy: the fish move about very little and rest on the surface close to the banks. When they swim their movement is sharp and disordered and their position in the water is abnormal. Certain fish girate and take on a dark colour. They develop exophthalmos (swollen eyes), anaemia characterized by a reduced haemoglobin rate and pale gills which turn pale grey immediately after death. The belly swells and a foul smelling yellowish liquid is found in the body cavity. The anus protrudes and there are slight haemorrhages in the muscles and on the swim bladder. The liver is grey brown and friable while the posterior parts of the kidney swell. A neutral or alkaline liquid is found in the stomach, the intestine turns red and is inflamed; muscular oedema may be present. The fins fray and there are sores on the skin.

Lipoid degeneration of the liver presents several signs identical with those of VHS. It is mainly a disease of hatchery salmonids and is due exclusively to poor feeding. It presents identical symptoms with those of VHS. Ghittino (1963) pointed out the major differences between the two conditions. With VHS the haemorrhage concerns the whole body and above all the muscles, swim bladder and the gills. In lipoid degeneration the liver turns yellowish-brown. If trout suffering from VHS have an affected liver it is spotted and turns wine red or grey in colour. The two diseases can exist at the same time.

Brunner (1961) and Bellet (1962) also speak of an intestinal form of furunculosis with signs identical to those of VHS. Furunculosis is caused by *Aeromonas salmonicida* and can be treated with sulphonamides, nitrofurans and antibiotics, but none of these are of any value in cases of VHS. Often only a full bacteriological examination can tell the difference between the two diseases.

Treatment. Currently there is no effective treatment of the VHS disease. Medicines are of little use. According to Schäperclaus it may probably be possible to obtain a certain measure of immunity by putting young fish in contact with the disease but there is a risk and the immunity will be very uncertain. The same writer also advises selection of resistant strains.

When farms are affected it is possible to reduce the virulence of the VHS disease and the death rate in the following way. Bringing in fish from uncontaminated farms; isolation of the diseased fish and removal of dead or seriously affected ones. Disinfect the ponds if the disease is very bad. Utilize good ponds with clean bottoms and a sufficient water supply. Avoid high stocking densities and over feeding. The food distributed should not be too fatty but rich in vitamins (A, B complex, E, T), containing appropriate protein and fillers.

If the disease is chronic, feeding must be temporarily stopped and then followed by the progressive distribution of light and varied food.

3   Infectious pancreatic necrosis of trout (IPN)

This virus brings sudden and widespread death to young trout at the start of artificial feeding.

Signs. Affected fish revolve on their longitudinal axis, swim around rapidly and in a disordered fashion and then become immobilized near the banks or on the bottom of the pond. Often they turn dark in colour and their eyes protrude (exophthalmos). Dissection shows a viscous whitish liquid in the stomach and intestine. The liver and the spleen are pale and the gall bladder shows severe necrosis.

Treatment. There is no known method of treating this highly contagious disease effectively. Prophylaxis has been tried in an effort to avoid the spread of the pathogenic germs. The feeding of povidone-iodine may be beneficial (Economon, 1963).

4   Furunculosis of salmonids

Furunculosis of salmonids is a bacterial disease caused by *Aeromonas salmonicida*. In fish culture it principally attacks brown trout and also brook trout, but rainbow trout appear to be more resistant. It also attacks grayling and many other species. The infection is transmitted through the digestive tract or through small wounds on the skin. The bacteria breed abundantly in the blood, the liver, the spleen and the kidneys. This disease can kill rapidly and on a large scale, above all among 2-year-old salmonids, though fry and small trout can also be attacked.

Signs. The name of the disease comes from the way in which it shows itself in certain cases as bloody boils of various sizes from a pea to a nut (Fig. 438) or by small lumps under the skin. Often, also, the centre of infection is hidden in the muscular fibres. In these forms the disease develops slowly.

Fig. 438 Brown trout affected by furunculosis. Above, sores; below, boils (internal and external views). (Hofer, 1904.)

In other cases the signs are very difficult to recognize. All of the internal organs are infected. This is the acute form, feared as soon as the mortality rate starts to grow greater progressively. A bacteriological examination is always necessary in order to obtain a sure diagnosis for all the symptoms: inflammation of the intestines, spotted liver, haemorrhages at the base of the fins, within the gill covers and in the muscle fibres, are very similar to other diseases. The disease is, above all, frequent in dirty water which contains organic matter.

Treatment. Infected ponds must be isolated and the dead fish carefully removed and destroyed. Care must be taken that they are not thrown into neighbouring ponds or water courses. After emptying, the ponds should be disinfected with quicklime or calcium cyanamide. All materials used should be disinfected. Handling and sudden changes of the environmental conditions should be avoided if the disease in its latent form is suspected.

If there is a danger of infection, fish must be raised in ponds fed with very pure water and in which the temperature is not too high and the amount of dissolved oxygen constantly high.

Furunculosis can be treated by mixing sulphonamides, nitrofurans and antibiotics with the food. These must be administered as soon as the disease is noticed.

According to McCraw (1952) and Deufel (1963) 100 kg (220 lb) of trout can take a daily dose of a mixture of 10 gm of sulphamerazine and 3 gm of sulphaguanidine over a period of 8 days. A second treatment, if necessary, can be given 4 weeks later. Gutsell (1947) advises a daily dose of 13·2 gm of sulphamerazine for 100 kg of trout over a period of at least 3 weeks and up to 10 days after the disease has disappeared. Bellet (1962) gives the following indications for antibiotics: 10 gm daily of chloramphenicol for 100 kg of fry over a period of 10 days; 5 gm of chloramphenicol for 100 kg of fingerlings or trout over a period of from 6 to 10 days. It is also possible to administer oxytetracycline.

According to Christensen (1966) possible infection of eyed eggs is prevented by disinfecting them with a 0·015 per cent solution of merthiolate or a 0·185 per cent solution of acriflavine. Egg disinfection with 1 per cent providone-iodine is highly effective, and is less toxic than other disinfectants to rainbow trout eggs (McFadden, 1969).

5 Bacterial gill disease

This gill disease, caused by myxobacteria, mainly affects young salmonids reared in intensive systems and fed a dry concentrated powder feed. Bad hygienic conditions favour the appearance of the disease: troughs or tanks with too many food wastes, too high density, insufficient oxygenation.

Signs. The young diseased fish stop feeding and often remain motionless near the water's surface. The gills are swollen and rather dark red in colour. The gill filaments secrete a lot of mucus and the opercula seem to be open.

Treatment. First of all, troughs and tanks should be kept clean and provided with sufficient water

Fig. 439 Roach showing wounds resulting from brutal manipulations and threatened with fungus infection.

Fig. 440 Dead pike eggs attacked by fungus.

renewal. The following treatments can be efficient: copper sulphate bath: 1 g in 2 litres of water for one minute, 1 g in 10 litres of water for 10 minutes, or 1 g in 100 litres of water for one hour; quaternary ammonium bath: 2 g of active ingredient in 1,000 litres of water for one hour.

II Fungal Diseases

1 *Saprolegnia* infection

Water mould or *Saprolegnia* infection is caused by fungi of the genera *Saprolegnia* and *Achlya* which develop in injured, weakened, diseased or dead fish. They also develop in dead eggs in incubation trays or jars and by contact can contaminate healthy eggs. These fungi are present in all fresh water, but above all in water rich in organic matter in which they find a favourable medium for development. In all media parasitic fungi live as saprophytes, on the remains of food and the bodies of dead fish.

It is not really correct to call disease the *Saprolegnia* infection for it is in fact a parasite quite inoffensive for healthy fish living in healthy surroundings, but attacking weakened fish.

Saprolegnia can affect all species of fish in all environments and at all ages. Particularly all external lesions open the way for the parasitic fungus.

Signs. Infection is characterized by woolly, grey-white or lightly brown blotches on the skin, fins, eyes, mouth or gills (Fig. 439). Affected eggs (Fig. 440) are completely covered by the fungus, including dead eggs and neighbouring healthy eggs which are stuck together.

Treatment. Action must be taken to avoid or treat the primary causes: injuries, long storage in cement tanks, brutal handling, diseases, weakness, unhealthy surroundings, bad quality water or too high stocking.

The best treatment likely to cure fungus disease includes baths for fish held in storage or eggs being incubated.

a Treatment of stored fish. These are treated by one of the following baths: potassium permanganate: 1 gm per 100 litres (22 gal) of water for between 60 to 90 minutes; salt baths: 10 gm ($\frac{1}{3}$ oz) per litre of water during 20 minutes for young fish, 25 gm (1 oz) per litre of water for 10 minutes for older fish; a copper sulphate bath: 5 gm in 10 litres (2 gal 2 pt) of water in a wooden vat until the fish show signs of weakness; a malachite green (zinc-free) bath prepared from a stock solution of 1 gm of malachite green in 450 ml (3 gills) of water; 1 to 2 ml of the solution are used per litre as a bath lasting 1 hour.

Malachite green can also be used for treating small ponds: 1 gm of malachite green for 5 to 10 m³ (6½ to 13 yd³) of water; in trout ponds a lower concentration is used.

b Treatment of incubating eggs. It is possible to stop fungus developing on trout eggs in incubation by giving them a daily formalin bath for 15 minutes. To each litre of water fed to the hatching troughs 1 to 2 ml of 30 per cent formol should be added (Steffens, 1962), by means of a recipient provided with a tap.

The malachite green (generally a copper oxalate) is currently used to treat fungi when they appear and develop in great numbers on incubating eggs. According to Bellet (1965) the concentration for jars and incubation troughs is 5 mg/litre or 1 gm of malachite green for 200 litres (44 gal) from $\frac{1}{2}$ to 1 hour. A stock solution is previously prepared for example with 10 gm of malachite green for 1 litre of water. The amount of water in the incubation troughs is measured and the supply is cut. A quantity of stock solution is taken to obtain a 5 mg/litre concentration or 5 mg/m³ in the troughs. This solution to which water should be added, is sprayed out evenly in the troughs. A trickle of water is allowed to run in order to assure the even spread. After $\frac{1}{2}$ or 1 hour the taps are turned on as before. If incubation is in jars the malachite green is tipped into the feeding reservoir. This treatment is carried out twice per week from the spawning to the first hatching.

Deufel (1965) suggests also a concentration of 5 gm/m³ during 45 minutes to be repeated every 5 or 6 days. The treatment must stop about 5 days before hatching. It is carried out in this way in the hatching troughs. A recipient with a pierced hole at its base is placed at the head of the trough through which the malachite green solution flows into the water supply. This recipient empties in 45 minutes so the volume of the solution it contains is of no importance. The concentration of the solution is calculated in such a way that the troughs receive 5 gm of malachite green per cubic metre of water. For treating eggs incubated in Zoug jars the water supply is stopped temporarily in order to remove later about half the water which is replaced by a solution containing 5 gm of malachite green per cubic metre of water. After having been well mixed with a feather the supply water taps are opened.

Gottwald (1961) suggests a concentration of 10 mg of malachite green per litre during 15 minutes every 2 days.

In the cases cited above the water must be cut off during treatment. It is also possible to treat the eggs without touching the normal water supply.

Deuffel (1957) suggests the following method for incubation troughs. First the volume of water to be treated in the troughs is measured. The stock solution to be used should contain 10 gm of malachite green per litre of water and the quantity needed equals $\frac{1}{20}$th of the volume of water in the troughs. (Ex: for an incubation trough of 2 × 0·50 × 0·20 m or 400 litres (88 gal) some 20 litres (4 gal 3 pt) of solution at a ratio of 10 gm per litre that is 200 gm (7 oz) are necessary.) If the water supply is 20 litres per minute then the solution is

Fig. 441 Installation for treating fish eggs in Zoug jars; use of malachite green against fungus.

poured in at the inlet of the troughs at a rate of 1 litre per minute (in the example given above 20 litres of solution will last 20 minutes). If the flow is 10 litres per minute 1 litre solution must be poured in every 2 minutes and so on.

Cappello (1967) describes a treatment method for eggs in jars which is also applicable to incubation troughs. A small diameter vertical tube provided with a tap is connected to the supply pipe of the water and installed beyond the tap of this pipe. The malachite green solution flows into the tube through a siphon, at a given speed. If the flow in the jars or troughs is regulated at 10 litres per minute (2 gal 1½ pt) there will be a flow of 10 ml per minute of solution containing 0·5 gm of malachite green per litre, during 45 minutes. This treatment is given twice weekly (Fig. 441).

Steffens *et al.* (1961) draw attention to the fact that the treatment of rainbow trout eggs with malachite green can cause certain genetic defects. For this reason they express certain reserve on the use of malachite green in fish culture.

It should not be forgotten that in commerce there are many basic salts sold as malachite green, of which the toxicity can vary for the same concentration. Preudhomme (1960) points out that one of the basic salts, a chlorozincate, should not be used in fish culture. A test is necessary before using any malachite green salt in fish culture.

2 Gill rot or branchiomycosis

Fungi of the genus *Branchyomyces* cause the gills to rot in most cyprinids and also in pike. On farms, particularly carp of 1 or 2 summers are attacked. The disease manifests itself principally in the summer in densely stocked ponds, rich in organic matter and with abundant phytoplankton.

Signs. At the start the fish have pale gills showing deep red coloured patches. Finally the gills are partially destroyed (Fig. 442), turn a yellow brown colour and are attacked by *Saprolegnia*. In warm weather the development of the disease can be rapid.

Treatment. One means of prevention is to avoid too dense stocking and when the weather is warm, a water supply too rich in organic matter such as lightly polluted water. When the weather is very hot it is necessary to reduce or even to suppress artificial feeding and if possible provide a good supply of fresh water.

When the disease is present, 200 kg (440 lb) of

Fig. 442   A carp of 2 summers having suffered from branchiomycosis (gill rot). The gills are partially rotted away. (Schäperclaus, 1954.)

finely ground quicklime should be spread per hectare taking care that the pH does not exceed 9·0. It is also possible to use, even as a preventative, an algicide, such as copper sulphate or possibly benzalkonium chloride. Schäperclaus advises the following doses of copper sulphate: 8 kg (18 lb) per hectare for ponds with an average depth of 0·50 m (1 ft 8 in); 12 kg (26 lb) per hectare for ponds with an average depth of 1 m (3 ft 3 in). These quantities are spread out in four monthly distributions of 2 or 3 kg (4 to 7 lb) each from May to August. Benzalkonium chloride may be used as a 1 hour bath containing 1 to 4 ppm of active ingredient. A copper sulphate bath: 1 gm in 10 litres (2 gal 1½ pt) of water for from 10 to 30 minutes will kill the parasites.

III   Diseases caused by Protozoa

Numerous protozoan parasites live on the bodies of the fish. They attack the skin both on the surface and underneath, the gills and they spare no organ. Muscle fibres and cartilaginous tissues may also be attacked.

1   Costiasis

Costiasis is a common disease which can attack all species of fish from the earliest age. In fish culture it is particularly important among salmonid fry. It is also dangerous for young carp. Above all it is a

disease which develops from storage and is generally noticed when the fish are too densely packed in the rearing troughs and in holding ponds and tanks. It is also found in wintering ponds or when the living conditions of the fish are unfavourable through lack of food or if the water is too acid. In salmonid culture outbreaks are most serious at the first warming of the water in spring. Within a few days the infection can become intense.

The disease is caused by the flagellate *Costia necatrix* measuring from 10 to 20 thousandths of a millimetre. They live in large numbers on the skin, the fins and the gills (Fig. 443).

They should be considered especially as a parasite favoured by debility.

Signs. The skin of the affected fish is covered with a light grey-blue film, while those parts which are seriously affected can show red patches. Affected gills turn brown and can be partially destroyed. The fish weaken and lose their appetite, while small carp swim about slowly. Finally the too-weakened fry and fish die.

Treatment. Preventive methods do not always succeed. Baths can be used to treat costiasis after which the infected fish must be placed in ponds rich in food but they must not be too densely stocked.

Formalin baths suggested as early as 1909 by Léger are effective and trout fry withstand them well: 20 to 25 ml (1·2 to 1·5 in³) of formalin in 100 litres of water over a period of from 30 to 45

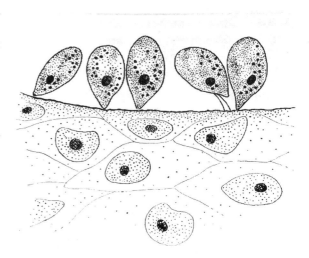

Fig. 443   *Costia necatrix* attached to the epidermis of a catfish. (Davis, 1956.)

minutes or 40 ml (2·4 in³) of formalin in 100 litres of water for 15 minutes.

For trout fry and small carp it is also possible to give salt baths: 10 gm of kitchen salt per litre of water for 20 minutes. The doses can be doubled for large fish and the bath should then take from 10 to 15 minutes.

Davis (1956) suggests a PMA (pyridil mercuric acetate) bath for 1 hour at a concentration of 1 gm for 500 litres (110 gal) of water, but PMA is toxic.

Tack (1956) points out that rainbow trout fingerlings coming from badly infected ponds have been cured by dividing them up between other ponds, that is, less densely.

2 Whirling disease

This disease is found all over Europe and North America and attacks salmonids. It is very dangerous in fish culture as it can lead to a high mortality rate and to severe economic loss, especially in rainbow trout and brook trout fingerlings. Brown trout fingerlings are more resistant but are not immune to the disease.

The causative agent is a protozoan *Myxosoma* (*Lentospora*) *cerebralis*. The spores are found in the mud on the pond bottom where they are very resistant, often over years, to drying out and frost so much so that from one year to another the crops are contaminated by the spores remaining in the pond bottom. Generally the fish are infected when young, probably through swallowing the spores. The disease is not transmitted by the eggs. Most often it is noticed some time after the stocking of the ponds. According to Putz and Hoffman (1966), in contaminated water small fry trout can be affected from the third day after hatching. Once in the intestines of the young fish, the spores liberate sporozoites which, carried by the blood, settle

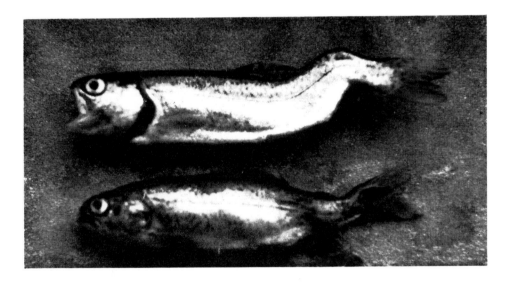

Fig. 444   Rainbow fingerlings suffering from whirling disease. The spinal column is deformed; the posterior section of the body sometimes turns black. (The bottom figure: Plehn, 1924.)

principally in the soft cartilage of the head, especially in the auditory region as well as in the spine and thus affects the equilibrium. When the young trout reach 7 or 8 cm (3 in) the cartilaginous tissues harden and the conditions become more difficult for development of the sporozoites. They then form spores. By this time the acute period of the disease is passed. The infected fish which have survived are now spore carriers even to 3 years of age. They are in no danger but their growth is generally poor. If the infected fish die in the pond then the spores return to the mud and the cycle can recommence the following year from the moment the fry are stocked. It is not impossible that spores can also be liberated with the excrement of the fish.

Signs. Fingerling trout whirl round and round in the same direction several times and then fall to the bottom. They come up again some time later, swim normally but start whirling again after an interval more or less long. This can go on for weeks. Certain fish can develop black tails caused by an infection in the region of the sympathetic nerve which controls the pigmentation of the caudal part of the body. Also certain malformations are often noticed (Fig. 444) in the infected fish, such as deformation of the spine, shortening of the jaws and the gill cover, bumps and small cavities in the head. These deformities in young trout are not always noticeable but they are recognized at a more advanced age. The first symptoms of whirling disease are easily noticeable from 40 to 60 days after the start of the infection. The disease can cause high losses among trout fingerlings of which more than 80 per cent can be infected.

Treatment. Steps must be taken to avoid infection as contaminated fish cannot be cured. Once a farm is infected it is difficult to rid it of the disease.

The nursery ponds and the fattening ponds must be strictly separate and the same ponds must always be used for the same farming operation. The dead fish must be removed and buried with quick lime.

It is recommended to empty the nursing ponds no later than the month of September. At this time the spores have not yet formed in the bodies of the young fish and cannot, therefore, aggravate the infection of the water and the bottom of the pond. It is also possible to reduce the loss percentage considerably by rearing the fingerlings trout in large ponds only slightly stocked and without artificial feeding.

On an infected, intensive production salmonid farm, only complete and careful disinfection can really combat the disease. All conduits, the incubation shed and its installation, the storage tanks and all the material used must be disinfected either with quicklime, calcium cyanamide, benzalkonium chloride, formalin or sodium hypochlorite. Ponds and banks can be disinfected with quicklime or better still with calcium cyanamide (Fig. 391). Tack (1951) advises the use of 0·5 to 1 kg (1 to 2 lb) of calcium cyanamide per square metre (5 to 10 tons per hectare) on the bottom, the banks and in the immediate vicinity of the pond. The first disinfection should be carried out in the autumn after emptying followed by winter drying and a second application should follow in the spring about 6 weeks before liberating the fry. After being put under water the pH should be controlled and it should not exceed 8·5. The toxicity of the water can be tested by keeping some fish in a holding net or in a basket.

The complete disinfection of a fish farm is both difficult and costly. According to Rasmussen (1965) Danish salmonid farmers avoid infection from an early age by rearing fry in concrete tanks over a period of at least 2 months and until they reach 5 or 6 cm (2 to $2\frac{1}{3}$ in). Some 35,000 fry are reared in tanks 6 m long, 0·70 m wide, 0·70 m deep (about 20 ft × 2 ft 4 in × 2 ft 4 in) or about 10 fry per cubic decimetre. The fingerling trout are graded and those exceeding 5 cm (2 in) are liberated in ordinary ponds with natural banks which offer less danger at that stage if they are infected.

A similar and general measure which can prevent infection is to feed the incubation shed with spring water which in practice can always be protected from contamination. The fry are fed until 6 or 8 weeks in troughs supplied by the same water; thus they pass the most dangerous period of risk of infection in spore-free water. Farming continues in circular ponds or in ponds of normal shape.

The complete rearing of fingerling trout in long shaped cement basins or in round tangential circulation tanks which can be easily disinfected is one of the best ways of fighting this dreaded whirling disease.

3 Ichthyophthiriasis

Ichthyophthirius multifiliis is a ciliated protozoan which affects the skin and the gills of most fish species of all age classes. Contamination can develop rapidly in water in which fish are closely packed.

The life cycle of the parasite is complex (Fig. 445).

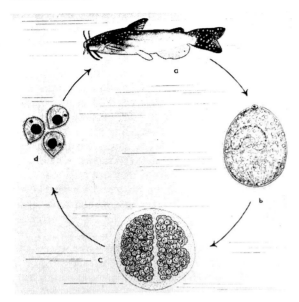

Fig. 445 Life cycle of *Ichthyophthirius multifiliis*: (a) parasites in the skin of catfish; (b) adult parasite swimming freely; (c) encysted parasite; (d) young parasites in search of a host. (Davis, 1956.)

The young parasites are very small and move about in the water seeking a host. If they meet a fish they attach themselves between the dermis and the epidermis. They raise the epidermal cells and grow rapidly until they are about 1 mm in size and are visible to the naked eye. When they reach full size they leave the host, fall to the bottom, encyst and then multiply by a cellular division, liberating swarms of young parasites which start looking for hosts.

Signs. Small white specks are noticeable on the skin and when the attack is serious these can change into white spots. The gills can also be attacked. The fish jump in the water and try to rid themselves of the parasites by rubbing themselves up against the bottom or against any submerged object. A severe attack can cause heavy losses among carp, trout and other young fish.

Treatment. It is difficult to get rid of parasites once they are under the epidermis. When they leave their hosts they can be eliminated by placing the affected fish for several days in storage tanks traversed by a strong current. It goes without saying that this water must not be emptied into ponds containing fish.

If there are enough ponds the fish can be changed from one pond to another every 2 or 4 days. As *Ichthyophthirius* cannot live without hosts they die after a few days in ponds which are not stocked.

All fish carrying parasites must be eliminated, particularly when the wintering and storage ponds are stocked, because here the fish are densely packed. If the infection is noticed from the start it can be fought by spreading the fish out over a greater area.

Deufel (1960) has drawn attention to the cure of infected trout and carp in ponds thanks to two or three distributions of 1 gm of malachite green for 10 m² (12 yd²) of water every 2 days. Amlacher (1961) also found that after 10 days the parasites died when the carp ponds were treated with malachite green at a concentration of 0·15 gm/m³. During that period the water supply must be cut off.

Tests by Rychlicki (1968) who spread carp nursing ponds with 3,000 kg (3 tons) of quicklime per hectare in 2 distributions with an interval of 2 days between, destroyed the parasites. The tests were undertaken in alkaline water with a large dose of quicklime; the fish were never in danger. The pH reached values of 8·5 to 9·0.

It is also advisable to give baths over several days: quinine (1 gm for 20 litres of water) (4 gal 3 pt), trypoflavine (1 gm for 100 litres of water) (22 gal), or chloramine (1 gm for 100 litres of water). During this operation the water must be well aerated. Repeated commercial formalin baths: 1/4000 during 1 hour, PMA (pyridyl mercuric acetate): 1/500 000 during 1 hour, and of common salt: 30 gm per litre (1·1 oz) until fish show signs of weakness are used in the U.S.A.

IV Diseases caused by Worms

The number of external and internal parasitic worms is very considerable and their form, structure and habits vary considerably.

1 Disease caused by fish leeches

The blood sucking leech (*Piscicola geometra*) is a very common external parasite found especially in calm water. It attacks all species of fish in ponds.

The blood sucking leech is a cylindrical, ringed worm about 2 or 3 cm (¾ to 1 in) in length and 1 mm in diameter. It has a sucker at each end of its body, thanks to which it attaches itself to any part of the body of the fish where it sucks the blood. When it has had its fill it detaches itself from the fish and

Fig. 446 Carp with fish leeches (*Piscicola geometra* L.). (Hofer, 1904.)

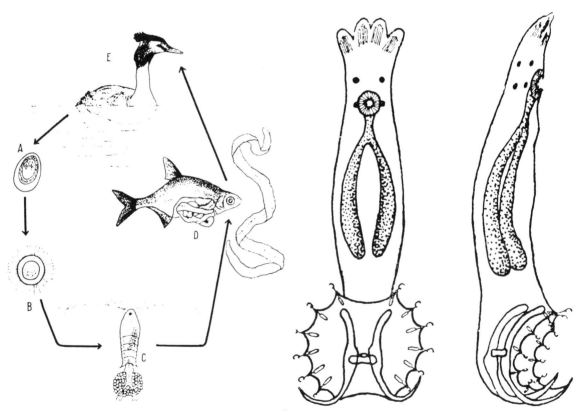

Fig. 447 Life cycle of *Ligula intestinalis*. A: egg; B: ciliated larvae; C: procercoid larvae in the body of *Diaptomus* (copepod); D: tapeworm in the body of a bream; E: sex ripe worm in the intestine of a teal. (Schäperclaus, 1954.)

Fig. 448 Ventral and lateral view of the gill fluke, *Dactylogyrus*. The head has four eyes. (Christensen, 1966.)

swims freely in the water with a undulating movement. When it is attached to the fish it moves like a caterpillar.

Signs. Affected fish may be almost covered with these parasites (Fig. 446). They look poor and are weakened through loss of blood. The incisions make it possible for them to be affected by other parasites and diseases such as fungus. The harm done depends on the number of parasites and this can be very high.

Treatment. If the parasites are very numerous the pond must be emptied and the fish bathed. With the aid of a slightly folded hand net the fish are dipped into a solution of lysol: 1 ml of lysol for 5 litres (1 gal) of water for 5 to 15 seconds. The fish must then be rinsed in fresh water. Lysol is a mixture of 50 per cent cresol and 50 per cent soap. A salt bath can also be given: 10 gm of kitchen salt for 1 litre of water for 20 minutes when the fish are young or 20 gm for 10 minutes when the fish are older.

A simple and rapid method permitting the treatment of a large quantity of fish is a lime bath. For a period of 5 seconds the fish are plunged, with the aid of a scoop net, into water containing 2 gm of quicklime per litre of water.

When there is an outbreak, after the drying out of the pond the bottom must be treated with quicklime in order to destroy the eggs and the leeches still living there.

Fish ponds can also be treated with trichlorfon: 0·4 g/m³ for carp, tench and eel, or 0·2 g/m³ for more sensitive species such as trout and pike. If the treatment has to be repeated, this is done two to three weeks later.

2 Ligulosis

This is caused by a long ribbon-like tapeworm, *Ligula intestinalis*, which is yellowish-white in colour and measures between 15 to 40 cm (6 in to 1 ft 4 in). (It can reach up to 2 ft 5 in.) It is 0·6 to 1·5 cm wide and its life cycle (Fig. 447) is complex.

The disease is met more frequently in open waters than in fish culture ponds. The worm attacks numerous species of fish but particularly cyprinids. In fish culture it is possible to prevent the disease by eliminating water fowl.

3 Disease caused by *Dactylogyrus*, **the gill fluke**

Dactylogyrus, of which there are many species is a small flat trematode worm (Fig. 448) which attacks the gills of nearly all fish, above all in ponds. One of the most common species found on carp is *Dactylogyrus vastator*, which can reach 1 mm in length. This parasite can be a great danger to young carp up to 5 or 6 cm (2 to 2½ in) especially if the fish are not in good condition and the attack is severe: several hundred parasites to one fry. Large fish rarely suffer from *Dactylogyrus*.

Signs. It is in the summer that the parasite is found in first carp nursing ponds. It is not easy to discover the disease if the attack is not severe. The gills swell and turn grey at the edges. Afterwards they can also be partially destroyed.

Treatment. The first nursing ponds must be well prepared and rich in food and the stocking density calculated so that the 5 to 6 cm length is reached within a few weeks. The spawning ponds should

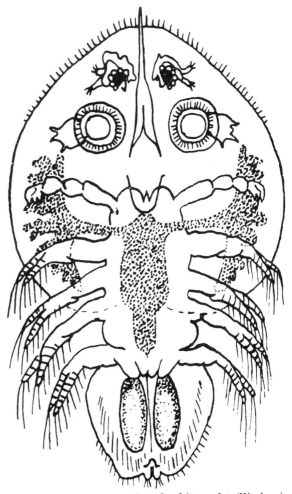

Fig. 449   The carp louse (*Argulus foliaceus* L.). (Wagler, in Amlacher, 1961.)

only be put under water during the spawning period and the first nursing ponds should be left dry during the winter.

Fish suffering from *Dactylogyrus* infection can be treated with baths: salt baths at a ratio of 25 gm (1 oz) per litre of water during 10 minutes for large fish and 15 gm per litre during 20 minutes for fry. Formalin baths can be given at 0·25 to 0·50 ml per litre of water for 30 minutes for fry and 1 ml per litre of water over 15 minutes for large fish.

One good precaution to take is to give the brood fish a salt (25 gm per litre for 10 minutes) or formalin bath before placing them in the spawning ponds.

Fish ponds can be treated with trichlorfon as mentioned for controlling blood sucking leeches.

V   Diseases caused by Crustaceans

1   Argulus

Argulus spp., or fish lice, are common parasites which attack fish. There are several species. *Argulus* is a small, flat crustacean (Fig. 449), green-yellow in colour and up to 8 mm in length in the case of *Argulus foliaceus* and 12 mm for *Argulus coregoni*. They attach themselves to the skin of the fish – especially at the base of the fins, by means of hooks and two suckers situated under the eyes. Between their antennae there is a kind of sharp pointed dart situated in front of the mouth which is shaped like a horn. The sting can cause wounds which open the way for other infections. In the same way as the blood sucking leech the harm done depends on the number of parasites. If the attack is minor then little harm will be done. In extreme cases the wound inflicted and the resulting anaemia can cause death.

Signs. The sting of fish lice causes red blotches on the skin. The fish show signs of nervousness and scratch themselves to get rid of the argulids.

Treatment. Treatment includes baths: lysol baths as for leeches or potassium permanganate: 1 gm in 1 litre of water during 40 seconds. The drying of the ponds followed by liming with quicklime is a good preventative.

In Israel (Sarig and Lahav, 1959), lindane baths are given to carp during transportation. For a journey lasting 2 or 3 hours 0·3 gm of lindane per cubic metre of water is put in tanks. The fish are tightly packed (1 kg of carp for 1·5 litres of water) ($2\frac{1}{4}$ lb of carp for $3\frac{1}{4}$ pt) and the water is well aerated. Lindane can also be used as a treatment in ponds without danger for the carp. The concentra-

tion should be 0·015 gm of lindane (active ingredient) per cubic metre of water. If a commercial product is used, preferably an emulsion containing 200 gm (7 oz) of lindane per litre, then the ratio should be 7·5 cm³ for 100 m³ of pond water. This treatment is also withstood by roach, tench and pike. There is a danger that the nutrient fauna may be partially destroyed.

Fish ponds can also be treated with trichlorfon: 0·5 g/m³ for carp, tench and eels, or 0·2 g/m³ for more sensitive species such as trout and pike. If the treatment has to be repeated, this is done two to three weeks later.

VI   Environmental Diseases

The principal physico-chemical characteristics of water normally used for fish culture have been discussed in Chapter XII, Section III. Certain fish are more sensitive than others to the chemical and physical deficiencies, for example salmonids to lack of oxygen in the water.

Too acid water. If the pH falls below 5·5 the water can, little by little, become toxic for most of the fish in the pond. From a pH equal to 5·0, mortality can start. The skin of the fish will be covered with a whitish film and secrete a lot of mucus; the edges of the gills will turn brownish. Certain fish will swim about slowly while others will die close to the banks and in a normal position. If, at the same time, the water is ferruginous the colloidal iron will settle on the gills and make breathing difficult or impossible, thus increasing the harmful effects of the acidity of the water. Colloidal iron may also favour the development of gill disease in young trout.

The pH of such ponds exposed to this danger must be tested regularly. If it falls to 5·5 then, without waiting 500 kg ($\frac{1}{2}$ ton) of calcium carbonate should be spread per hectare. As far as it is possible water from melting snow or run-off water should not be allowed to enter the ponds, particularly if they are surrounded by conifer woods.

Too alkaline water. A pH above 9·0 should be considered dangerous for the fish. This can be caused by pollution and also in concrete tanks if the concrete is too fresh. It can also follow the distribution of quicklime or results from a biogenic decalcification which liberates some quicklime. This can be caused by too strong an assimilation of an abundant submerged vegetation in bright sunshine.

Fig. 450 Brown trout dead from asphyxiation. Note the large open mouth and the widely lifted gill cover.

The gills of the fish will be burned and their fins injured. Biogenic decalcification can be avoided by previous liming in such a way as to obtain a sufficiently high alkalinity (SBV) (1·5 or more) and thereby a better buffered water. Submerged vegetation can also be controlled and reduced.

Temperature variations. Damage caused by strong temperature variations is real but has sometimes been exaggerated. Throughout their existence the fish withstood differences in temperature between 10 and 12°C very well if the change is gradual. It is necessary, however, to be very careful with water below 10°C.

Lack of oxygen. Dangers of an insufficient dis-

solved oxygen content – the trouble this can cause and its remedies – were discussed in Chapter XII, Section III. Fish dead from suffocation have wide open mouths, raised gill covers and widely separated gills (Fig. 450).

VII Nutritional Diseases

Artificial feeding plays an important part on many farms especially those which raise trout intensively. Over-feeding is often the cause of disease and death. This danger developed with the general use of pellets though without doubt it was even greater

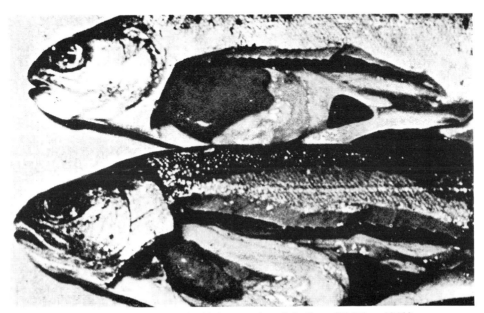

Fig. 451 Rainbow trout affected by lipoid degeneration of the liver. (Ghittino, 1966.)

when sea fish and meat of doubtful quality formed the principal food for salmonids.

In salmonid culture *lipoid hepatic degeneration* (Fig. 451) is essentially a nutritional disease of which many symptoms are identical with viral haemorrhagic septicaemia. It is characterized by a yellow-brown colour of the liver. Treatment includes avoiding over-feeding, letting the fish go without food from time to time and the periodical distribution of fresh food such as beef liver and lean fish.

Enteritis is also a feeding disease. If the abdomen is pressed slightly a yellow-red liquid will flow from the anus. An examination of living or moribund fish shows red, congested and severely inflamed intestines. Normally the colour is white or slightly pink, and the capillary blood vessels scarcely noticeable. When they are inflamed the finest blood vessels can be seen. This inflammation is caused by feeding errors.

Hepatoma of rainbow trout which shows itself on the body as an external hard tumour behind the pectoral fins is believed by American specialists also to be due to feeding, and has been shown to be caused by aflatoxin.

Treatment. In salmonid culture a lot of troubles arising from faulty diets can be avoided by taking the following precaution. Food must be distributed in good condition and even dry concentrates should not be stocked or distributed in too large quantities at one time. Food must be rich in vitamins, not too fatty or too salty (not more than 2 per cent salt) and should include sufficient fillers. The ration advised by the manufacturers should not be exceeded, and in general the daily ration should not be more than 2·5 per cent of the weight of the fish. Feeding should

Fig. 452   Abnormally constituted bream; the upper jaw is short (so-called "Mops-head" or "pug-head").

be reduced or even stopped when the weather is too hot or too cold (water temperature under 6°C). One ration should preferably be divided up and fed twice or even more, this is better than a single feeding. The intestine and the liver of the fish should be examined from time to time, and if there are any doubts feeding should be suspended for several days and then restarted gradually. The fish should go without food one day per week and from time to time dry concentrates should be replaced by fresh food.

VIII Constitutional Anomalies

These phenomena (Fig. 452) can be caused by disease, whirling disease for example, but they are often caused by rachitis.

Treatment. Vigorous and healthy parents should be chosen; feeding, especially for the young, should be complete and varied; badly formed fish should be eliminated at the time of grading.

Part Four

Chapter XVI

HARVESTING THE FISH

SECTION I

FISHING OUT THE POND

THE capture of farmed fish can be carried out in two ways: drying out (emptying the ponds) or not drying out. Whichever method is used the pond will have to be dried out to permit new, controlled stocking. Ponds which cannot be dried cannot be considered to be real farm ponds.

In order to reduce considerable losses or avoid them altogether at the time of capture, grading and handling, the following precautions should be taken: 1. Feeding should be suspended 2 or 3 days before emptying. 2. Only harvest when the weather is cool, preferably in the morning, except when the weather is cloud covered or rainy; no harvest should be done during thundery weather. 3. A perfected grading installation is necessary; long transportation should be avoided. 4. Fish should not be heaped up in scoop nets or transport receptacles–particularly young fish, salmonids or other delicate fish.

I Catching Fish by Draining the Ponds

1 The period of draining

The best times to drain vary according to the nature of the farming. In European cyprinid culture the second nursing ponds, second year carp ponds and growing ponds are generally emptied in the autumn, from September to November. During the winter the ponds remain dry. Wintering ponds are harvested in the spring from March to May. In salmonid culture, fingerlings ponds are dried out in the autumn. In order to avoid loss through cannibalism emptying should not be delayed or postponed. Trout fattening ponds are emptied according to the demands of the market, throughout the year in practice, so the salmonid farmer organizes his establishment accordingly.

Except in special cases ponds are not emptied during the warm season. For trout it is absolutely essential to avoid great heat. If emptying has to be carried out during a hot period, then it should be done early in the morning when the weather is cooler. Frost should also be avoided for it is harmful for the fish and leaves lesions on the skin.

2 Methods of draining

Draining well-designed and built ponds is quite easy, but it is that much more difficult if the pond leaves much to be desired.

All emptying must be very carefully prepared and carried out. The successive order in which the ponds should be emptied should be studied in advance. All the tools (nets, scoop nets, waders, sorting tables, baskets, cans, basins and other transport receptacles, platform scales) must be ready, in good condition and in their places. There must be abundant clean water available; its temperature should be as close as possible to that of the pond being emptied. Work can only be done with experienced personnel with each one knowing what he has to do.

Draining must be slow and regular so that the fish can follow the lowering of the water. If draining is too fast some of the fish will not follow the lowering water level and will be imprisoned in the floating or submerged vegetation, get buried in the mud and will be lost. They may also be swept up against the screens and be injured. This could kill them. The draining of large ponds takes several days, and can take up to several weeks.

The sluice boards of the monk are progressively lifted. In order to avoid taking too long over draining, two or three boards are lifted at a time at first. The screen and wire netting are selected according to the size of the fish in order to prevent their escape. If the monk has two grooves the screen will occupy all of the first groove or the lower part only. In the latter case, it is surmounted by a series of boards that overtop the level of the water, for this arrangement permits easy removal of boards in the second groove which control the level of the water. The screens must be cleared regularly of leaves, twigs and mud carried by the current and which will rapidly become an obstruction. This happens especially towards the end of the operation.

If there are too many water weeds, and particularly if it is a nursery pond, part of the vegetation should be cleared away before harvesting. While draining the pond it is often necessary to go over the pond in order to catch fish remaining in the principal and secondary emptying ditches. The day following the draining the pond must be examined in order to capture any fish which may have escaped the first day.

The harvesting of fish is carried out especially at the termination of the draining. The greatest precautions must be taken in capturing them in order to ensure their good condition. This operation calls for the greatest care if the fish are delicate (e.g. fry and brood fish). Capture can be carried out behind the monk and the dike of the pond or in the pond itself in front of the monk.

During draining, if the fish are not to be transported immediately to holding tanks or ponds, where they will remain until sold or being restocked, they should be held temporarily in storage tanks installed at the side or nearby where they were captured. These tanks must be fed by fresh water, generally drawn off the pond by-pass (Fig. 454). To help the subsequent removal of the fish from these tanks it is a good idea to install them previously with hanging nets. The fish clean themselves in the tanks and get rid of the mud covering them and found in their gills.

If there are no provisional storage tanks available near to the point of capture, the fish can be placed in the by-pass in which the depth is assured by means of a dam, either permanent or provisional, which is installed there. The fish are placed in small holding nets (Fig. 481) if they are not numerous or in larger nets (Fig. 476) if the quantities are greater. The nets are fixed to the sides of the by-pass and supported by poles laid across it. These same poles divide the nets up into separated pockets.

a Catching fish behind the monk and the dike
For this system the fish are allowed to cross the monk and the emptying pipe and are taken behind the dike in the fishing-out device which is in stone, cemented or in concrete if the size of the pond justifies it. If the pond is small, a mobile catching box can be used.

A permanent fishing-out device (Figs. 454 and 455), is a construction with a concrete or cemented bottom and the dimensions should accord with the size of the pond. Down-stream there are grooves into which screens are inserted to stop the fish escaping during the draining but permitting the evacuation of the water. It is a good thing to have two successive

grooves so that one can be lifted for cleaning after the other has been cleaned and put back in place. If the screens are cleaned with a brush or are scoured with a rake while in place, then it is best that the bars should be in a vertical and not an horizontal position. It is sometimes useful to slip a sluice board into the posterior groove to get a sufficient depth of water in the fishing-out device.

To facilitate harvesting it is a good idea to suspend, in the fishing-out device, a net more or less the same size as the device itself (Fig. 455). The net is fixed at the sides by hooks embedded in the wall for that purpose or by simpler hooks just stuck into the ground. In front the net is fixed to the back section of the draining pipe by a sleeve which is part of the net. Nets help the removal of the fish and are not obstructed so quickly as the vertical screen of the monk which can be removed when the net is in place. The mesh must be selected to suit the size of the fish.

There is a great advantage in allowing fresh water to flow into the fishing-out section. In this way the end of the draining can be carried out under the best conditions, for it is at this moment that the fish arrive in a mass and the water is very muddy. This is quite possible in diversion ponds (Fig. 454). The water supply for the fishing-out device is linked into what is the properly called "by-pass" of the pond (Fig. 3). This cannot be done when the ponds are barrage ponds.

Mobile catching boxes (Fig. 456) are about 2 × 0·60 × 0·60 m (6 ft 6 in × 2 ft × 2 ft). They have strong wooden frames. The bottom is made up of smooth wood planks while the sides and back have wire netting whose mesh size is suitable to that of the fish to be caught. The top is open, as is the forward end which is provided with a solid jute canvas sleeve attached to the back end of the draining pipe.

If a catching box is not available, a net can be placed at the back end of the evacuation pipe stretched on two horizontal poles. This method is rather rudimentary.

Carried by the water, the fish reach the place of capture without being handled. They are picked up progressively and any filth or mud brought by the water is eliminated. If at some time the fishing-out device gets clogged up with fish or dirt, draining is stopped temporarily by slipping a sluice board into the monk until the draining can restart.

The system of capturing fish behind the monk and dike works very well and is to be recommended, especially for delicate and small fish. The catching of fish in fishing-out sections, outside of the pond, makes for more convenient handling than the system described below and, in particular, it helps the cleaning of the fish.

b Capture of fish in front of the monk

This method is only recommended for such fish as cyprinids and catfish which can withstand a long stay in agitated muddy water, such as occurs in a capturing ditch, before the monk, because of the movement of the fish and the method of capture.

Capture only starts when most of the fish are assembled in a prepared ditch in front of the monk. The object of the operation is also to collect those fish which are still in the drainage ditches. The size of the capturing ditch must be sufficient to hold, at one time, all the fish which are in the pond.

The bottom of this ditch should be from 10 to 20 cm (4 to 8 in) below that of the principal drainage ditch. On the other hand it should be higher than the level of the emptying pipe which in any case must ensure the complete drying of the pond. In no case should there be a pool of water in front of the monk, which cannot be emptied by gravity.

In order to reduce turbidity at the time of emptying, and to facilitate capture, the ditch can have a hard bottom either in stone, cemented or in concrete and the drainage ditches must be well maintained (Fig. 457) and cleaned after drying.

Capture in front of the monk must be utilized when the topographical conditions – lack of space or insufficient slope – hinder the employment of the preceding method. It can be used also when ponds are large and hold a great number of fish.

When a large number of fish have to be harvested at the time of emptying, to facilitate their capture a net can be used in that part of the pond which remains under water at the termination of draining (Fig. 458). A seine net is used. The fish are lifted with scoop nets and loaded onto a van which transports them to the ponds or storage reservoirs if they are not to be held in temporary holding tanks at the side of the pond.

When drainage is extensive, and a large number of fish are handled, transport from the pond to the van can be helped by the use of an elevator. This is an endless belt, the foot of which is placed near the point of capture while the top hangs over the vehicle (Fig. 459).

When a pond has to be emptied during high temperatures (as is the case in warm temperate and intertropical regions), it is necessary at the time of emptying the pond to supply clean water sprayed upwards for aeration in front of the monk where the fish collect (Fig. 460). It is, in fact, an operation similar to that of bringing fresh water to the fishing-out section when fish are harvested behind the monk.

II Catching Fish without Draining

1 Application

This method of capture is used in the following cases: intermediate fishing, fishing out carp fry in spawning ponds, the daily capture of trout for consumption, trial fishing to test the success of reproduction or to control growth, or for fishing ponds which cannot be emptied. Several different methods are used.

The opportunity for intermediate fishing and its uses were discussed earlier (Chapter XI). This fishing can be practised with a variety of different equipment; lines, trap nets, seines, cast nets or drop nets, and even bamboo lattice has been used. This last is normal practice in Indonesia and is possible if the ponds are not too deep.

Carp fry hatched in Dubisch spawning ponds are caught with the aid of scoop nets made up of stretched very fine muslin (Fig. 200). It is first necessary to concentrate the fry in front of the monk. This can be done by lowering the level of the water. This method of capture may also be used for the fry of other species which are too small to be caught any other way.

In salmonid culture the daily fishing of small quantities of trout for eating does not involve drying out the pond. It can be done with a scoop net, a drop net, a cast net or by electricity (Fig. 467). A part of the fish population is concentrated in one section of the pond – which is quite possible by distributing food at those spots where they generally eat (Fig. 468).

Control fishing is sometimes carried out in fry and growing ponds. The method is currently used in carp culture and also for other species. A cast net (Fig. 421) or drop net is employed. If the fish are accustomed to being fed, then groundbait is used at the point of capture.

Fishing in ponds which cannot be dried out is done with lines, trap nets and nets.

2 Principal equipment for capture

Various devices and equipment are used for capture.

They differ according to the nature of the pond as well as the species and the age of the fish.

1 *Scoop Nets.* There are numerous models (Fig. 461). They are adapted to existing conditions and the mesh is proportionate to the size of the fish. If delicate fish are being handled the net is shallow. Scoop nets can be round in shape. They have a straight lower hem when used for storage tanks or for fishing-out sections where it is necessary to be able to work at angles and in corners. In any case the net is not fixed outside the frame but inside it. To this end it is possible to use a double frame (Fig. 461) or if it is a single frame, holes are made and the net is attached by means of string or rings (Fig. 462). A hoop net (Fig. 122) is a large scoop net with a very solid frame handled by personnel in the water. It is used in shallow running water to capture brood trout.

2 *Trap Nets and Nets.* This equipment can be rigid or pliable, mobile or fixed. There is considerable variation. All the equipment in fish culture is used only in shallow water. The size of the mesh varies according to the size of the fish.

Trap nets are rigid, made in wicker and vary considerably both in dimensions, shape and handling. They are generally cylindrical in the lower part and like a truncated cone in the upper part. The interior of the trap net is provided with one or two cone-shaped necks which are a sort of addition forming a funnel and stopping the escape of the fish caught in the net. The interior opening has a diameter close to the sizes of most fish currently captured. Trap nets are particularly effective when they are placed in spaces left free by the wattle fencing in the shallow parts lining the weedy places and the passages which cut it.

Fyke nets (Fig. 232) are the same as trap nets though they are supple and formed by a net fixed to wooden hoops. They are shaped like cones and are held between three pegs, two of which are driven into the soil each side of the mouth, and a third planted right behind, which holds the net taut. Inside, a fyke net is the same as a trap net with one or more necks, but they are more supple than trap nets and are carried more easily when the hoops are drawn together. They are used for catching pike spawners, for example, on spawning grounds. To secure the fish the bottom of the pocket is untied.

A *drum* is a cylindrical fyke net with an opening provided with a neck at each end.

The effectiveness of trap nets, fyke nets and drum nets is aided considerably by placing rigid or supple wings made from wicker or from fixed nets separating the mouth of the trap net or the fyke net (Fig. 325). The fish bump up against the wing, swim along the side and enter the capturing net.

A *drop net* comprises, essentially, a net mounted squarely and suspended at its corners by two supple hoops tied in a cross at the top and attached to a pole. It is pulled up and down from the banks of the water (Fig. 463) or from a boat. When it is pulled up the net forms a deep pocket which holds the fish. Its dimensions vary considerably.

A *cast net* (Fig. 421) is a cone-shaped net cast from the edge of the water or from a boat or in the water itself if it is not too deep. When it opens it forms a kind of circular sheet which imprisons the fish. This cloth is bordered with a large hem to which leads are attached and which carry the net to the bottom. The hem forms purses which are closed by strings. A play of cords pulls in the net filled with all the fish within its reach. When closed it can be 1·50 to 4 m (5 ft to 13 ft) high. The handling of a cast net requires training in order to spread it right out when it falls onto the water.

A *seine* (Fig. 466) is a net used to encircle a part of the water believed to contain a lot of fish. It is drawn by its ends. Sizes vary considerably. It is generally rectangular in shape and is sometimes provided with a medium purse which increases its catching potential. The net is mounted between a wire or balk with floats, and a lower balk provided with leads in order to keep it stretched out vertically in the water. The seine is often used in running water and, in fish culture, at the time of emptying large ponds (Fig. 458) or for test fishing to control growth and reproduction (Fig. 276).

A *bamboo screen* (Fig. 464), used in the Far East in shallow ponds, is manipulated and handled like a seine. It is possible to combine a bamboo screen and a drop net (Fig. 465).

The maintenance of nets and scoop nets. It is important to ensure that nets and scoop nets are kept in good condition.

Nowadays, nets and scoop nets are generally made in nylon and without knots. This is a very solid material, scarcely rots and demands far less attention than cotton. Such nets should be kept out of reach of rodents and be protected from the sun.

To ensure good maintenance, cotton nets and scoop nets must be washed carefully after each use in order to rid them of mud and mucus which may cover them. They are suspended in shaded spots to dry, protected from rain and dew. In no case should

Fig. 453 A rudimentary method of draining ponds and harvesting reared fish in muddy water. Not advised.

Fig. 454 Catching fish behind the dyke in the fishing-out section seen on the left. On the right, two storage tanks in which the fish are kept during the emptying; they are given clean fresh water from a pipe connected with the diversion canal. Flamizoulle ponds, Luxembourg prov., Belgium.

Fig. 455 Harvesting fish with a net stretched out in the fishing-out device behind the dyke. The net is hooked to the back end of the drainage pipe. Bokrijk fish culture station, Limburg prov., Belgium.

Fig. 456 Catching box for fish harvesting behind the dyke. This device can be used for small ponds.

Fig. 457   Catching fish in front of the monk, in a catching ditch with a cement bottom and at a level slightly superior of that of the draining pipe. Bussche ponds, "Mission piscicole du Katanga (1947)" Lubumbashi, Zaire.

Fig. 458   Catching fish in front of the monk and the dyke in a large carp culture pond. The level of the water is lowered considerably and the fish are brought together with a seine net. Nasice fish farm, Yugoslavia.

Fig. 459   Lifting and loading fish by means of an elevator. In the pond the fish are loaded into the elevator vat which lifts them above the dyke where they are tipped into a transport lorry. Ma'agan Michael fish farm, Israel.

Fig. 460   Aeration of pond water at the moment of draining, by supplying sprayed water at the point where the fish are assembled. Draining at the time of a high summer temperature. On left, the elevator. Sassa fish farm, Upper Galilee, Israel.

 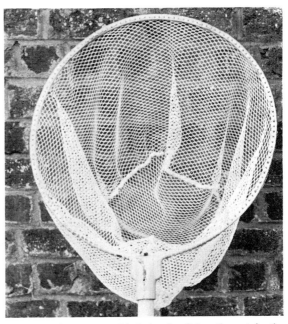

Fig. 461   Above, models of several scoop nets with double frame. Below, scoop nets with perforated frames. Groenendaal, Belgium.

Fig. 462   Scoop net with holes for fixing the net in the perforated frame.

Fig. 463   Drop net for catching reared bait fish. Ben Adams Minnow Hatchery, Celina, Ohio, U.S.A. (Photo: Cl. F. Clark.)

Fig. 464 A partial catching with linked bamboo screen which progressively encircles the fish. Ponds at Bodjongloa, Bandung, Indonesia.

Fig. 465 Partial catching in a pond combining a bamboo screen and a large drop net. The principal fish are tilapias. Lembang, Indonesia.

Fig. 466   Partial catching in a cyprinid pond by manoeuvring a seine net. Szarvas fish culture station, Hungary.

Fig. 467   Fishing with electricity in a trout pond. Canton of Aarau, Switzerland. (Photo: Hey.)

Fig. 468   Catching reared trout with a scoop net after distributing the feed. The fish are transported in a tank mounted on a small rail truck. Ligneuville fish farm, Belgium.

they be exposed to the sun directly. Nets should never be left in a heap.

The conservation of cotton nets can be helped considerably by impregnations which can be carried out by the farmer himself. Creosote and other substances derived from tar must be avoided. Vegetable extracts are used such as catechou and quebracho. The nets are placed in a cauldron containing 1 kg (2¼ lb) of vegetable extract for 40 litres (about 8 gal) of water which is sufficient to treat 10 kg (22 lb) of net. The operation is carried out at 60°C and lasts from 24 to 48 hours.

3   *Electric Equipment.* Electric equipment can be used for fishing in ponds and water courses which are neither wide nor deep. The current is provided by a generator (Fig. 123) or from batteries. The electric current is weak, about 1½ amperes. Voltage varies according to depth, distance and the conductivity of the water. Salmonids are more sensitive than cyprinids, and large fish more than small ones. The voltage is between 150 and 250 volts. Direct current can be used as well as alternating or interrupted current.

Two insulated cables carry the current to the poles between which an electric field is created. With direct current the negative electrode is often a metal plate and the positive electrode can be transformed into a scoop net.

It is best to use direct current or interrupted direct currents which are less dangerous and by which the fish are drawn to the positive pole. With alternating current the fish that find themselves within the electric field, receive a shock followed by numbness for a few moments during which they are seized.

This method of fishing can be used for catching brood trout in open water or trout for the table in growing ponds (Fig. 467). If the intensity and the voltage of the current are well regulated the fish soon recover from the shock in the electric field.

SECTION II
SORTING AND GRADING THE FISH

The aim of sorting and grading is to separate the fish by species and then by sizes. Sorting must be carried out at least annually when ponds are emptied. In salmonid culture it is more frequent, notably when first year trout are being reared in order to ensure that the most developed do not eat the smallest.

Sorting and grading precedes final storage. According to the circumstances of each case, this operation is done immediately after capture of the fish or after the provisional storage which might follow.

1   Grading first year trout and other young fish
The aim of grading first year trout is not only to avoid cannibalism but also to permit the smallest fish to eat normally without being chased away by the stronger.

In salmonid culture grading fish is very important in order to reduce loss, especially when raising fingerling trout the first year. If the farming operation is to raise fish from 1 to 3 months in rearing troughs, then the small fingerlings are sorted at the end of this period before being released in ponds and a new grading will not be necessary before the autumn emptying. If the fry are released directly into the ponds when their yolk sacs are absorbed, sorting will be necessary in July-August, but it is not always possible because of the high temperatures at this time of the year.

Grading must be done very carefully so that those subjects which grow quickly, in particular, can be separated and kept as future brood fish.

Buschkiel suggests the following categories for summerlings; category 0: less than 5 cm (2 in), I: from 5 to 7·5 cm (2 in to 3 in), II: from 7·5 to 10 cm (3 to 4 in), III: from 10 to 12·5 cm (4 to 5 in), IV: over 12·5 cm (over 5 in). It is also current practice to employ units of classification using 2 cm and 3 cm brackets.

The grading of small fish must be done more or less automatically. All the fish caught are brought to the handling shed from the rearing ponds and tanks. The work is done in a long trough or tank traversed by a strong current (Fig. 471). A grader is plunged into the trough. This is a kind of screened basket, the mesh of which is adapted to the size of the fish to be graded. Small fish, frightened by the movement which tosses them around in the screened basket, escape while the rest remain. If several screened grading baskets are used with

Fig. 469   Bachmeyer apparatus for grading fish. The space between the bars can be regulated between 3 and 30 mm.

Fig. 470   Automatic fish grader for salmonids. U.S. fish culture station, Lamar, Pennsylvania, U.S.A.

Fig. 471   Grading trout fingerlings with a Bachmeyer grader. Canton of Aarau, Switzerland. (Photo: Hey.)

Fig. 472 Automatic grader for table trout. Canizzano fish farm, Venice, Italy.

Fig. 473 Sorting table divided into three compartments.

Fig. 474 Sorting fish by species and sizes. Undesirable species are eliminated.

Fig. 475   Graded, V-shaped board for measuring fish.

different size mesh it is possible to grade rapidly into as many categories as desired.

Instead of a set of graders with fixed mesh dimensions it is possible to use a Bachmayer or other similar apparatus (Figs. 469 and 470) of which the bottom has tubes which can be spaced between 3 and 30 mm.

After the autumn grading of trout fingerlings, another is useful and even necessary in the spring if feeding has gone on during the winter and if the first year trout have shown an appreciable growth during that period.

During the second year there are more gradings but the number and sizes depend on the local conditions of the market.

2   Sorting and grading other fish

The sorting and grading of second year and older fish is done in different ways.

1   Sorting Species. If several species of fish are farmed or captured at the same time it is necessary to separate them when emptying. This is done at the time of drainage itself, or immediately after the provisional storage if this is necessary.

The work is done on a sorting table. This (Figs. 473 and 474) is a smooth table covered with zinc and comprising a bottom and raised sides about 10 cm (4 in) in height. One end terminates in a narrow neck. The table stands on a support or on

mobile legs or wheels and is periodically cleaned of detritus and mud which the fish bring with them.

Fish brought from the pond can be tipped directly onto the sorting table. If they are very dirty they are placed at first in basins filled with water or on provisional storage nets where they clean themselves. They are then tipped onto the sorting table.

2   Grading the Sizes. This is also done on the sorting table for resistant species such as cyprinids and other warm water fish. It is also possible to use mechanical graders or nets.

Mechanical graders are used for large fish (salmonids, cyprinids and others) and are the same models as those used for young fish, but the spaces between the bars are adapted to the sizes of the fish being graded (Fig. 472). Marek (1958) describes a grader suitable for carp. The space between the bars varies from 20 to 50 mm.

It is also possible to grade fish by using large superimposed *nets* with different size mesh. The nets are plunged into the storage tanks or into narrow storage ponds. Fish to be graded are tipped into the top net which has the largest mesh. The smaller fish pass through the net and are held by the lower one which has a smaller mesh. This method is used especially for sorting of small fish such as roach.

The sorted and graded fish are divided into species and by sizes in different buckets, two thirds filled with clean water. When these are filled sufficiently with fish they are taken to the growing ponds, storage ponds, or wintering ponds. Before being released they are counted and weighed.

If at the time of grading it is necessary to measure the fish, the simplest way is to use a graduated V-shaped measuring board (Fig. 475). The graduations are in centimetres and are marked clearly every 5 cm (2 in).

SECTION III
STORING THE FISH

During or after emptying it is almost always indispensable to keep the harvested fish for a longer or shorter time until they are sent to their final destination.

Storing can precede sorting or follow it. There can be a first storage before sorting and grading and a second storage after it.

Storing can be of short duration (provisional storing) or long duration (prolonged storage). Provisional storing is often used at the time the fish are caught. They are stored immediately pending later sorting. After sorting the fish can be sent immediately to their final destination, that is restocking ponds or water courses, or for eating. In other cases storing following sorting can be of either long or short duration.

Precautions must be taken to ensure that fish of various origin do not constitute a danger by introducing diseases or parasites onto the farm.

1   Conditions for good storage
1.   The fish must remain in good condition.
2.   The quality of their flesh must improve. The latter is often poor in the case of cyprinids living in ponds with muddy bottoms and for salmonids fattened with rather poor quality food. Satisfactory storage over a few days should improve the quality of the flesh.
3.   Loss of weight should be at a minimum.

Conditions cited above can be realized by placing the fish in water with an adequate flow to ensure a rich oxygen supply. But it should not be excessive in order to reduce swimming efforts to a minimum. The water should be fresh in order to reduce the metabolism, but not too cold (not below 4°C). The fish should not be fed if the storage is of short duration.
4.   Fish must be protected from their natural enemies and from thieves. To ensure good supervision they should be kept close to the living quarters of the farmer.

2   Principal techniques for storing
The storage of small quantities of fish is possible in small holding nets, floating boxes or aquaria. Storing large numbers of fish can be done with nets or better still in storage tanks if they are long enough. To hold a large number a rather long time the fish should be placed in ponds (holding ponds).

1   *Nets.* Holding nets suspended in the water are suitable if the fish are to be kept for a short time only, for example during emptying (Fig. 480). If several divisions (partitions) are needed, pockets can be made by means of poles on which the nets rest (Fig. 478). The water in which the nets are suspended must be clean and well aired such as water from the by-pass of the pond which is being emptied. It is best that the installation should be near the monk or the fishing-out section (Fig. 476).

If the place (by-pass or some other) in which the nets are suspended is traversed by an abundance of clean water then a large number of fish can be held. They will clean themselves very well.

2   *Small Ponds.* If the species is delicate, and especially if the fish are to be kept for a long time, then they must be placed in small ponds known as holding ponds (Fig. 477). Their use for carp for eating has been described in Chapter IV, Section IV (Figs. 213 and 214).

3   *Storage Tanks.* Dimensions vary: 2 to 6 m × 1 to 2 m × 1 m (6 ft 7 in to 20 ft × 3 ft 4 in to 6 ft 7 in × 3 ft 4 in). The walls are in stone or cemented brick or even in concrete. The walls are smooth in order to avoid injury to the fish. It is essential that the tanks should be fed with a fairly abundant supply of water. It must be possible to regulate the level, for example by a series of boards which can be slipped into grooves in the same way as for the monk.

The handling of fish in storage tanks is easier if nets are suspended inside (Fig. 478) as described for the catching of fish in the fishing-out sections behind the monk. The recovery of the fish is

Fig. 476   Counting and storing temporarily small carp on a net placed over the by-pass channel, the level of which is raised by a dam. Bokrijk fish culture station, Limburg prov., Belgium.

Fig. 477   Small storage ponds. Bokrijk fish culture station, Limburg prov., Belgium.

Fig. 478 Concrete storage tanks through which a current of pure water flows. The flow may be regulated. The fish are stored on suspended nets in the tanks. Bokrijk fish culture station, Limburg prov., Belgium.

Fig. 479 Floating boxes for storing small quantities of fish in a water course, a diversion canal or a pond. The plastic wire mesh is very fine. Bokrijk fish culture station, Limburg prov., Belgium.

Fig. 480 Holding nets suspended from poles or stretched by cords. Limal fish culture station, Belgium.

Fig. 481 Small holding net with iron rings suspended from an inflated tube.

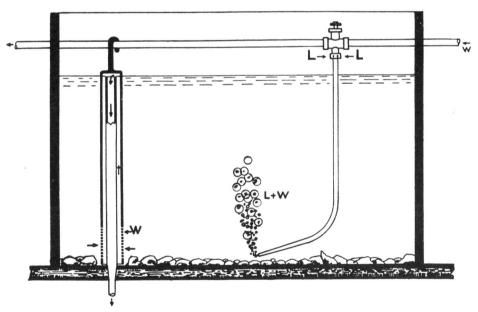

Fig. 482 Schematic drawing of an aquarium with water supply (W) and air (L) at the bottom and water overflow at the surface by means of a double conduit which permits evacuation of water from the bottom. (Demoll-Steinmann, 1949.)

accelerated and helped by using the nets, which are simply lifted out, whereas in tanks without nets the level of the water must be lowered. This takes more time.

Storage tanks can be built under a cover shed and this helps supervision.

Before using new tanks in cement or concrete they should be placed under water for about 15 days. The water is then emptied out.

4 *Floating Boxes, Crates and Small Holding Nets.* If only small quantities of fish are involved then floating crates are enough. These are much used by professional fishermen fishing in lakes and rivers.

Floating boxes (Fig. 479) are solidly built in wood on which are fixed slats and wire netting or perforated metal sheet adapted to the dimensions of the fish to be kept. The floating crates partially emerge. The lids can be padlocked in order to avoid theft and the escape of the fish. Crates are attached by a chain to a pole sunk into the water or to a ring embedded in the wall. They can also have arms attached to help in carrying them on land.

Small holding nets can also be used. They are as good as floating crates. This netting is sometimes in metal but is always supple and is used by anglers in which to keep their catch. An excellent receptacle along the same lines (Fig. 481) is made of nylon netting mounted on metal hoops and supported by a tube inflated by air. Thanks to the latter the basket floats and partially emerges, which avoids the escape of the fish which are quite comfortable behind the netting where the water is of course regularly renewed. The fish keep better in these small nets than in crates though the latter can be padlocked and are secured against theft.

5 *Aquaria.* These are used by restaurants to keep fish, and permit the customers to choose the fish they want. If there is plenty of water then nothing need be done about oxygenation, other-wise air must be fed by a compressor or simply by suction if the aquaria is not too densely stocked.

Both tanks and aquaria should be so arranged that the current of water crosses the mass, care being taken that it does not just renew the top level. One system adopted brings the water from the bottom and evacuates it at the top (Fig. 482) or another has the water supply on the surface and evacuation at the bottom. If the evacuation is in depth, then there is a system permitting the admission of air at the same time as the water. This reduces the consumption of water. Care must be taken not to supply the tanks directly with spring water which is generally poor in oxygen. Aquaria may have a stopcock at the bottom for emptying and cleaning.

3 Quantities of fish which can be stored

These quantities are high but very variable according to the species and the sizes of the fish, the dimensions of the tanks, the abundance of the water, its oxygen content and its temperature. The quantity of fish which can be stocked is the greater as the breathing demands of the fish become small.

From this fact there is a numerous variety of different stocking conditions and it is practically impossible to give precise indications as to the quantities of fish which can be stocked in the different storage devices.

By way of example, it is possible to mention that according to Wiesner a tank containing 150 litres (33 gal) with a water supply of $\frac{1}{2}$ litre per second of well-aerated water at $10°C$ can keep over several weeks: 8 to 12 kg (18 to 26 lb) of trout from 200 to 300 gm (7 to 11 oz) (up to 80 kg (176 lb) per cubic metre), 6 to 8 kg (14 to 18 lb) of trout of 100 gm ($3\frac{1}{2}$ oz), 1,000 trout fingerlings from 8 to 10 cm (3 to 4 in).

According to Schäperclaus (1933) it is possible to keep 150 kg (330 lb) of K_3 per cubic metre in storage tanks.

SECTION IV

TRANSPORT OF FISH

I Transport of Live Fish

The transportation of fish is of particular importance in fish culture.

On a farm, resistant fish can be transported dry in buckets without water, or in baskets. Others are carried in cans, pans, vats and other receptacles carried by hand or on vehicles of all kinds: wheelbarrows, barrows, vans or trailers. On large farms such as certain cyprinid and salmonid farms, narrow gauge railways were once used and, in some cases, are still used (Fig. 468). These same narrow gauge

railways can carry food for the fish and fertilizers.

Outside the farm, transportation and food costs
have been reduced considerably since the use of
motor vans became general. Aeroplanes provide
an excellent service for the transportation of small
quantities of fish such as brood fish, eggs or fry of
species to be introduced to new surroundings far
from their original habitat. It is thanks to the
satisfactory solution of transport problems that the
development of fish culture is largely indebted.

1 Receptacles for transport

The receptacles used in transport are varied and are
adapted to the size, but above all to the species and
the quantity of fish to be transported.

1 *Dry Transport.* The transportation over short
distances of large resistant fish such as carp, can be
dry. It is possible to use baskets and hampers (Fig.
483) or round or oval-shaped buckets with perforated bottoms.

2 *Small Cans or "Goujonnières."* If transport
involves only small quantities of fry or a few average
size fish over short distances, small cans known
as "goujonnières", similar to models used by anglers
for carrying live bait when fishing for voracious
fish can be used.

3 *Cans.* For transporting larger quantities of
fry or fish of average size, cans with flat bottoms
have been and are still being used. They are either
circular or oval (Fig. 483). They generally have a
capacity of from 20 to 40 litres (35 to 70 pt) and are
adapted to the quantity of fish to be carried. The
cans are in galvanized sheet metal (Fig. 485), zinc
or plastic. If the water is acid the zinc cans must be
lined inside with a protective layer, in the same way
as incubation trays, to avoid the fish, particularly
fry, being poisoned. If they are painted in light
colours inside then the fish can be more easily seen.
Certain cans include two apertures with a removable meshing. They are closed by two stoppers. The
lower stopper allows the can to be emptied while
the upper one makes possible the partial renewal
of water when the fish are already installed.

For transport in mountains there is a special
model which can be carried by one man on his
back. For transporting fry with absorbed yolk sacs
the cans should be small: 15 litres (about 26 pt)
which are more practical than large cans.

In the Far East fish are often transported in
water-proofed bamboo receptacles (Fig. 324)
whether fry or fish for eating which are generally
small. This form of transport is generally attached
on the back by means of a shoulder balance; at the
extremities of the bamboo pole there are two fish
cans. In these regions trade in restocking fry and
fish for eating is very prosperous (Fig. 484).

4 *Polythene Bags.* For carrying small quantities
over land or by aeroplane polythene bags are now
in current usage because of their reduced weight
and because this light packing does not have to be
returned. It comprises a solid polythene bag one
third or one quarter filled with water. The fish to be
transported are installed in it. Then by means of a
bottle of oxygen and a rubber pipe, oxygen is let in
above the water, blowing the bag up and stretching
it (Fig. 487). The upper part of the bag, about 6 to 8
cm (about 3 in) is twisted, folded and closed with a
solid elastic band wound round the twisted neck of
the bag. It is best to place the first bag inside a
second bag which is closed just as carefully. This
can be done separately or at the same time as the
first. One or more bags are then placed in a carton
for shipping (Fig. 488). The oxygen above the water
diffuses slowly during transport which ensures
good oxygenation and avoids the danger of an
excess of oxygen.

5 *Barrels and Vats.* Before the general use of
motor vans, significant quantities of both average
and large fish were carried in barrels and vats.
Barrels are now less used and old type wooden vats
are being replaced by metal or plastic of the same
design and carried by vans.

Barrels (Fig. 489) are made of wood and have
oval shaped sections, not round (in which case they
would roll) and they lie lengthwise on one side.
They are provided with solid handles on either side.
While they are being transported, thanks to their
shape, they have a slight swaying movement thereby
stirring the water a little. This helps oxygenation. The shape makes them best suited for automatic aeration when being carried. The opening must
be large enough to allow easy handling. The covers
of the barrels are shaped like truncated pyramids.
They are hollow. The bottom is pierced with holes
from which the water from melting ice, which can
be placed in the truncated cover, flows slowly.

Vats (Fig. 490) are shaped either like truncated
cones or cylinders. They rest on their flat end which
is either round or oval in circumference. They
are used generally for transporting with oxygen.
These vats are of varying capacities (50 to 150 litres
average) (11 to 33 gal) and contain a bottle of compressed oxygen of 3 to 7 litres (5 pt to 1 gal 4 pt) at

130 to 150 atmospheres, inserted vertically and fixed with two collars. The bottle can also be fixed outside the vat. These receptacles have an average capacity of 80 litres (about 17 gal) and are more easily handled and preferable to those of larger size. A pressure reducing valve with a tube carries the oxygen to a spray which is installed on the bottom of the vat and is protected by a cover. The first vats were built in wood and then in metal. They can be transported on a wheelbarrow with either one or two pneumatic tyred wheels (Fig. 491).

The handles and hoops of barrels and vats in wood must be protected against rust. It is also important to see that they do not dry too quickly while not in use during hot periods. Before being used for the first time wood receptacles must be submerged for 1 or 2 weeks in running water.

6 *Special Vehicles.* When large quantities of fish are transported over long distances and periods, up-to-date fish farms have motor lorries mounted with tanks of different shapes and generally made in metal (Figs. 494 and 496). Sometimes these tanks are covered with insulating materials.

For the most part transport tanks are provided with an aeration system. There are several kinds. Often bottles of oxygen or compressed air are used (Fig. 492). Compressed air takes up more space than oxygen which means that larger bottles are necessary and this increases transport costs.

The distribution of the gas is ensured by the installation of sprays at the bottom of the tanks. Sprays are in either hard porous materials or comprise a simple plastic perforated tube fixed to a metal frame on the bottom of the tank (Fig. 493). Air pumps powered by the batteries or the motor of the lorry are also used. Certain vehicles are equipped with hydraulic pumps which suck in water from the bottom of the tank and then spray it in atomized form from the top. If an injector is included in the closed circuit of an hydraulic pump then the water jet will be rich in air bubbles. The pumps are worked either by an independent petrol motor or by an electric motor fed off the battery of the vehicle. Sometimes the pumps are provided with a filter to clean the water.

To stop fish injuring themselves against the sides of the tank, due to inevitable bumps during transport, nets can be suspended in the tanks (Fig. 495) as is done for storage tanks.

Large quantities of fish can be shipped by railway and even in barges equipped in the same way as are the motor lorries.

2   General rules for transportation

The transport of fish, and fry in particular, must be carried out very carefully if it is to be successful. If it is badly organized the fish will die after being released and the result will be a waste of effort, time and money.

The primary condition to ensure successful transport of living fish is to maintain sufficient oxygen in the receptacle throughout the whole of the journey. This is not always easy considering the large number of fish held in a small volume of water.

The best ways of assuring sufficient oxygen content in the water are: to keep the water at a low temperature, to renew the water, to ensure moderate movement of the receptacle and to secure adequate aeration by a supply of diffused air or oxygen.

Low temperatures automatically ensure better oxygenation of the water, and the same is true for the renewal of the water, either partial or complete, at the different stops and stages. Moderate movement caused by the moving vehicle during transport produces the same result. The diffusion of oxygen or air permits a considerable increase in the stocking of the receptacle.

The principal factors which influence transport of living fish are as follows and the stocking of the receptacle must be proportionate to these factors.

(*a*) The species of fish. Oxygen needs vary considerably according to species; salmonids are the most demanding.

(*b*) Age and size of the fish. The larger the size of each individual fish the greater its breathing requirements. However, the smaller the fish the greater are the breathing needs high, corresponding to their total weight.

(*c*) Relative resistance of the fish. Fish fed artificially are less resistant than those fed naturally. At the time of reproduction, spawning fish withstand transpórtation badly.

(*d*) The temperature of the water. Transport must be carried out in water whose temperature has been progressively lowered. Respiratory requirements are then less and the natural oxygen content of the water is higher. A high temperature will mean reduced stocking. Nevertheless the temperature should not fall below 4°C even for trout. For the latter, the best temperatures are between 4 and 10°C.

(*e*) Length of time of transport. The shorter the time taken the greater the stocking rate can be.

Fig. 483 Different models of transport cans for fish. Basket (hamper) for transporting resistant fish able to be carried without water (carp and tilapias). Jonkershoek fish culture station, South Africa.

Fig. 484 General view of the morning market in fry at Tjisäät, Soekabumi, Indonesia. The fry are transported in baskets water-proofed with latex, grouped in pairs suspended at the end of poles and carried across the shoulders.

Fig. 485   Transport can in galvanized sheet metal convenient for transporting small quantities of fish. Much used before the introduction of polythene bags.

Fig. 486   Partly perforated pail for transporting and releasing in running water trout fry with absorbed yolk sacs.

Fig. 487   Oxygen inflated polythene bags for transporting small quantities of fish.

Fig. 488   Polythene bags placed in a carton container.

Fig. 489   Oval shaped barrels for the transport of live fish. These were much used at one time.

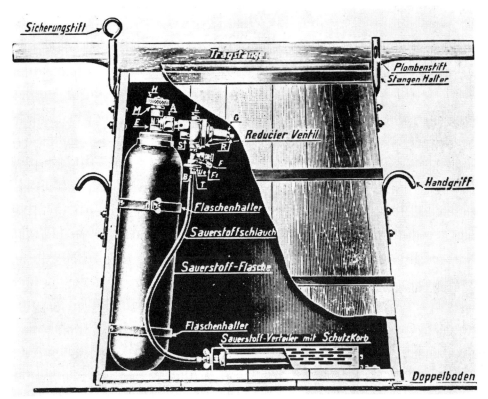

Fig. 490   Wooden, cone shaped vat equipped with an oxygen container and diffuser. (Schäperclaus, 1933.)

Fig. 491   Small reservoir with an oxygen container for transporting trout inside a fish farm. Augst fish farm, Switzerland.

Fig. 492   Lorry equipped with oxygen containers and tanks with diffusion apparatus.

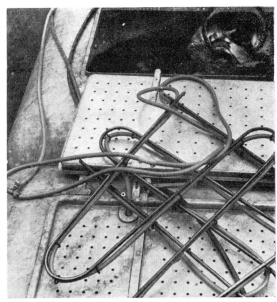

Fig. 493   Perforated plastic oxygen diffuser fixed to a metal frame and installed on the bottom of transport tanks.

Fig. 494 Aluminium tanks for transporting fish. Dimensions: 95 cm square (about 3 ft square). Limal fish farm, Belgium.

Fig. 495 Net for suspending inside the metal transport tanks.

Fig. 496 Lorry specially equipped for transporting fish in tanks with oxygen or compressed air diffusion. The tanks are thermally insulated. Wijnbelt fish farm, the Netherlands.

(*f*) Means of transport and lengths of stops. The faster and the easier the transport (good communications) and the shorter the stops, the better the chances of success.

(*g*) The nature of the transport receptacles. Those in wood warm up more slowly than metal but the latter can be thermically insulated.

(*h*) Climatic conditions. These influence the temperature of the receptacles and also the oxygen content.

To increase the number of fish transported, tranquillizers (e.g. MS 222, propoxate) have been used. These reduce the metabolism of the fish and their oxygen intake. This technique is not yet in current usage but trials are showing promise.

Lamarque (1962) has described the following tests. Before being placed in polythene bags for transportation salmonids are placed in a MS 222 bath with a concentration of 1/10,000 until the fish lose their equilibrium and rest inanimate on the bottom. They continue to breathe. As soon as they are removed from the MS 222 solution they are washed before being placed in the polythene bags. About 2 litres ($3\frac{1}{2}$ pt) of water are needed for 10 trout of 100 gm ($3\frac{1}{2}$ oz) or 2 litres for 1 kilo ($2\frac{1}{4}$ lb) of fish. This process permits a stay in the bags of about 6 hours.

Instead of using the anaesthetics previously described, a tranquillizer can be placed in the transport receptacle. In this case Lamarque suggests a concentration of MS 222 which should not exceed 1/100,000.

3   Preparation of transport

1. Transport must be undertaken in weather that much cooler as the fish need oxygen. In warm weather, transportation must be carried out at night or in the early morning.

2. The transport must be carefully prepared in advance. Important shipments and delicate fish always travel in escorted convoys. The moment the merchandise arrives at its destination it is taken over. If the railway is used, attention must be paid to the time-table.

3. The fish fast when travelling in order to avoid their excrement soiling the water. They should also fast for at least 1 or 2 days in advance.

4. For transport in cans or barrels receptacles that can be handled are used. Cans should not be completely filled, no more than 70 to 80 per cent. The space left aids aeration. Carriers made in wood are filled with water 1 day in advance so that the apertures can close.

4   Transportation

1. Everything possible must be done to prevent any rapid warming of the water. To this end receptacles should be placed in the shade out of the direct rays of the sun.

When the weather is very hot the cans or barrels are kept cool by placing blocks of ice in the cover of the cans, or in small cases placed on the top of the barrels. Ice should never be placed directly in the water, but only in the perforated covers of the receptacles.

In hot countries one way is to cover the cans with sacks or cloth soaked with water. This packing is kept soaked constantly for the evaporation of the water it contains will maintain freshness inside the can.

2. The water should be changed whenever necessary, notably during prolonged stops and always over a long journey when the weather is hot and thundery. It will be realized that the fish have an urgent need of fresh water when, instead of remaining at the bottom or in the body of the water they come to the surface to breathe. If water is not changed immediately they will suffocate and perish. As long as they stay on the bottom they are in good condition.

Changing water is a rather delicate operation which must not be carried out all at once, especially when the fish have been held a long time in the same water. Precautions are necessary not only because of temperature differences which can only be a few degrees, but also because of the differences in the composition of the original and the new water. First, half of the carrier is emptied and is replaced with fresh water; after 5 or 10 minutes the whole of the water is changed. When the fish are carried in cans the renewal of the water is helped if the receptacle is provided with a screened overflow or a grilled open tap for draining the water.

The choice of water to be used is important. The new supply should be fresher than that in the carrier and should come, for example, from a rheocrene spring (a fountain) or better still from a clear stream. It must be pure (unpolluted), not too acid (peaty water). Poorly aerated water (cisterns, reservoirs and wells) should be avoided.

In case of urgency, if renewal of the water is not possible, then a small quantity can be taken from any receptacle and then poured from a height of about 50 cm (1 ft 8 in). This will aerate the water although it will not freshen it. It is also possible to pump air into the water.

3. Jolting should be avoided for this will throw the fish up against the side of the carrier. Repeated shocks of this kind are bad for them. They injure the fish, which may die immediately or a few days later.

4. On arrival the fish must be acclimatized slowly and progressively to their new environment. Instead of transferring them directly and brusquely from the carrier water into the water ready to receive them, the same precautions must be taken as those described above for renewing water during transport.

If the fish are transported in cans they should be placed under a tap. The water should be drained away by means of the screened overflow. If there is no overflow then the can should be covered with a loose meshed cloth on which the water is allowed to trickle for a certain time.

Similar precautions must be taken with other transport equipment.

5 Quantities of fish transported

Quantities vary considerably according to the type of transport, i.e. with or without diffused oxygen.

The quantity of fish transported depends on many factors, and in particular: the species of the fish, size, the length of time taken, the temperature of the water, the temperature and atmospheric pressure and the length of the stops.

Under these conditions it is practically impossible to give norms for transportation. The following calculations are given as examples corresponding to predetermined circumstances. (Table *a*)

a   Transport without diffused oxygen

Quantity of water necessary, in litres, for transport in barrels without diffused oxygen for 1 kg (2½ lb) of K$_2$ (individual weight 250 to 500 gm).

(According to Schäperclaus, 1961)

Temperature of the air °C	Transport in hours									
	2	4	6	8	10	12	14	16	18	20
− 5	3·1	3·5	3·9	4·3	4·7	5·1	5·5	6·0	6·5	7·0
0	3·3	3·7	4·1	4·5	5·0	5·5	6·0	6·5	7·0	7·6
+ 5	3·6	3·9	4·4	5·0	5·6	6·2	6·8	7·4	8·0	8·6
+ 10	3·9	4·3	5·0	5·6	6·4	7·1	7·8	8·5	9·3	10·0
+ 15	4·2	5·0	5·8	6·6	7·5	8·4	9·3	10·2	11·2	12·2

Instead of 1 kg of K$_2$ of the above chart it is possible, according to the same author to transport:

1·2 kg (2⅔ lb) of carp or tench for eating.
1 kg (2¼ lb) of tench of 100 to 125 gm (3½ to 4½ oz).
0·25 kg (9 oz) of trout (200 gm) (8 oz), pike or pike-perch.
15 K$_1$ of 9 cm (3½ in) and less.
10 K$_1$ of 9 to 12 cm (3½ to 4¾ in).
6 K$_1$ over 12 cm (4¾ in).
70 S$_1$ of 4 to 9 cm (1½ to 3½ in).
25 S$_2$ from 9 to 18 cm (3½ to 7 in).
20 trout of 5 to 7 cm (2 to 2¾ in).
5 trout of 10 to 13 cm (4 to 5½ in).

The *transport of fry in cans* demands the following

quantities of water for transport lasting from 6 to 10 hours at average temperatures of 8 to 10°C (according to Schäperclaus, 1933 and 1961).

1,000 K$_0$ —   5 to 10 litres (1 gal 1 pt to 2 gal 2 pt)
1,000 K$_v$ —   —   150 to 300 litres (33 to 66 gal)
1,000 coregonids fry 1 to 2 litres (1¾ to 3½ pt)
1,000 pike fry   —   1·5 to 3 litres (2¾ to 5¼ pt)
1,000 trout fry   —   7 to 20 litres (1 gal 4 pt to 4 gal 3 pt).

According to Hofmann (1967) the *transport of carp in a 100 litres (22 gal) container* can be carried out under the following conditions:

	Time of transport h	K_3 kg-number	K_2 kg-number	K_1 9 to 12 cm	K_1 6 to 9 cm	K_v
Without diffusion of oxygen	5	30 (66 lb)-20	20 (44 lb)-50	500	600	1,000
	15	20 (44 lb)-15	15 (33 lb)-35	350	400	800
With diffusion of oxygen	5	50 (110 lb)-30	40 (88 lb)-100	1,000	1,200	2,000
	10	35 (77 lb)-25	25 (55 lb)-60	700	800	1,600
	20	25 (55 lb)-15	15 (33 lb)-40	500	700	1,200

Charpy (1945) gives the following charges for *transporting in cans of 50 to 60 litres (88 to 105 pt) in cool weather.*

Duration of transport time: 5 hours				Duration of transport time: 36 to 48 hours		
Length	Roach	Tench	Carp	Length	Roach	Carp and Tench
3 to 5 cm (1 to 2 in)	2,000 to 2,500	3,000	—	—	—	—
5 to 8 cm (2 to 3 in)	1,800 to 2,000	2,000	1,000	5 to 10 cm (2 to 4 in)	250	300
8 to 10 cm (3 to 4 in)	1,000	1,500	700	10 to 15 cm (4 to 6 in)	70 to 80	150
10 to 12 cm (4 to 5 in)	500	1,000	500	15 to 20 cm (6 to 8 in)	35 to 40	100
12 to 15 cm (5 to 6 in)	200	250	200 to 250	20 to 25 cm (8 to 10 in)	25	
Large Fish	5 kg (11 lb)	5 kg (11 lb)	5 kg (11 lb)	Large Fish	3 to 4 kg (7 to 9 lb)	3 to 4 kg (7 to 9 lb)

b Transport with diffused oxygen or compressed air
Transportation with diffused oxygen or compressed air is becoming more and more generalized, and is usually replacing the preceding method except over short distances, for small quantities or for fry.

One of the preceding charts, according to Hofmann, gives transport indications for carp in a receptacle with diffused oxygen.

According to Schäperclaus, Wiesner and Buschkiel, it is possible to carry in a receptacle with a capacity of 120 litres (26 gal), over a period of 15 hours the following:

Temperature of the water	5 to 6°C	10°C	Relationship between weight of fish and weight of water
Marketable trout	25 kg (55 lb)	20 kg (44 lb)	1 : 5-6
8 cm (3 in) trout fingerlings	3,000	2,000 to 2,500	1 : 6-7
Carp	50 kg (110 lb)	40 kg (88 lb)	1 : 2·5-3

According to Wohlfarth *et al.* (1961), in Israel it is possible to transport from 4 to 8 hours in a poly-thene bag with oxygen, 1 kg of small carp with an average weight of 30 gm (1 oz) in from 2 to 2·5 litres ($3\frac{1}{2}$ to $4\frac{1}{2}$ pt) of water.

The diffusion of oxygen and compressed air must be carefully controlled. It depends on the quantity of fish transported and other factors given above. Excessive oxygen can be harmful to the health of the fish, particularly of the fry, for it produces gas embolism in them and a too strong mucus secretion. During transport with oxygen, foam often forms on the surface of the water.

II   Transport of Freshly Killed Fish

The transportation of freshly caught and killed fish is possible but there are difficulties because of the speed at which their flesh decomposes. Besides, when this is done, it is necessary to ensure that the transported fish will find a ready market when they reach their destination.

One way of dealing with the danger of decomposition is to transport at low temperature. This will stop the development of bacteria. The temperature should be between 0 and 4°C, and to this end the fish are placed in crushed ice.

The type of packing used more or less depends on the value of the fish to be transported. In the case of trout, for example, "burning" of the skin, resulting from direct contact with the ice, should be avoided. The fish can be protected by a greaseproof paper wrapping. Normally the fish are laid out in lots side by side, alternatively head to tail.

The fish should be lightly wiped before being set out. The packed fish are covered with ice and the whole pack then wrapped in greaseproof paper and placed in a basket or a box on some kind of bed able to absorb water from the melting ice; dry moss for example. The carriers should not be too deep for, to avoid crushing, a large number of fish cannot be superimposed one layer on another. A satisfactory size is 50 × 30 × 25 cm (1 ft 8 in × 12 in × 10 in). If a long time is to be taken on the journey then the fish should be refrigerated first in order to prevent the ice from melting too rapidly.

Packing should be carefully carried out and the fish sorted for weight and size. Before being packed the fish should be killed. The simplest way to kill a trout for eating is to give it a sharp crack on the head with a stick. There also exist special pincers and electric equipment for killing fish.

Since the transportation of live trout has become more general, the shipping of dead fish has diminished in commercial fish culture.

SECTION V

CONSERVATION OF FISH

1 Conservation by cold

At the right temperatures cold will permit the conservation of dead fish in a fresh condition for quite a long time. Refrigeration will not permit fish to be kept for a period exceeding 2 weeks. Deep freezing will allow conservation for several months.

1 Refrigeration. The aim is to conserve the fish at around 0°C which will prevent decomposition for a limited time. The most widely used system is that of ice carried in appropriate receptacles. The fish and the ice are mixed, that is to say, in alternate layers of fish and crushed ice or artificial snow, in small cases, baskets or barrels which allow the water from the melting ice to flow easily. For warehousing, the receptacles are kept in crushed ice or artificial snow at temperatures around 0°C.

2 Deep Freezing. Deep freezing of fish calls for a temperature equal or −18°C, or inferior which will permit conservation for a long time. Only the freshest fish in good condition are deep frozen. Before freezing the small fish are washed and the heads of large fish are removed. They are also gutted and washed. Fillets only can also be frozen.

2 Salting, drying, and smoking of fish

1 Salting. Salting means the partial dehydration of the fish by osmosis with sodium chloride. If intense, this will kill the microbes and stop diastasis. Fish with their heads removed, when gutted and washed, are salted as soon as possible after being caught, especially if the temperature is high.

Salting can be carried out: (*a*) dry (alternate layers of fish and salt); (*b*) in brine (light – 16 per cent salt; average: 16 to 20 per cent; strong: + 25 per cent); (*c*) in brine after the fish have been dipped in salt; (*d*) cold by spreading salt and crushed ice on the fish (10 to 12 kg (22 to 26 lb) of salt for 100 kg (220 lb) of ice).

Fig. 497   Tilapias of 1 lb in weight, scaled and gutted, placed in brine before freezing. San Samuel, Hadera, Israel.

For light and cold salting, maturation and conservation must take place in a cold room (2 to 3°C); for average salting the temperature can rise to 16°C and for strong salting the maturation and conservation can take place in normal atmospheric conditions.

2 Drying. Drying fish can be carried out: (*a*)

in the open air if the atmospheric conditions permit, (*b*) artificially. Immediately after catching, the fish to be dried are gutted and have their heads removed. They are then cut lengthwise and a large part of their spine is removed. They are washed and dried if simple drying is desired or salted and dried if a dry salted fish is required.

Natural drying needs a dry, well-aired climate, with sunshine which is not too hot. The fish are suspended or laid out on a lattice or something else, or even on the ground. For artificially dried fish special dryers are used, the best known being a labyrinth type gallery traversed by a current of hot air.

3 *Smoking*. Smoking means submitting the fish to the action of wood smoke (small faggots of wood, shavings or sawdust). By impregnating the flesh of the fish with the empyreumatic products contained in the smoke the conservation capacity is reinforced. This permits the preparation of delicate specialities.

The composition of the smoke depends on the wood. Conifer wood must not be used. The smoke must be conditioned, that means, its temperature regulated and the rate of circulation controlled.

Smoking can be hot or cold. For cold smoking the fish are first dried and then lightly salted and exposed to a smokeless fire (temperature not exceeding 38°C) after which the real smoking takes place at a temperature not above 28°C. Hot smoking is carried out on fresh fish or on fish rapidly salted in brine. First the fish is submitted to a high temperature of 130°C near a strong fire. Smoking follows at a temperature of around 40°C.

There is a difference between artisanal and industrial smoking installations. The first can simply be a chimney provided with a hood, a hearth and twigs used for the smoking, or a small stone construction. Industrial curing includes galleries with a special smoking installation and a system which forces the smoke to circulate.

APPENDIX TO CHAPTER XVI

I Results and Accounting

Once the raising of the fish is completed it is necessary to draw technical and financial conclusions.

From the technical point of view, it is a good idea to keep a register including a page or a formulary for each pond. This will include all useful indications, notably:

The designation of each pond by a number corresponding to a plan;

size;

depth and water supply;

the date of stocking with details such as, per species: quantities in numbers and (or) weight, length and individual average weights;

similar entries for the harvest;

total losses also given as percentages;

increases in total and individual weights of each species;

total production of the pond and production per unit of area (hectare or other);

the nature and quantities of fertilizers used;

the characteristics of the food distributed including its type and the frequency of its distribution; etc.

It is also necessary to take note for the farming in general or for each pond in particular of full information on temperature, chemical composition of the water, diseases, accidents, etc.

From the financial point of view an account should be kept for each group of ponds, or for each type of farming (rearing fry, fish for eating); of all earnings and expenditure, which will permit conclusions to be drawn on the manner of the farming. It is also good to include graphs.

Buschkiel (1932) estimates that for a fish farm to pay, the interest of the capital invested in the purchase of the land and the construction of the pond, or the rent of the fish farm, should not be more than one third of the revenue; another third should cover running costs and the final third should include salaries and profits covering risks.

II Marking Cultivated Fish

It is useful to know the growth, age, sex and behaviour of certain individual fish of any particular species cultivated in a single pond. The identification of these individual fish is made by marking. This can be done either by collective or individual marking. The first will permit recognition of a determined group of individuals and the second the identification of an individual fish.

A simple way of *collective marking* is by clipping a fin or part of a fin with scissors (Figs. 500 and 501). This technique can be used for nearly all species and sizes of fish. Generally speaking, experienced personnel can recognize the marking without too much difficulty even after one growing season. The regrowth of a clipped fin is that much slower or its deformation that much more pronounced, as it is cut near to its point of implantation. It is important not to injure the fish by allowing it to bleed when the fin is clipped. A partially clipped fin after complete regrowth can still be recognized thanks to the deformed fin rays at the point of clipping. This deformation can be recognized more easily by transparency. Numerous tests made in ponds have shown that fish scarcely suffer from this marking technique.

Either the right or left ventral or pectoral fin, or the top or lower part of the caudal fin can be clipped. In the case of salmonids the adipose fin can be clipped. This permits several groups of individuals in the same pond to be distinguished. It is also possible to increase the number of groups by clipping two fins.

There are other systems of collective markings such as tattooing or using coloured subcutaneous injections but these are less practised than those given above.

Individual marking is generally by metal or plastic numbered tags. These can be obtained in several models. They are fixed to different parts of the body. Some tags are fixed to the jaw (Fig. 502) or to the gill cover though these are only suitable for rather large fish and species with a pronounced lower jaw such as trout or perch, or with a sufficiently strong gill cover such as in carp. These tags often cause injury and those attached to the jaw can hinder feeding.

For sufficiently large fish a tag fixed by a metal wire or some synthetic material passed through the muscular fibres near the dorsal fin, is often used. Losses of tags can be rather high. It is the same for nylon thread fixed to the large spiny ray of the dorsal fin of carp.

Another method of marking is to brand the skin of the fish. This applies only when the fish have few or no scales such as mirror carp or leather carp as well as fish with very small scales. This can be done with a silver nitrate pencil or a heated metal wire such as a copper wire heated on a small portable stove, or a platinum or chrome-nickel wire heated electrically. Moav et al. (1960) has described an apparatus which simply uses an automobile battery. The method permits collective marking if only one sign is used, a cross for example. It also permits individual marking if a number (Fig. 503) is used. The method is simple and rapid and practically without danger for the fish.

Collective marking, and above all individual marking, is greatly helped by the use of an *anaesthetic*. There are several anaesthetics available, and MS 222 or propoxate are commonly used at present. First a stock solution with 20 gm of MS 222 per litre of water is prepared. The fish are then placed in a bath containing 5cm³ of the solution, per litre of water. Propoxate may be used for the anaesthesia of salmonids at a level of 1 to 2 ppm and for certain other freshwater and marine species at a level of 1 to 4 ppm of active ingredient (Bergström, 1967; Thienpont, 1965 a, b). The moment the fish turn on their sides they can be removed from the bath and marked.

Fig. 498   Carp, pike and European catfish in slices. Poljana fish farm, Croatia, Yugoslavia.

Fig. 499   Grey mullet and eels being smoked with wood. Lago di Paola lagoon, Sabaudia, Lazio, Italy.

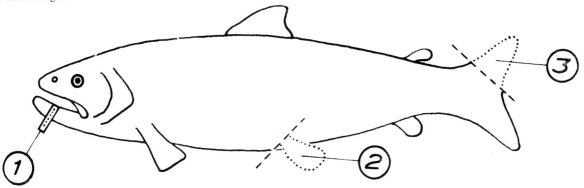

Fig. 500 Schematic drawing of different methods for marking fish. 1: Metallic jaw tag: 2: Clipping of left ventral fin: 3: Clipping of the upper part of the caudal fin.

Fig. 501 Roach marked by removal of the right ventral fin, 1 year after marking.

Fig. 502 Trout with metal, numbered jaw tag.

Fig. 503 Four-year-old carp with a number marked by a silver nitrate pencil.

GENERAL APPENDIX

Lists of principal cultivated fish included in the Textbook of Fish Culture

1 *Scientific and common names in French and English*

No.	Family	Species	Common Names French	Common Names English
1	**ACIPENSERIDAE**	*Acipenser sturio* L.	Esturgeon	Sturgeon
2		*Acipenser ruthenus* L.	Sterlet	
3	**CLUPEIDAE**	*Alosa alosa* (L.)	Grande alose	Allis shad
4		*Alosa sapidissima* (Wilson)		American shad
5		*Chanos chanos* (Forskal)		Milkfish
6	**SALMONIDAE**	*Hucho hucho* Berg (*Salmo hucho* L.)	Huchon	
7		*Oncorhynchus gorbuscha* (Walbaum)	Saumon du Pacifique	Pink salmon
8		*Oncorhynchus keta* (Walbaum)	(n° 7 à 12)	Chum salmon
9		*Oncorhynchus kisutch* (Walbaum)		Coho salmon
10		*Oncorhynchus masou* (Brevoort)		
11		*Oncorhynchus nerka* (Walbaum)		Sockeye salmon
12		*Oncorhynchus tshawytscha* (Walbaum)		Chinook salmon
13		*Salmo clarkii* Richardson	Truite des Montagnes Rocheuses	Cutthroat trout
14		*Salmo gairdneri* Richardson	Truite arc-en-ciel	Rainbow trout
15		*Salmo irideus* Gibbons	Truite arc-en-ciel	Steelhead trout
16		*Salmo shasta* Jordan		
17		*Salmo salar* L.	Saumon atlantique	Atlantic salmon
18		*Salmo trutta* L.	Truite de mer	Sea trout
19		*Salmon trutta fario* L.	Truite de rivière	Brown trout
20		*Salmo trutta lacustris* L.	Truite de lac	Lake trout (Europe)
21		*Salmo trutta caspius* Kessler		
22		*Salmo trutta labrax* Pallas		
23		*Salmo trutta macrostigma* (Dumeril)		
24		*Salmo trutta marmoratus* Cuvier		
25		*Salvelinus alpinus* (L.)	Omble chevalier	Arctic char
26		*Salvelinus fontinalis* (Mitchill)	Saumon de fontaine	Brook trout
27		*Salvelinus namaycush* (Walbaum)	Omble américain	Lake trout (America)
28	**COREGONIDAE**	*Coregonus albula* (L.)	Petite marène	Lake whitefish
29		*Coregonus lavaretus* (L.)	Lavaret	
30		*Coregonus lavaretus maraena* Bloch	Grande marène	
31	**THYMALLIDAE**	*Thymallus signifer* Richardson		Arctic grayling
32		*Thymallus thymallus* (L.)	Ombre de rivière	Grayling
33	**ESOCIDAE**	*Esox lucius* L.	Brochet	Northern pike
34		*Esox masquinongy* Mitchill	Muskellunge	Muskellunge
35	**OSTEOGLOSSIDAE**	*Heterotis niloticus* (Ehrenbaum)	Heterotis	Heterotis
36	**CHARACIDAE**	*Brycon guatemalensis* Regan		
37	**CYPRINIDAE**	*Alburnus alburnus* (L)	Ablette commune	Bleak
38		*Campostoma anomalum* (Rafinesque)		Stoneroller

No.	Family	Species	Common Names	
			French	English
39	**CYPRINIDAE**	*Carassius auratus* (L.)	Carassin doré	Goldfish
40		*Carassius carassius* (L.)	Carassin	Crucian carp
41		*Carassius gibelio* (Bloch)	Gibèle	
42		*Catla catla* (Hamilton Buchanam)	Catla	Catla
43		*Cirrhina molitorella* (Valenciennes)		Mud carp
44		*Cirrhina mrigala* (Hamilton Buchanam)	Mrigal	Mrigal
45		*Ctenopharyngodon idella* Valenciennes	Carpe herbivore	Grass carp
46		*Cyprinus carpio* L.	Carpe	Carp
47		*Hypophthalmichthys molitrix* (Valenciennes)	Carpe argentée	Silver carp
48		*Hypophthalmichthys nobilis* (Richardson)	Carp marbrée	Big head
49		*Idus idus* (L.) (*Idus orfus* Roule)	Orfe ou Ide rouge	Golden ide
50		*Labeo calbasu* (Hamilton Buchanam)	Calbasu	Calbasu
51		*Labeo rohita* (Hamilton Buchanam)	Rohu	Rohu
52		*Leucaspius delineatus* (Heckel)	Able de Heckel	
53		*Mylopharyngodon piceus* (Richardson)		Black carp
54		*Notemigonus crysoleucas* (Mitchill)		Golden shiner
55		*Osteochilus hasseltii* (Valenciennes)		Nilem
56		*Phoxinus phoxinus* (L.)	Vairon	Minnow (Europe)
57		*Pimephales notatus* Rafinesque		Bluntnose minnow
58		*Pimephales promelas* Rafinesque		Fathead minnow
59		*Puntius javanicus* (Bleeker)		Tawes
60		*Rhinichthys atratulus* Hermann		Blacknose dace
61		*Rutilus rutilus* (L.) [*Leuciscus rutilus* (L.)]	Gardon	Roach
62		*Scardinius erythrophthalmus* (L.)	Rotengle	Rudd
63		*Semotilus atromaculatus* (Mitchill)		Creek chub
64		*Tinca tinca* (L.)	Tanche	Tench
65	**CATOSTOMIDAE**	*Catostomus commersonii* (Lacepede)		White sucker
66		*Ictiobus cyprinellus* Valenciennes		Bigmouth buffalo
67	**ICTALURIDAE**	*Ictalurus catus* (L.)		White catfish
68		*Ictalurus furcatus* (Lesueur)		Blue catfish
69		*Ictalurus melas* (Rafinesque)	Poisson chat	Black bullhead
70		*Ictalurus nebulosus* (Lesueur)		Brown bullhead
71		*Ictalurus punctatus* (Rafinesque)		Channel catfish
72	**SILURIDAE**	*Silurus glanis* L.	Silure glane	Danubian wels
73	**CLARIIDAE**	*Clarias batrachus* (L.)		Catfish
74		*Clarias macrocephalus* Günther		
75	**SCHILBEIDAE**	*Pangasius larnaudi* Bocourt		
76		*Pangasius micronemus* Bleeker		
77		*Pangasius pangasius* (Hamilton Buchanam)		
78		*Pangasius sutchi* Fowler		
79	**ANGUILLIDAE**	*Anguilla anguilla* (L.) (*Anguilla vulgaris* Slem.)	Anguille européenne	European eel
80		*Anguilla japonica* Temminck & Schlegel	Anguille japonaise	Japanese eel
81		*Anguilla rostrata* (Lesueur)	Anguille américaine	American eel
82	**MUGILIDAE**	*Mugil capito* Valenciennes		
83		*Mugil cephalus* L.	Muge ou Mulet	Grey mullet
84		*Mugil tade* Forskal		
85	**ATHERINIDAE**	*Odontesthes bonariensis* (Valenciennes)	Pejerrey	Silverside
86		*Chirostoma estor* Jordan		
87	**CENTROPOMIDAE**	*Lates niloticus* (L.)	Capitaine	Nile perch

No.	Family	Species	Common Names	
			French	English
88	**PERCIDAE**	*Perca flavescens* (Mitchill)	Perche américaine	Yellow perch
89		*Perca fluviatilis* L.	Perche	Perch
90		*Sander lucioperca* (L.) (*Lucioperca lucioperca* Berg)	Sandre	Pike-perch
91		*Stizostedion vitreum* (Mitchill)	Walleye	Walleye
92	**CENTRARCHIDAE**	*Ambloplites rupestris* (Rafinesque)		Rock bass
93		*Eupomotis gibbosus* (L.)	Perche soleil	Pumpkinseed
94		*Lepomis cyanellus* Rafinesque		Green sunfish
95		*Lepomis macrochirus* Rafinesque	Bluegill	Bluegill
96		*Lepomis microlophus* (Günther)		Redear sunfish
97		*Micropterus dolomieu* Lacépède	Black-bass à petite bouche	Smallmouth bass
98		*Micropterus punctulatus* (Rafinesque)	Black-bass tacheté	Spotted bass
99		*Micropterus salmoides* (Lacépède)	Black-bass à grande bouche	Largemouth bass
100		*Pomoxis annularis* Rafinesque		White crappie
101		*Pomoxis nigromaculatus* (Lesueur)		Black crappie
102	**CICHLIDAE**	*Astatoreochromis alluaudi* (Pellegrin)		
103		*Etroplus suratensis* (Bloch)		Pearl spot
104		*Haplochromis mellandi* (Boulenger)		
105		*Hemichromis fasciatus* Peters		
106		*Tilapia andersonii* (Castelnau)	Tilapia (no 106 à 121)	Tilapia (nr 106-121)
107		*Tilapia aurea* (Steindachner)		
108		*Tilapia esculenta* Graham		
109		*Tilapia galilaea* (Artedi)		
110		*Tilapia heudelotii* Dumeril		
111		*Tilapia hornorum* Trewavas		
112		*Tilapia leucosticta* Trewavas		
113		*Tilapia macrochir* Boulenger		
114		*Tilapia melanopleura* Dumeril		
115		*Tilapia mossambica* (Peters)		
116		*Tilapia nigra* (Günther)		
117		*Tilapia nilotica* (L.)		
118		*Tilapia rendalli* (Boulenger)		
119		*Tilapia sparrmani* Smith		
120		*Tilapia tholloni* (Sauvage)		
121		*Tilapia zillii* (Gervais)		
122	**ANABANTIDAE**	*Helostoma temmincki* Cuvier		Kissing gourami
123		*Osphronemus goramy* (Lacépède)	Gourami	Gourami
124		*Trichogaster pectoralis* (Regan)		Sepat siam

2 *Alphabetical list in French*

The number is the same as that given in the list of scientific names

	No.		No.
Able de Heckel	52	Muge	83
Ablette commune	37	Mulet	83
Anguille américaine	81	Muskellunge	34
Anguille européenne	79	Omble américain	27
Anguille japonaise	80	Omble chevalier	25
Black-bass à grande bouche	99	Ombre de rivière	32
Black-bass à petite bouche	97	Orfe	49
Black-bass tacheté	98	Pejerrey	85
Bluegill	95	Perche	89
Brochet	33	Perche américaine	88
Calbasu	50	Perche soleil	93
Capitaine	87	Petite marène	28
Carassin	40	Poisson chat	69
Carassin doré	39	Rohu	51
Carpe	46	Rotengle	62
Carpe argentée	47	Sandre	90
Carpe herbivore	45	Saumon atlantique	17
Carpe marbrée	48	Saumon de fontaine	26
Catla	42	Saumon du Pacifique	7 à 12
Esturgeon	1	Silure glane	72
Gardon	61	Sterlet	2
Gibèle	41	Tanche	64
Gourami	123	Tilapia	106 à 121
Grande alose	3	Truite arc-en-ciel	14-15
Grande marène	30	Truite de lac	20
Heterotis	35	Truite de mer	18
Huchon	6	Truite des Montagnes Rocheuses	13
Ide rouge	49	Truite de rivière	19
Lavaret	29	Vairon	56
Mrigal	44	Walleye	91

3   *Alphabetical list in English*

The number is the same as that given in the list of scientific names

	No.		No.
Allis shad	3	Kissing gourami	122
American eel	81	Lake trout (America)	27
American shad	4	Lake trout (Europe)	20
Arctic char	25	Lake whitefish	28
Arctic grayling	31	Largemouth bass	99
Atlantic salmon	17	Milkfish	5
Big head	48	Minnow (Europe)	56
Bigmouth buffalo	66	Mrigal	44
Black bullhead	69	Mud carp	43
Black carp	53	Muskellunge	34
Black crappie	101	Nilem	55
Blacknose dace	60	Nile perch	87
Bleak	37	Northern pike	33
Blue catfish	68	Pearl spot	103
Bluegill	95	Perch	89
Bluntnose minnow	57	Pike	33
Brook trout	26	Pike-perch	90
Brown bullhead	70	Pink salmon	7
Brown trout	19	Pumpkinseed	93
Calbasu	50	Rainbow trout	14
Carp	46	Redear sunfish	96
Catfish	73	Roach	61
Catla	42	Rock bass	92
Channel catfish	71	Rohu	51
Chinook salmon	12	Rudd	62
Chum salmon	8	Sea trout	18
Coho salmon	9	Sepat siam	124
Creek chub	63	Silver carp	47
Crucian carp	40	Silverside	85
Cutthroat trout	13	Smallmouth bass	97
Danubian wels	42	Sockeye salmon	11
European eel	79	Spotted bass	98
Fathead minnow	58	Steelhead trout	15
Golden ide	49	Stoneroller	38
Golden shiner	54	Sturgeon	1
Goldfish	39	Tawes	59
Gourami	123	Tench	64
Grass carp	45	Tilapia	106-121
Grayling	32	Walleye	91
Green sunfish	94	White catfish	67
Grey mullet	83	White crappie	100
Heterotis	35	White sucker	65
Japanese eel	80	Yellow perch	88

BIBLIOGRAPHY

I General bibliography

A. GENERAL WORKS BEFORE 1940

1921. BENECKE, B. - VON DEBSCHITZ, H. - Die Teichwirtschaft. Paul Parey, Berlin, 6. Aufl., 172 p., 83 fig.

1931. BUSCHKIEL, A. L. - Salmonidenzucht in Mitteleuropa. In DEMOLL-MAIER: Handbuch der Binnenfischerei Mitteleuropas, Bd. IV, Lf. 2: 161-348, 72 fig.

1932. BUSCHKIEL, A. L. - De Teelt van Karpers en de beginselen der Vischteelt in Nederlandsch Oost-Indië. Archipel Drukkerij, Buitenzorg (Java), 135p., 40 fig.

1926. DIESSNER - ARENS - Die künstliche Zucht der Forelle. J. Neumann, Neudamm, 3. Aufl., 285 p. 120 fig.

1921. DOLJAN - HAEMPEL. - Handbuch der modernen Fischereibetriebslehre. Verlag Wilhelm Frick, Wien und Leipzig, 176 p., 27 fig., 5 pl.

1923. GUENAUX, G. - Pisciculture. J.-B. Baillière et fils, Paris, 2e édit., 488 p., 125 fig.

1931. MAIER, H. N. und HOFMANN, J. - Grundzüge der Karpfenteichwirtschaft für kleinere und mittlere Betriebe. (Weihenstephaner Schriftsammlung für praktische Landwirtschaft, Heft 24.) Datterer & Cie, Freising-München, 125 p., 116 fig., 2 pl.

1934. MEHRING, H. - Karpfenzucht. In DEMOLL-MAIER: Handbuch der Binnenfischerei Mitteleuropas, Bd IV, Lf. 3: 349-406, 13 fig.

1914. RAVERET-WATTEL, C. - La pisciculture industrielle. G. Doin et Cie, Paris, 408 p., 74 fig.

1932. ROULE, L. - Manuel de Pisciculture. Librairie Hachette, Paris, 158 p., 45 fig.

1933. SCHÆPERCLAUS, W. - Lehrbuch der Teichwirtschaft. Paul Parey, Berlin, 289 p., 71 fig.

1927. TAURKE, Fr. - Die Fischzucht und Fischhaltung. M.-H. Schaper, Hannover, 2. Aufl., 272 p.

1937. VIOSCA, P. - Pondfish Culture. The Pelican Publishing Company, New Orleans, 260 p., 68 fig.

1928. WALTER, E. - Karpfennutzung in kleinen Teichen. J. Neumann, Neudamm, 4. Aufl., 144 p., 30 fig.

1934. WALTER, E. - Grundlagen der allgemeinen fischereilichen Produktionslehre einschliesslich ihrer Anwendung auf die Fütterung. In DEMOLL-MAIER: Handbuch der Binnenfischerei Mitteleuropas, Bd. IV, Lf. 5: 483-662, 100 fig.

1937. WIESNER, R. - Lehrbuch der Forellenzucht und Teichwirtschaft. J. Neumann, Neudamm, 253 p., 12 pl., 17 fig.

B GENERAL WORKS AFTER 1940

1956. DAVIS, H. S. - Culture and Diseases of Game Fishes. University of California Press, Berkeley and Los Angeles, 332 p., 55 fig.

1924/62. DEMOLL, R. und MAIER, H. N. - Handbuch der Binnenfischerei Mitteleuropas. E. Schweizerbart'sche Verlagsbuchhandlung, Stuttgart. 6 volumes.

1969. GHITTINO, P. - Piscicoltura e Ittiopatologia, Vol. I: Piscicoltura. Edizione Rivista di Zootechnica, Torino, 333 p., 274 fig.

1963. GIERALTOWSKI, M. et al. - Hodowla Ryb W Stawach. Panstwowe Wydawnictwo Rolnicze i Lesne, Warszawa, 636 p., 233 fig.

1949. HALL, C. B. - Ponds and Fish Culture. Faber and Faber, London, 244 p.

1947. HEY, D. - The Culture of Freshwater Fish in South Africa. Inland Fisheries Department, Stellenbosch, 2d edit., 124 p., 48 fig.

1962. HICKLING, C. F. - Fish Culture. Faber and Faber, London, 295 p., 66 fig.

1967. HOFMANN, J. - Der Teichwirt. P. Parey, Hamburg und Berlin, 2. Aufl., 248 p., 166 fig.

1962. HORA, S. L. and PILLAY, T. V. R. - Handbook on fish culture in the Indo-Pacific Region. FAO Fisheries Biology Technical Paper, 14, 204 p.

1949. HUET, M. - Appréciation de la valeur piscicole des eaux douces. Trav. Stat. Rech. Eaux et Forêts, Groenendaal, Sér. D, 10, 55 p., 41 fig.

1956. HUET, M. - Aperçu de la pisciculture en Indonésie. Bul. Agric. Congo belge, 47 (4): 901-957, 66 fig. (Trav. Stat. Rech. Eaux et Forêts, Groenendaal, Sér. D. 19.)

1960. KOCH, W. - Fischzucht. Paul Parey, Berlin und Hamburg, 3. Aufl., 322 p., 216 fig., 11 pl.

1961. KOSTOMAROV, B. - Die Fischzucht. VEB Deutsche Landwirtschaftsverlag, Berlin, 375 p., 135 fig.

1951. KREUZ, A. - Teichbau und Teichwirtschaft. Neumann Verlag, Radebeul und Berlin, 2. Aufl., 167 p., 51 fig.

1956. LAGLER, K. F. - Freshwater Fishery Biology WM. C. Brown Company, Dubuque, Iowa, 2d edit., 421 p., 172 fig.

1967. LIVOJEVIC, Zl. et BOJCIC, C. - Prirucnik za Slatkovodno Ribarstvo. Agronomski glasnik, 704 p., fig.

1967/68. PILLAY, T. V. R. - Proceedings of the world symposium on warm-water pond fish culture. (Rome, 1966.) FAO Fisheries Report, 44, Vol. I, 55 p., Vol. II, 174 p., Vol. III, 423 p., Vol. IV, 492 p., Vol. V, 411 p.

1949. SCHÄPERCLAUS, W. - Grundriss der Teichwirtschaft. Paul Parey, Berlin und Hamburg, 236 p., 119 fig.

1961. SCHÄPERCLAUS, W. - Lehrbuch der Teichwirtschaft. Paul Parey, Hamburg und Berlin, 2. Aufl., 582 p., 290 fig.

1966. SOMMANI, E., VOLPE, R. – La Trota. Come si alleva. Ramo Editoriale degli Agricoltori, Roma, 150 p., 43 fig.

1961. VIBERT, R. et LAGLER, K. F. – Pêches continentales. Biologie et Aménagement. Dunod, Paris, 720 p., 164 fig.

1953. VILLALUZ, D. K. – Fish Farming in the Philippines. Bookman, Inc., Manila, 336 p.

1954. VIVIER, P. – La Pisciculture. Presses Universitaires de France, Paris, 127 p., 6 fig.

1963. WUNDSCH, H. H. und andere. – Fischereikunde. Neumann Verlag, Radebeul und Berlin, 2. Aufl., 351 p.

II   Special bibliography[1]

(Also consult general bibliography.)

CHAPTER I

CONSTRUCTION AND LAYOUT OF PONDS

1964. ARRIGNON, J. – Note technique sur l'emploi du Bentonil dans l'étanchement des barrages. *Bul. Off. Inform. Cons. Sup. Pêche*, 56: 79-81.

1963. BANK, O., KRUSCH, A. – So baut man Teiche. Paul Parey, Hamburg-Berlin, 123 p., 27 fig.

1948. BARCA, Gh. I. – Constructii si amenajeri piscicole. Institut de Cercetari piscicole al R.P.R., Bucaresti, 264 p.

1954. DE BONT, A. F. – La construction d'étangs de pisciculture au Congo belge. Ministère des Colonies, Bruxelles, 116 p., 38 fig.

1926. DIESSNER - ARENS – Die künstliche Zucht der Forelle. *

1925/26. EVRARD, J. B. – La construction des pièces d'eau dans les parcs publics et jardins privés. *Ann. Trav. Publ. Belgique*, 26 (6): 931-947; 27 (1): 7-61, 66 fig.

1960. HASKELL, D. C., DAVIES, R. O., RECKAN, J. – Factors in hatchery pond design. *New York Fish and Game Journal*, 7 (2): 112-129, 13 fig.

1948. HUET, M. – Construction et Aménagement piscicole des étangs. Ministère des Colonies, Bruxelles, 50 p., 41 fig.

1951. KREUZ, A. – Teichbau und Teichwirtschaft. *

1966. MAAR, A., MORTIMER, M. A. E., VAN DER LINGEN, I. – Fish Culture in Central East Africa. *(VI, 5).

1931. MAIER - HOFMANN. – Grundzüge der Karpfenteichwirtschaft. *

1965. RICHARDSON, W. M. – Electric rotary fish screens. *Progressive Fish-Culturist*, 27 (1): 20-22, 2 fig.

1933. SCHÄPERCLAUS, W. – Lehrbuch der Teichwirtschaft. *

1963. SCHÄPERCLAUS, W. – Teichbau, 86-94, 14 fig. *in* WUNDSCH, H. H. – Fischereikunde. *

1956. VIBERT, R. – Aperçu sur le problème des grilles et de leur nettoyage. *Bul. franç. Pisciculture*, 29 (183): 45-64, 16 fig.

1955. X. – Comment construire un étang de pisciculture familiale. Inspection générale des Eaux et Forêts, Brazzaville, 27 p., 44 fig.

CHAPTER II

NATURAL FOOD AND GROWTH OF CULTIVATED FISH

1937. BROCHER, F. – L'Aquarium de chambre. Les Naturalistes Belges, Bruxelles, 3e édit., 410 p., 186 fig.

1967. ENGELHARDT, W. – Was lebt in Tümpel, Bach und Weiher? Franckh'sche Verlagshandlung, Stuttgart, 4. Aufl., 258 p., 420 fig.

1910. FRANCÉ, R. H. – Die Kleinwelt des Süsswassers. Verlag Theod. Thomas, Leipzig, 160 p., 322 fig.

1957. GERMAIN, L. et SEGUY, E. – La faune des lacs, des étangs et des marais de l'Europe occidentale. Editions Paul Lechevalier, Paris, 2e édit., 549 p., 272 fig., 32 pl.

1927. GEYER, D. – Unsere Land- und Süsswasser-Mollusken. K. G. Lutz-Verlag, Stuttgart, 224 p., XXXIII pl.

1949. HUET, M. – Appréciation de la valeur piscicole des eaux douces. *

1937. JOHANNSEN, O. A. – Aquatic Diptera. Part III, Chironomidae. Cornell University, Agric. Exper. Sta., Memoir 205, June 1937, 84 p., XVIII pl., 274 fig.

1937. JOHANNSEN, O. A., THOMSEN, L. C. – Aquatic Diptera. Part IV-V, Chironomidae, Ceratopogonidae. Cornell University, Agric. Exper. Sta., Memoir 210, December 1937, 80 p., XVIII pl., 133 fig.

1934. KARNY, H. H. – Biologie der Wasserinsekten. Fritz Wagner, Wien, 311 p., 160 fig.

1925. LAMPERT, K. – Das Leben der Binnengewässer. Chr. Herm. Tauchnitz, Leipzig, 3. Aufl., 892 p., 17 pl., 286 fig.

1934. MIALL, L. C. – The Natural History of Aquatic Insects. Macmillan and Co, London, 395 p., 116 fig.

1962. NEEDHAM, J. G. and NEEDHAM, P. R. – A Guide to the Study of Fresh-Water Biology. Holden-Day, Inc., San Francisco, 5th edit., 107 p., 15 pl.

1915. ROUSSEAU, E. – Les poissons d'eau douce indigènes et acclimatés de la Belgique. (*In* La Pêche fluviale en Belgique.) Pêche et Pisciculture, Bruxelles, 192 p., 102 fig.

[1]For references marked with an asterisk, the complete reference is given either in the general bibliography (*) or in the special bibliography relative to the chapter indicated in Roman numerals by the side of the asterisk [*(IV)].

1921. ROUSSEAU, E. – Les larves et nymphes aquatiques des insectes d'Europe. Office de Publicité, Bruxelles, 967 p., 344 fig.

1933. SCHÄPERCLAUS, W. – Lehrbuch der Teichwirtschaft. *

1930. SCHOENEMUND, Ed. – Eintagsfliegen oder Ephemeroptera. *In* Die Tierwelt Deutschlands, Gustav-Fischer, Jena, 19. Teil, 106 p., 186 fig.

1913. WALTER, E. – Einführung in die Fischkunde unserer Binnengewässer. Quelle und Meyer, Leipzig, 364 p., 62 fig.

1939. WESENBERG-LUND, C. – Biologie der Süsswassertiere. Julius Springer, Wien, 817 p., 1138 fig.

1943. WESENBERG-LUND, C. – Biologie der Süsswasserinsekten. Julius Springer, Berlin, Wien, 682 p., 501 fig.

1936. WUNDER, W. – Physiologie der Süsswasserfische Mitteleuropas. *In* DEMOLL-MAIER: Handbuch der Binnenfischerei Mitteleuropas. Bd. IIB, 340 p., 213 fig.

1926. WUNDSCH, H. H. – Die Reinhaltung unserer Fischgewässer. *In* DEMOLL-MAIER: Handbuch der Binnenfischerei Mitteleuropas, Bd. VI, Lfg. 2: 146-222, 60 fig.

CHAPTER III

BREEDING AND CULTIVATION OF SALMONIDS

1977. BERGER, M. – Silox-Fischzuchtanlage. Ein neues Verfahren zue Intensiv-produktion von Speisefischen. *Fischer und Teichwirt*, 28 (9): 116-118.

1955. BURROWS, R. E. and PALMER, D. D. – A vertical egg and fry incubator. *Progressive Fish-Culturist*, 17 (4): 147-155, 4 fig.

1931. BUSCHKIEL, A. L. – Salmonidenzucht in Mitteleuropa. *

1961. BUSS, K. and FOX, H. – Modifications for the jar culture of trout eggs. *Progressive Fish-Culturist*, 23 (3): 142-144, 2 fig.

1965. CALDERON, E. G. – L'élevage de la Truite fario et de la Truite arc en-ciel dans les eaux à température très élevée. FAO Stud. Rev. Gen. Fish. Counc. Medit., 30, 25 p.

1967. CAPPELLO, G. – Un semplice "embrionatore" per uova di trota. *Riv. Ital. Piscicoltura e Ittiopatologia*, 2 (3): 60-61, 4 fig.

1941. CHARPY, R. – De la construction et de l'aménagement d'un établissement de trutticulture orienté en vue de la production d'alevins et truitelles de repeuplement. *Bul. franç. Pisciculture*, 13 (122): 5-87, 32 fig., 26 pl.

1965. CUINAT, R. – Quelques conseils pour l'aménagement et l'exploitation des rigoles d'alevinage. *Bul. Off. Inform. Cons. Sup. Pêche*, 59: 12-19.

1956. DAVIS, H. S. – Culture and Diseases of Game Fishes. *

1949. DEMOLL, R. und STEINMANN, P. – Praxis der Aufzucht von Forellen-Besatzmaterial. E. Schweizerbart'sche Verlagsbuchhandlung, Stuttgart, 98 p., 40 fig.

1926. DIESSNER - ARENS – Die künstliche Zucht der Forelle. *

1921. DOLJAN - HAEMPEL. – Handbuch der modernen Fischereibetriebslehre. *

1951. DORIER, A. – Action du liquide cœlomique sur les spermatozoïdes de Truite arc-en-ciel. *Trav. Labor. Hydr. et Piscic. Univ. Grenoble*, XLIe et XLIIe années (1949 et 1950), 69-73.

1951. DORIER, A. – Conservation de la vitalité et du pouvoir fécondant des spermatozoïdes de Truite arc-en-ciel. *Trav. Labor. Hydr. et Piscic. Univ. Grenoble*, XLIe et XLIIe années (1949 et 1950), 75-85.

1974. GERARD, J.-P. – Sur l'utilisation du formol en thérapeutique piscicole. *Bul. franç. Pisciculture*, 47 (254): 18-19.

1960. GREENBERG, D. B. – Trout farming. Chilton Company, Philadelphia, 197 p., 158 fig.

1967. GRÜNSEID, G. – Forellenzucht im Umlaufverfahren. *Oesterreichs Fischerei*, 20 (2/3): 21-28, 4 fig.

1965/66. GUYARD, H. – Eléments de génétique et d'embryologie piscicoles. *Bul. franç. Pisciculture*, 38 (219): 66-73, (220): 81-100, (221): 134-145.

1955. HILDEBRAND, S. F. – The Trouts of North America. U.S. Fish and Wildlife Service, Fishery Leaflet 355, 13 p.

1949. HUET, M. – Appréciation de la valeur piscicole des eaux douces. *

1921. JUILLERAT, E. – L'élevage industriel des Salmonidés. Librairie Delagrave, Paris, 203 p., 42 fig.

1960. KOCH, W. – Fischzucht. *

1951. KREUZ, A. – Teichbau und Teichwirtschaft. *

1932. LÉGER, L. – La pratique du déversement d'alevins dans les cours d'eau. Allier père et fils, Grenoble, 2e édit., 39 p., 4 fig.

1934/45. LÉGER, L. – Petite salmoniculture fermière. *Trav. Labor. Hydr. et Piscic. Univ. Grenoble*, 1934 (Année 1932), 1-32 et 1935 (Années 1933-1934), 1-103, 7 fig.

1959. LEITRITZ, E. – Trout and Salmon Culture (Hatchery Methods). Fish Bulletin No 107, State of California, Department of Fish and Game, 169 p., 58 fig.

1956. LINDROTH, A. – Salmon stripper, egg counter, and incubator. *Progressive Fish-Culturist*, 18 (4): 165-170, 10 fig.

1959. LOUCHET, Cl. – La pisciculture salmoniculture. Yvert et Cie, Amiens, 198 p.

1949. MEDEM, F., RÖTHELI, A., ROTH, H. – Biologischer Nachweis von Befruchtungsstoffen bei Felchen und Flussforelle. *Schweiz. Zeit. f. Hydrologie*, 11 (3/4): 361-377.

1967. ROTH, H. und NEF, W. – Intensive Zucht von Besatzfischen in Rundtrog. *Schweiz. Zeit. f. Hydrologie*, 29 (1): 251-268.

1968. ROTH, H. und GEIGER, W. – Aufzucht von Besatzfischen in Trögen. Ver. d. Eidg. Amtes f. Gewässerschutz u.d. Eidg. Fischereiinspektion, Bern, 25, 40 p., 32 fig.

1933. SCHÄPERCLAUS, W. – Lehrbuch der Teichwirtschaft. *

1936. SURBER, E. W. – Circular rearing pools for trout and bass. *Progressive Fish-Culturist*, 21: 1-14.

1952/53. TACK, E. – Ueber die Aufzucht von Forellensetzlingen in Naturteichen. *Archiv. f. Fischereiwissenschaft*, 4: 70-86, 2 fig.

1963. THUMANN, M. E., STEFFENS, W. – Forellenteichwirtschaft, *in* WUNDSCH, H. H. – Fischereikunde, 144-153. *

1959. VIBERT, R. – Dispositf vertical d'incubation en masse. *Bul. franc. Pisciculture*, 31 (192): 104-115, 6 fig.

1975. VIBERT, R. – Repeuplement des eaux à truites. *Pisc. franç.*, 42: 25–48.

1937. WIESNER, R. – Lehrbuch der Forellenzucht und Teichwirtschaft. *

CHAPTER IV

BREEDING AND CULTIVATION OF CYPRINIDS

1966. ALIKUNHI, K. H. – Synopsis of biological data on Common Carp (*Cyprinus carpio* L.). FAO Fisheries Synopsis N° 31.1, 83 p.

1968. ANTALFI, A., TÖLG, Is. – Növenyevo Halak. Mezogazdasagi Kiado, Budapest, 155 p., 44 fig.

1968. ANTALFI, A. und TÖLG, I. – Fortpflanzung und kunstliche Vermehrung der chinesischen pflanzenfressenden Fische. *Allg. Fisch. Zeitung*, 93 (13): 402-404.

1967. BANK, O. – Der Gras- und Silberkarpfen, seine Lebensweise, Haltung und Züchtung. *Der Fischwirt*, 17 (6): 141-146.

1921. BENECKE, B. - VON DEBSCHITZ, H. – Die Teichwirtschaft. *

1935. BURDA, V. und WALTER, E. – Grundlagen der Karpfenzucht. J. Neumann, Neudamm, 3. Aufl., 63 p., 4 fig.

1965. CHIBA, K. – Studies on the Carp culture in running water pond. I. *Bul. Fresh. Fish. Res. Lab.* (Japan), 15 (1): 13-33.

1958. CHIMITS, P. – Les Carpes chinoises. *Bul. franç. Pisciculture*, 30 (188): 84:91.

1923. GUENAUX, G. – Pisciculture. *

1967. HOFMANN, J. – Der Teichwirt. *

1956. HUET, M. – Aperçu de la pisciculture en Indonésie. *

1966. HUET, M. et TIMMERMANS, J.-A – Het kweken van pootvis in de viskwekerij van Bokrijk. Cypriniden en roofvis. 1958-1963. – Production de cyprins et de voraces de repeuplement à la pisciculture de Bokrijk, de 1958 à 1963. Trav. Stat. Rech. Eaux et Forêts, Groenendaal, Sér. D, 38, 68 p., 30 fig.

1960. KOCH, W. – Fischzucht. *

1961. KOSTOMAROV, B. – Die Fischzucht. *

1951. KREUZ, A. – Teichbau und Teichwirtschaft. *

1931. MAIER, H. N. und HOFMANN, J. – Grundzüge der Karpfenteichwirtschaft für kleinere und mittlere Betriebe. *

1934. MEHRING, H. – Karpfenzucht. *

1933. SCHÄPERCLAUS, W. – Lehrbuch der Teichwirtschaft. *

1961. SCHÄPERCLAUS, W. – Lehrbuch der Teichwirtschaft. *

1967. v. SENGBUSCH, R., MESKE, Ch., SZABLEWSKI, W. u. LÜHR, B. – Gewichtszunahme von Karpfen in Kleinstbehältern, zugleich ein Beitrag zur Ausklärung des Raumfaktors. *Zeit. f. Fischerei*, 15 (1/2): 45-60.

1969. STEFFENS, W. – Der Karpfen. Die Neue Brehm-Bücherei, A. Ziemsen Verlag, Wittenberg Lutherstadt, 3. Aufl., 156 p., 84 fig.

1927 TAURKE, Fr. – Die Fischzucht und Fischhaltung. *

1956. TIMMERMANS, J. A. – Deux appareils simples et pratiques en pisciculture. Trav. Stat. Rech. Eaux et Forêts, Groenendaal, Sér. D, 20, 6 p., 8 fig.

1928. WALTER, E. – Karpfennutzung in kleinen Teichen. *

1964. WOYNAROVICH, E. – Ueber die künstliche Vermehrung des Karpfens und Erbrütung des Laiches in Zuger-Gläsern. *Wasser und Abwasser. Beiträge zur Gewasserforschung*, IV: 210-217.

1949. WUNDER, W. – Fortschrittliche Karpfenteichwirtschaft. E. Schweizerbart'sche Verlagsbuchhandlung, Stuttgart, 385 p., 219 fig.

1953. WUNDSCH, H. H. u. a. – Fischereikunde. *

1967. WURTZ-ARLET, J. – La reproduction artificielle des poissons d'étang. *La pisciculture française*, 9: 17-23.

Selection of carp

1967. BEUKEMA, J. J. – Hengelproeven met verschillende karperrassen. O.V.B. Jaarverslag 1966-1967: 70-74.

1968. KIRPITSCHNIKOV, V. S. – Efficiency of mass selection and selection for relatives in fish culture. (World symposium on warm-water pond fish culture, Rome, 1966.) FAO Fisheries Report, 44 (4): 179-194.

1968. MOAV, R. and WOHLFARTH, G. W. – Genetic improvement of yield in carp. (World symposium on warm-water pond fish culture, Rome, 1966.) FAO Fisheries Report, 44 (4): 12-29.

1961. O. V. B. – De selectie in de visteelt in het algemeen en in de pootkarperteelt in het bijzonder. Jaarverslag 1960-1961, 61-80.

1980. WOYNAROVICH, E. and HORVATH, L. – The artificial propagation of warm-water finfishes – a manual for extension. FAO Fish. Tech. Pap. (201): 183 pp.

1961. WUNDER, W. – Durchführung von Leistungsprüfungen in der Karpfenzucht. Arbeiten der DLG, DGL-Verlags-GMBH, Frankfurt (Main), Bd. 67, 56 p., 25 fig.

CHAPTER V

SPECIAL TYPES OF FISH CULTIVATION FOR RESTOCKING

1 Cultivation of pike

1967. ANWAND, K. und GROBMANN, G. – Besatzversuche mit Hechtbrut in Karpfenteichen. *Zeit. f. Fischerei*, 14 (5/6): 383-391.

1938. DORIER, A. – A propos de l'œuf et de l'alevin de Brochet. *Bul franç. Pisciculture*, 10 (110): 61-73, 3 pl.

1968. VAN DRIMMELEN, D. E. – Personal communication.

1968. GRAFF, D. L. – The successful feeding of a dry diet to Esocids. *Progressive Fish-Culturist*, 30 (3): 152.

1940. HEUSCHMANN, O. – Die Hechtzucht. *In* DEMOLL-MAIER: Handbuch der Binnenfischerei Mitteleuropas, Bd. IV, Lfg. 7: 749-787, 20 fig., 1 pl.

1959. HUET, M. et TIMMERMANS, J. A. – Esociculture. Production de brochetons de sept semaines. Trav. Stat. Rech. Eaux et Forêts, Groenendaal, Sér. D, 24, 18 p., 4 fig.

1958. JOHNSON, L. D. – Pond Culture of Muskellunge in Wisconsin. Technical Bulletin N° 17, Wisconsin Conservation Department, Madison, 54 p.

1960. KOCH, W. – Fischzucht. *

1946. LINDROTH, A. – Zur Biologie der Befruchtung und Entwicklung beim Hecht. Mitt. d. Anst. f. Binnenfischerei, Stockholm, 24, 173 p., 53 fig., 3 pl.

1958. O. V. B. – Organisatie ter Verbetering van de Binnenvisserij, Jaarverslag, 70 p.

1950. ROTH, H., MEDEM, F., RÖTHELI, A. – Biologischer Nachweis von Befruchtungsstoffen bei Hecht und Aesche. *Schweiz. Zeit. f. Hydrologie*, 12 (1): 67-68.

1976. STEFFENS, W. – Hechtzucht. *Z. Binnenfischerei* DDR, 23 (11): 327-343; 23 (12): 360-371.

1966. TONER, E. D. – *Esox lucius* L. FAO Fisheries Synopsis, 30, 39 p.

1938. VOUGA, M. – Le rôle du Brochet dans l'économie piscicole des lacs et des rivières. *Pêche et Pisciculture*, N° 5, 6, 7, 13 fig.

1948. VOUGA, M. et PRUDHOMME, J.-G. – Repeuplons nos eaux en Brochets. Editions Camille Rousset, Paris, 31 p., 17 fig.

1955. WILLEMSEN, J. – De waarde van pootsnoekjes voor uitzetting. *Visserij-Nieuws*, 7 (10): 150-151.

2 Cultivation of coregonids

1948. AMMANN, Ed. und STEINMANN, P. – Die Verbesserung der Methoden in der Felchenzucht. Kommission für die Erforschung fischereiwirtschaftlicher Fragen, Zurich, 56 p.

1958. LIBOSVARSKY, J. – Einiges über die Zucht der Grossen Maräne in der Tschechoslowakei. *Deutsche Fischerei-Zeitung*, 5 (6): 174-176.

1967. QUARTIER, A. A. – La pisciculture des Corégones. *Le Pêcheur et le Chasseur suisses*, 31 (6): 222-223.

1950/51. STEINMANN, P. – Monographie der schweizerischen Koregonen. *Schweiz. Zeit. f. Hydrologie*, 12: 109-189 and 340-491; 13: 54-155, 67 fig.

3. Cultivation of shad

1932. ROULE, L. – Manuel de Pisciculture. *

4 Cultivation of grayling

1956. DAVIS, H. S. – Culture and Diseases of Game Fishes. *

1958. SVETINA, M. – L'Ombre et sa reproduction artificielle. *Bul. franç. Pisciculture*, 31 (191): 59-65.

1958. VIVIER, P. – L'Ombre commun (*Thymallus thymallus* L.). Sa reproduction et son élevage. *Bul. franç. Pisciculture*, 31 (191): 45-58.

5 Cultivation of sturgeon

1962. MAGUIN, Et. – Recherches sur la systématique et la biologie des Acipenséridés. *Ann. Stat. Cent. Hydrob. Appl.*, 9: 9-242, 54 fig.

1969. MOJDEHI, M. – Personal communication.

1961. ROSTAMI, Is. – Biologie et exploitation des Acipenséridés caspiens. Thèse Fac. Sc. Univ. Paris, 210 p.

1955. ROUSSOW, G. – Les Esturgeons du fleuve Saint-Laurent en comparaison avec les autres espèces d'Acipenséridés. Montréal, Office de Biologie, 124 p., 25 fig.

1939. UNGER, E. – Der Sterlet (*Acipenser ruthenus* L.) in Karpfenteichen. *In* DEMOLL-MAIER: Handbuch der Binnenfischerei Mitteleuropas, Bd. IV, Lf. 6-7: 745-746.

6 Cultivation of pejerrey or silversides

1967. CONROY, D. A. – The raising of silversides, with an appendix on diseases. *Rev. Ital. Piscicoltura e Ittiopatologia*, 2 (3): 48-51.

1946. MAC DONAGH, E. J. – Piscicultura del Pejerrey. *Revista de la Facultad de Agronomia*, La Plata, 26: 33-51.

CHAPTER VI

BREEDING AND CULTIVATION OF PERCIFORMES

1 Cultivation of perch

2 Cultivation of pike-perch

1961. KOSTOMAROV, B. – Die Fischzucht. *

1961. SCHÄPERCLAUS, W. – Lehrbuch der Teichwirtschaft. *

1960a. STEFFENS, W. – Ernährung und Wachstum des jungen Zanders (*Lucioperca lucioperca* L.) in Teichen. *Zeit. f. Fischerei*, 9 (3/4): 161-271, 47 fig.

1960b. STEFFENS, W. – Zanderzucht in Karpfenteichen. *Deutsche Fischerei-Zeitung*, 7: 82-89, 8 fig.

1966. TÖLG, Is. und PENZIS, B. – Ueber den Zander, sein Leben und die Voraussetzungen für erfolgreichen Besatz. *Oesterreichs Fischerei*, 19 (5/6): 71-74, 4 fig.

1961. WILLEMSEN, J. – Over de biologie van jonge snoekbaars en de mogelijkheid van kunstmatige teelt *Visserij-Nieuws*, 13 (12): 189-193.

1960a. WOYNAROVICH, E. – Aufzucht der Zanderlarven bis zum Raubfischalter. *Zeit. f. Fischerei*, 9 (1/2): 73-83.

1960b. WOYNAROVICH, E. – Erbrütung von Fischeiern im Sprühraum. *Arch. f. Fischereiwiss.*, 10 (3): 179-189, 6 fig.

1939. UNGER, E. – Die Zucht des Zanders in Karpfenteichwirtschaften und in freien Gewässern. *In* DEMOLL-MAIER: Handbuch der Binnenfischerei Mitteleuropas, Bd. 4, Lf. 6/7: 723-742.

3   Cultivation of walleye

1956.   DAVIS, H. S. – Culture and Diseases of Game Fishes. *

1959.   NIEMUTH, W., CHURCHILL, W. and WIRTH, Th. – The Walleye. Life History, Ecology and Management. Wisconsin Conservation Department, Publication Nº 227, 14 p.

4   Cultivation of black bass and other centrarchids

1956.   DAVIS, H. S. – Culture and Diseases of Game Fishes. *

1955.   DAVISON, V. E. – Managing Farm Fishponds for Bass and Bluegills. U.S. Dep. Agric. Farmers Bulletin Nº 2094, 17 p.

1929.   HESEN, H. D. – A trap for the capture of large-mouthed bass fry and fingerlings. *Trans. Am. Fish. Soc.*, 59: 118-124.

1947.   HEY, D. – The Culture of Freshwater Fish in South Africa. *

1961.   MRAZ, D., KMIOTEK, St. and FRANKENBERGER, L. – The Largemouth Bass. Its Life History, Ecology and Management. Wisc. Conserv. Department, Publication Nº 232, 15 p.

1950.   SWINGLE, H. S. – Relationships and Dynamics of Balanced and Unbalanced Fish Populations. Ala. Poly. Inst. Agr. Exp. Stat., Bul. 274, 74 p.

1952.   WURTZ-ARLET, J. – Le Black-Bass en France.

Esquisse monographique. *Ann. Stat. Cent. Hydrob. Appl.*, Paris, 4: 203-286, 8 pl.

5   Cultivation of tilapias and other African cichlids

1955.   CHIMITS, P. – Le Tilapia et son élevage. Bibliographie préliminaire. *Bulletin des pêches de la FAO*, 8 (1): 1-35.

1957.   CHIMITS, P. – Les Tilapias et leur élevage. Seconde étude et bibliographie. *Bulletin des pêches de la FAO*, 10 (1): 1-27, 7 fig.

1948.   DE BONT, A. F., HALAIN, C., HUET, M., HULOT, A. – Premières directives pour l'élevage de poissons en étangs au Katanga. Pisciculture des *Tilapia*. Ministère des Colonies, Bruxelles, 24 p., 25 fig.

1959.   GRUBER, R. et MATHIEU, Y. – *Hemichromis fasciatus* et la pisciculture. *Bul. Agric. Congo belge*, 50 (2): 421-436.

1957.   HUET, M. – Dix années de Pisciculture au Congo belge et au Ruanda-Urundi. Compte rendu de mission piscicole. Ministère des Colonies, Bruxelles, 154 p., 110 fig.

1968.   LEMASSON, J. et BARD, J. – Nouveaux poissons et nouveaux systèmes pour la pisciculture en Afrique. *(IX, 5)

1966.   MAAR, A., MORTIMER, M. A. E., VAN DER LINGEN, I. – Fish Culture in Central East Africa. FAO, Rome, 160 p., 73 fig.

1967.   MESCHKAT, A. – The Status of warm-water fish culture in Africa. *(IX, 5)

CHAPTER VII

BREEDING AND CULTIVATION OF CATFISH

1   Breeding and cultivation of Danubian wels

1961.   KOSTOMAROV, B. – Die Fischzucht. *

1955.   WOYNAROVICH, E. – Die künstliche Zucht des Welses. *Deutsche Fischerei-Zeitung*, 2: 361-362.

2   Breeding and cultivation of American catfish

1960.   CRAWFORD, B. – Propagation of Channel Catfish (*Ictalurus punctatus*). Arkansas Game and Fish Commission, Centerton, Arkansas, 2d edit., 10 p., 6 fig.

1956.   DAVIS, H. S. – Culture and Diseases of Game Fishes. *

1967.   TIEMEIER, O. W., DEYOE, C. W. – Producing Channel

Catfish. Kansas State University, Manhattan, Kansas, Bull. 508, 23 p.

1951.   TOOLE, M. – Channel Catfish Culture in Texas. *Progressive Fish-Culturist*, 13 (1): 3-10, 11 fig.

3   Breeding and cultivation of clarias in the Far East

1962.   HORA, S. L., PILLAY, T. V. R. – Handbook of fish culture in the Indo-Pacific Region. *

1968.   SIDTHIMUNKA, A., SANGBERT, J., PAWAPOOTANON, A. – The Culture of Catfish (*Clarias* sp.) in Thailand. (World symposium on warm-water pond fish culture, Rome, 1966.) FAO Fisheries Reports, 44 (5): 196-204.

CHAPTER VIII

SPECIAL TYPES OF FISH BREEDING AND CULTIVATION

1   Cultivation of eels

1968.   DEELDER, C. L. – Over het kweken van aal in het buitenland. *Visserij-Nieuws*, 21 (2): 85-86.

1968.   FUJI, A. R. – A short note on eel culture in Japan. Personal communication.

1962.   HORA, S. L. and PILLAY, T. V. R. – Handbook on fish culture in the Indo-Pacific Region. *

1966.   KOOPS, H. – Die Aalproduktion in Japan. *Archiv. f. Fischereiwissenschaft*, 17 (1): 44-50.

1967.   KOOPS, H. – Ergebnisse der Aalwirtschaft im Ausland und Folgerungen für unsere einheimische Aalwirtschaft. *Der Fischwirt*, 17 (2): 29-40.

1980.   KOOPS, H. and KUHLMANN, H. – Eel farming in the thermal effluents of a conventional power station in the harbour of Emden. *EIFAC-Symposium, Stavanger, May 1980*, 16 pp.

1968.   MEYER-WAARDEN, P. F. und KOOPS, H. – Qualitätsbeurteilungen von gefütterten Teichaalen. *Der Fischwirt*, 18 (3): 57-61.

2 Fish cultivation in rice fields

1958. ARDIWINATA, R. O. – Fish culture on paddy fields in Indonesia. *Proc. Indo-Pac. Fish Counc.*, 7 (II-III): 119-162.

1953. CHEN, T. P. – The Culture of Tilapia in Rice Paddies in Taiwan. Chinese-American Joint Commission on Rural Reconstruction, Fish. Ser., 2, 29 p.

1967. COCHE, A. G. – Fish Culture in Rice Fields. A World-wide Synthesis. *Hydrobiologia*, 30 (1): 1-44.

1962. HICKLING, C. F. – Fish Culture. *

1962. HORA, S. L. and PILLAY, T. V. R. – Handbook on fish culture in the Indo-Pacific Region. *

1956. HUET, M. – Aperçu de la pisciculture en Indonésie. *

3 Fish cultivation in brackish water

1952. CHEN. T. P. – Milkfish culture in Taiwan. Chinese-American Joint Commission on Rural Reconstruction, Fish. Ser., 1, 17 p., 12 fig.

1962. HORA, S. L. and PILLAY, T. V. R. – Handbook on fish culture in the Indo-Pacific Region. *

1956. HUET, M. – Aperçu de la pisciculture en Indonésie.*

1967. JENSEN, K. W. – Saltwater rearing of Rainbow trout and Salmon in Norway. FAO EIFAC Technical Paper; 3: 43-48.

1962. PILLAI, T. G. – Fish farming methods in the Philippines, Indonesia and Hong-Kong, FAO Fisheries Biol. Techn. Paper, 18, 68 p., 74 fig.

1960. SCHUSTER, W. H. – Synopsis of Biological Data on Milkfish *Chanos chanos* (FORSKAL), 1775 - FAO Fish Biol. Synopsis, 4, 65 p.

1966. YASHOUV, A. – Breeding and Growth of Grey Mullet (*Mugil cephalus* L.). *Bamidgeh*, 18 (1): 3-13.

Appendix: Cultivation of salmonids in sea water

1978. EDWARDS;, D. J. – Salmon and trout farming in Norway. Fishing News Books Limited, Farnham, Surrey, England, 195 pp., 78 fig.

4 Fish cultivation in floating cages

1978. COCHE, A. G. – Revue des pratiques d'élevage de poissons en cages dans les eaux continentales. *Aquaculture*, 13: 157–189.

1975. DAHM, E. – Die Verwendung von Netzkäfigen zur Fischintensivsucht. *Protokolle zur Fischereitechnik*, 63; 319–358.

1974. MANN, H. – Auswirkungen der Intensivhaltung von Forellan auf das Gewässer am Beispiel eines Baggersees. *Der Fischwirt*, 24 (2): 7–10.

5 Fish cultivation in heated water

1978. COCHE, A. C. – Revue des pratiques d'élevage de poissons en cages dans les eaux continentales. *Aquaculture*, 13: 157–189.

1974. WESTIN, D. T. – Nitrate and nitrite toxicity to salmonid fishes. *Progressive Fish-Culturist*, 36 (2): 86–89.

CHAPTER IX

REGIONAL FRESH WATER FISH CULTURE

1 Fish culture in Europe

1967. GAUDET, J. L. – The status of warm-water fish culture in Europe. (World symposium on warm-water pond fish culture, Rome, 1966.) FAO Fisheries Reports, 44 (2): 70-87.

1961. MANN, H. – Fish cultivation in Europe. *In* BORGSTROM, G. Fish as food, Academic Press, New York and London, Vol. I: 77-102.

2 Fish culture in North America

1960. CLARK, Cl. F. – Fish propagation, artificial and natural. Ohio Dept. Nat. Res., Div. Wildlife, 16 p.

1956. DAVIS, H. S. – Culture and Diseases of Game Fishes. *

1956. DOBIE, J. *et al.* – Raising bait fishes. U.S. Fish and Wildlife Service, Circular 35, 124 p.

1967. MALOY, Ch. R. – Status of fish culture in the North American region. (World symposium on warm-water pond-fish culture, Rome, 1966.) FAO Fisheries Reports, 44 (2): 123-134.

1939. MARKUS, H. C. – Propagation of bait and forage fish. U.S. Fishery Circular N° 28, 19 p.

1953. PRATHER, E. E. *et al.* – Production of bait minnows in the Southeast. Agric. Exper. Stat. Auburn, Alabama, Circ. N° 112, 71 p., 33 fig.

1940. SURBER, Th. – The propagation of Minnows. Minnesota Department of Conservation, Div. Fish and Game, 22 p.

1951. WASCKO, H. and CLARK, Cl. F. – Pond propagation of Bluntnose and Blackhead Minnows. Ohio Dept. Nat. Res., Wildlife Conservation Bul., 4, 16 p.

3 Fish culture in the Near and Middle East

1967. JOB, T. J. – Status of fish culture in the Near East region. (World symposium on warm-water pond fish culture, Rome, 1966.) FAO Fisheries Reports, 44 (2): 54-69.

4 Fish culture in the Indo-Pacific region

1957. CHAUDHURI, H. and ALIKUNHI, K. H. – Observations on the spawning in Indian carps by hormone injection. *Current Science*, 26 (12): 381-382.

1953. HOFSTEDE, A. E., ARDIWINATA, R. O. and BOTKE, F. – Fish Culture in Indonesia. FAO Indo-Pacific Fisheries Council, Special Publications, N° 2, 129 p.

1962. HORA, S. L. and PILLAY, T. V. R. – Handbook on fish culture in the Indo-Pacific Region. *

1956. HUET, M. – Aperçu de la pisciculture en Indonésie. *

1968. KURONUMA, K. – New systems and new fishes for culture in the Far East. (World symposium on warm-water pond fish culture, Rome, 1966.) FAO Fisheries Reports, 44 (5): 123-142.

1967. TUBB, J. A. – Status of fish culture in Asia and the Far East. (World symposium on warm-water pond fish culture, Rome, 1966.) FAO Fisheries Reports, 44 (2): 45-53.

5　Fish culture in Africa

1956.　DAGET, J. et D'AUBENTON, F. – *Heterotis niloticus* peut-il être un poisson de pisciculture? 2e Symposium sur l'hydrobiologie et la pêche en eaux douces en Afrique, Brazzaville, Public. N° 25 C.C.T.A./C.S.A., 109-111.

1968.　LEMASSON, J. et BARD, J. – Nouveaux poissons et nouveaux systèmes pour la pisciculture en Afrique. (World symposium on warm-water pond fish culture, Rome, 1966.) FAO Fisheries Reports, 44 (5): 182-195, 2 fig.

1967.　MESCHKAT, A. – The status of warm-water fish culture in Africa. (World symposium on warm-water pond fish culture, Rome, 1966.) FAO Fisheries Reports, 44 (2): 88-122.

1963.　OLANIYAN, C. I. O. and ZWILLING, K. K. – The suitability of *Heterotis niloticus* EHRENBAUM as a fish for cultivation; with a note on its spawning behaviour. *Bul. de l'I.F.A.N.*, 25, A (2): 513-525.

1964.　REIZER, Ch. – Comportement et reproduction d'*Heterotis niloticus* en petits étangs. *Bois et Forêts des Tropiques*, (95): 49-60.

6　Fish culture in Latin America

1967.　MILES, C. – Posibilidades de la piscicultura en estanques en America Latina. (World symposium on warm-water pond fish culture, Rome, 1966,) FAO Fisheries Reports, 44 (2): 34-44.

Appendix: Restricted fish cultivation

1961.　CHIMITS, P. – La salmoniculture fermière dans le cadre d'une petite ferme de montagne. *Plaisirs de la Pêche*, 54-55: 98-102; 56-57: 158-160.

1956.　HUET, M. – Aperçu de la pisciculture en Indonésie. *

1934/45.　LÉGER, L. – Petite salmoniculture fermière. *Trav. Labor. Hydr. et Piscic. Univ. Grenoble*, 1934 (Année 1932): 1-32 et 1935 (Années 1933-1934): 1-103, 7 fig.

1931.　MAIER, H. N. und HOFMANN, J. – Grundzüge der Karpfenteichwirtschaft für kleinere und mittlere Betriebe. *

1921.　WALTER, E. – Kleinteichwirtschaft. J. Neumann, Neudamm, 2. Aufl., 146 p., 25 fig.

CHAPTER X
NATURAL PRODUCTIVITY OF PONDS

1962.　BARD, J. – Theory and practice of fish culture in central Africa. Centre Technique Forestier Tropical, Nogent-sur-Marne, 44 p. (stencilé).

1949.　HUET, M. – Appréciation de la valeur piscicole des eaux douces. *

1964.　HUET, M. – The evaluation of the fish productivity in fresh waters (The coefficient of productivity k). *Verh. Internat. Verein. Limnol.*, 15: 524-528.

1968.　HUET, M. – Méthodes biologiques d'accroissement de la production piscicole (Europe et Afrique). *(XI)

1910.　LÉGER, L. – Principes de la méthode rationnelle du peuplement des cours d'eau à Salmonides. *Trav. Labor. Piscic. Univ. Grenoble*, 1910: 531-568.

1945.　LEGER, L. – Economie biologique et productivité de nos rivières à Cyprinides. *Bul franç. Pisciculture*, 18 (139): 49-69.

CHAPTER XI
BIOLOGICAL MEANS OF INCREASING PRODUCTION

1961.　EPPEL. – Ertragssteigerung der Speisefischproduktion durch Entenfreihaltung in der Binnenfischerei. *Deutsche Fischerei-Zeitung*, 8 (8): 246-251.

1950.　HICKLING, C. F. – *Tilapia* culture in Singapore. Comptes rendus conf. pisc. anglo-belge, Elisabethville 1949, 287-292.

1962.　HICKLING, C. F. – Fish culture. *

1968.　HUET, M. – Méthodes biologiques d'accroissement de la production (Europe et Afrique). (World symposium on warm-water pond fish culture, Rome, 1966.) FAO Fisheries Reports, 44 (4): 289-327.

1968.　RABANAL, H. R. – Stock manipulation and other biological methods of increasing production of fish through pond fish culture in Asia and the Far East. (World symposium on warm-water pond fish culture, Rome, 1966.) FAO Fisheries Reports, 44 (4): 274-288.

1960.　STEFFENS, W. – Zanderzucht in Karpfenteichen. *Deutsche Fischerei-Zeitung*, 7 (3): 82-89.

1950.　SWINGLE, H. S. – Relationships and dynamics of balanced and unbalanced fish populations. *(VI, 4)

1968.　SWINGLE, H. S. – Biological means of increasing productivity in ponds. (World symposium on warm-water pond fish culture, Rome, 1966.) FAO Fisheries Reports, 44 (4): 243-257.

1959.　YASHOUV, A. – Studies on the productivity of fish ponds. Carrying capacity. *Proc. Gen. Fish. Coun. Medit.*, 5: 409-419.

1968.　YASHOUV, A. – Mixed fish culture, an ecological approach to increase pond productivity. (World symposium on warm-water pond fish culture, Rome, 1966.) FAO Fisheries Reports, 44 (4): 258-273.

CHAPTER XII
MAINTENANCE AND IMPROVEMENT OF PONDS

1　Control of excessive aquatic vegetation

1958.　BANK, O. – Stickstoffdüngermittel zur Bekämpfung unerwünschter Pflanzen. *Der Fischbauer*, 9 (117): 473-475.

1962.　BANK, O. – Algen, ihre Bekämpfung mit chemischen Mitteln. *Der Fischwirt*, 12 (9): 260-269.

1961.　BAUER, K. – Studien über Nebenwirkungen von Pflanzenschutzmitteln auf Fische und Fischnährtiere.

Mitt. Biol. Bundesanstalt f. Land- und Forstwirtschaft, Berlin-Dahlem, (105), 72 p.

1961. SCHÄPERCLAUS, W. – Lehrbuch der Teichwirtschaft. *

1954/56. SCHUBERTH, A. – Ueber den gegenwartigen technischen Stand der Unterwassermähmaschinen. *Deutsche Fischerei-Zeitung,* 1 (3): 78-81, (5): 137-143, (9): 255-258; 2 (11): 336-342, (12): 371-374; 3 (7): 203-206.

1958. SCHUBERTH, A. – Ein Hilfsgerät zum abraumen geschnittenen Schilfes von flachen Teichen. *Deutsche Fischerei-Zeitung,* 5 (9): 242-247.

1960. SCHUBERTH, A. – Aus der Entwicklung der Schilfschneidemaschinen "Erpel", eine Neukonstruktion mit Amphibiencharakter. *Deutsche Fischerei-Zeitung,* 7 (4): 112-116.

1961. TIMMERMANS, J. A. – Bestrijding der overtollige waterplanten. Lutte contre la végétation aquatique envahissante. Trav. Stat. Rech. Eaux et Forêts, Groenendaal, Sér. D, N° 31, 41 p., 13 fig.

1955. X. – L'hélice antiherbes à grand rendement Dynophyta. *Bul. Off. Inform. Cons. Sup. Pêche,* (22): 95.

2   Improvement and restoration of pond bottoms

1960. BANK, O. – Der Schwimmbagger, Saug- und Spülbagger als Entlandungsgerät in Teichen. *Allg. Fischerei-Zeitung,* 85 (4): 83-85.

1966. BANK, O. – Bodenbearbeitung unter Wasser. *Der Fischwirt,* 16 (8): 199-201.

1965. X. – (Aspiro-dragueuse Biver). *Bul. Off. Inform. Cons. Sup. Pêche,* (62): 102-105.

3   Control of the principal physico-chemical factors of fish production

1936. BANDT, H. J. – Der für Fische "tödliche ph-Wert" im alkalischen Bereich. *Zeit. f. Fischerei,* 34: 359-361.

1941. HUET, M. – pH et réserves alcalines (SBV). Notion, détermination, importance piscicole. Comm. Stat. Rech. Eaux et Forêts, Groenendaal, Sér. D, 1, 20 p.

1968. HUET, M. – Méthodes biologiques d'accroissement de la production. *(XI)

4-5   Liming and fertilization of the ponds

1965. BRÜNING, D. und MÜLLER, W. – Die Düngung der Teiche. *In* Handbuch der Pflanzenernährung und Düngung, Bd. III: Düngung der Kulturpflanzen Springer-Verlag, Wien-New York, 1517-1539.

1931. DEMOLL, R. – Die Düngung der Teiche. *In* Handbuch der Pflanzenernährung und Düngerlehre, Bd. II, Julius Springer, Berlin, 861-875, 5 fig.

1967. GOOCH, B. C. – Appraisal of North American fish culture fertilization studies. (World symposium on warm-water pond fish culture, Rome, 1966.) FAO Fisheries Reports, 44 (3): 13-26.

1962. HEPHER, B. – Ten years of research in fish ponds fertilization in Israel. *Bamidgeh,* 14 (2): 29-38.

1967. HEPHER, B. – Some limiting factors affecting the dose of fertilizers added to fishponds with special reference to the Near East. (World symposium on warm-water pond fish culture, Rome, 1966.) FAO Fisheries Reports, 44 (3): 1-6.

1954. MORTIMER, C. H. and HICKLING, C. F. – Fertilizers in Fishponds. A Review and Bibliography. Colonial Office, Fishery Publications, N°. 5, 155 p.

1946. PREVOST, G. – Quatrième rapport de l'Office de Biologie. Province de Québec, Ministère de la Chasse et des Pêcheries, 91 p., 18 fig.

1967. PROWSE, G. A. – A review of the methods of fertilizing warm-water fish ponds in Asia and the Far East. (World symposium on warm-water pond fish culture, Rome, 1966.) FAO Fisheries Reports, 44 (3): 7-12.

1963. SUMMERS, M. W. – Managing Louisiana Fish Ponds. Louisiana Wildlife and Fisheries Commission, 64 p.

1922. WALTER, E. – Kleiner Leitfaden der Teichdüngung. Verlag J. Neumann, Neudamm, 140 p.

1924/33 WALTER, E. – Die Versuche in der bayerischen teichwirtschaftlichen Versuchsanstalt Wielenbach 1923-1932. Sammlung fischereilicher Zeitfragen, J. Neumann, Neudamm.

1967. WOLNY, P. – Fertilization of warm-water fish ponds in Europe. (World symposium on warm-water pond fish culture, Rome, 1966.) FAO Fisheries Reports, 44 (3): 64-81.

1956. WUNDER, W. – Düngung in der Teichwirtschaft. Tellus-Verlag, Essen, 75 p., 13 fig.

CHAPTER XIII

ARTIFICIAL FEEDING OF FISH

1967. BANK, O. – Keine Rückstände in den Muskeln von Karpfen nach Dauerverfütterung antibiotikahaltigen Trockenfutters (Carpi). *Der Fischwirt,* 17 (1): 11-13.

1964. DEUFEL, J. – Ueber die Forellenernährung mit Trockenfutter. *Oesterreichs Fischerei,* 17 (2): 19-22.

1965. EINSELE, W. – Futter und Fütterung. *Oesterreichs Fischerei,* 18 (5): 75-83.

1969. GAUDET, J. L. – Symposium sur les récents développements dans la nutrition de la Carpe et de la Truite. FAO, EIFAC Technical Paper, (9): 213 p.

1967. HASTINGS, W. H. – Warm-water fish nutrition. Fish Farming Experimental Station, Stuttgart, Arkansas, 10 p. (mimeographed).

1968. JENSEN, K. W. and GAUDET, J. L. – Bibliography on nutritional requirements of salmonoid fishes. FAO, EIFAC Occasional Paper N° 2, 28 p.

1967. LING, S. W. – Feeds and feeding of warm-water fishes in ponds in Asia and the Far East. (World symposium on warm-water pond fish culture, Rome, 1966.) FAO Fisheries Reports, 44 (3): 291-309.

1967. MANN, H. – Weitere Versuche zur Frage der Antibiotikarückstände im Fischfleisch. *Der Fischwirt,* 17 (1). 13-14.

1966. MESKE, Ch. – Karpfenaufzucht in Aquarien. *Der Fischwirt,* 18 (12): 309-316.

1966. PENZIS, B. et TÖLG, Is. – Etude de la croissance et de l'alimentation de la "Grass Carp" (*Ctenopharyngodon idella*) en Hongrie. *Bul. franç. Pisciculture*, 39 (223): 70-75.

1963. PHILLIPS, A. M., TUNISON, A. V., BALZER, G. C. – Trout feeds and feeding. United States Department of the Interior, Fish and Wildlife Service, Circular 159, 38 p.

1964. PHILLIPS, A. M., Jr, HAMMER, G. L. and PYLE, E. A. Dry concentrates as complete trout foods. *Progressive Fish-Culturist*, 26 (1): 21-24.

1967. SHELL, E. W. – Feeds and feeding of warm-water fish in North America. (World symposium on warm-water pond fish culture, Rome, 1966.) FAO Fisheries Reports, 44 (3): 310-325.

1934. WALTER, E. – Richtlinien zur Karpfenfütterung. Sammlung Fischereilicher Zeitfragen, J. Neumann, Neudamm, Hf. 12, 2. Aufl., 20 p.

1965. X. – Elektrische Fischfütterungsautomaten. *Allg. Fischerei-Zeitung*, 90 (15): 463-464.

CHAPTER XIV

TOTAL PRODUCTIVITY AND STOCKING OF PONDS

1962. BARD, J. – Theory and practice of fish culture in Central Africa. *(x)

1957. CHARPY, B. – *Tilapia macrochir* et *Tilapia melanopleura*. Leur élevage en pisciculture. Centre Technique Forestier Tropical, Nogent-sur-Marne, 24 p.

1931. CONTAG, E. – Der Einfluss verschiedener Besatzstärken auf die natürliche Ernährung zweisömmerigen Karpfen und auf die Zusammensetzung der Tierwelt ablassbarer Teiche. *Zeit. f. Fischerei*, 20: 569-596.

1952. DE BONT, A. – La production de poisson de consommation au Congo belge. Méthode préconisée. *Bul. Agric. Congo belge*, 43 (4): 1053-1068.

1962. HICKLING, C. F. – Fish Culture. *

1949. HUET, M. – Appréciation de la valeur piscicole des eaux douces. *

1964. HUET, M. – The evaluation of the fish productivity in fresh waters (The coefficient of productivity k). *(x)

1968. HUET, M. – Méthodes biologiques d'accroissement. de la production piscicole (Europe et Afrique). *(xi)

1959. LESSENT, P. – Rapport sur les recherches piscicoles entreprises sur la Station de pisciculture de Bouaké (Côte d'Ivoire), d'avril 1958 à avril 1959. Notes Pêche et Pisc. Centre Tech. For. Trop. (D.G.), 8, 49p.

1933. SCHÄPERCLAUS, W. – Lehrbuch der Teichwirtschaft. *

1960a. SCHÄPERCLAUS, W. – Erhöhte Karpfenproduktion durch Steigerung der Fütterung. *Deutsche Fischerei-Zeitung*, 7 (3): 75-79.

1960b. SCHÄPERCLAUS, W. – Durch intensive Fütterung zur Erhöhung der Marktprodukten von Karpfen. *Deutsche Fischerei-Zeitung*, 7 (11): 341-345.

1963. SCHÄPERCLAUS, W. – Das Verhältnis von Naturzuwachs zu Fütterzuwachs im Karpfenabwachsteich sowie seine Auswirkung auf den Gesamtzuwachs, den Fütterquotienten und die Beschaffenheit des Karpfen. *Zeit. f. Fischerei*, 11 (3/4): 265-300.

1967. VAN DER LINGEN, M. I. – Some preliminary remarks on stocking rate and production of *Tilapia* species at the fisheries research center. *In* Proc. first fisheries day in Southern Rhodesia, Governt. printer, Salisbury, 54-68.

1931. WALTER, E. – Arbeiten aus der Bayerischen teichwirtschaftlichen Versuchsstation Wielenbach aus den Jahren 1928 bis 1930. Sammlung fischereilicher Zeitfragen, J. Neumann, Neudamm, Hf. 22, 56 p.

1934. WALTER, E. – Grundlagen der allgemeinen fischereilichen Produktionslehre. *

1959. YASHOUV, A. – Studies on the productivity of fish ponds. Carrying capacity. *Proc. Gen. Fish. Coun. Medit.*, 5: 409-419.

CHAPTER XV

ENEMIES AND DISEASES OF FISH

1 General bibliography

1961. AMLACHER, E. – Taschenbuch der Fischkrankheiten. VEB Gustav Fischer Verlag, Jena, 286 p., 195 fig.

1968. CHRISTENSEN, N. O. – Maladies des poissons, 1966. (Adaptation française de P. BESSE.) Syndicat des Pisciculteurs salmoniculteurs de France, Paris, 97 p., 42 fig.

1956. DAVIS, H. S. – Culture and Diseases of Game Fishes. *

1926. DIESSNER - ARENS. – Die künstliche Zucht der Forelle. *

1967. GHITTINO, P. – La Piscicoltura in Europa. *Riv. Ital. Piscicoltura e Ittiopatologia*, 1 (2): 23-46, 43 fig.

1906. HOFER, B. – Handbuch der Fischkrankheiten. E. Schweizerbart'sche Verlagsbuchhandlung, Stuttgart, 2. Aufl., 359 p., 222 fig.

1924. PLEHN, M. – Praktikum der Fischkrankheiten. E. Schweizerbart'sche Verlagsbuchhandlung, Stuttgart, 179 p., 173 fig.

1966. REICHENBACH-KLINKE, H. H. – Krankheiten und Schädigungen der Fische. Gustav Fischer Verlag, Stuttgart, 389 p., 338 fig.

1961. SCHÄPERCLAUS, W. – Lehrbuch der Teichwirtschaft. *

1954. SCHÄPERCLAUS, W. – Fischkrankheiten. Akademie-Verlag, Berlin, 3. Aufl., 708 p., 389 fig.

1961. STOLK, A. – Visziekten. Diergeneeskundig Memorandum, 8 (4-5): 173-251, 49 fig.

Prophylaxis, hygiene and disinfection in fish culture

1965. BANK, O. – Zur Anwendung des Kochsalzbades bei Karpfen. *Oesterreichs Fischerei*, 18 (2): 23-30.
1964. BESSE, P. et DE KINKELIN, P. – Modes d'emballage de poissons destinés aux examens de laboratoire. *Bul. franç. Pisciculture*, 37 (214): 36-37.
1963. EINSELE, W. – Kochsalzbäder zur Heilbehandlung und zur allgemeinen Kräftigung von Fischen. *Oesterreichs Fischerei*, 16 (3/4): 50-53.
1974. GERARD, J.-P. – Sur l'emploi des iodophores en pisciculture. *Bul. franç. Pisciculture*, 47 (254): 13-15.
1974. GERARD, J.-P. – Sur la désinfection en pisciculture. *Bul. franç. Pisciculture*, 47 (254): 20-21.
1966. OTTE, E. – Richtlinien für eine sachgemäsze Einsendung von Fischen und Wasserproben. *Oesterreichs Fischerei*, 10 (10): 151-153.

2 Special bibliography for certain fish diseases
(*First consult the general bibliography* p. 425)

Infectious abdominal dropsy of carp

1967. BANK, O. – Keine Rückstände in den Muskeln von Karpfen nach Dauerverfütterung antibiotikahaltigen Trockenfütters. *(XIII)
1967. MANN, H. – Weitere Versuche zur Frage der Antibiotikarückstände im Fischfleisch. *(XIII)
1972. OTTE, E. – Die Schwimmblasenkrankheit eine Form der IBW? *Münchner Beiträge zur Abwasser-Fischerei- und Flussbiologie*, Bd 20: 56–63.
1956. SCHÄPERCLAUS, W. – Die Bauchwassersucht des Karpfens, eine bakterielle Infektionskrankheit, und neue Methoden zu ihrer erfolgreichen Heilung und Bekämpfung durch antibiotische Mittel. *Archiv. f. Fischereiwissenschaft*, 7 (1): 9-17.
1966. TOMASEC, I. – Considérations générales sur le problème de l'étiologie de l'Hydropisie infectieuse de la Carpe. *Bul. Off. int. Epiz.*, 65 (5-6): 721-730.
1964. TOMASEC, I., BRUDNJAK, Z., FIJAN, N. und KUNST, Lj. – Weiterer Beitrag zur Aetiologie der Infektiösen Bauchwassersucht des Karpfens, Jugoslavenska Akademija Znanosti i Umjetnosti, Zagreb, 35-44.

Viral haemorrhagic septicaemia of trout

1962. BELLET, R. – La furonculose de la truite ou "Septicémie hémorragique". *Bul. franç. Pisciculture*, 35 (207): 45-66.
1961. BRUNNER, G. – Ueber ein seuchenhaftes Auftreten der Viruskankheit "INuL", und der Furonkulose bei importierten Regenbogenforellen. *Allg. Fischerei-Zeitung*, 86: 694-695.
1963. ECONOMON, P. P. – Experimental treatment of infectious pancreatic necrosis of brook trout with polyvinyl-pyrrolidone-iodine. *Trans. Am. Fish Soc.* 92: 180-182.
1963. GHITTINO, P. – Differential Diagnose der bei Regenbogenforellen vorkommenden "lipoiden Leberdegeneration" und "infectiösen Nierenschwellung und Leberdegeneration" auf Grund histopathologisch-anatomischer Untersuchungen. *Zeit. f. Fischerei*, 11 (7/8): 549-556.

1965. RASMUSSEN, C. J. – A biological study of the Egtved disease (INuL). *Annals of the New York Academy of Science*, 126 (1): 427-460, 10 fig. ·
1965. ZWILLENBERG, L. O., JENSEN, M. H., ZWILLENBERG, H. H. L. – Electronmicroscopy and classification of the virus of viral haemorrhagic septicaemia of rainbow trout. Second symposium of the permanent commission of the O.I.E. on diseases of fish, Munich, pp. 20-24.

Furunculosis of salmonids

1963. DEUFEL, J. – Zur Bekämpfung der Furunkulose in der Forellenzucht. *Der Fischwirt*, 13 (5): 143-145.
1947. GUTSELL, J. S. – Furunculosis and its treatment. *Progressive Fish-Culturist*, 9 (1): 13-20.
1952. McCRAW, B. M. – Furunculosis of fish. Special Scientific Report Fisheries N° 84, U.S. Dept. of Interior, Fish and Wildlife Service, 87 p.
1969. McFADDEN, T. W. – Effective disinfection of trout eggs to prevent egg transmission of *Aeromonas liquefaciens*. *J. Fish. Res. Bd Can.* 26: 2311-2318.

Saprolegnia infection

1965. BELLET, R. – La prophylaxie en alevinage. *La pisciculture française*, 1 (2): 20-22.
1967. CAPPELLO, G. – Un semplice "embrionatore" per uova di trota. *(III)
1957. DEUFEL, J. – Bekämpfung der Verpilzung von Fischeiern mit Malachitgrün. *Der Fischwirt*, 7 (6): 153-156.
1965. DEUFEL, J. – Malachitgrün in der Fischzucht. *Oesterreichs Fischerei*, 18 (1): 5-8.
1961. GOTTWALD, S. – Die Anwendung von Malachitgrün und Kochsalz beim Erbrüten von Fischeiern und Hältern von Laichfischen in Polen. *Deutsche Fischerei-Zeitung*, 8: 48-52.
1960. PREUDHOMME, J. G. – Note sur les verts malachite. *Bul. franç. Pisciculture*, 32 (197): 164-165.
1961. STEFFENS, W., LIEDER, U., NEHRING, D., HATTOP, H.-W. – Möglichkeiten und Gefahren der Anwendung von Malachitgrün in der Fischerei. *Zeit. f. Fischerei*, 10 (8-10): 745-771.
1962. STEFFENS, W. – Verhütung des Saprolegnia-Befalls von Forelleneiern durch Formalin. *Deutsche Fischerei-Zeitung*, 9 (9): 287-289.

Costiasis

1933. LÉGER, L. – Maladies épidémiques au cours du premier âge dans l'élevage de la Truite. Procès-verbaux de la Société Dauphinoise d'Etudes Biologiques, N° 216, 8 p.
1956. TACK, E. – Neue Erfahrungen bei der Bekämpfung von Costia. *Deutsche Fischerei-Zeitung*, 3: 313-316.

Whirling disease

1966. PUTZ, R. E., HOFFMAN, G. L. – Earliest susceptible age of rainbow trout to whirling disease. *Progressive Fish-Culturist*, 28 (2): 82.
1965. RASMUSSEN, C. J. – La lutte contre la maladie du tournis dans les stations d'élevage de truites au Danemark. FAO, EIFAC Technical Paper N° 2: 15-20.

1951. TACK, E. – Bekämpfung der Drehkrankheit mit Kalkstickstoff. *Der Fischwirt*, 1 (5): 123-130, 3 fig.

Ichthyophthiriasis

1960. DEUFEL, J. – Malachitgrün zur Bekämpfung von Ichthyophthirius bei Forellen. *Der Fischwirt*, 10 (1:) 13-14.
1968. RYCHLICKI, Z. – Eradication of *Ichthyophthirius multifiliis* in carp. (World symposium on warm-water pond fish culture, Rome, (1966.) FAO Fisheries Reports, 44 (5): 361-364.

Argulus

1959. SARIG, S. and LAHAV, M. – The treatment with lindane of carp and fish ponds infected with the fish-louse, Argulus. *Proc. Gen. Fish. Coun. Medit.*, 5: 151-156.

CHAPTER XVI

HARVESTING THE FISH

1 Fishing out the pond

(*See general bibliography* pp. 425-426)

2 Sorting and grading the fish

1958. MAREK, M. – A modified fish grader with an extendable grille. *Bamidgeh*, 10 (2): 29-30.

3-4 Storing and transport of fish

1945. CHARPY, B. – Vente et transport des alevins. *L'action forestière et piscicole*, oct. 1945.
1949. DEMOLL, R. und STEINMANN. P. – Praxis der Aufzucht von Forellenbesatzmaterial. *(III)
1967. HOFMANN, J. – Der Teichwirt. *
1964. LAMARQUE, P. – Anesthésie et transport. *Bul. Off Inform. Cons. Sup. Pêche*, (55): 1-3.
1960. NORRIS, K. S., BROCATO, F. CALANDRINO, F,. McFARLAND, W. N. – A survey of fish transportatiou methods and equipment. *California Fish and Game*, 46 (1): 5-33, 7 fig.
1961. WOHLFARTH, G., LAHMAN, M., MOAV, R. – Transporting live carp in polyethylene bags. *Bamidgeh*, 13 (3/4): 74-75.

5 Conservation of fish

1965. BORGSTROM, G. – Fish as food. Academic Press, New York and London, Vol. III, 489 p.
1953. PENSO, G. – Les produits de la pêche. Vigot Frères, Paris, 418 p., 334 fig.

Appendix: Marking cultivated fish

1967. BERGSTRÖM, E. – Propoxate as an anaesthetic for salmon (*Salmo salar* L.). Swedish Salmon Res. Inst. Rept. LFI Medd. 7/1967.
1962. LÜHMANN, M. und MANN, H. – Ueber die Wanderungen von Fischen in der Elbe nach Markierungsversuchen. *Der Fischwirt*, 12 (12): 353-365.
1960. MOAV, R., WOHLFARTH, G. and LAHMAN, M. – An electric instrument for brandmarking fish. *Bamidgeh*, 12 (4): 92-95.
1962. MÜLLER, W. – Erste Erfahrungen mit der Höllensteinmarkierung bei Karpfen. *Deutsche Fischerei-Zeitung*, 9 (6): 178-180.
1965a. THIENPONT, D. – L'anesthésie des poissons. "*Zoo*" (S. R. Zool. d'Anvers), 32: 28-30.
1965b. THIENPONT, D. – Immobilisation et anesthésie des poissons. *Bull. Soc. Roy. Zool. Anvers*, 35: 11-18.
1940. WUNDER, W. – Die Markierung des Karpfens in der Teichwirtschaft. *Fischerei-Zeitung*, 43 (5): 25-28, 4 fig.

GENERAL APPENDIX

Lists of principal cultivated fish

1968. BLANC, M. et BANARESCU, P. – Rapport du groupe de travail sur les noms de poissons d'eau douce européens. FAO, EIFAC, Cinquième Session, Gen-4, 32 p.
1962. HORA, S. L. and PILLAY, T. V. R. – Handbook on fish culture in the Indo-Pacific Region. *
1965. GOLVAN, Y.-J. – Catalogue systématique des poissons actuels. Masson et Cie, Paris, 227 p.
1965. LADIGES, W. und VOGT, D. – Die Süsswasserfische Europas. Verlag Paul Parey, Hamburg und Berlin, 250 p., 425 fig.
1964. THYS VAN DEN AUDENAERDE, D. – Révision systématique des espèces congolaises du genre Tilapia (Pisces, Cichlidae). Musée Royal de l'Afrique Centrale, Annales, Sciences Zoologiques, N° 124, 24 fig., 11 pl.
1961. SPILLMANN, C. J. – Faune de France. 65, Poissons d'eau douce. Ed. Paul Lechevalier, Paris, 304 p., 102 fig., 11 pl.
1961. TORBETT, H.D. – The Angler's Freshwater Fishes. Putnam, London, 352 p., 145 fig.
1960. X. – A list of Common and Scientific Names of Fishes from the United States and Canada. American Fisheries Society, Special Publication, N° 2, 2d edit., 102 p.

Other books published by
Fishing News Books Ltd

Free catalogue available on request

Advances in aquaculture
Advances in fish science and technology
Aquaculture practices in Taiwan
Aquaculture training manual
Atlantic salmon: its future
Better angling with simple science
British freshwater fishes
Business management in fisheries and
   aquaculture
Commercial fishing methods
Control of fish quality
The crayfish
Culture of bivalve molluscs
Echo sounding and sonar for fishing
The edible crab and its fishery in
   British waters
Eel capture, culture, processing and
   marketing
Eel culture
Engineering, economics and
   fisheries management
European inland water fish: a
   multilingual catalogue
FAO catalogue of fishing gear designs
FAO catalogue of small scale fishing
   gear
Fibre ropes for fishing gear
Fish and shellfish farming in
   coastal waters
Fish catching methods of the world
Fisheries oceanography and ecology
Fisheries of Australia
Fisheries sonar
Fishermen's handbook
Fishery development experiences
Fishing boats and their equipment
Fishing boats of the world 1
Fishing boats of the world 2
Fishing boats of the world 3
The fishing cadet's handbook
Fishing ports and markets
Fishing with electricity
Fishing with light
Freezing and irradiation of fish

Freshwater fisheries management
Glossary of UK fishing gear terms
Handbook of trout and salmon diseases
How to make and set nets
Introduction to fishery by-products
The lemon sole
A living from lobsters
Making and managing a trout lake
Marine fisheries ecosystem
Marine pollution and sea life
Marketing in fisheries and aquaculture
Mending of fishing nets
Modern deep sea trawling gear
More Scottish fishing craft and
   their work
Multilingual dictionary of fish
   and fish products
Navigation primer for fishermen
Netting materials for fishing gear
Pair trawling and pair seining
Pelagic and semi-pelagic trawling gear
Penaeid shrimps – their biology
   and management
Planning of aquaculture development
Power transmission and automation
   for ships and submersibles
Refrigeration on fishing vessels
Salmon and trout farming in Norway
Salmon fisheries of Scotland
Scallop and queen fisheries in the
   British Isles
Scallops and the diver-fisherman
Seine fishing
Squid jigging from small boats
Stability and trim of fishing vessels
The stern trawler
Study of the sea
Training fishermen at sea
Trends in fish utilization
Trout farming handbook
Trout farming manual
Tuna: distribution and migration
Tuna fishing with pole and line